RISK
ASSESSMENT *AND*
DECISION
ANALYSIS *WITH*
BAYESIAN
NETWORKS
Second Edition

RISK ASSESSMENT *AND* DECISION ANALYSIS *WITH* BAYESIAN NETWORKS

Second Edition

NORMAN FENTON

MARTIN NEIL

CRC Press
Taylor & Francis Group
Boca Raton London New York

CRC Press is an imprint of the
Taylor & Francis Group, an **informa** business

A CHAPMAN & HALL BOOK

CRC Press
Taylor & Francis Group
6000 Broken Sound Parkway NW, Suite 300
Boca Raton, FL 33487-2742

Library of Congress Cataloging-in-Publication Data

Names: Fenton, Norman E., 1956- author. | Neil, Martin D., author.
Title: Risk assessment and decision analysis with bayesian networks / by Norman Fenton, Martin Neil.
Description: Second edition. | Boca Raton, Florida : CRC Press, [2019] | Includes bibliographical references and index.
Identifiers: LCCN 2018009541| ISBN 9781138035119 (hardback : alk. paper) | ISBN 9781315269405 (e-book)
Subjects: LCSH: Bayesian statistical decision theory. | Decision making. | Risk management.
Classification: LCC QA279.5 .F46 2019 | DDC 519.5/42--dc23
LC record available at https://lccn.loc.gov/2018009541

Visit the Taylor & Francis Web site at
http://www.taylorandfrancis.com

and the CRC Press Web site at
http://www.crcpress.com

For our parents, wives and children

Contents

Foreword

Probabilistic models based on directed acyclic graphs have a long and rich tradition, beginning with work by the geneticist Sewall Wright in the 1920s. Variants have appeared in many fields. Within statistics, such models are known as directed graphical models; within cognitive science and artificial intelligence, such models are known as Bayesian networks. The name honors the Reverend Thomas Bayes (1702–1761), whose rule for updating probabilities in the light of new evidence is the foundation of the approach. The initial development of Bayesian networks in the late 1970s was motivated by the need to model the top-down (semantic) and bottom-up (perceptual) combination of evidence in reading. The capability for bidirectional inferences, combined with a rigorous probabilistic foundation, led to the rapid emergence of Bayesian networks as the method of choice for uncertain reasoning in AI and expert systems, replacing earlier ad-hoc rule-based schemes. Perhaps the most important aspect of Bayesian networks is that they are direct representations of the world, not of reasoning processes. The arrows in the diagrams represent real causal connections and not the flow of information during reasoning (as in rule-based systems or neural networks). Reasoning processes can operate on Bayesian networks by propagating information in any direction. For example, if the sprinkler is on, then the pavement is probably wet (prediction); if someone slips on the pavement, that also provides evidence that it is wet (abduction, or reasoning to a probable cause). On the other hand, if we see that the pavement is wet, that makes it more likely that the sprinkler is on or that it is raining (abduction); but if we then observe that the sprinkler is on, that reduces the likelihood that it is raining. It is the ability to perform this last form of reasoning—called explaining away—that makes Bayesian networks so powerful compared to rule-based systems or neural networks. They are especially useful and important for risk assessment and decision-making.

Although Bayesian networks are now used widely in many disciplines, those responsible for developing (as opposed to using) Bayesian network models typically require highly specialist knowledge of mathematics, probability, statistics, and computing. Part of the reason for this is that, although there have been several excellent books dedicated to Bayesian Networks and related methods, these books tend to be aimed at readers who already have a high level of mathematical sophistication—typically they are books that would be used at graduate or advanced undergraduate level in mathematics, statistics, or computer science. As such they are not accessible to readers who are not already proficient in those subjects. This book is an exciting development because it addresses this problem. While I am sure it would be suitable for undergraduate courses on probability and risk, it should be understandable by any numerate reader interested in risk assessment and decision making. The book provides sufficient motivation and examples (as well as the mathematics and probability where needed from scratch) to enable readers to understand the core principles and power of Bayesian networks. However, the focus is on ensuring that readers can build practical Bayesian network models, rather than understand in depth the underlying algorithms and theory. Indeed readers are provided with a tool that performs the propagation, so they will be able to build their own models to solve real-world risk assessment problems.

Judea Pearl

UCLA Computer Science Department

Los Angeles, California

Preface

The era of "big data" offers enormous opportunities for societal improvements. There is an expectation—and even excitement—that, by simply applying sophisticated machine learning algorithms to "big data" sets, we may automatically find solutions to problems that were previously either unsolvable or would incur prohibitive economic costs.

Yet, the clever algorithms needed to process big data cannot (and will never) solve most of the critical risk analysis problems that we face. Big data, even when carefully collected, is typically unstructured and noisy; even the "biggest data" typically lack crucial, often hidden, information about key causal or explanatory variables that generate or influence the data we observe. For example, the world's leading economists failed to predict the 2008–2010 international financial crisis because they relied on models based on historical statistical data that could not adapt to new circumstances even when those circumstances were foreseeable by contrarian experts. In short, analysts often depend on models that are inadequate representations of reality—good for predicting the past but poor at predicting the future.

These fundamental problems are especially acute where we must assess and manage risk in areas where there is little or no direct historical data to draw upon; where relevant data are difficult to identify or are novel; or causal mechanisms or human intentions remain hidden. Such risks include terrorist attacks, ecological disasters and failures of novel systems and marketplaces. Here, the tendency has been to rely on the intuition of "experts" for decision-making. However, there is an effective and proven alternative: the *smart data* approach that combines expert judgment (including understanding of underlying causal mechanisms) with relevant data. In particular *Bayesian Networks (BNs)* provide workable models for combining human and artificial sources of intelligence even when big data approaches to risk assessment are not possible.

BNs describe networks of causes and effects, using a graphical framework that provides rigorous quantification of risks and clear communication of results. Quantitative probability assignments accompany the graphical specification of a BN and can be derived from historical data or expert judgment. A BN then serves as a basis for answering probabilistic queries given knowledge about the world. Computations are based on a theorem by the Reverend Thomas Bayes dating back to 1763 and, to date, provides the only rational and consistent way to update a belief in some uncertain event (such as a decline in share price) when we observe new evidence related to that event (such as better than expected earnings). The problem of correctly updating beliefs in the light of new evidence is central to all disciplines that involve any form of reasoning (law, medicine, and engineering as well as finance and indeed AI). Thus, a BN provides a general approach to reasoning, with explainable models of reality, in contrast to big data approaches, where the emphasis is on prediction rather than explanation, and on association rather than causal connection.

BNs are now widely recognized as a powerful technology for handling risk, uncertainty, and decision making. Since 1995, researchers have incorporated BN techniques into software products, which in turn have helped develop decision support systems in many scientific and industrial applications, including: medical diagnostics, operational and financial risk, cybersecurity, safety and quality assessment, sports prediction, the law, forensics, and equipment fault diagnosis.

A major challenge of reasoning causally is that people lacked the methods and tools to do so productively and effectively. Fortunately, there has been a quiet revolution in both areas. Work by Pearl (Turing award winner for AI), has provided the necessary philosophical and practical instruction on how to elicit, articulate and manipulate causal models. Likewise, our work on causal idioms and dynamic discretization has been applied in many application areas to make model building and validation faster, more accurate and ultimately more productive. Also, there are now software products, such as AgenaRisk, containing sophisticated algorithms, that help us to easily design the BN models needed to represent complex problems and present insightful results to decision makers. Compared to previous generations of software these are more powerful and easier to use—so much so that they are becoming as familiar and accessible as spreadsheets became in the 1980s. Indeed, this big leap forward is helping decision makers think both graphically, about relationships, and numerically, about the strength of these relationships, when modelling complex problems, in a way impossible to do previously.

This book aims to help people reason causally about risk and uncertainty. Although it is suitable for undergraduate courses on probability and risk, it is written to be understandable by other professional people generally interested in risk assessment and decision making. Our approach makes no assumptions about previous probability and statistical knowledge. It is driven by real examples that are introduced early on as motivation, with the probability and statistics introduced (from scratch) as and when necessary. The more mathematical topics are separated from the main text by comprehensive use of boxes and appendices. The focus is on applications and practical model building, as we think the only real way to learn about BNs is to build and use them.

Many of the examples in this book are influenced by our academic research but also by our experience in putting the ideas into practice with commercial and government decision and policy makers. Together we have consulted and supplied software to a wide variety of commercial and government organizations, including Milliman LLP, Exxon Mobil, Medtronic, Vocalink, NATO, Sikorsky, World Agroforestry Centre, Virgin Money, TNO, Royal Bank of Canada, Bosch, KPMG, QinetiQ, RGA, GTI, EDF Energy, Boeing and Universities worldwide. We are firm believers in technology transfer and putting useful decision-making theory into the hands of those at the sharp end.

Although purists have argued that only by understanding the algorithms can you understand the limitations and hence build efficient BN models, we overcome this by providing pragmatic advice about model building to ensure models are built efficiently. Our approach means that the main body of the text is free of the intimidating mathematics that has been a major impediment to the more widespread use of BNs.

Acknowledgments

Numerous people have helped in various ways to ensure that we achieved the best possible outcome for the second edition of this book.

For general support we thank:

Nic Birtles, Rob Calver, Neil Cantle, Anthony Constantinou, Paul Curzon, Christian Dahlman, Phil Dawid, Shane Cooper, Eugene Dementiev, Stephen Dewett, Kevin Korb, Paul Krause, Evangelina Kyrimi, Dave Lagnado, Helge Langseth, Peng Lin, Paul Loveless, Eike Luedeling, David Mandel, Amber Marks, David Marquez, William Marsh, Scott McClachlan, Takao Naguchi, Anne Nicholson, Magda Osman, Judea Pearl, Toby Pilditch, Elena Perez-Minana, Thomas Roelleke, Lukasz Radlinksi, Keith Shepherd, Nigel Tai, Ed Tranham, Maggie Wang, Patricia Wiltshire, Rob Wirszycz, Barbaros Yet, Yun Zhou, Jacob de Zoete.

For recommendations on modelling and AgenaRisk we thank:

Steven Frazier, Sune Hein, Chris Hobbs, Ignatius Prashanth, Venkatasubramanian Ramakrishnan, Kurt Schulzke, Charles Warnky.

For finding errors in the first edition we thank:

Bayan Al Mutawa, Richard Austin, Thomas Barnett, Mark Blackburn, Fredrik Bökman, Michael Corning. Daniel Keith Farr, Jesus Rubio Garcia, Robbie Hearmans, Robbie Heremans, Chris Hobbs, Sandra Johnson, Paul King, Ernest Lever, Felipe Oliveira, Shem Malmquist, Hashmat Mastin, Ian McDavid, Dave Palmer, Andrew Rattray, Wendell Rylander, Andrew Steckley, Erik Vestergaard, Jaime Villalpando, Jun Wang, Yun Wang, Vince Whelan, Mark Woolley, Teresa Zigh.

Authors

Norman Fenton is a professor in risk information management at Queen Mary, the School of Electronic Engineering and Computer Science University of London and also a founding Director of Agena, a company that specializes in risk management for critical systems. His current work on quantitative risk assessment focuses on using Bayesian networks. Norman's experience in risk assessment covers a wide range of application domains such as software project risk, legal reasoning (he has been an expert witness in major criminal and civil cases), medical decision-making, vehicle reliability, embedded software, football prediction, transport systems, and financial services. From 2014–2018 Norman was the holder of a European Research Council Advanced Grants (BAYES-KNOWLEDGE). Norman has a special interest in raising public awareness of the importance of probability theory and Bayesian reasoning in everyday life (including how to present such reasoning in simple lay terms) and he maintains a website dedicated to this and also a blog focusing on probability and the law. In March 2015 Norman presented the award-winning BBC documentary Climate Change by Numbers. In 2016 Norman led a 6-month Programme on Probability and Statistics in Forensic Science at the Isaac Newton Institute for Mathematical Sciences, University of Cambridge. In addition to his research on risk assessment, Norman is renowned for his work in software engineering (including pioneering work on software metrics); the third edition of his book *Software Metrics: A Rigorous and Practical Approach* was published in November 2014. He has published over 200 articles and five books on these subjects, and his company Agena has been building Bayesian network-based decision support systems for a range of major clients in support of these applications.

Martin Neil is a professor in computer science and statistics at the School of Electronic Engineering and Computer Science, Queen Mary, University of London. He is also a joint founder and Director of Agena, Ltd. and is visiting professor in the Faculty of Engineering and Physical Sciences, University of Surrey. Martin has over twenty years experience in academic research, teaching, consulting, systems development, and project management and has published or presented over 100 papers in refereed journals and at major conferences. His interests cover Bayesian modeling and risk quantification in diverse areas, operational risk in finance, systems and design reliability (including software), software project risk, decision support, simulation, AI and statistical learning. Martin earned a BSc in mathematics, a PhD in statistics and software metrics, and is a chartered engineer.

1

Introduction

In December 2017, the Royal Statistical Society (RSS) announced the winner of its International Statistic of the Year. The citation[1] announced it as follows:

Winner: International Statistic of the Year

69

This is the annual number of Americans killed, on average, by lawnmowers—compared to two Americans killed annually, on average, by immigrant Jihadist terrorists.

The figure was highlighted in a viral tweet this year from Kim Kardashian in response to a migrant ban proposed by President Trump; it had originally appeared in a Richard Todd article for the *Huffington Post*.

Todd's statistics and Kardashian's tweet successfully highlighted the huge disparity between (i) the number of Americans killed each year (on average) by "immigrant Islamic Jihadist terrorists" and (ii) the far higher average annual death tolls among those "struck by lightning," killed by "lawnmowers," and in particular "shot by other Americans."

Todd and Kardashian's use of these figures shows how everyone can deploy statistical evidence to inform debate and highlight misunderstandings of risk in people's lives.

Judging panel member Liberty Vittert said: "Everyone on the panel was particularly taken by this statistic and its insight into risk—a key concept in both statistics and everyday life. When you consider that this figure was put into the public domain by Kim Kardashian, it becomes even more powerful because it shows anyone, statistician or not, can use statistics to illustrate an important point and illuminate the bigger picture."

The original Kim Kardashian tweet is shown in Figure 1.1.

While the announcement was met with enormous enthusiasm, one significant dissenter was Nassim Nicolas Taleb—a well-known expert on risk and "randomness." He exposed a fundamental problem with the statistic, which he summed up in the tweet of Figure 1.2.

Indeed, rather than "inform debate and highlight misunderstandings of risk in people's lives," as stated by the RSS, this example does exactly the opposite. It provides a highly misleading view of risk because it omits crucial causal information that explains the statistics observed and that is very different for the two incomparable numbers. One of the

Contrary to the statement of the Royal Statistical Society citation, the figures directly comparing numbers killed by lawnmower with those killed by Jihadist terrorists do *not* "highlight misunderstandings of risk" or "illuminate the bigger picture." They do the exact opposite, as we explain in this book.

---

[1] https://www.statslife.org.uk/news/3675-statistic-of-the-year-2017-winners-announced. All models in this chapter are available to run in AgenaRisk, downloadable from www.agenarisk.com.

Because of the particular 10-year period chosen (2007–2017), the terrorist attack statistics do not include the almost 3,000 deaths on 9/11 and also a number of other attacks that were ultimately classified as terrorist attacks.

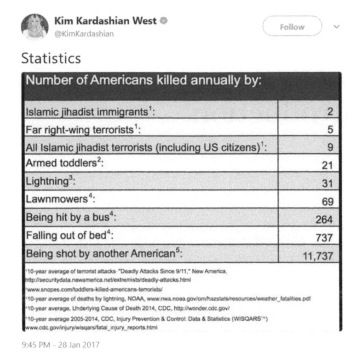

Figure 1.1   Tweet by Kim Kardashian that earned "International Statistic of the Year" 2017.

objectives of this book is to help readers understand how to see through such statistics and build models that incorporate the necessary causal context.

Informally, Taleb's argument is that there is a key difference between risks that are *systemic*—and so can affect more than one person (such as a terrorist attack)—and those that are not (such as using a lawnmower), which can be considered *random*. The chances that the number of people who die from a nonsystemic risk, like using a lawnmower, will double next year are extremely unlikely. But this cannot be said about the number of people dying from systemic risks like terrorist attacks and epidemics. The latter can be "multiplicative," whereas the former cannot. It is impossible for a thousand people in New York City to die from using lawnmowers next year, but it is not impossible for a thousand to die from

Figure 1.2   Taleb's response to the RSS announcement.

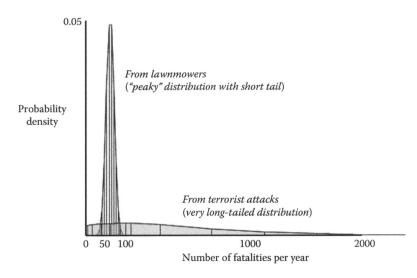

Figure 1.3 Comparing the probability distributions of number of fatalities per year.

terrorist attacks. Systemic and nonsystemic risks have very different *probability distributions*, as shown in Figure 1.3.

Using the number of deaths per year to compare different types of "risk" fails to consider the range of factors that affect the true risk to particular individuals or groups. For example, the probability of being killed by a lawnmower in New York City is especially low because relatively few people there have lawns to mow. In fact, death by lawnmower is essentially impossible for those not using a lawnmower, whereas there is a greater risk to gardeners. Residents of major cities are at greater risk from terrorists than residents of the countryside.

Crucially, there are also causal factors that *explain* the number of terrorist deaths that need to be considered. Most obviously, there are extensive security measures in place to stop terrorist attacks; without these, deaths from terrorist attacks would drastically increase. Also, terrorist cells can be responsible not just for multiple deaths in a single attack, but also multiple attacks, so deaths in terrorist attacks can be related by a *common cause*. These types of causal influences and relations—summarized in Figure 1.4—are the focus of much of this book.

An especially concerning part of the RSS citation was the implication that the relatively low number of terrorist deaths suggested that new measures to counter terrorism were unnecessary because of the "low risk." To make such reasoning explicit, we would have to perform a cost-benefit and trade-off analysis (Figure 1.5). Imposing new measures to counter terrorist threats involves both a financial cost and a human rights cost. But they also involve potential benefits, not just in terms of lives saved but also in reduction of other existing (secondary) security costs and improved quality of life. The implication from the RSS was that the costs were greater than the benefits.

But even if this trade-off analysis had been made explicit (which would involve putting actual numbers to all the costs and benefits as well

Figure 1.5 is an example of an extended type of Bayesian network, called an influence diagram, which we discuss in Chapter 11.

In the lawnmower case, Fred and Jane are killed by different lawnmowers. This is what makes them independent ... in the absence of common lawnmower design flaws (such as a controller bug inserted by a terrorist designer).

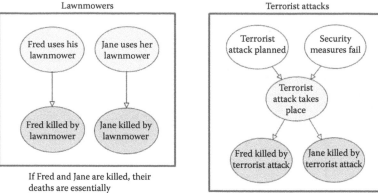

Lawnmowers

Fred uses his lawnmower   Jane uses her lawnmower

Fred killed by lawnmower   Jane killed by lawnmower

If Fred and Jane are killed, their deaths are essentially "independent."

Terrorist attacks

Terrorist attack planned   Security measures fail

Terrorist attack takes place

Fred killed by terrorist attack   Jane killed by terrorist attack

If Fred and Jane are killed, their deaths may be the result of the same terrorist attack or same group.

Figure 1.4   Causal view of lawnmower versus terrorist attack deaths.

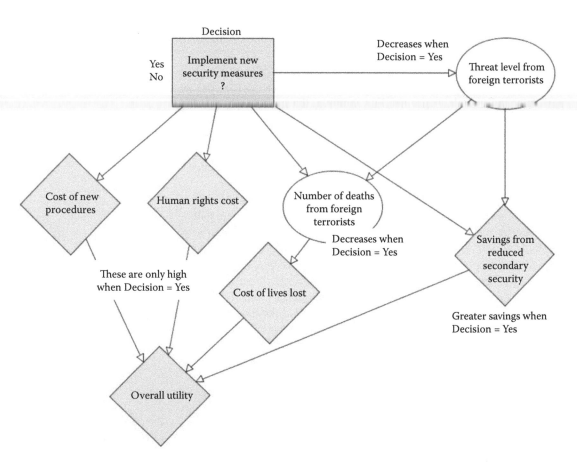

Figure 1.5   The kind of cost-benefit trade-off analysis required for informed decision-making.

as the number of expected deaths from jihadis who would otherwise not have entered the United States), there is a fundamental flaw in relying only on previous years' fatalities data. The number of fatalities depends (among other things) on the security measures that are put in place. The *Thames barrier* provides a very good analogous example:

> In several decades prior to 1974 the number of deaths in London due to the River Thames flooding was zero. Based on these data (and applying the RSS reasoning) what possible justification could there have been for the British Government to decide to build a flood barrier (which cost £650 million before its completion in May 1984) rather than do nothing? The decision on the Thames Barrier was made because steadily rising water levels were already causing expensive (but non-fatal) flooding and reliable measurements predicted catastrophic flooding within 50 years if no barrier was in place. In this case the simplistic counts of past number of fatalities were clearly insufficient for rational risk assessment and decision-making.

While the Thames Barrier decision still made use of historical data (namely monthly water levels, cost of flood damage, etc.), the key point is that we need to go beyond the simplistic data and consider contextual and situational factors. Moreover, in many risk scenarios, "triggers" and "threats" that are analogous to rising water levels in this example might require expert judgments and models in addition to data. Without an explanatory or causal model, the data alone would be meaningless from an inferential or decision-making perspective.

Completely novel risks (such as crashing civilian planes into skyscrapers prior to 9/11) can only be quantified using expert imagination and judgment. Indeed, the 9/11 scenario had previously been considered seriously by security experts (and movie scriptwriters), and terrorist "chatter" suggested the threat was increasing. However, the probability of such an event was considered sufficiently low not to merit additional security measures that could have been put in place to avert it. Had security measures—which are now routine at all airports in the world—been put in place before 9/11, there would have been no mass fatalities on 9/11.

Yet we find the same flawed "datacentric" reasoning being applied yet again. The context and motivation for the choice of the 2017 RSS International Statistic of the Year was the partial immigration ban proposed by President Trump in 2017. Opponents to the proposed measures argued that they were unnecessary because everyday risks from, for example, lawnmowers are greater than the risk from jihadis. This demonstrates again the problems of relying solely on historical fatality data.

One of our own areas of research—predicting system reliability, which is covered in Chapter 13—is especially prone to these kinds of misconceptions about oversimplistic past data. A "system" could be a software program, a physical device or component thereof (phone, TV, computer, microprocessor), or even a process (such as a method for manufacturing steel). It is standard to measure a system's reliability in terms of the frequency with which it fails—both before releasing the system for general use and also after release. But using this data alone is problematic. Why? For instance, consider a system where, after two years,

The methods described in this book enable us to build models that incorporate expert subjective judgment with data (when available) in order to provide fully quantified risk assessment in terms of probabilities of specific events and overall utilities.

The methods described in this book—based on Bayesian networks—are currently accessible only to mathematical and statistical specialists. The aim of this book is to make them accessible to anybody with an interest in risk assessment who has not forgotten their high school math.

there are very few or zero reports of failures. At first glance, this might suggest that the system is very reliable. But there is another possible explanation—the cause of the low number of failures may well be that the system was so bad that it was rarely or never used. So, here we have competing causal explanations that are very different but give rise to the same observable data.

In many areas of life, past data is a good indicator of future behavior and may be sufficient for good decision-making. Based on average temperatures in previous years, we can be pretty confident that if we are going to Cairo in June, we will not need a fur coat to keep warm. You don't even need the past data to be "constant." A company that has seen a steady year-on-year increase in sales of widgets can be confident of next year's sales based on simple regression models. The same is true in many industries. In both of these examples, we do not use the data alone. We use it with a (often implied) model to interpret and make inferences, either using other relevant circumstances connected to weather or customer demand. But as soon as there are novel circumstances and factors, this type of model for decision-making is likely to be poor.

While the above example of misuse of statistics for risk assessment might be considered harmless, the same cannot be said of the financial crisis of 2008–9, which brought misery to millions around the world. The armies of analysts and statisticians employed by banks and government agencies had failed to predict either the event or its scale until far too late. Similarly, the results of major elections in 2016 (in the USA and the UK Brexit vote) were contrary to what pollsters were consistently and almost uniformly predicting. Yet the methods that could have worked—that are the subject of this book—were largely ignored. Moreover, the same methods have the potential to transform risk analysis and decision-making in all walks of life, including medicine and the law as well as business.

Examples of the kind of problems we want to be able to solve include the following:

- Medical—Imagine you are responsible for diagnosing a medical condition and for prescribing one of a number of possible treatments. You have some background information about the patient (some information is objective, like age and number of previous operations, but other information is subjective, like "depressed" and "stressed"). You also have some prior information about the prevalence of different possible conditions (for example, bronchitis may be ten times more likely in a certain population than cancer). You run some diagnostic tests about which you have some information of the accuracy (such as the chances of the test outcome positive for a condition that is not present and negative for a condition that is present). You also have various bits of information about the success rates of the different possible treatments and their side effects. On the basis of all this information, how do you arrive at a decision of which treatment pathway to take? And how would you justify that decision if something went wrong? If something went wrong, you may be open to negligence. The issue is about justifying your actions if your contingent diagnosis turns out to have been the wrong one.

- Legal—As a judge or member of a jury, you hear many pieces of evidence in a trial. Some of the evidence favors the prosecution hypothesis of guilty, and some of the evidence favors the defense hypothesis of innocence. Some of the evidence is statistical (such as the match probability of a DNA trace found at a crime scene) and some is purely subjective, such as a character witness statement. It is your duty to combine the value of all of this evidence to arrive at a probability of innocence. If the probability value you arrive at is sufficiently small ("beyond reasonable doubt"), you must return a guilty verdict. How would you arrive at a decision? Similarly, before a case comes to trial, how should a member of the criminal justice system, the police, or a legal team determine the value of each piece of evidence and then determine if, collectively, the evidence is sufficient to proceed to trial?

- Safety—A transport service (such as a rail network or an air traffic control center) is continually striving to improve safety but must nevertheless ensure that any proposed improvements are cost effective and do not degrade efficiency. There is a range of alternative competing proposals for safety improvement, which depend on many different aspects of the current infrastructure (for example, in the case of an air traffic control center, alternatives may include new radar, new collision avoidance, detection devices, or improved air traffic management staff training). How do you determine the "best" alternative, taking into account not just cost but also impact on safety and efficiency of the overall system? How would you justify any such decision to a team of government auditors?

- Financial—A bank needs sufficient liquid capital readily available in the event of exceptionally poor performance (either from credit or market risk events, or from catastrophic operational failures of the type that brought down Barings in 1995 and threatened Societe Generale in 2007). It has to calculate and justify a capital allocation that properly reflects its "value at risk." Ideally, this calculation needs to take account of a multitude of current financial indicators, but given the scarcity of previous catastrophic failures, it is also necessary to consider a range of subjective factors, such as the quality of controls in place at different levels of the bank hierarchy and business units. How can all of this information be combined to determine the real value at risk in a way that is acceptable to the regulatory authorities and shareholders?

- Reliability—The success or failure of major new products and systems often depends on their reliability, as experienced by end users. Whether it is a high-end digital TV, a software operating system, or a complex military vehicle, like a tank, too many faults in the delivered product can lead to financial disaster for the producing company or even a failed military mission, including loss of life. Hence, pre-release testing of such systems is critical. But no system is ever perfect and a perfect system delivered after a competitor gets to the market first may be worthless. So how do you determine when a system is "good enough" for release or how much more testing is needed? You may have hard data in the form of a sequence of test results, but this has to be considered along with subjective data about the quality of testing and the realism of the test environment.

What is common about all of the aforementioned problems is that a "gut-feel" decision based on doing all the reasoning "in your head" or on the back of an envelope is fundamentally inadequate and increasingly unacceptable. Nor can we base our decision on purely statistical data of "previous" instances, since in each case the "risk" we are trying to calculate is essentially unique in many aspects. The aim of this book is to show how it is possible to do rigorous analysis of all of the above types of risk assessment problems using Bayesian networks.

2

Debunking Bad Statistics

The aim of this chapter is to introduce, by way of examples and stories, some of the basic tools needed for the type of risk assessment problems introduced in Chapter 1. We do this by introducing and analysing the traditional statistics and data analysis methods that have previously been used. In doing so you should be able to understand both their strengths and limitations for risk assessment. In particular, in discussing their limitations, we also provide a gentle introduction to the power of causal models (which are often implemented as Bayesian networks).

2.1 Predicting Economic Growth: The Normal Distribution and Its Limitations

Table 2.1 contains the (annualized) growth rate figures for the UK for each quarter from the start of 1993 to the end of 2007 (which was just prior to the start of the international economic collapse). So, for example, in the fourth quarter of 2007 the annual growth rate in the UK was 2.36%.

Data such as this, especially given the length of time over which it has been collected, is considered extremely valuable for financial analysis and projections. Since so many aspects of the economy depend on the growth rate, we need our predictions of it for the coming months and years to be very accurate. So imagine that you were a financial analyst presented with this data in 2008. Although it would be nice to be able to predict the growth rate in each of the next few years, the data alone gives you little indication of how to do that. If you plot the growth over time as in Figure 2.1 there is no obvious trend to spot.

But there is a lot that you can do other than making "point" predictions. What financial institutions would really like to know is the answer to questions like those in Sidebar 2.1.

Indeed, economic analysts feel that the kind of data provided enables them to answer such questions very confidently. The way they typically proceed is to "fit" the data to a standard curve (also called a *statistical distribution*). The answers to all the aforementioned questions can then be answered using standard statistical tables associated with that distribution.

Sidebar 2.1

The kind of things financial institutions would really like to know:

- What are the chances that the next year's growth rate will be between 1.5% and 3.5% (stable economy)?

- What are the chances that the growth will be less than 1.5% in each of the next three quarters?

- What are the chances that within a year there will be negative growth (recession)?

Table 2.1

Quarterly Annualized UK Growth Rate Figures 1993-2007 Adjusted for Inflation

Quarter	Annual GDP%	Quarter	Annual GDP%
1993 Q1	1.42	2000 Q4	3.04
1993 Q2	2.13	2001 Q1	3.08
1993 Q3	2.50	2001 Q2	2.31
1993 Q4	2.82	2001 Q3	2.27
1994 Q1	3.29	2001 Q4	2.19
1994 Q2	4.28	2002 Q1	1.79
1994 Q3	4.83	2002 Q2	1.95
1994 Q4	4.70	2002 Q3	2.19
1995 Q1	3.95	2002 Q4	2.44
1995 Q2	3.01	2003 Q1	2.29
1995 Q3	2.76	2003 Q2	2.83
1995 Q4	2.51	2003 Q3	2.88
1996 Q1	3.14	2003 Q4	3.23
1996 Q2	3.03	2004 Q1	3.58
1996 Q3	2.54	2004 Q2	3.22
1996 Q4	2.84	2004 Q3	2.57
1997 Q1	2.70	2004 Q4	2.45
1997 Q2	3.15	2005 Q1	1.81
1997 Q3	3.48	2005 Q2	1.97
1997 Q4	3.90	2005 Q3	2.50
1998 Q1	3.85	2005 Q4	2.40
1998 Q2	3.67	2006 Q1	3.16
1998 Q3	3.52	2006 Q2	2.71
1998 Q4	3.39	2006 Q3	2.59
1999 Q1	3.13	2006 Q4	2.70
1999 Q2	3.29	2007 Q1	2.59
1999 Q3	3.72	2007 Q2	2.88
1999 Q4	3.74	2007 Q3	2.91
2000 Q1	4.37	2007 Q4	2.36
2000 Q2	4.55	2008 Q1	1.88
2000 Q3	3.73		

Source: UK Office of National Statistics.

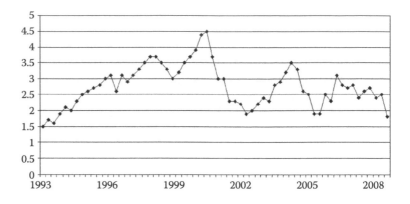

Figure 2.1 Growth rate in GDP (%) over time from first quarter 1993 to first quarter 2008.

In most cases the analysts assume that data of the kind seen here can be fitted by what is called a *Normal distribution* (also called a *bell curve* because that is its shape as shown in Figure 2.2).

The key thing about a Normal distribution is that it is an "idealized" view of a set of data. Imagine that, instead of trying to model annual growth rate, you were trying to model the height in centimeters of adults. Then, if you took a sample of, say, 1,000 adults and plotted the frequency of their heights within each 10 centimeter interval you would get a graph that looks something like Figure 2.3. As you increase the sample size and decrease the interval size you would eventually expect to get something that looks like the Normal distribution in Figure 2.4.

The Normal distribution has some very nice mathematical properties (see Box 2.1), which makes it very easy for statisticians to draw inferences about the population that it is supposed to be modelling.

Unfortunately, it turns out that, for all its nice properties the Normal distribution is often a very poor model to use for most types of risk assessment. And we will demonstrate this by returning to our GDP growth rate data. In the period from 1993 to 2008 the average growth rate was 2.96% with a standard deviation of 0.75.

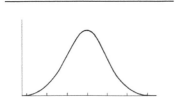

Figure 2.2 Normal distribution (bell curve).

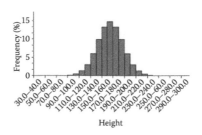

Figure 2.3 Histogram of people's height (centimetres).

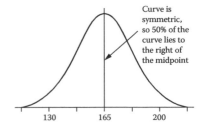

Figure 2.4 Normal distribution model for people's height.

Box 2.1   Properties of the Normal Distribution

- The distribution is symmetric around the midpoint, which is called the *mean* of the distribution. Because of this exactly half of the distribution is greater than the mean and half less than the mean. If the model is a good fit of the real data then the mean of the distribution should be close to the mean of the data from which it is drawn (that is, just the average calculated by adding all the data and dividing by the number of points). So, from Figure 2.4, we can infer that there is a 50% chance that a randomly selected adult in the UK will be taller than 165 cm.
- The "spread" of the data (which is the extent to which it varies from the mean) is captured by a single number called the *standard deviation*. Examples of Normal distributions with different standard deviations are shown in Figure 2.5. Once we know the mean and the standard deviation there are tables and calculators that tell us what proportion of the distribution lies between any two points.

  So, for example, the distribution that supposedly models height in Figure 2.4 has a mean of 165 and a standard deviation of 14. Approximately 20% of such a distribution lies between 178 and 190, so if this distribution really is an accurate model of height we can conclude that there is a 20% chance a randomly selected adult will be between 178 and 190 cm tall.
- As shown in Figure 2.6 it is always the case that 95% of the distribution lies between the mean and plus or minus 1.96 times the standard deviation. So (by symmetry) in the height example, 2.5% of the distribution lies above 195 cm. This means there is a 2.5% chance a randomly selected person is taller than 195cm.
- The Normal distribution approaches zero frequency in both directions, towards plus and negative infinity, but never reaches it. So, no matter how far we go away from the mean the curve never quite touches zero on the frequency axis (mathematicians say it is *asymptotic*). However, as we move away from the mean we very quickly get into tiny regions of the curve. For example, less than 0.001% of the distribution lies beyond the mean plus four standard deviations. So, in our height example less than 0.001% of the distribution lies to the right of 224 cm. This means that there is less than 1 in a 100,000 chance of an adult being more than 224 cm tall according to the model. Although the infinite tails of the Normal distribution are very useful in the sense that the model poses no limit on the possible values that can appear, it also leads to "impossible" conclusions. For example, the model implies there is a nonzero (albeit very small) chance that an adult can be less than zero centimeters tall, and a nonzero chance that an adult can be taller than the Empire State Building.

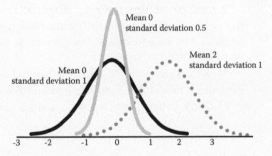

Figure 2.5   Examples of different Normal distributions.

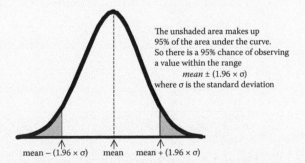

Figure 2.6   Normal distribution: 95% lies between the mean ±1.96 standard deviations.

Sidebar 2.2

Answers to original questions if we assume Normal distribution:

- What are the chances that the next quarter growth rate will be between 1.5% and 3.5%? Answer based on the model: approximately 72%.
- What are the chances that the growth will be less than 1.5% in each of the next three quarters? Answer: about 0.0125%, which is 1 in 8,000.
- What are the chances that within a year there will be negative growth (recession)? Answer: about 0.0003%, which is less than 1 in 30,000.

Table 2.2

Quarterly Annualized UK Growth Rate Figures from 2008 Q2-2010 Adjusted for Inflation

Quarter	Annual GDP%
2008 Q2	1.04
2008 Q3	−0.40
2008 Q4	−2.75
2009 Q1	−5.49
2009 Q2	−5.89
2009 Q3	−5.28
2009 Q4	−2.87

Source: UK Office of National Statistics.

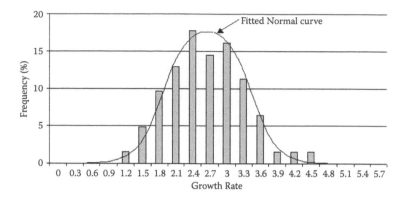

Figure 2.7   Histogram of annualized GDP growth rate from 1993 to 2008.

Following the approach as described earlier for the height example, we can create an appropriate histogram of the growth rate data and fit a Normal distribution to it as shown in Figure 2.7.

The fitted Normal curve has a mean of 2.96 and a standard deviation of 0.75. Using standard tables (plus a little of the kind of probability that you will learn about in Chapters 4 and 5) this enables us to answer the original questions that we posed in Sidebar 2.2.

Things turned out very differently from these optimistic predictions, as the actual data (between 1993 and 2009) shown in Figure 2.8 and Table 2.2 clearly demonstrate.

Within less than a year the growth rate was below −5%. According to the model a growth rate below −5% would happen considerably less frequently than once every 14 billion years (i.e., the estimated age of the universe).

Actual predictions made by financial institutions and regulators in the period running up to the credit crunch were especially optimistic because they based estimates of growth on the so-called Golden Period of 1998–2007.

So what went wrong? Clearly the Normal distribution was a hopelessly inadequate model. It looked like a reasonable fit for the data in the period 1993–2008 (when there was not a single period of negative growth;

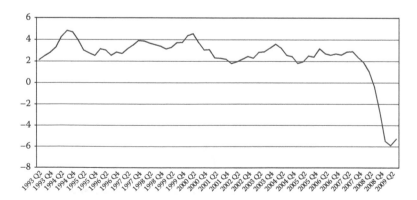

Figure 2.8   Growth rate in GDP (%) over time from 1993 to 2009.

indeed growth was never below 1.8%). But although the tails are infinite they are *narrow* in the sense we explained in Box 2.1: observations a long way from the mean are almost impossible. *Hence, a Normal distribution is inherently incapable of predicting many types of rare events.*

Whereas analysts like to focus on the most recent data as being the most relevant, especially during periods of prolonged stability such as had been experienced, a look at the longer term data reveals much greater potential for volatility. This can be seen in Figure 2.9, which charts the growth rate for the period 1956–2017.

When we plot these quarterly growth rates as a histogram (Figure 2.10) it looks very different from the histogram we saw in Figure 2.7 of the

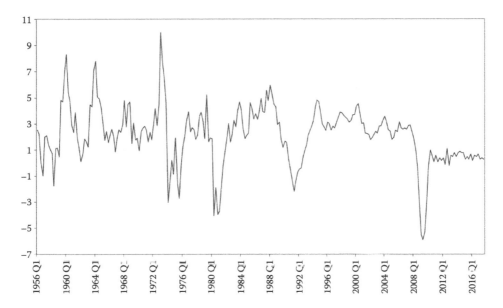

Figure 2.9   Growth rate in GDP (%) over time from 1956 to 2017.

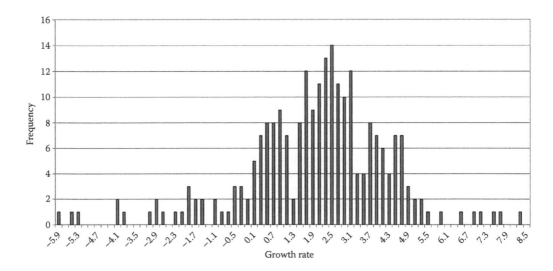

Figure 2.10   Histogram of annualized GDP growth rate from 1956 to 2017.

School Number	Score
38	175
43	164
44	163
25	158
31	158
47	158
11	155
23	155
48	155
40	153
7	151
30	151
6	150
9	149
33	149
19	148
10	147
12	147
32	147
2	146
27	146
42	146
28	145
35	145
49	145
45	144
46	143
1	142
18	142
22	141
26	141
4	140
14	140
29	140
39	139
8	138
5	136
17	136
34	136
3	134
24	133
36	131
37	131
15	130
21	130
16	128
13	120
20	116
41	115

period 1993–2008. Not only is the spread of the distribution much wider, but it is clearly not "Normal" because it is not symmetric.

Unfortunately, while basing predictions on the longer term data may have led to slightly more realistic results (in the sense of being less optimistic) even this data (and any amount of previous data that we may have collected) would still have been insufficient to predict the scale of the collapse in growth. The conditions prevailing in 2008 were unlike any that had previously been seen. The standard statistical approaches inevitably fail in such cases.

2.2 Patterns and Randomness: From School League Tables to Siegfried and Roy

Take a look at Table 2.3. It shows the scores achieved (on an objective quality criteria) by the set of state schools in one council district in the UK. We have made the schools anonymous by using numbers rather than names. School 38 achieved a significantly higher score than the next best school, and its score (175) is over 52% higher than the lowest ranked school, number 41 (score 115). Tables like these are very important in the UK, since they are supposed to help provide informed "choice" for parents. Based on the impressive results of School 38 parents clamour to ensure that their child gets a place at this school. Not surprisingly, it is massively oversubscribed. Since these are the only available state schools in this district, imagine how you would feel if, instead of your child being awarded a place in School 38, he or she was being sent to school 41. You would be pretty upset, wouldn't you?

You should not be. We lied. The numbers do not represent schools at all. They are simply the numbers used in the UK National Lottery (1 to 49). And each "score" is the *actual number of times* that particular numbered ball had been drawn in the first 1,172 draws of the UK National Lottery. So the real question is: Do you believe that 38 is a "better" number than 41? Or, making the analogy with the school league table more accurate:

Do you believe the number 38 is more likely to be drawn next time than the number 41? (Since the usual interpretation of the school league table is that if your child attends the school at the top he or she will get better grades than if he or she attends the school at the bottom.)

The fact is that the scores are genuinely random. Although the "expected" number of times any one ball should have been drawn is about 144 you can see that there is a wide variation above and below this number (even though that is still the average score).

What many people fail to realise is that this kind of variation is *inevitable*. It turns out that in any sequence of 1,172 lottery draws there is about a 50% chance that at least half the numbers will be chosen either less than 136 times or more than 152 times. That indeed is roughly what happened in the real sample. Moreover, the probability that at least one number will be chosen more than 171 times is about 45%. You may find it easier to think of rolling a die 60 times. You would almost certainly not get each of the six numbers coming up 10 times. You might get 16 threes and only 6 fours. That does that not make the number three "better" than the number four. The more times you roll the die, the closer in relative terms will be the frequencies of each number (specifically, if you roll the die n times the frequency of each number will get closer to n divided by 6 as n gets bigger); but in absolute terms the frequencies will not be exactly the same. There will inevitably be some numbers with a higher count than others. And one number will be at the "top" of the table while another will be "bottom."

We are not suggesting that all school league tables are purely random like this. But, imagine that you had a set of genuinely equal schools and you ranked them according to a suitable criteria like average exam scores. Then, in any given year, you would inevitably see variation like the earlier table. And you would be wrong to assume that the school at the top was better than the school at the bottom. In reality, there may be inherent quality factors that help determine where a school will come in a league table. But this does not disguise the fact that much of the variation in the results will be down to nothing more than pure and inevitable chance. See Box 2.2 for another example.

Box 2.2 Sporting Form: Quality or Luck?

The English premier league consists of 20 teams. The 20 competing in 2016–17 are shown in Table 2.4. This table also shows the results after each team has played every other team once where we have used 2 to represent victory for the first named team, 1 to represent draw, and 0 to represent defeat for the first named team.

Table 2.4
Premiership Results

	Ars	Bou	Bur	Che	CP	Ev	Hull	Lei	Liv	MC	MU	Mid	Sou	Sto	Sun	Swa	Tot	Wat	WB	WH
Arsenal		1	2	1	0	1	0	1	1	1	0	0	0	1	0	1	2	1	0	0
Bournemouth			0	2	1	0	0	2	2	1	0	2	1	1	1	0	1	2	0	2
Burnley				2	0	1	0	1	0	0	1	2	2	1	0	1	1	2	2	0
Chelsea					1	0	0	2	2	1	2	2	2	2	0	2	0	2	1	0
Crystal Palace						0	1	1	0	2	1	0	0	2	0	2	2	0	1	2
Everton							1	1	2	2	2	2	2	0	1	2	2	2	0	1
Hull City								2	0	2	2	0	2	1	1	1	2	1	0	2
Leicester									2	1	0	1	0	1	0	0	2	2	0	1
Liverpool										2	2	2	1	0	1	2	0	1	2	2
Man City											2	0	2	0	1	0	2	2	2	0
Man United												0	2	1	1	0	0	1	1	1
Middlesbrough													2	2	1	2	1	2	2	2
Southampton														0	1	0	0	0	0	1
Stoke City															1	1	2	0	0	0
Sunderland																1	0	1	0	2
Swansea City																	1	2	0	0
Tottenham																		1	0	0
Watford																			1	2
West Brom																				2
West Ham																				

So, for example, the 2 in the cell with row Arsenal and column Bur (Burnley) means that Arsenal defeated Burnley.

This data is not real. We generated random results in each result cell using the Excel function RANDBETWEEN(0,2). Based on these randomly generated results we determine the league table as shown in Table 2.5a using the premier league convention of 3 points for a win, 1 for a draw, and 0 for defeat.

It seems difficult to understand when you look at this table that there really is no difference in "form" between Everton (leaders with 40 points) and Leicester (bottom with 15 points). However, the career of many team managers will depend on perceptions of their performance. How then can we discriminate between those managers that are merely lucky and those that are genuinely competent?

What is striking about this table is how, *in its distribution of points*, it looks little different from the actual premier league table as it stood in week 19 of the season (January 2017) when all teams had played each other once (Table 2.5b).

Table 2.5
Premiership Table (a)

Everton	40
Middlesbrough	39
West Brom	37
Hull City	36
Liverpool	31
Crystal Palace	30
Chelsea	29
Burnley	28
Sunderland	29
Swansea	27
Bournemouth	26
West Ham	25
Man City	24
Stoke City	24
Tottenham	23
Watford	22
Man United	20
Southampton	16
Arsenal	15
Leicester City	15

[a] Based on random results

Table 2.5
Premiership Table (b)

Chelsea	49
Liverpool	43
Arsenal	40
Tottenham	39
Man City	39
Man United	36
Everton	27
West Brom	26
Bournemouth	24
Southampton	24
Burnley	23
West Ham	22
Watford	22
Stoke City	21
Leicester City	20
Middlesbrough	18
Crystal Palace	16
Sunderland	14
Hull City	13
Swansea	12

[b] Real table after 19 games Jan 2017

Unfortunately, many critical decisions are made based on wrongly interpreting purely random results exactly like these, even though the randomness was entirely *predictable* (indeed, we will show in Chapters 4 and 5 that probability theory and statistics are perfectly adequate for making such predictions).

In fact, most serious real-world "risks" are not like lotteries, and a very good illustration of this is to contrast the predictable risks confronting casinos (whose job is to run different types of lotteries) with less predictable and far more serious ones.

The risk for a casino of making massive losses to gamblers is entirely predictable because of our understanding of the "mechanical" uncertainty of the games played. For example, the "true" probability that a roulette wheel throw ends with the ball on a specific number (from 1 to 36) is not 1 in 36 as suggested by the winning "odds" provided by the casino but 1 in 38 because (in the USA) there are also two zero slots in addition to the 1-36 numbers (in UK casinos there is only one zero slot so the true probability of winning is 1 in 37). Hence their risk of ruin from losing money at the roulette wheel is easily controlled and avoided. Simply by placing a limit on the highest bet means that in the long run the casinos cannot lose really large sums because of the odds in their favor. And to minimize the risk of losses to cheating (i.e., where gamblers use techniques that swing the odds in their own favor) the casinos spend significant sums on security even though, again, such risks are also foreseeable.

Yet, the casinos can be blind to the kind of risks that could really bring them down, and there was no better example of this than what happened at the Mirage Resort and Casino on 3 October 2003. For many years the illusionists Siegfried and Roy had been the biggest draw in Las Vegas with their nightly show (Figure 2.11). A key part of the show was Roy's interaction with tigers. Roy had lived and even slept with tigers for some 40 years without any incident. Yet, after over 5,000 performances, on that fateful night Roy was mauled by one of his beloved tigers, causing life-threatening injuries. The show was immediately and permanently closed leading to the dismissal of 267 staff and performers and other losses (from ticket sales, hotel bookings, and legal costs) of hundreds of millions of dollars, making it the single worst loss in Las Vegas history. The casino managers—who you would think are the ultimate experts in risk assessment—were "beaten" by a risk they had not even considered. The magnitude of the resulting losses from that risk dwarfed the largest possible loss they could ever have suffered from the very predictable risks they had spent millions protecting themselves against.

Figure 2.11 Siegfried and Roy (pre-2003), illustrated by Amy Neil, aged 12.

This example was cited in an excellent book by Nassim Taleb whose title, *The Black Swan*, is the expression used to describe highly unpredictable events (for centuries all swans were assumed to be white because nobody had yet seen a black swan; the sighting of a black swan reversed all previous beliefs about the colour of swans). For the Mirage what happened on 3 October 2003 was a devastating black swan event.

Traditional statistical methods, which, we will see, rely on ideas like ideal lotteries, repeated observations and past data, cannot predict black swan events. Some, including Taleb himself, argue that no methods can predict such "unknowns" and "unknowables." We believe the methods advocated in this book allow for management and assessment of risks that include black swan events. We will see this in Chapter 13 on operational risk.

2.3 Dubious Relationships: Why You Should Be Very Wary of Correlations and Their Significance Values

The close "correlation" between *per capita cheese consumption* and *number of people who died by becoming tangled in their bedsheets* in

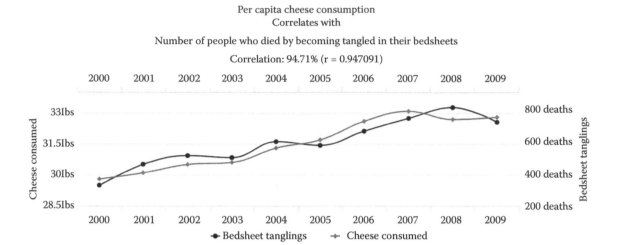

Figure 2.12   Example of spurious correlations published on website http://tylervigen.com. (Data sources: U.S. Department of agriculture and centers for disease control and prevention)

the United States, as shown in Figure 2.12, is one of many spurious correlations published on the website http://tylervigen.com.

Correlations and significance values (also called *p-values*) are the standard techniques that statisticians use to determine whether there are genuine relationships between different variables. In the approach to probability that we espouse in this book these traditional techniques (along with their first-cousins *regression analysis* and *confidence intervals,* which we will look at in Chapter 12) are superseded by techniques that we feel are simpler and more intuitive. But the acceptance and entrenchment of these ideas are so widespread across all empirical disciplines that you need to be aware of what they are in order to appreciate the damage they can do to rational decision making and risk analysis.

Look at Table 2.6. This gives (a) the average temperature and (b) the number of automobile crashes resulting in fatalities in the United States in 2008 broken down by month (source: U.S. Department of Transport 2008). We can plot this data in a scatterplot graph as shown in Figure 2.13.

From a quick view of the chart there is a relationship between temperature and fatalities. There seem to be more fatalities as the temperature increases. Statisticians use a formula—called the *correlation coefficient* (see Box 2.3)—that measures the extent to which the two sets of numbers are related. You do not need to know what this formula is because any spreadsheet package like Excel will do the calculation for you. It so happens that the correlation coefficient in this case is approximately 0.869. Using standard tables this turns out to be "highly significant" (comfortably passing the criteria for a *p*-value of 0.01 that is also explained in Box 2.3). Statisticians would normally conclude from this data that the number of road fatalities and the minimum temperature on any given day are significantly related (although note that we have severe concerns about the limitations of *p*-values as explained in Box 2.3).

Table 2.6
Temperature and Fatal Automobile Crashes

Month	Average Temperature	Total Fatal Crashes
January	17.0	297
February	18.0	280
March	29.0	267
April	43.0	350
May	55.0	328
June	65.0	386
July	70.0	419
August	68.0	410
September	59.0	331
October	48.0	356
November	37.0	326
December	22.0	311

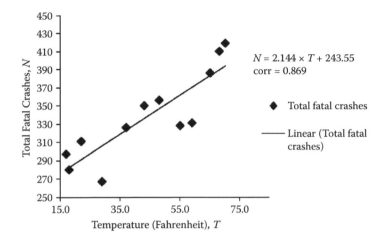

Figure 2.13 Scatterplot of temperature against road fatalities (each dot represents a month).

Box 2.3   Correlation Coefficient and *p*-Values: What They Are and Why You Need to Be Very Wary of Them

The correlation coefficient is a number between –1 and 1 that determines whether two paired sets of data (such as those for *height* and *intelligence* of a group of people) are related. The closer to 1 the more "confident" we are of a positive linear correlation and the closer to –1 the more confident we are of a negative linear correlation (which happens when, for example, one set of numbers tends to decrease when the other set increases as you might expect if you plotted a person's age against the number of toys they possess). When the correlation coefficient is close to zero there is little evidence of any relationship.

Confidence in a relationship is formally determined not just by the correlation coefficient but also by the number of pairs in your data. If there are very few pairs then the coefficient needs to be very close to 1 or –1 for it to be deemed "statistically significant," but if there are many pairs then a coefficient closer to 0 can still be considered "highly significant."

The standard method that statisticians use to measure the "significance" of their empirical analyses is the *p*-value. Suppose we are trying to determine if the relationship between height and intelligence of people is significant and have data consisting of various pairs of values (height, intelligence) for a set of people; then we start with the "null hypothesis," which, in this case is the statement "height and intelligence of people are unrelated." The *p*-value is a number between 0 and 1 representing the probability that the data we have arisen if the null hypothesis were true. In medical trials the null hypothesis is typically of the form that "the use of drug X to treat disease Y is no better than not using the drug."

The calculation of the *p*-value is based on a number of assumptions that are beyond the scope of this discussion, but people who need *p*-values can simply look them up in standard statistical tables (they are also computed automatically in Excel when you run Excel's regression tool). The tables (or Excel) will tell you, for example, that if there are 100 pairs of data whose correlation coefficient is 0.254, then the *p*-value is 0.01. This means that there is a 1 in 100 chance that we would have seen these observations if the variables were unrelated.

A low *p*-value (such as 0.01) is taken as evidence that the null hypothesis can be "rejected." Statisticians say that a *p*-value of 0.01 is "highly significant" or say that "the data is significant at the 0.01 level."

A competent researcher investigating a hypothesized relationship will set a *p*-value in advance of the empirical study. Typically, values of either 0.01 or 0.05 are used. If the data from the study results in a *p*-value of less than that specified in advance, the researchers will claim that their study is significant and it enables them to reject the null hypothesis and conclude that a relationship really exists.

In their book *The Cult of Statistical Significance* Ziliak and McCloskey expose a number of serious problems in the way *p*-values have been used across many disciplines. Their main arguments can be summarized as:

- Statistical significance (i.e., the *p*-value) is arbitrarily set and generally has no bearing on what we are really interested in, namely impact or magnitude of the effect of one or more variables on another.
- By focusing on a null hypothesis all that we are ever considering are existential questions, the answers to which are normally not interesting. So, for example, we might produce a very low *p*-value and conclude that road deaths and temperature are not unrelated. But the *p*-value tells us nothing about what we are really interested in, namely the nature and size of the relationship.
- Researchers sometimes wrongly assume that the *p*-value (which, remember, is the chance of observing the data if the null hypothesis is true) is equivalent to the chance that the null hypothesis is true given the data. So, for example, if they see a low *p*-value of say 0.01 they might conclude that there is a 1 in a 100 chance of no relationship (which is the same as a 99% chance that there is a relationship). This is, in fact, demonstrably false (we will show this in Chapter 6)—the *p*-value tells us about the probability of observing the data if the null hypothesis is true, and this may be very different from the probability of the hypothesis given the data; it is an example of one of the most pernicious and fundamental fallacies of probability theory that permeates many walks of life (called the *fallacy of the transposed conditional*). For example, in 2013 the 5th report of the Intergovernmental Panel on Climate (IPCC) Summary for Politicians asserted that "there is a 95% certainty that at least half the warming in the last 60 years is man-made." In fact, what the IPCC report actually showed was that the null hypothesis "Less than half the warming in the last 60 years is man-made" could be rejected at the 5% level (*p*-value 0.05), i.e. that there was less than a 5% probability of observing the actual data under the null hypothesis. That is very different from the assertion in the summary report.
- In those many studies (notably medical trials) where the null hypothesis is one of "no change" for some treatment or drug, the hypothesis comes down to determining whether the arithmetic mean of a set of data (from those individuals taking the treatment/drug) is equal to zero (supposedly representing status quo). In such cases, we have the paradox that, as we substantially increase the sample size, we will inevitably find that the mean of the sample, although approximately close to and converging to zero, will be significantly different from zero, even when the treatment genuinely has no effect. This is covered in Chapter 12 and is known as Meehl's conjecture.
- The choice of what constitutes a valid *p*-value is arbitrary. Is 0.04 radically different from 0.05? A treatment or putative improvement that yields a *p*-value that just misses the 0.05 target may be completely rejected and one that meets the target may be adopted.

Ziliak and McCloskey cite hundreds of examples of studies (all published in highly respected scientific journals) that contain flawed analyses or conclusions arising from the aforementioned misunderstandings. They give the following powerful hypothetical example of a fundamental weakness of using *p*-values:

Suppose we are interested in new drugs for reducing weight in humans. Two candidate drugs (called *Precision* and *Oomph* respectively) are considered. Neither has shown any side effects and their cost is the same. For each drug we conduct a study to test the null hypothesis "taking the drug leads to no weight loss." The results are:

- For drug *Precision* the mean weight loss is 5 lb and every one of the 100 subjects in the study loses between 4.5 lb and 5.5 lb.
- For drug *Oomph* the mean weight loss is 20 lb and every one of the 100 subjects in the study loses between 10 lb and 30 lb.

Since the objective of weight loss drugs is to lose as much weight as possible, any rational, intuitive review of these results would lead us to recommend drug *Oomph* over *Precision*. Yet the *p*-value test provides the opposite recommendation. For drug *Precision* the *p*-value is much lower (i.e. more significant) than the *p*-value for drug *Oomph*. This is because *p*-values inevitably "reward" low variance more than magnitude of impact.

The inevitable temptation arising from such results is to infer causal links such as, in this case, higher temperatures cause more fatalities. Indeed, using the data alone and applying traditional statistical regression techniques to that data you will end up with a simple model like that shown in Figure 2.14. Here the equation $N = 2.144 \times T + 243.55$ is the best linear fit for the data calculated using Excel's regression analysis tool. Using this equation we can predict, for example (by simply substituting the temperature values), that at 15°F we might expect to see 275 fatal crashes, per month while at 80°F we might expect to see 415 fatal crashes per month.

Such an analysis could lead to an apparently counterintuitive (and dangerous?) newspaper headline:

New research proves that driving in winter is actually safer than at any other time of the year.

What is happening in this example is that there are other underlying factors (such as *number of journeys made* and *average speed*) that contribute to an explanation of the number of road fatalities on any given day (we will return to this example in Chapter 3 to motivate the need for more intelligent causal models for risk analysis).

There are many well-known examples of similar dubious correlations that expose the dangers and weaknesses of this standard statistical method. Some are shown in Sidebar 2.3. The folklore belief that babies are delivered by storks is strongly supported by analysis of real statistical data. In 2001 Matthews showed that the stork population and number of births in European cities were correlated to a very high level of significance (*p*-value 0.008). But, of course, the correlation misses the explanation of a third common factor: population size. Obviously cities with larger populations have more births, but they also attract more storks.

Similarly, studies have shown that height and intelligence (as measured by an IQ test) of people are highly correlated. But, as illustrated in Figure 2.15, any attempt to explain intelligence causally by height misses the fact that the relationship is almost entirely due to a third factor, age; many people in the study were children between the ages of 4 to 16.

Bayesian Networks

You can consider the diagrams in Figure 2.14 and Figure 2.15 as our first examples of Bayesian networks. Each node (i.e. bubble) represents some variable of interest and an arc between two nodes represents some kind of influential (or even causal) relationship between the corresponding variables.

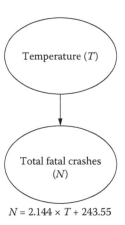

$N = 2.144 \times T + 243.55$

Figure 2.14 Simple regression model for monthly automobile fatal crashes.

Sidebar 2.3

Examples of purely coincidental (but strong) correlations:

- Level of beer drinking in the United States and child mortality in Japan.
- Solar radiation and the London Stock Exchange index.
- Sunspots and the lynx population density.
- Per capita consumption of mozzarella cheese and the number of civil engineering doctorates awarded.
- The website tylervigen.com provides many similar examples

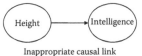

Inappropriate causal link

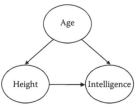

Correct influential relationship through underlying common cause

Figure 2.15 Spurious relationship resulting from failure to consider underlying common factor.

2.4   Spurious Correlations: How You Can Always Find a Silly "Cause" of Exam Success

Although the preceding examples illustrate the danger of reading too much into dubious correlations between variables, the relationships we saw there did not arise purely by chance. In each case some additional common factors helped explain the relationship.

But many studies, including unfortunately many taken seriously, result in claims of causal relationships that are almost certainly due to nothing other than pure chance.

Although nobody would seriously take measures to stop Americans drinking beer in order to reduce Japanese child mortality, barely a day goes by when some decision maker or another somewhere in the world takes just as irrational a decision based on correlations that turn out to be just as spurious.

For example, on the day we first happened to be drafting this section (16 March 2009) the media was buzzing with the story that working night shifts resulted in an increased risk of breast cancer. This followed a World Health Organization study and it triggered the Danish government to make compensation awards to breast cancer sufferers who had worked night shifts. It is impossible to state categorically whether this result really is an example of a purely spurious correlation. But it is actually very simple to demonstrate why and how you will *inevitably* find a completely spurious correlation in such a study—which you might then wrongly claim is a causal relationship—if you measure enough things.

Example 2.1   The Shotgun Fallacy

Let us suppose that we are interested in possible "causes" of student exam success. To make our example as simple as possible let us assume that exam scores are measured on a scale of 1 (worst) to 10 (best).

Now let us think of a number of possible "causes" of exam success. These could include plausible factors like *coursework score* and *class attendance*. But we could also throw in some implausible factors like the *number of sexual partners, number of football matches attended*, or *number of potatoes eaten on 12 January*. In fact, to effectively demonstrate the point let us only consider a set of totally implausible factors. For simplicity we will assume that, like the exam score, they can all be measured on a scale of 1 to 10.

Now although these factors—suppose we think of 18—are completely silly, let's actually remove any possibility that they are in any way valid factors by generating the results for them *purely randomly*. You can do this yourself. Create an Excel spreadsheet and type the entry =RANDBETWEEN(1,10) into cell A1. This will generate a random number between 1 and 10. By copying and pasting this entry create a set of random numbers like the set shown in Figure 2.16. There are 18 columns (A through to R) that we can think of as being the 18 silly factors associated with the students. We have also added column S, which represents the student exam score, again generated randomly in the same way.

Now, using Excel's built-in data analysis package, run a correlation analysis for each column (A through R) against the exam score

A	B	C	D	E	F	G	H	I	J	K	L	M	N	O	P	Q	R	S
10	9	5	3	10	4	7	4	3	9	5	4	5	9	4	4	10	5	5
10	10	4	2	3	10	6	1	8	5	8	8	8	7	6	3	8	3	1
5	7	9	9	3	2	5	2	6	6	7	8	2	9	10	3	8	2	1
9	3	6	10	3	1	5	8	2	9	5	8	7	4	8	8	2	7	7
2	6	1	1	10	8	8	5	8	7	10	4	7	9	7	4	3	7	3
10	3	6	7	1	10	9	9	6	2	8	5	8	3	9	9	2	2	7
1	1	1	7	5	1	4	9	1	6	9	8	9	9	4	1	2	7	5
3	5	8	4	2	4	6	2	7	9	5	2	2	5	4	3	2	1	1
1	8	8	10	6	4	10	7	6	6	5	7	3	7	10	7	4	9	8
4	4	8	8	3	1	1	9	1	9	10	9	10	2	8	1	3	4	10
9	3	5	3	3	2	4	4	3	10	4	9	8	7	3	10	2	8	4
2	3	1	1	6	7	10	5	5	1	4	4	3	10	9	5	7	1	6
10	9	1	3	10	6	7	7	8	1	9	4	3	7	3	3	10	3	7
4	2	5	10	9	9	2	4	9	8	9	7	5	7	6	6	1	7	2
1	7	3	5	5	8	8	10	2	10	7	10	2	10	4	8	5	2	8
9	8	8	1	4	2	8	7	10	1	6	8	1	1	9	6	4	1	2
4	4	3	2	4	7	5	3	1	3	5	10	2	5	2	6	7	8	9
4	7	3	10	5	10	7	3	6	6	6	5	10	8	1	6	2	7	5
10	9	6	5	9	7	4	8	10	2	8	6	3	5	9	9	6	3	2
7	5	5	3	5	4	8	4	3	4	4	2	1	9	6	4	7	6	9

Figure 2.16   Randomly generated numbers.

(column S). If the correlation coefficient is higher than 0.561 then the correlation is considered to be highly significant (the *p*-value is 0.01).

In fact, because of the number of factors we are trying here *it is very likely that you will find at least one column for which there is a significant correlation with S*. In Figure 2.16 the correlation coefficient of H and S is 0.59. Since column H is just as likely to represent *number of potatoes eaten on 12 January* as any other factor, would we be correct in concluding that eating potatoes on 12 January is the key to exam success?

In fact, because of the number of factors, it is also almost certain that among the 18 factors themselves you will also find at least two pairs that have a significant correlation. For example, in this case columns B and Q have a correlation coefficient of 0.62, which apparently might lead us to conclude that you can increase the number of football matches you attend by taking on more sexual partners.

What is clear from Example 2.1 is that if you measure enough different things about your subjects you will inevitably find at least one that is significantly correlated with the specific factor of interest. This may be the most likely explanation for night-shift work showing up as a significant "causal factor" in breast cancer.

This should put you on your guard when you next hear about a recommendation for a lifestyle change that results from a statistical study. It should also make you extremely wary of correlation coefficients and *p*-values.

2.5   The Danger of Regression: Looking Back When You Need to Look Forward

Suppose that you are blowing up a large balloon. After each puff you measure the surface area and record it as shown in Figure 2.17. So, after the 23rd puff the surface is 181 sq cm. What will the surface area be on the 24th puff? On the 50th puff?

As opposed to the data on growth rates in Section 2.1, there is no doubt that the data here exhibits a clear trend. When presented with this kind of problem professionals often try to find lines that best fit the historical trend. As we saw in Section 2.3 this is an example of *regression analysis*. As in the road fatalities example there, the simplest (and most common) approach is to assume a simple straight line fit (called linear regression), producing a line such as line A shown in Figure 2.18. Alternatively, we might decide that the relative slow down of increase toward the end of the data is indicative of a *curve* such as line B (this is an example of *nonlinear* regression). The lines provide

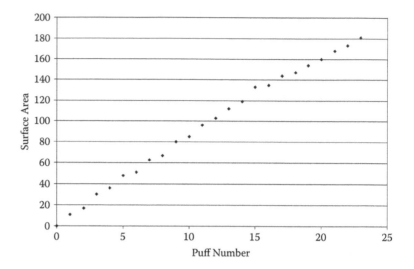

Figure 2.17   Increasing surface area of balloon.

us with a method of predicting future values. The line A fit results in a prediction of 186 for the 24th puff, whereas the line B fit results in a prediction of 183.

It is also common for analysts to apply what are called *time-series* adjustments into their prediction to take account of the fact that there are local sequential differences; in this case the even-numbered puffs tend to result in lower increases than the odd-numbered puffs (for the simple reason that we blow harder on alternative puffs). Factoring in the time-series analysis results in an adjusted prediction of 184 for puff 24 in the linear regression and 182 in the quadratic regression. Predictions further ahead, such as at puff 30, are farther apart (235 for line A and 185 for line B).

Unfortunately the balloon burst after 24 puffs (Figure 2.19).
Neither model was able to predict this.

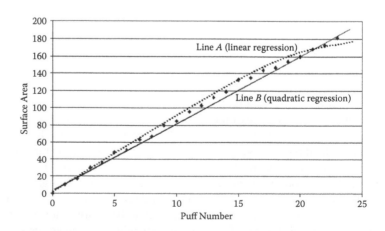

Figure 2.18   Lines of best fit for the data.

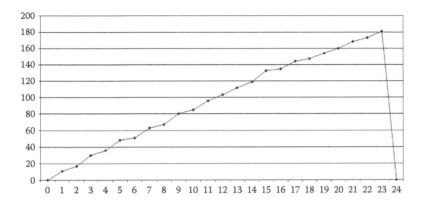

Figure 2.19   Balloon bursts on puff 24.

As we saw in Section 2.1 it was for reasons quite similar to this that the traditional statistical models were unable to predict the collapse of the banking sector in 2008 that ushered in a major worldwide recession. Although the models can incorporate millions of historical data to produce highly complex—and accurate—predictions over the short term during periods of growth, they were predicated on a set of barely articulated overoptimistic assumptions. The most basic knowledge about balloons would have indicated that a complete burst was inevitable, but traditional statistical models cannot incorporate this knowledge. Failure to do so is an example of what is commonly called the *fallacy of induction*. A similar example is highlighted in the Sidebar 2.4.

Whereas methods that rely purely on past data cannot predict these kinds of events, the methods advocated in this book at least provide the possibility of predicting them by enabling us to incorporate expert judgement about assumptions, missing variables, the structure of causal relations, and our uncertainty about these.

Sidebar 2.4

Does the Data Tell the Full Story?

Suppose a government collects data on terrorist attacks on its territory as shown in Figure 1.20. Traditional statistical modelling predicts that in year 17 the number of attacks will be 15% fewer than in year 16. This makes the threat sufficiently low that a range of expensive security measures can now be lifted.

But what if the decreasing number of attacks is the result not just of a reduced threat but also of the increasingly sophisticated counterterrorist measures? The causal impact of these measures is not incorporated in the statistical model and is therefore wrongly ignored.

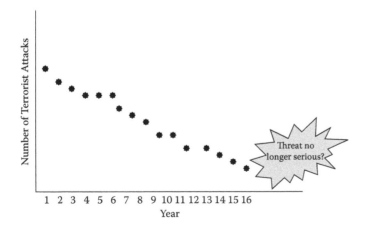

Figure 2.20   Charting attacks over time.

Table 2.7

Course Results

	Fred	Jane
Year 1 average	50	40
Year 2 average	70	62
Overall average	**60**	**51**

Table 2.8

Revised Score Information

	Fred	Jane
Year 1 total	350 (7 × 50)	80 (2 × 40)
Year 2 total	210 (3 × 70)	496 (8 × 62)
Overall total	560	576
Real overall average	**56**	**57.6**

Table 2.9

Overall

	Pre-natal care	
	Yes	No
Survives Yes	93	90
No	7	10
Survival rate	**93%**	90%

Table 2.10

Clinic 1

	Pre-natal care	
	Yes	No
Survives Yes	8	80
No	2	10
Survival rate	80%	**88%**

2.6   The Danger of Averages

Fred and Jane study on the same course spread over two years. To complete the course they have to complete 10 modules. At the end, their average annual results are as shown in Table 2.7. Jane's scores are worse than Fred's every year. So how is it possible that Jane got the prize for the student with the best grade? It is because the overall average figure is an average of the year averages rather than an average over all 10 modules. We cannot work out the average for the 10 modules unless we know how many modules each student takes in each year.

In fact:

- Fred took 7 modules in Year 1 and 3 modules in Year 2
- Jane took 2 modules in Year 1 and 8 modules in Year 2.

Assuming each module is marked out of 100, we can use this information to compute the total scores as shown in Table 2.8. So clearly Jane did better overall than Fred.

This is an example of *Simpson's paradox*. It seems like a paradox—Fred's average marks are consistently higher than Jane's average marks but Jane's overall average is higher. But it is not really a paradox. It is simply a mistake to assume that you can take an average of averages without (in this case) taking account of the number of modules that make up each average.

Look at it the following way and it all becomes clear: In the year when Fred did the bulk of his modules he averaged 50; in the year when Jane did the bulk of her modules she averaged 62. When you look at it that way it is not such a surprise that Jane did better overall.

This type of instance of Simpson's paradox is particularly common in medical studies. Consider the example shown in Tables 2.9–2.11 (based on a simplified version of a study described in Bishop et al., 1975) in which the indications from the overall aggregated data from a number of clinics (Table 2.9) suggest a positive association between pre-natal care and infant survival rate. However, when the data are analysed for each individual clinic (Tables 2.10–2.11) the survival rate is actually lower when pre-natal care is provided *in each case*. Bishop et al. concluded:

"If we were to look at this [combined] table we would erroneously conclude that survival was related to the amount of care received."

Pearl 2000 notes that:

"Ironically survival *was* in fact *related* to the amount of care received ... What Bishop et al. meant to say is that looking uncritically at the combined table, we would erroneously conclude that survival was *causally* related to the amount of care received."

We will look at a more profound and troubling example of Simpson's paradox later in the chapter. In fact, such examples provide a very convincing motivation for why causal models (implemented by Bayesian networks) are crucial for rational decision making. But first we have to address more fundamental concerns about the use of averages.

2.6.1 What Type of Average?

When we used the average for the exam marks data above we were actually using one particular (most commonly used) measure of average: the *mean*. This is defined as the sum of all the data point values divided by the number of data points.

But it is not the only measure of average. Another important measure of average is the *median*. If you put all the data point values in order from lowest to highest then the median is the value directly in the middle, that is, it is the value for which half the data points lie below and half lie above.

Since critical decisions are often made based on knowledge only of the average of some key value, it is important to understand the extent to which the mean and median can differ for the same data. Take a look at Figure 2.21. This shows the percentage distribution of salaries (in $) for workers in one city.

Note that the vast majority of the population (83%) have salaries within a fairly narrow range ($10,000–$50,000). But 2% have salaries in excess of $1 million. The effect of this asymmetry in the distribution is that the *median* salary is $23,000, whereas the mean is $137,000. By definition half of the population earn at least the median salary; but just 5% of the population earn at least the mean salary.

Of course, the explanation for this massive difference is the "long tail" of the distribution. A small number of very high earners massively skew the mean figure. Nevertheless, for readers brought up on the notion that most data is inherently bell-shaped (i.e. a Normal distribution in the sense explained in Section 2.1) this difference between the mean and median will come as a surprise. In Normal distributions the mean and median are always equal, and in those cases you do not therefore need to worry about how you measure average.

The ramifications in decision making of failing to understand the difference between different measures of average can be devastating.

Example 2.2   Using the Mean When You Really Need the Median

Suppose you are the mayor of the city mentioned earlier. To address the problem of unequal wealth distribution you decide to introduce a modest

Table 2.11
Clinic 2

	Pre-natal care	
	Yes	No
Survives Yes	85	10
No	5	0
Survival rate	94%	100%

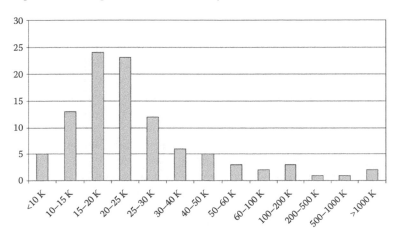

Figure 2.21   Percentage distribution of salaries for a large group of workers.

redistribution package. Every worker earning above "average" salary will pay a tax of $100 while every worker earning below average will receive an extra $100. You feel that this will prove not only popular, but crucially will be tax neutral overall; it will not cost the city a penny. Unfortunately, by basing your calculations on the mean ($137,000) rather than the median ($23,000) just 5,000 workers pay the extra $100 in tax, while 95,000 benefit from the extra $100. The move certainly proves popular but it bankrupts the city since you have to find $9 million of extra cash.

Example 2.3   Using the Median When You Really Need the Mean

Again suppose you are the mayor of the aforementioned city. This time you have to raise $100 million from taxpayers to fund a major new transport project. It is agreed that all workers will contribute a fixed proportion of their salary to pay for the project. What should the fixed percentage be? Stung by your unfortunate experience at wealth redistribution, this time you base your calculation on an "average" salary of $23,000. You work out that the necessary new "tax" is a whopping 4.3% for each of the 100,000 workers; this is because you believe that an "average" salary of $23,000 yields $1,000 and multiplying this by the total number of workers gets you to the magical $100 million. But this would only make sense if the mean salary was $23,000. In fact, because the mean salary is $137,000 the tax of 4.3% actually yields close to $600 million. The city makes an incredible profit, but unfortunately you are voted out of office because it is rightly perceived as an unnecessarily harsh tax. Basing your calculations on the mean salary of $137,000 requires a far more modest (and politically acceptable) 0.73% rate to make the target.

2.6.2   When Averages Alone Will Never Be Sufficient for Decision Making

Whereas Simpson's paradox and skewed distributions alert us to the need to be very careful in how we use averages, there are some fundamental reasons why, in many cases of critical decision making and risk analysis, averages should be avoided altogether.

If you were going on holiday to a particular location in July, then knowing that the average July temperature there (however you measure it) is 27°C does not provide you with sufficient information to know what clothes to pack; your decision would be very different if the temperature *range* was 10°C to 40°C compared to a range of 22°C to 29°C. Similarly, if you were a poor swimmer, it is doubtful that you would be willing to wade across a river if you were told that the average depth was 5 feet, even if you were 6 feet tall.

Some decision makers avoid these problems by insisting that they have what is called a *three-point estimate* for each key value.

So, in the temperature example above the three-point estimate might be {10, 27, 40}.

Such three-point estimates are very widely used by decision makers in critical applications (the military is especially keen on the approach). Unfortunately, although the three-point estimate seems an attractively simple way to describe the range for a value, it will generally be insufficient for rational decision making.

To see why, consider again the salary distribution in Figure 2.21. What is the three-point estimate here? We have already seen that the "average

A three-point estimate is simply three numbers:

{lowest possible value, average value, highest possible value}.

value" will be very different depending on whether we use the mean or median, and we clearly need at least both as the examples showed. But there are also serious problems with the lowest and highest possible values. In this case the lowest possible value is something very close to zero since there will be at least a small number of workers earning almost nothing. At the other end there is almost certainly at least one person earning over $20 million, so this will be the highest possible value. Neither the three-point estimate {0, 23,000, 20,000,000} nor {0, 137,000, 20,000,000} is sufficiently informative even to help us solve the problems in Examples 2.2 and 2.3. When confronted with this issue, proponents of three-point estimates will often propose that the lowest and highest values are replaced with what are called *percentiles*, typically 10% and 90% where the *n*% percentile is the value for which *n*% of the data items lie below. In the salary case this is more informative but still insufficient, whereas in the holiday example (where we were interested in temperature) it obscures the information we really need. To solve this we end up having to add additional percentiles, giving us not a three-point estimate but a five-, seven, or nine-point estimate. But no matter what we chose we can always find examples where the number of points may be insufficient.

Fortunately, there is a simple way out. We can just use the full distribution such as that shown in Figure 2.21. When decision makers use either an average or a 3-point estimate what they are trying to do is characterize the whole distribution in as simple way as possible. But truly rational decision making often requires us to consider the full distribution, rather than a crude simplification of it. It turns out that such a distribution is precisely what probability theory and Bayesian networks provide us with for all values of interest. And in some cases there may even be a very small number of values (called parameters) that enable you to determine the whole distribution (this is something we will explain properly in Chapter 5).

> The Normal distribution is an example where just two parameter values—the mean and the variance—determine the whole distribution.

2.7 When Simpson's Paradox Becomes More Worrisome

Consider the following more troubling example of Simpson's paradox (based on one from Pearl, 2000):

> A new drug is being tested on a group of 800 people (400 men and 400 women) with a particular disease. We wish to examine the effect that taking the drug has on recovery from the disease. As is standard with any randomised controlled trial, such as this clinical trial, half of the people (randomly selected) are given the drug and the other half are given a placebo. The results in Table 2.12 show that, of the 400 given the drug, 200 (i.e. 50%) recover from the disease; this compares favourably with just 160 out of the 400 (i.e. 40%) given the placebo who recover.

Therefore, clearly we can conclude that the drug has a positive effect. Or can we? A more detailed look at the data results in *exactly the opposite* conclusion. Specifically, Table 2.13 shows the results when broken down into male and female subjects.

Table 2.12
Drug Trial Results

Drug Taken	No	Yes
Recovered		
No	240	200
Yes	160	200
Recovery rate	40%	50%

Table 2.13
Drug Trial Results with Sex of Patient Included

Sex	Female		Male	
Drug Taken	No	Yes	No	Yes
Recovered				
No	210	80	30	120
Yes	90	20	70	180
Recovery rate	30%	20%	70%	60%

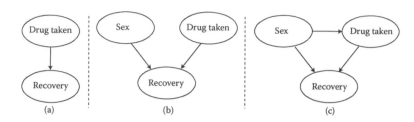

What we have in Figure 2.22 are three more examples of Bayesian networks. In this case we know not just the graphical structure of the network, but also the underlying "statistical" content. For the initial model Table 2.12 provides us with the necessary information about the outcome of "recovery" (yes or no) given the information about "drug taken" (yes or no).

For the revised causal model Table 2.13 provides us with the necessary information about the outcome of "recovery" (yes or no) given the different combinations of information about sex ("male" or "female") and "drug taken" (yes or no). It also provides us with the necessary information about the outcome of drug taken given sex. You need to get used to these kinds of tables because they, together with the graphical model, are exactly what you have to specify to complete a Bayesian network. You will learn how to do that, but we leave that until Chapter 7 when we will return to these same models and show how they easily explain and overcome Simpson's paradox.

There have been many well-known cases where Simpson's paradox has clouded rational judgement and decision making. Many of these cases are in medicine, but the most famous occurred at Berkeley University, which was (wrongly) accused of sex discrimination on the grounds that its admissions process was biased against women. Overall the data revealed a higher rate of admissions for men, but no such bias was evident for any individual department. The overall bias was explained by the fact that more women than men applied to the more popular departments (i.e. those with a high rejection rate).

Figure 2.22 Explaining Simpson's paradox using a causal model. (a) Initial model; (b) Revised causal model; (c) Causal model with additional information.

Focusing first on the men we find that 70% (70 out of 100) taking the placebo recover, but only 60% (180 out of 300) taking the drug recover. Therefore, *for men, the recovery rate is better without the drug.*

With the women we find that 30% (90 out of 300) taking the placebo recover, but only 20% (20 out of 100) taking the drug recover. Therefore, *for women, the recovery rate is also better without the drug.* **In every subcategory the drug is worse than the placebo.**

The process of drilling down into the data this way (in this case by looking at men and women separately) is called *stratification.* Simpson's paradox is simply the observation that, on the same data, stratified versions of the data can produce the opposite result to non-stratified versions. Often, there is a *causal* explanation. In this case men are much more likely to recover naturally from this disease than women. Although an equal number of subjects overall were given the drug as were given the placebo, and although there were an equal number of men and women overall in the trial, the drug was ***not*** equally distributed between men and women. More men than women were given the drug. Because of the men's higher natural recovery rate, overall more people in the trial recovered when given the drug than when given the placebo.

Unfortunately, as explained in Box 2.4 things can get even worse.

The difference between the types of data analysis is captured graphically in Figure 2.22. In the initial model we only have information about whether the drug is taken to help us determine whether a subject recovers. The revised causal model tells us that we need information about the subject's sex in addition to whether they take the drug to help us better determine whether the subject recovers. The final model introduces the further dependence, which is relevant for this particular case study namely that sex influences drug taken because **men are much more likely in this study** to have been given the drug than women.

2.8   How We Measure Risk Can Dramatically Change Our Perception of Risk

The way we measure risk can dramatically change our perception of risk. A good example surrounds the claim that flying is the safest form of transport. What is the basis for this claim? It is based on measuring

Box 2.4   Can we avoid Simpson's paradox?

The answer to this question is yes, but only if we are certain that we know every possible variable that can impact the outcome variable. If we are not certain—and in general we simply cannot be—then Simpson's paradox is theoretically unavoidable.

First, let us see how we can avoid the paradox in previous drug example of Section 2.8. Another way of looking at the paradox in that example is that, although the number of men and women in the study is the same, the drug is not equally distributed between men and women. The variable "sex" *confounds* the recovery rate. Confounding is the bias that arises when the treatment (drug) and the outcome (recovery) share a common cause—as illustrated in Figure 2.22(c); confounding is often viewed as the main shortcoming of such studies. To avoid it, we need an equal number of subjects for each state of the confounding variable (in this case there are two "states" of "sex" namely male and female) for each state of the other dependent variable ("drug taken").

So, in the example studied it is not sufficient to simply divide the subjects into two equal size "control groups" (400 taking the drug and 400 taking the placebo) *even if the total number of males and females are equal*. We actually need four equal size control groups corresponding to each state combination of the variables, that is:

- 200 subjects who fit the classification ("drug," "male")
- 200 subjects who fit the classification ("drug," "female")
- 200 subjects who fit the classification ("placebo," "male")
- 200 subjects who fit the classification ("placebo," "female")

Therefore, let us suppose that in a new study for some different drug we ensure that our 800 subjects are assigned into equal size control groups and that the results are as shown in Table 2.14.
Note the following:

- All four control groups have 200 subjects.
- The overall recovery rate is 63% with the drug compared with 52% with the placebo
- The recovery rate among men is 72% with the drug compared with 58% with the placebo
- The recovery rate among women is 54% with the drug compared with 46% with the placebo

Therefore, in contrast to the previous example, the drug is more effective overall and more effective in every sub-category. So surely we can recommend the drug and cannot possibly fall foul of Simpson's Paradox in this case?

Unfortunately, it turns out that we really can fall foul of the paradox—as soon as we realise there may be another confounding variable that is not explicit in the data. Consider the variable *age*, and for simplicity let us classify people with respect to this variable into just two categories "<40" and "40+," Even if we are lucky enough to have exactly 400 of the subjects "in each category" we may have a problem. Look at the results in Table 2.15 when we further stratify the data of Table 2.14 by age.

Since it is the *same* data obviously none of the previous results are changed, that is a higher proportion of people overall recover with the drug than the placebo, a higher proportion of men overall recover with the drug

Table 2.14
New Drug Trial Results with Sex of Patient Included

			Sex	Female		Male	
Drug taken	No	Yes	Drug taken	No	Yes	No	Yes
Recovered			Recovered				
No	192	148	No	108	92	84	56
Yes	208	252	Yes	92	108	116	144
Recovery rate	52%	63%	Recovery rate	46%	54%	58%	72%
Overall result: Favours drug			In each subcategory: Favours drug				

Table 2.15

New Drug Trial Results with Sex and Age of Patient Included

Age	40+				<40			
Sex	Female		Male		Female		Male	
Drug taken	No	Yes	No	Yes	No	Yes	No	Yes
Recovered								
No	96	28	80	24	12	64	4	32
Yes	64	12	80	16	28	96	36	128
Recovery rate	40%	30%	50%	40%	70%	60%	90%	80%

than the placebo, and a higher proportion of women overall recover with the drug than the placebo A. However, we can now see:

- *The proportion of young men who recover with the drug is 80% compared with 90% with the placebo*
- *The proportion of old men who recover with the drug is 40% compared with 50% with the placebo*
- *The proportion of young women who recover with the drug is 60% compared with 70% with the placebo*
- *The proportion of old women who recover with the drug is 30% compared with 40% with the placebo*

Therefore, *in every single subcategory* the drug is actually *less effective* than the placebo. This is despite the fact that in every single super-category the exact opposite is true. How is this possible? Because, just as in the earlier example, the size of the control groups at the lowest level of stratification are not equal; more young people were given the drug than old people (750 against 80). And young people are generally more likely to recover naturally than old.

The only way to be sure of avoiding the paradox in this case would be to ensure we had eight equal size control groups:

- 100 subjects who fit the classification ("drug," "male," "young")
- 100 subjects who fit the classification ("drug," "female," "young")
- 100 subjects who fit the classification ("placebo," "male," "young")
- 100 subjects who fit the classification ("placebo," "female," "young")
- 100 subjects who fit the classification ("drug," "male," "old")
- 100 subjects who fit the classification ("drug," "female," "old")
- 100 subjects who fit the classification ("placebo," "male," "old")
- 100 subjects who fit the classification ("placebo," "female," "old")

Even then we cannot be sure to have not missed another confounding variable.

deaths per distance travelled for the different modes of transport as shown in Figure 2.23.

While this is very comforting for those about to fly away on their holidays, things are not so simple. In terms of distance travelled, a single plane journey from London to the popular holiday location Majorca is the same as about 110 car journeys from West to East London. However, each such car journey, like the flight to Majorca, takes about 90 minutes. When we think of safety and risk what we are really interested in is *surviving a journey* and for this it is not fair to equate a single plane trip with 110 car trips. So let's look at safety by deaths per billion journeys rather than distance travelled. The results are shown in Figure 2.24.

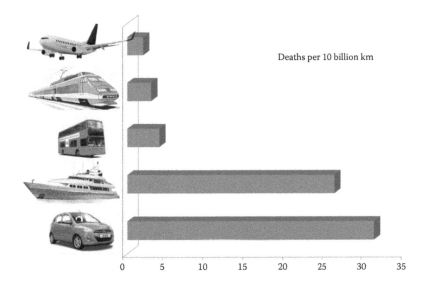

Figure 2.23 Safest form of travel? Travel by airplane is "20 times safer" than travel by car when safety is measured by deaths per distance travelled.

On this measure an airplane journey is three times as "risky" as a car journey. But, even with this measure the aircraft is a very safe way of travelling compared with modes of transport not yet considered. Indeed, if we change the scale in Figure 2.23 we get the results in Figure 2.25.

The astronomically higher "risk" of the space shuttle is based on the fact there were only 138 journeys resulting in 14 deaths.

It's also the case that simple differences in the way we measure risk can completely change our understanding and attitudes. Figure 2.26 shows a typical and widely reported recent news story on medical risk.

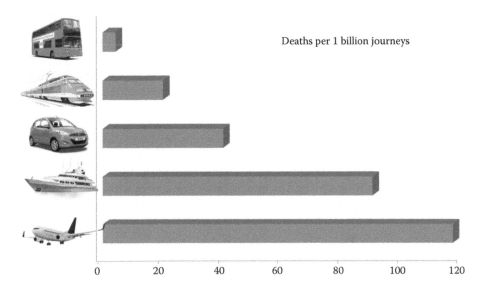

Figure 2.24 Safest form of travel? Travel by car is "3 times safer" than travel by airplane when safety is measured by deaths per number of journeys.

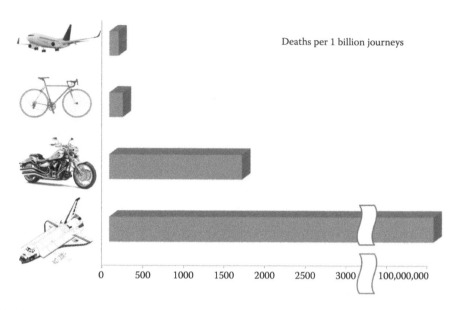

Deaths per 1 billion journeys

Figure 2.25 Safest form of travel? Measured by deaths per 1 billion journeys, the airplane is safer than a bicycle or motor-bike. By far the least safe form of travel is the space shuttle.

Drinking two glasses of wine a day triples risk of developing mouth cancer

Figure 2.26 Typical newspaper headline on medical risk. Beware of the difference between relative and absolute risk.

The story reported that drinking wine regularly (about two glasses at night) triples the risk of mouth cancer. It sounds like a devastating finding but in reality it is not, because what is being reported here is *relative risk* whereas what is of most interest is *absolute risk*. There are actually few deaths from mouth cancer. In fact, for every 200,000 deaths in the United Kingdom, about eight are from mouth cancer. Assuming the results of the study are reliable (i.e. the "tripling of risk") then what it means is that about six out of every eight people who die of mouth cancer drank wine regularly and two did not. That is where the "tripling" of risk comes from.

However this relative measure of "risk" is simply the ratio of drinkers to non-drinkers among those who die of mouth cancer. What we are really interested in knowing is the actual chance of dying from mouth cancer if we drink wine regularly compared with the chance of dying if we do not. To calculate this absolute risk we need to know what the proportion of regular wine drinkers is among those who did. Suppose the proportion is 15%. So for every 200,000 deaths about 30,000 are regular wine drinkers. That means about six out of 30,000 who are regular wine drinkers die of mouth cancer—i.e. 0.02%; this compares to two out of 170,000 who are not regular drinkers but who also die of mouth cancer—i.e. 0.0012%. So the absolute risk of dying from mouth cancer increases from 0.0012% to 0.002% for those who drink wine regularly. That is an increase of just 0.0008%. Interesting, but hardly the story implied.

2.9 Why Relying on Data Alone Is Insufficient for Risk Assessment

The last decade has seen an explosion of interest in "big data" and sophisticated algorithms for analysing such data. The popular belief is that, with sufficiently "big" data and increasingly powerful "machine learning" algorithms it should be possible, by using purely automated methods applied to the data, to discover all of the properties and relationships of interest for both improved prediction and decision-making. For example, such methods have been applied to large databases of supermarket customers to understand and predict the buying patterns of customers and to determine the optimal time to release new products. In areas such as healthcare the hope is that, given large patient databases, such methods can be used to understand both the causes of particular diseases and the optimum treatments. Unfortunately, in most areas of critical decision making there is limited relevant data (e.g. in medicine doctors do not always record what they do), while in other areas even very large databases will never provide the required answers. Nor does "big data" necessarily mean good quality data.

For example, a popular and important area for such machine learning is the use of "credit scoring" by banks to determine the risk associated with making loans to customers. The kind of database used by banks for this purpose is shown in Table 2.16, where each record (i.e. row) corresponds to a customer who was previously granted a loan.

Since too many people "default" on loans, the bank wants to use machine learning techniques on this database to help decide whether or not to offer credit to new applicants. In other words they expect to

Table 2.16

Typical Bank Database of Customers Given Loans

Customer	Age	Marital Status	Employment Status	Home Owner	Salary	Loan	...	Defaulted
1	37	M	Employed	Y	50,000	10,000	...	N
2	45	M	Self-employed	Y	60,000	5000	...	N
3	26	M	Self-employed	Y	30,000	20,000	...	Y
4	29	S	Employed	N	50,000	15,000	...	N
5	26	M	Employed	Y	90,000	20,000	...	N
6	35	S	Self-employed	N	70,000	20,000	...	Y
7	32	M	Self-employed	Y	40,000	5000	...	N
8	37	M	Employed	Y	25,000		...	Y
9	18	S	Unemployed	N	0	50,000	...	N
10	40	M	Employed	Y	65,000	45,000	...	N
11	21	S	Employed	N	20,000	10,000	...	Y
12	30	S	Employed	N	40,000	5000	...	N
13	22	M	Self-employed	N	30,000	10,000	...	Y
14	35	M	Unemployed	Y	0	3000	...	Y
15	19	S	Unemployed	N	0	100,000	...	N
...	...	...	...	...	...	...	...	...
100001	34	M	Employed	Y	45,000	1000	...	N
100002	28	S	Self-employed	N	25,000	2000	...	N
100003	19	S	Unemployed	N	0	25,000	...	N
...	...	...	...	...	...	...	...	...

"learn" when to refuse loans on the basis that the customer profile is too "risky."

The fundamental problem with such an approach is that the database contains only records of those who were granted loans. Analysis of such a database can learn nothing about those customers who were refused credit precisely because the bank decided they were likely to default. Any causal knowledge about such (potential) customers is missing from the data.

Suppose, for example, that the bank normally refuses credit to people under 20, unless their parents are existing high-income customers known to a bank manager. Such special cases (like customers 9, 15, 100003 above) show up in the database and they never default. Any pure data-driven learning algorithm will "learn" that unemployed people under 20 never default—the exact opposite of reality in almost all cases. Simplistic machine learning will therefore recommend giving credit to people known most likely to default.

2.10 Uncertain Information and Incomplete Information: Do Not Assume They Are Different

Consider the following assertions:

1. Oliver Cromwell spoke more than 3,000 words on 23 April 1654.
2. O.J. Simpson murdered his wife.
3. You (the reader) have an as-yet undiagnosed form of cancer.
4. England will win the next World Cup.

The events in assertions 1 and 2 either happened or did not. Nobody currently knows whether the assertion in statement 1 happened. Only O.J. Simpson knows for certain whether assertion 2 happened. Assertion 3 describes a fact that is either true or false. Assertion 4 is different because it describes the outcome of an event that has not yet happened.

While all four assertions are very different what that all have in common is that our knowledge about them is uncertain (unless we happen to be O.J. Simpson). In this book the way we reason about such uncertainty is the same whether the events have happened or not and whether they are unknown or not. Unfortunately, many influential people do not accept the validity of this approach. We have an obligation to demonstrate why those influential people are wrong. To do this we will consider the simple

---

Box 2.5   Uncertain versus Incomplete Information

Suppose you ask your friend Naomi to roll a die without letting you see the result, but before she rolls it you have to answer the following:

Question 1: Will the number rolled be a 3?

Having rolled the die Naomi must write down the result on a piece of paper (without showing you) and place it in an envelope.

> Now answer:
> Question 2: Is the number written down a 3 (i.e. was the number rolled a 3)?

scenario in Box 2.5 that captures the key differences between uncertain information and incomplete information.

Most people would be happy to answer Question 1 with something like the following (which, as we will see in Chapter 3 is an example of a *probabilistic* statement): There is a one in six chance of it being 3. Yet, there are many people who are convinced that such a probabilistic statement is meaningless for Question 2. Their reasoning is as follows:

1. There is no uncertainty about the number because it is a "fact"—it is even written down (and is known to Naomi).
2. The number either is a 3 (in which case there is 100% chance it is a 3) or it is not a 3 (in which case there is 0% chance it is a 3).

So some people are happy to accept that there is genuine uncertainty about the number *before* the die is thrown (because its existence is "not a fact"), but not *after* it is thrown. This is despite the fact that our knowledge of the number after it is thrown is as *incomplete* as it was before.

Nowhere is this type of distinction more ingrained than in the law: A defendant stands trial for a crime that, of course, *has already been committed*. Because the crime has already been committed the defendant either is or is not guilty of that crime.

In most cases the only person who knows for certain whether the defendant is guilty is the defendant. However, the defendant is not the one who has to determine guilt. Although the law implicitly endorses probabilistic reasoning when it talks about "balance of probabilities" and "beyond reasonable doubt" it often abhors any explicit probabilistic reasoning about innocence and guilt in court based on the same irrational argument as above. As an eminent lawyer told us:

> Look, the guy either did it or he didn't do it. If he did then he is 100% guilty and if he didn't then he is 0% guilty; so giving the chances of guilt as a probability somewhere in between makes no sense and has no place in the law.

What is curious about the rejection of the probabilistic answer to Question 2, is that we can *prove* that this rejection leads to irrational decision making as follows. This type of argument is commonly known as the Dutch Book:

The scenario in which people are convicted on the basis of crimes that they are *predicted* to commit remains the domain of pure science fiction like the Hollywood movie *Minority Report*.

> Suppose that you ask 60 people to each bet $1 on the number written down by Naomi (and you can assume Naomi is not one of the 60 betting). You have to set the odds and must choose one of the following options:

Calculating the Break-Even Odds for Guessing the Correct Die Number Written Down

Out of 60 people we can expect about 10 to choose the number 1, 10 to choose the number 2, 10 to choose the number 3, and so on. Of course, in practice these actual numbers will vary (as we saw in Section 2.2), but as this is the most likely outcome, it would be irrational to make any other assumption.

So, whatever number is written down we can expect about 10 people to win and 50 people to lose. Using Option A this results in us taking $50 from the losers and paying out $40 to the winners. So we expect to win $10 overall. Using Option C this results in us taking $50 from the losers and paying out $60 to the winners. So we expect to lose $10 overall. Only using Option B do we expect to break even (taking $50 from the losers and paying out $50 to the winners). Of course a bookie, if he wanted to stay in business, would offer Option A.

- Option A—If they choose the correct number you pay them $4 plus their $1 stake. Otherwise you win their $1 stake.
- Option B—If they choose the correct number you pay them $5 plus their $1 stake. Otherwise you win their $1 stake.
- Option C—If they choose the correct number you pay them $6 plus their $1 stake. Otherwise you win their $1 stake.

The twist to the scenario is that *your life depends on getting as close as possible to breaking even.* In that case, whatever your views about the uncertainty or otherwise of the number being 3, you will surely do the kind of calculations shown in the sidebar to choose Option B rather than Option A or Option C. But that means that you accept that the chances of the number being a 3 must be closer to 1 in 6 than to either 1 in 5 or 1 in 7. And if you accept this then it is irrational to reject a probabilistic answer to Question 2. Moreover, by accepting the validity of the statement "There is a one in six chance of it being 3" you have just almost certainly saved your life. Not accepting this statement means you will probably die (actually there is a 2 in 3 chance you will die because you should have no preference between any of the three options).

What really lies at the heart of people's concerns about using probabilities to describe incomplete information is that people with different levels of knowledge about the information will have different probabilities. So, whereas we should accept as reasonable the probability of a one in six chance of Naomi's number being a 3, Naomi really does have reason to reject it because she knows the chance is 100% (if it is a 3) and 0% if it is not. If Naomi told her friend Hannah that the number written down is an odd number, then Hannah's personal probability for the number being a 3 should be a one in three chance (because 3 of the possible numbers are odd).

By the same argument, for the defendant in court there is no uncertainty about guilt. But that does not remove the obligation from the jury to make a probabilistic assessment of guilt based on the incomplete information made available to them.

What is clear from the preceding discussion is that:

- If our knowledge of an event that has already happened is incomplete then we need to reason about it in the same way as we reason about uncertain events yet to happen. The failure to recognize that uncertainty and incomplete knowledge have to be handled in the same way leads to irrational decision making in some of the most critical situations.
- Different people will generally have different information about the same event (and this applies both to past and future events). Because of this people will generally have their own

personal probability assessment of the event. In economics this difference in knowledge is often called "information asymmetry": for example, when buying a used automobile you may not know whether it is a "peach" or a "lemon," but the salesman selling the car has no such uncertainty and his probabilities will be very different from that of his customers.

This notion of "personal probabilities" is central to the Bayesian reasoning in this book. It is a property of the mind and not of the object, hence the contrasting use of the labels "subjective" and "objective." We will explore it further when we define probability formally in Chapter 5. Unfortunately, as the next section demonstrates, it turns out that correct probabilistic reasoning can be very difficult and seem counter-intuitive.

2.11   Do Not Trust Anybody (Even Experts) to Properly Reason about Probabilities

Try answering the puzzle in Box 2.6.

Box 2.6   Birthdays Puzzle

In a class of 23 children the chances that at least two children share the same birthday is:

    a. Approximately 1 in 16
    b. Approximately 1 in 10
    c. Approximately 1 in 5
    d. Approximately 1 in 3
    e. Approximately 1 in 2

If you have not already seen the puzzle then you may be surprised to know that the "correct" answer here is *e* (in fact, the chances are slightly better than 1 in 2). You don't need to know why at this point; a proper explanation will be provided in Chapter 5.

Box 2.7   The Monty Hall Problem

Let's Make a Deal, a classic American '60s game show, hosted by Monty Hall, involved contestants choosing one of three doors. Behind one of the doors was a valuable prize such as a new car. Behind the other two doors was something relatively worthless like a banana.

After the contestant chooses one of the three doors Monty Hall (who knows which door has the prize behind it) always reveals a door (other than the one chosen) that has a worthless item behind it. He now poses the question to the contestant:

"Do you want to switch doors or stick to your original choice?"

Most people assume that there is no benefit in switching; they feel that by sticking to their original choice they have a 50% chance of winning, the same as if they switch.

In fact they are wrong. It turns out that, by switching, you have a 2 in 3 chance of winning. We will give a simple explanation why once we have formally defined probability in Chapter 5, and will also describe a Bayesian network solution in Chapter 7.

But if you answered *a*—on the instinctive basis that it is the closest to 23/365—then although you are completely wrong you are at least in good company. It is easily the most common answer. People are similarly stumped by the classic Monty Hall problem described in Box 2.7.

Not knowing the probability that children share the same birthday (or even the probability of winning the Monty Hall game) is hardly going to affect your life. But these problems are strikingly similar to many problems that can and do affect lives. In fact, even highly intelligent people, like world-leading barristers, scientists, surgeons, and businessmen misunderstand probability and risk, as the following examples indicate.

Example 2.4   The Harvard Medical School Question

In a classic and much referenced study by Casscells and colleagues the following question was put to students and staff at Harvard Medical School:

One in a thousand people has a prevalence for a particular heart disease. There is a test to detect this disease. The test is 100% accurate for people who have the disease and is 95% accurate for those who don't (this means that 5% of people who do not have the disease will be wrongly diagnosed

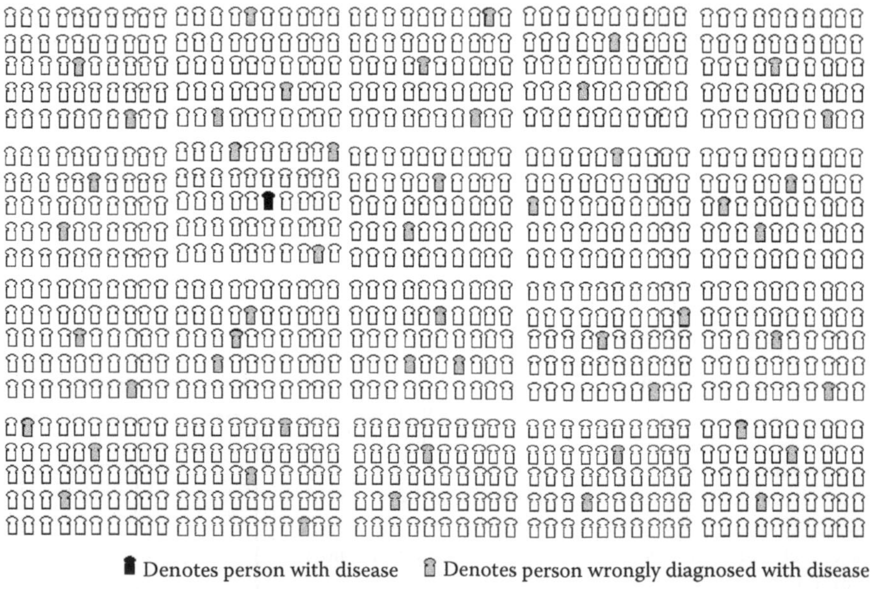

🯅 Denotes person with disease   🯄 Denotes person wrongly diagnosed with disease

Figure 1.27 In 1,000 random people about 1 has the disease but about 50 more are wrongly diagnosed as having the disease. So about 1 in 51 people who test positive for the disease actually have the disease, i.e. less than 2%.

as having it). If a randomly selected person tests positive what is the probability that the person actually has the disease?

Almost half gave the response 95%. The "average" answer was 56%. In fact, as we will explain formally, in Chapter 6, the correct answer is just below 2%. Figure 2.27 provides an informal visual explanation.

Think, for a moment, of the implications of this. If you test positive it is still extremely unlikely that you have the disease. But there are doctors who would believe you almost certainly have the disease and would proceed accordingly. This could result not just in unnecessary stress to you but even unnecessary surgery.

Example 2.5   The Prosecutor's Fallacy

Suppose a crime has been committed and that the criminal has left some physical evidence, such as some of his blood at the scene. Suppose the blood type is such that only 1 in every 1,000 people has the matching type. A suspect, let's call him Fred, who matches the blood type is put on trial. In court the prosecutor argues as follows:

> The chances that an innocent person has the matching blood type is 1 in a 1,000. Fred has the matching blood type. Therefore the chances that Fred is innocent is just 1 in a 1,000.

When an eminent prosecutor makes a statement like this, backed by forensic evidence, it is clear that its influence on the jury could be profound. Yet, as we will explain in Chapter 4, the prosecutor's conclusion generally massively understates the true probability that Fred is innocent. Figure 1.28 provides an informal visual explanation of this in the case where the number of potential suspects is 10,000 (in this case the chances that Fred is innocent is about 91%). And, as we will show in Chapters 6 and 15, mistakes exactly like this continue to be made by lawyers and forensic scientists in courtrooms throughout the world. The result is that

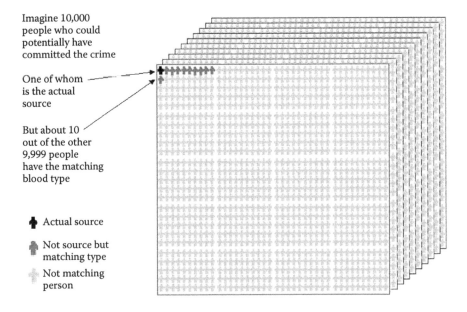

Figure 2.28   The potential source population.

the value of evidence is misunderstood, and juries are influenced to make poor decisions.

You would expect that where critical decisions need to be made, the probabilities are calculated correctly. Unfortunately, they are usually not. In medical and legal situations lives are affected as a result. In business, companies can go bust and in many everyday financial cases the public's general inability to understand probabilities is cunningly used against them.

And it is not just about relying on other people's ability to calculate probabilities properly. Every day you make decisions that, consciously or not, depend on probability assessments. Whether it is deciding which way to travel to work, deciding if it is worth taking out a particular insurance, deciding if you should proceed with a major project, or just improving your chances of winning at cards or on a sporting bet, the ability to do accurate probability calculations is the only way to ensure that you make the optimal decisions. One of the main challenges of this book is to help improve the way you do it.

Key points covered include:

- Uncertainty is a function of the lack of information and differences in certainty between individuals that reflect differences in personal experience and beliefs.
- Real risks do not often behave like lotteries.
- Beware spurious correlations purporting to reveal causal connections.
- Averages can be dangerous because they provide false security.
- Risks are not necessarily distributed normally nor are they symmetric.
- Big data and machine learning will not help in most risk assessment problems.
- Beware when people quantify risk as relative risk when they should be using absolute risk.
- Beware of the fallacy of induction.

2.12 Chapter Summary

The aim of this chapter was to introduce, by way of motivating examples, the key ideas of risk assessment and causal modelling that are the focus of the book. To appreciate why causal modelling (implemented by Bayesian networks) is such an effective method for risk assessment and decision analysis you first need to understand something about the traditional statistics and data analysis methods that have previously been used for this purpose. In introducing such methods, we have exposed a number of misconceptions and identified the most important limitations. In particular, we have demonstrated why these techniques provide little support for real practical risk assessment. To address these issues we now need to turn to causal models (Bayesian networks) and a different approach to probability than is typically used by statistical analysts.

Further Reading

Adams, J. (1995). *Risk*, Routledge.
BAYES-KNOWLEDGE (2018) http://bayes-knowledge.com/
Bickel, P. J., Hammel, E. A., and O'Connell, J. W. (1975). 'Sex bias in graduate admissions: Data from Berkeley.' *Science* 187, 398–404.
Casscells, W., Schoenberger, A., and Graboys, T. B. (1978). 'Interpretation by physicians of clinical laboratory results.' *New England Journal of Medicine* 299, 999–1001.
Eastaway, R., and Wyndham J. (1998). *Why Do Buses Come in Threes? The Hidden Mathematics of Everyday Life*, John Wiley & Sons.
Fenton, N. E. (2012). 'Making Sense of Probability: Fallacies, Myths and Puzzles.' www.deecs.qmul.ac.uk/~norman/papers/probability_puzzles/Making_sense_of_probability.html.

Gigerenzer, G. (2002). *Reckoning with Risk: Learning to Live with Uncertainty*, Penguin Books.

Haigh, J. (2003). *Taking Chances: Winning with Probability*, Oxford University Press.

Haldane, A. G. (2009). 'Why banks failed the stress test.' Marcus-Evans Conference on Stress Testing, 9-10 Feb 2009. London, Bank of England, www.bankofengland.co.uk/publications/speeches/2009/speech374.pdf.

Hubbard, D. W. (2009). The Failure of Risk Management: Why It's Broken and How to Fix It, Wiley.

Hubbard, D. W. (2010). How to Measure Anything: Finding the Value of Intangibles in Business, 2nd Edition, Wiley.

Kendrick, M. (2015). Doctoring data : how to sort out medical advice from medical nonsense., Columbus Publishing.

Lewis, H. W. (1997). *Why Flip a Coin? The Art and Science of Good Decisions*, John Wiley & Sons.

Matthews, R. (2001). 'Storks deliver babies (p = 0.008).' *Teaching Statistics* 22(2), 36–38.

Pearl, J. (2000). *Causality: Models, Reasoning and Inference*, Cambridge University Press.

Pearl, J., and Mackenzie, D. (2018). *The book of why: the new science of cause and effect*, Basic Books, New York.

Piatelli-Palmarini, M. (1994). *Inevitable Illusions: How Mistakes of Reason Rule Our Minds*, John Wiley & Sons.

Taleb, N. N. (2007) "Fooled by Randomness: The Hidden Role of Chance in Life and in the Markets", Penguin Books.

Taleb, N. N. (2007). *The Black Swan: The Impact of the Highly Improbable*, Random House.

Vigen, T. Spurious Correlations: (Amazon Books, 2015).

Ziliak, S. T., and McCloskey, D. N. (2008). *The Cult of Statistical Significance*, The University of Michigan Press.

The Need for Causal, Explanatory Models in Risk Assessment

3.1 Introduction

The aim of this chapter is to show that we can address some of the core limitations of the traditional statistical approaches exposed in Chapter 2 by introducing causal explanations into the modeling process. The causal models are examples of Bayesian Networks (BNs).

Although we will not formally define BNs until Chapter 6 (because such a definition requires an understanding of Bayesian probability that we explain in Chapter 5), our intention here is to provide a flavor of their power and flexibility in handling a range of risk-assessment problems.

In Section 3.2 we use the automobile crash example from Chapter 2 to explain the need for and structure of a causal BN. In Section 3.3 we explain why popular methods such as risk registers and heat maps are insufficient to properly handle risk assessment. We describe the causal approach to risk assessment in Section 3.4, showing how it overcomes the limitations of the popular methods.

We describe a simple causal approach to risk assessment that overcomes basic flaws and limitations of other commonly used approaches.

3.2 Are You More Likely to Die in an Automobile Crash When the Weather Is Good Compared to Bad?

We saw in Chapter 2, Section 2.3 some data on fatal automobile accidents. The fewest fatal crashes occur when the weather is at its worst and the highways are at their most dangerous. Using the data alone and applying the standard statistical regression techniques to that data we ended up with the simple regression model shown in Figure 3.1.

But there is a grave danger of confusing prediction with risk assessment. For risk assessment and management the regression model is useless, because it provides no explanatory power at all. In fact, from a risk perspective this model would provide irrational, and potentially dangerous, information. It would suggest that if you want to minimize your chances of dying in an automobile crash you should do your driving when the highways are at their most dangerous—in the winter months.

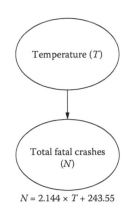

$$N = 2.144 \times T + 243.55$$

Figure 3.1 Simple regression model for automobile fatalities.

All models in this chapter are available to run in AgenaRisk, downloadable from www. agenarisk.com.

Table 3.1

Risk of Fatal Crash per Billion Miles Traveled in the Northeastern States of the United States in 2008

Month	Total Fatal Crashes	Miles Traveled (millions)	Crash Rate
January	297	34241	8.67
February	280	31747	8.82
March	267	36613	7.29
April	350	36445	9.60
May	328	38051	8.62
June	386	37983	10.16
July	419	39233	10.68
August	410	39772	10.31
September	331	37298	8.87
October	356	38267	9.30
November	326	34334	9.49
December	311	37389	8.32

Source: U.S. Department of Transportation, 2008.

In Risk Assessment Normalized Measures Are Usually Better Than Absolute Measures

Knowing that there are 20 fatal crashes in one year in city A compared to 40 in city B does not necessarily mean it is safer to drive in city A. If we know that 1 million miles were traveled in city A compared to 10 million in city B, then the crash rate is 20 per million miles for city A compared to 4 per million miles in city B. As obvious as this seems, the failure to "normalize" data in this way continues to lead to common errors in risk analysis. Often, low absolute numbers (such as service failures) may simply be because nobody uses the service rather than because it is high quality; indeed poor quality may be the very reason why the service is rarely used.

Instead of using just the total number of fatal crashes to determine when it is most risky to drive, it is better to factor in the number of miles traveled so that we can compute the *crash rate* instead, which is defined as the number of fatal crashes divided by miles traveled. Fortunately we have ready access to this data for the northeastern states of the United States in 2008 on a monthly basis as shown in Table 3.1.

As explained in the sidebar, the crash rate seems to be a more sensible way of estimating when it is most risky to drive.

However, when we graph the crash rate against the temperature, as shown in Figure 3.2, there still seems to be evidence that warmer

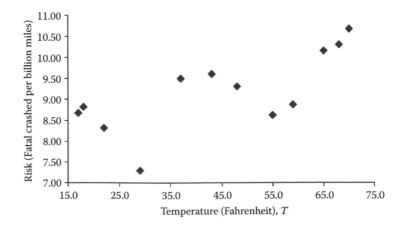

Figure 3.2 Scatterplot of number of crashes per billion miles traveled by temperature.

weather is "riskier" (although the correlation is weaker than when we considered simply the total number of fatal crashes).

Since common sense suggests that we should expect the risk to increase during winter (when road conditions are most dangerous) we must look elsewhere for an explanation.

What we know is that, in addition to the number of miles traveled (i.e., journeys made), there are other underlying causal and influential factors that might do much to explain the apparently strange statistical observations and provide better insights into risk. With some common sense and careful reflection we can recognize the following:

- The temperature influences the highway conditions (they will be worse as the temperature decreases).
- But temperature also influences the number of journeys made; people generally make more journeys in spring and summer, and will generally drive less when weather conditions are bad. This means the miles traveled will be less in winter.
- When the highway conditions are bad people tend to reduce their speed and drive more slowly. So highway conditions influence speed.
- The actual number of crashes is influenced not just by the number of journeys but also the speed. If relatively few people are driving and taking more care, we might expect fewer fatal crashes than we would otherwise experience.

The influence of these factors is shown in Figure 3.3, which is an example of a BN. To be precise, it is only the graphical part of a BN.

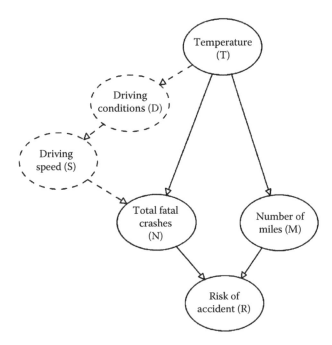

Figure 3.3 Causal model for fatal crashes.

In the highway example we have information in a database about temperature, number of fatal crashes, and number of miles traveled. These are therefore often called *objective* factors. If we wish our model to include factors for which there is no readily available information in a database we may need to rely on expert judgment. Hence, these are often called *subjective* factors. Just because the driving speed information is not easily available does not mean it should be ignored. Moreover, in practice (as we shall see in Chapter 4) the distinction between what is objective and what is subjective can be very blurred.

People are known to adapt to the perception of risk by tuning the risk to tolerable levels. For example, drivers tend to reduce their speed in response to bad weather. This is formally referred to as *risk homeostasis*.

To complete the BN model we have to specify the strength of the relationships between linked factors, since these relationships are generally uncertain. It turns out we need probabilities to do this (we actually need a probability table for each of the nodes). The details of what this all means are left until Chapter 6. But you do not need to know such details to get a feel for the benefits of such a model. The crucial message here is that the model no longer involves a simple single causal explanation; instead it combines the statistical information available in a database (the objective factors as explained in the sidebar) with other causal subjective factors derived from careful reflection.

The objective factors and their relations are shown with solid lines and arrows in the model, and the subjective factors are shown using dotted lines. Furthermore, these factors now interact in a nonlinear way that helps us to arrive at an explanation for the observed results. Behavior, such as our natural caution to drive slower when faced with poor road conditions, leads to lower accident rates. Conversely, if we insist on driving fast in poor road conditions then, irrespective of the temperature, the risk of an accident increases and so the model is able to capture our intuitive beliefs that were contradicted by the counterintuitive results from the simple regression model.

We could extend this process of identifying and attributing causes to help explain the change in risk to include other factors, such as increased drink driving during the summer and other holiday seasons. This is shown in Figure 3.4.

The role played in the causal model by driving speed reflects human behavior. The fact that the data on the average speed of automobile drivers was not available in a database explains why this variable, despite

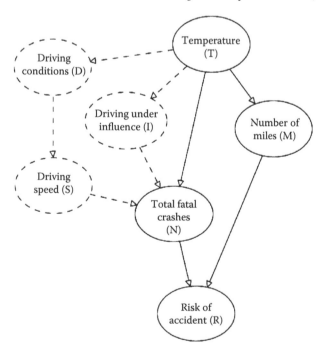

Figure 3.4 Extended causal model for fatal highway crashes.

its apparent obviousness, did not appear in the statistical regression model.

By accepting the naïve statistical model we are asked to defy our senses and experience and actively ignore the role unobserved factors play in the model. In fact, we cannot even explain the results without recourse to factors that do not appear in the database. This is a key point: With causal models we seek to dig deeper behind and underneath the data to explore richer relationships than might be admitted by oversimplistic statistical models. In doing so we gain insights into how best to control risk and uncertainty. The original regression model, based on the idea that we can predict automobile crash fatalities based on temperature, fails to answer the substantial question: How can we control or influence behavior to reduce fatalities? This at least is achievable; control of weather is not.

> The situation whereby a statistical model is based only on available data, rather than on reality, is called *conditioning on the data*. This enhances convenience but at the cost of accuracy.

3.3   When Ideology and Causation Collide

One of the first things taught in an introductory statistics course is that correlation is not causation. As we have seen, a significant correlation between two factors A and B (where, for example A is yellow teeth and B is cancer) could be due to pure coincidence or to a causal mechanism such that:

a. A causes B
b. B causes A
c. Both A and B are caused by C (where in our example C might be smoking) or some other set of factors.

The difference between these possible mechanisms is crucial in interpreting the data, assessing the risks to the individual and society, and setting policy based on the analysis of these risks. However, in practice causal interpretation can collide with our personal view of the world and the prevailing ideology of the organization and social group, of which we will be a part. Explanations consistent with the ideological viewpoint of the group may be deemed more worthy and valid than others, irrespective of the evidence. Discriminating between possible causal mechanisms A, B, and C can only formally be done if we can intervene to test the effects of our actions (normally by experimentation). But we can apply commonsense tests of causal interaction to, at least, reveal alternative explanations for correlations.

Box 3.1 provides an example of these issues at play in the area of social policy, specifically regarding the provision of prenatal care.

Box 3.1   The Effect of Prenatal Care Provision: An Example of Different Causal Explanations According to Different Ideological Viewpoints

Thomas Sowell (1987) cites a study of mothers in Washington, DC. The study found a correlation between prenatal care provision and weight of newborn babies. Mothers receiving low levels of prenatal care were disproportionately black and poor. Sections of the U.S. media subsequently blamed society's failure to provide enough prenatal care to poor black women and called for increased provision and resources. They were using a simple

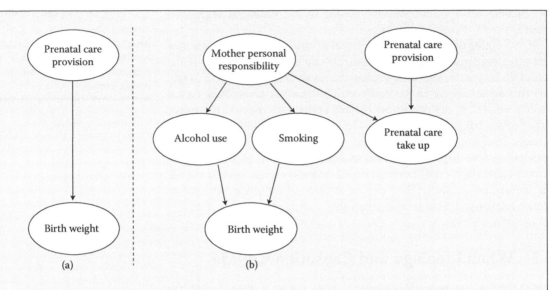

Figure 3.5 Ideological perspectives and their causal claims as applied to prenatal care provision. (a) Liberal. (b) Conservative.

causal explanation: lack of prenatal care provision causes low birth weight with consequent detrimental effects on infant health. However, a closer look at the data revealed that, independently of race and income, smoking and alcohol use were respectively twice and six times more prevalent among mothers who did not get prenatal care support. Sowell argues that smoking and alcohol abuse are symptoms of another factor not addressed in the study—personal responsibility. Among mothers who did not smoke or drink alcohol, there was no correlation between prenatal care provision and birth weight. A deficit in personal responsibility could explain the failure to seek prenatal care and also explain the increased propensity to smoke and drink alcohol.

Clearly ideology plays a role in these explanations. What might be characterized as the liberal perspective will tend to blame the problem on an absence of state provision of prenatal care, whereas the conservative perspective looks for an explanation based on individual behavior and personal responsibility. Figure 3.5 shows the difference in causal model from these two perspectives.

Simplistic causal explanations (such as the liberal perspective in Box 3.1) are usually favored by the media and reported unchallenged and taken as axiomatic. This is especially so when the explanation fits the established ideology, helping to reinforce ingrained beliefs. Picking apart oversimplistic causal claims and deconstructing and reconstructing them into a richer, more realistic causal model helps separate ideology from reality and determines whether the evidence fits reality. The richer model may also help identify more effective possible policy interventions.

Another example where ideology and causation collide is in the so-called conspiracy of optimism explained in Box 3.2.

The time spent analyzing risks must be balanced by the short-term need to take action and the magnitude of the risks involved. Therefore, we must make judgments about how deeply we model some risks and how quickly we use this analysis to inform our actions. There is a trade-off between efficiency and thoroughness here, called the efficiency-thoroughness trade-off (ETTO) (coined by Hollnagel, 2009): too much

---

Box 3.2   The Conspiracy of Optimism

The conspiracy of optimism refers to a situation in which a can-do attitude, where action is favored to the exclusion of risk analysis, leads to the underestimation of risk. A classical example of this occurs in tender bids to supply a new system to a disengaged customer who has not taken the care to identify their needs. A bidder that points out the risks and takes a realistic approach is likely to lose the tender to a competitor that shows a positive attitude and plays down future problems. The competitor and the customer are storing up problems for later (such as higher costs, delays, and even failure), but for the winning bidder this looks a much better short-term position to be in than the realistic, but losing, bidder. The smart approach for a realistic contractor who is determined to win the tender is to structure the risk mitigants and controls in such a way that when the risks are revealed during the project the contractor is protected and the customer ends up paying for them.

---

analysis can lead to inaction and inefficiency, and too little analysis can lead to inefficient actions being undertaken.

Consider, for example, a doctor attempting a diagnosis. The doctor may be tempted to undertake many diagnostic tests, at considerable expense but for little additional certainty. In the meantime the disease may progress and kill the patient. The doctor has purchased near certainty but at a devastating cost. Trade-offs such as these are made all of the time but the challenge is to be explicit about them and search for better options.

The flip side of the conspiracy of optimism is the conspiracy of pessimism (or paralysis by analysis).

3.4   The Limitations of Common Approaches to Risk Assessment

The previous sections provided some insight into both why standard statistical techniques provide little help when it comes to risk assessment and why causal models (BNs) help. In this section we show that when it comes to quantifying risk, again, the traditional techniques are fundamentally flawed, while BNs provide an alternative solution. This section uses some terms from probability that we will define properly in Chapters 4 and 5, so do not worry too much if there are some things you do not fully understand on first reading. You should still get the gist of the argument.

3.4.1   Measuring Armageddon and Other Risks

By destroying the meteor in the film *Armageddon*, Bruce Willis saved the world. Both the chance of the meteor strike and the consequences of such a strike were so high, that nothing much else mattered except to try to prevent the strike. In popular terminology what the world was confronting (in the film) was a truly massive risk.

But if the NASA scientists in the film had measured the size of the risk using the standard approach in industry they would quickly have discovered such a measure was irrational, and it certainly would not have explained to Bruce Willis and his crew why their mission made sense.

Bruce takes to space.

Before we explain why, let's think about more mundane risks, like those that might hinder your next project. These could be:

- Some key people you were depending on become unavailable.
- A piece of technology you were depending on fails.
- You run out of funds or time.

Whether deliberate or not, you will have "measured" such risks. The very act of listing and then prioritizing risks, means that mentally at least you are making a decision about which risks are the biggest.

What you probably did, at least informally, is what most standard texts on risk propose. You decompose risks into two components:

- *Probability* (or likelihood) of the risk
- *Impact* (or loss) the risk can cause

Figure 3.6 Standard impact-based risk measure. Typically risk analysts using this approach measure both probability and impact on a scale of 1 to 5 or 1 to 10.

Risk assessors are assumed to be at the leading edge of their profession if they provide quantitative measures of both probability and impact, and combine them to give an overall measure of risk. The most common such measure is to multiply your measure of probability of the risk (however you happen to measure that) with your measure of the impact of the risk (however you happen to measure that) as in Figure 3.6.

The resulting number is the size of the risk; it is based on analogous utility measures. This type of risk measure is quite useful for prioritizing risks (the bigger the number, the greater the risk), but it is normally impractical and can be irrational when applied blindly. We are not claiming that this formulation is wrong, and indeed we cover its use in later chapters. Rather, we argue that it is normally not sufficient for decision making.

One immediate problem with the risk measure of Figure 3.6 is that, normally, you cannot *directly* get the numbers you need to calculate the risk without recourse to a much more detailed analysis of the variables involved in the situation at hand (see Box 3.3 for the *Armageddon* example).

Box 3.3   Limitations of the Impact-Based Risk Measure Using the *Armageddon* **Example**

According to the standard risk measure of Figure 3.6, we have a model like the one in Figure 3.7. The problems with this are:

- *We cannot get the probability number*—According to the NASA scientists in the film, the meteor was on a direct collision course with Earth. Does that make it a certainty (i.e., a 100% chance, or equivalently a probability of 1) of it striking Earth? Clearly not, because if it was then there would have been no point in sending Bruce Willis and his crew up in the space shuttle (they would have been better off spending their last few remaining days with their families). The probability of the meteor striking Earth is *conditional* on a number of other control events (like intervening to destroy the meteor) and trigger events (like being on a collision course with Earth). It makes no sense to assign a direct probability without considering the events it is conditional on. In general it makes no sense (and would in any case be too difficult) for a risk manager to give the unconditional probability of every risk irrespective of relevant controls, triggers, and mitigants. This is especially significant when there are, for example, controls that have never been used before (like destroying the meteor with a nuclear explosion).

- *We cannot get the impact number*—Just as it makes little sense to attempt to assign an (unconditional) probability to the event "meteor strikes Earth," so it makes little sense to assign an (unconditional) number to the *impact* of the meteor striking. Apart from the obvious question, "impact on what?" we cannot say what the impact is without considering the possible mitigating events such as getting people underground and as far away as possible from the impact zone.
- *Risk score is meaningless*—Even if we could get round the two aforementioned problems what exactly does the resulting number mean? Suppose the (conditional) probability of the strike is 0.95 and, on a scale of 1 to 10, the impact of the strike is 10 (even accounting for mitigants). The meteor risk is 9.5, which is a number close to the highest possible 10. But it does not measure anything in a meaningful sense.
- *It does not tell us what we really need to know*—What we really need to know is the probability, given our current state of knowledge, that there will be massive loss of life if (a) we do nothing and (b) we attempt to destroy the meteor.

Hence, the analysis needs to be coupled with an assessment of the impact of the underlying variables, one on another, and in terms of their effect on the ultimate outcomes being considered. To put it another way, the accuracy of the risk assessment is crucially dependent on the fidelity of the underlying model used to represent the risk; the simple formulation of Figures 3.6 and 3.7 is insufficient. Unfortunately, much risk analysis involves going through the motions to assign numbers without actually doing much thinking about what lies underneath.

Figure 3.7 Risk of meteor strike.

We provide a solution to this problem in Section 3.5.

3.4.2 Risks and Opportunities

COSO (Committee of Sponsoring Organizations of the Treadway Commission, 2004) defines risk as follows:

Risk is an event that can have negative impact. Conversely an event that can have a positive impact is an *opportunity*.

In fact, although risk analysts often fail to recognize it (because they are focused on the potential negative aspects) risks and opportunities are inevitably intertwined. In many situations you do not even need to distinguish between whether an event is a risk or an opportunity. Its consequences, which may be positive or negative, determine whether it is a risk or an opportunity.

Decision Theory

Classical decision theory assumes that any event or action has one or more associated outcomes. Each outcome has an associated positive or negative utility. In general you choose the action that maximizes the total expected utility. That way you explicitly balance potential loss (risk) against potential gain (opportunity). The consequence of choosing to drive excessively fast is that it exposes you and others to increased risk (which could be measured as per Figure 3.6) as probability of car crash times impact of that event (cost of injury or loss of life). In most circumstances the risks of driving excessively fast far outweigh the opportunity (generally, getting to work five minutes earlier is not going to save a life). The hard part is that in many decision-making situations a specific action might expose one party to risks and a second party to opportunities. As a society we prefer not to favor individual persons making such decisions (hence the imposition of speed limits).

Example 3.1

Consider the event "drive excessively fast." According to the COSO definition, this is certainly a risk because it is an event that can have a negative impact such as "crash." But it can also be an opportunity since its positive impact might be delivering a pregnant woman undergoing a difficult labor to hospital or making it to a life-changing business meeting.

In general, risks involving people tend to happen because people are seeking some reward (which is just another word for opportunity). Indeed, Adams (1995) argued that most government strategies to risk are fundamentally flawed because they focus purely on the negative aspects of risk while ignoring the rewards. Adams also highlights the important role of people's risk appetite that is often ignored in these strategies. For example, as better safety measures (such as seat belts and airbags) are introduced into cars, drivers' appetite for risk increases, since they feel the rewards of driving fast may outweigh the risks.

3.4.3 Risk Registers and Heat Maps

The obsession with focusing only on the negative aspects of risk also leads to an apparent paradox in risk management practice. Typically, risk managers prepare a risk register for a new project, business line, or process whereby each risk is scored according to a formula like in Figure 3.6. The cumulative risk score then measures the total risk. The paradox involved in such an approach is that the more carefully you think about risk (and hence the more individual risks you record in the risk register) the higher the overall risk score becomes. Since higher risk scores are assumed to indicate greater risk of failure it seems to follow that your best chance of a new project succeeding is to simply ignore or underreport any risks.

There are many additional problems with risk registers:

- Different projects or business divisions will assess risk differently and tend to take a localized view of their own risks and ignore that of others. For example, to counter the risk of losing out to a competitor's new product, a company's marketing department may impose a product release deadline that creates major new quality risks for the production department. Such "externalized" risks are easy to ignore if other departments' interests are not represented when constructing the register.
- A risk register does not record opportunities or serendipity, and so does not deal with upside uncertainty, only downside. Hence, risk managers become viewed as doomsayers.
- Risks are not independent. For example, in most manufacturing processes cost, time, and quality will be inextricably linked (see Figure 3.8); you might be able to deliver faster but only by

sacrificing quality. Yet "poor quality" and "missed delivery" will appear as separate risks on the register giving the illusion that we can control or mitigate one independently of the other.

Given the difficulty of quantifying risk, and the lack of time usually given over to addressing it thoroughly, the probability and impact numbers needed on a risk register are often replaced by labels (like low, medium, high) and the items in the risk register plotted on a *heat map* as shown in Figure 3.9. Here we have three risks, 1, 2, and 3, where both probability and impact are scored on a scale of low, medium, and high. The green colored sector on a heat map indicates low risk (risk 3 being an example), amber is more risky (risk 1 being an example), and red might mean dangerously high risk (risk 2 being an example).

Conventional wisdom dictates that we worry less about risks 3 and 1 and focus more of our energies on reducing risk 2. This strategy would be fine, all things being equal, if the risks were completely independent of each other. But if they are actually dependent then reducing risk 2 to a green status does not guarantee that the overall risk will reduce, because not all risks are additive: should all of the risks occur together or in some other combination, they may give rise to an unforeseen red, event, albeit it with small probability but with much higher impact than either risk alone (as explained in the sidebar 3.1 in relation to the 2008 subprime loan crisis).

Risk is therefore a function of how closely connected events, systems, and actors in those systems might be. If you are a businessman the last

Figure 3.8 The iron triangle: choose any two but do not expect all three.

Sidebar 3.1

In the subprime loan crisis of 2008–2009 there were three risks: (1) extensive defaults on subprime loans, (2) growth in novelty and complexity of financial products, and (3) failure of AIG (American International Group Inc.) to provide insurance to banks when customers default. Individually if these risks were plotted on a heat map by the regulator they would each have been placed in the green sector. However, as we know from bitter experience, when they occurred together the total risk was much larger than the individual risks. In fact, it never made sense to consider the risks individually at all.

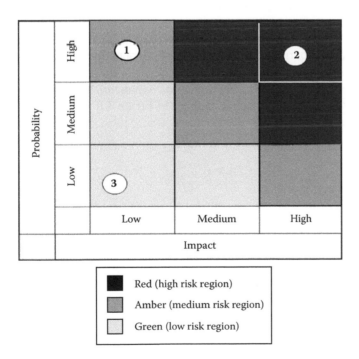

Figure 3.9 Heat map.

thing you should do is look at the risks to your business separately using a risk register or heat map. Instead, you need to adopt a holistic outlook that embraces a causal view of interconnected events. Specifically:

To get rational measures of risk you need a causal model. Once you do this, measuring risk starts to make sense. However, it requires an investment in time and thought.

3.5   Thinking about Risk Using Causal Analysis

It is possible to avoid all these problems and ambiguities surrounding the term *risk* by considering the causal context in which both risks and opportunities happen. The key thing is that a risk (and, similarly, an opportunity) is an *event* that can be characterized by a causal chain involving (at least):

- The event itself
- At least one consequence event that characterizes the impact (so this will be something negative for a risk event and positive for an opportunity event)
- One or more *trigger* (i.e., initiating) events
- One or more *control* events that may stop the trigger event from causing the risk event (for risk) or *impediment* events (for opportunity)
- One or more *mitigating* events that help avoid the consequence event (for risk) or *impediment* event (for opportunity)

This approach (which highlights the symmetry between risks and opportunities) is shown in the example of Figure 3.10.

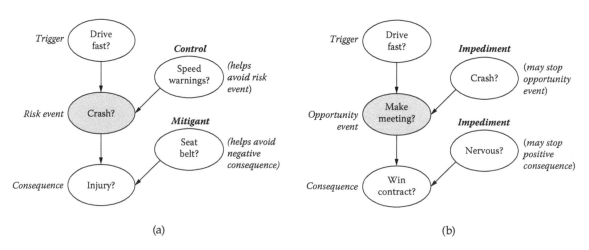

(a)   (b)

Figure 3.10   Causal taxonomy of risk/opportunity. (a) Causal view of risk. (b) Causal view of opportunity.

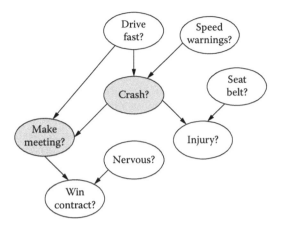

Figure 3.11   Unified model with risk events and opportunity events.

In practice, we would not gain the full benefits of building a causal model, unless we combine the risk events and opportunity events in a single model as in Figure 3.11.

In many situations it is actually possible to use completely neutral language (neither risk nor opportunity) as explained in the example of Box 3.4.

Box 3.4   Using Neutral Language to Combine Risk/Opportunity

Companies undertake new projects because ultimately they feel the rewards outweigh the risks. As shown in Figure 3.12 the *project delivery* (whether it is late or not) and *project quality* are examples of key events associated with any new project. If the quality is bad or if the delivery is late then these represent risk events, whereas if the quality is good or the delivery is on time or even early, these represent opportunity events. The income is one of the

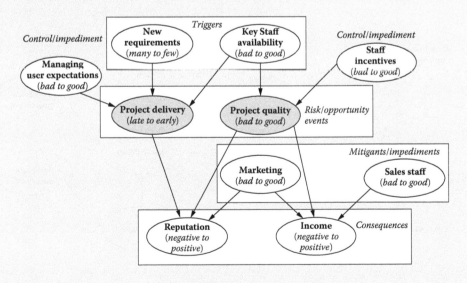

Figure 3.12   Neutral causal taxonomy for risk/opportunity.

---

consequence events of quality. It can be positive (in the event of a good quality project) or negative (in the event of a poor quality project). The company reputation is one of the consequences of both delivery and quality. It can be positive (if the quality is good and the delivery is on time) or negative (if either the quality is bad or the delivery is late).

Among the triggers, the risk/opportunity events have a common trigger, namely, key staff availability. If key staff become unavailable then the effect will be negative (risk), whereas if key staff are available then the effect will be positive (opportunity). The negative (positive) effect of key staff availability is controlled (impeded) by good (bad) staff incentives.

Among the mitigants/impediments there is a common factor, *marketing*. If the marketing is bad then even a good quality project delivered on time could lead to loss of reputation, whereas if it is good then even a poor quality project could lead to enhanced reputation.

With this causal perspective of risk, a risk is therefore actually characterized not by a single event but by a set of events. These events each have a number of possible outcomes (to keep things as simple as possible in the examples in this chapter we will assume each has just two outcomes: true and false). The uncertainty associated with a risk is not a separate notion (as assumed in the classic approach). Every event (and hence every object associated with risk) has uncertainty that is characterized by the event's *probability distribution* (something we will cover in depth in Chapter 5).

Clearly risks in this sense depend on stakeholders and perspectives, but the benefit of this approach is that once a risk event is identified from a particular perspective, there will be little ambiguity about the concept and a clear causal structure that tells the full story. For example, consider the risk of Flood in Figure 3.13(a). Because the risk event "Flood" takes the central role the perspective must be of somebody who has responsibility for both the associated control and mitigant. Hence, this is the perspective the local authority responsible for amenities in the village

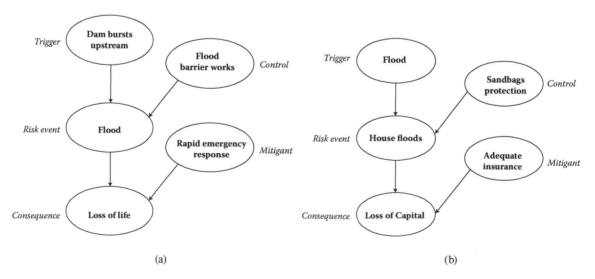

(a)   (b)

Figure 3.13 Risk from different perspectives. (a) Flood risk from the local authority perspective. (b) Flood risk from the householder perspective.

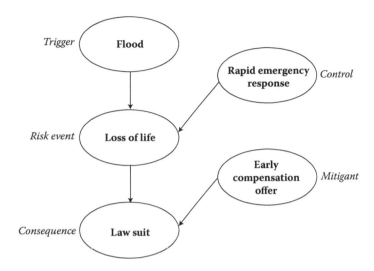

Figure 3.14   Interchangeability of concepts depending on perspective.

(rather than, for example, a householder in the village). A householder's perspective of risk would be more like that shown in Figure 3.13(b).

What is intriguing is that the types of events are all completely interchangeable depending on the perspective. Consider the example shown in Figure 3.14. The perspective here might be of the local authority lawyer. Note that:

- The risk event now is "Loss of life." This was previously the consequence.
- "Flood" is no longer the risk event, but the trigger.
- "Rapid emergency response" becomes a control rather than a mitigant.

It is not difficult to think of examples where controls and mitigants become risk events and triggers. This interchangeability stresses the symmetry and simplicity of the approach.

This ability to decompose a risk problem into chains of interrelated events and variables should make risk analysis more meaningful, practical, and coherent. In Chapter 11 we will explain in detail how the BN causal approach can accommodate decision making as well as measures of utility but before then we make some simplifying assumptions to keep the material digestible:

- We model all variables as chance events rather than decision/actions that a participant might take.
- The payoffs for different decisions/actions are obvious and do not need to be assigned a utility value. So a "Law suit" or "Flood" is obviously a worse outcome (of lower utility than "No law suit" or "No flood") but we are not attempting here to measure the utility in order to identify the optimum decision to take.

The Fallacy of Perfect Risk Mitigation

Causes of risk are viewed as uncertain, whereas mitigation actions to reduce impact are assumed to operate perfectly. This fallacy is widespread. Many risk management standards and guidelines assume that once a mitigation action is put in place that it will never degrade, be undermined, and hence will be invulnerable. Likewise, even sophisticated thinkers on risk can make the same mistake. For example, in his book *The Black Swan*, Taleb claims that high impact risk events cannot be readily foreseen and the only alternative answer is to mitigate or control the consequences of such risk events *after the fact*. How then do we guarantee perfect mitigation?

We take the view that the performance of mitigants is itself uncertain since they will need to be maintained and supported and if such maintenance and support is not forthcoming they will degrade.

3.6 Applying the Causal Framework to *Armageddon*

We already saw in Section 3.4.1 why the simple impact-based risk measure was insufficient for risk analysis in the *Armageddon* scenario. In particular, we highlighted:

1. The difficulty of quantifying (in isolation) the probability of the meteor strike.
2. The difficulty of quantifying (in isolation) the impact of a strike.
3. The lack of meaning of a risk measure that is a product of (isolated measures of) probability and impact.

To get round these problems we apply the causal framework to arrive at a model like the one shown in Figure 3.15 (if we want to stick to events with just true or false outcomes then we can assume "Loss of life" here means something like loss of at least 80% of the world population).

The sensible risk measures that we are proposing are simply the probabilities you get from executing the BN model. Of course, before you can execute it you still have to provide some probability values (these are the strengths of the relationships). But, in contrast to the classic approach, the probability values you need to supply are relatively simple and they make sense. And you never have to define vague numbers for impact.

If you are not comfortable with basic probability then you may wish to return to the next few paragraphs after reading Chapters 4 and 5.

To give you a feel of what you would need to do, in the BN for the uncertain event "Meteor strikes Earth" we still have to assign some probabilities. But instead of second guessing what this event actually means in terms of other conditional events, the model now makes it explicit and it becomes much easier to define the necessary *conditional* probability. What we need to do is define the probability of the meteor strike given each combination of parent states as shown in Figure 3.16.

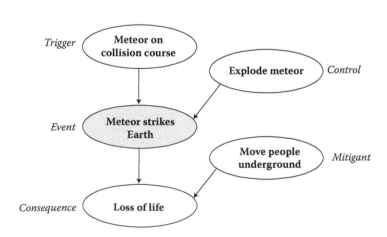

Figure 3.15 Meteor strike risk.

Meteor on collision course	False		True	
Explode meteor	False	True	False	True
False	1.0	1.0	0.0	0.8
True	0.0	0.0	1.0	0.2

Figure 3.16 Conditional probability table for "Meteor strikes Earth." For example, if the meteor is on a collision course then the probability of it striking the Earth is 1 if it is not destroyed, and 0.2 if it is. In completing such a table we no longer have to try to factor in any implicit conditioning events like the meteor trajectory.

There are some events in the BN for which we do need to assign unconditional probability values. These are represented by the nodes in the BN that have no parents; it makes sense to get unconditional probabilities for these because, by definition, they are not conditioned on anything (this is obviously a choice we make during our analysis). Such nodes can generally be only triggers, controls, or mitigants. An example, based on dialogue from the film, is shown in Figure 3.17.

The wonderful thing about BNs is that once you have supplied the initial probability values (which are called the *priors*) a Bayesian inference engine (such as the one in AgenaRisk) will run the model and generate all the measures of risk that you need. For example, when you run the model using only the initial probabilities the model (as shown in Figure 3.18) computes the probability of the meteor striking Earth as

We are not suggesting that assigning the probability tables in a BN is always easy. You will generally require expert judgment or data to do it properly (this book will provide a wealth of techniques to make the task as easy as possible). What is important is that it is easier than the classic alternative. At worse, when you have no data, purely subjective values can be supplied.

False	0.001
True	0.999

Figure 3.17 Probability table for "Meteor on collision course with Earth."

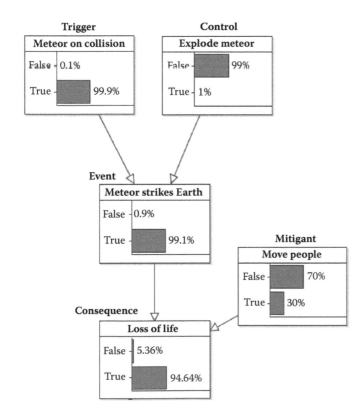

Figure 3.18 Initial risk of meteor strike.

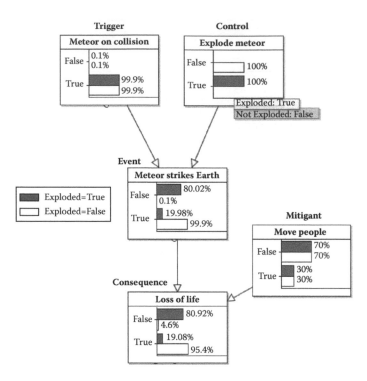

Figure 3.19   The potential difference made by Bruce Willis and crew.

99.1% and the probability of loss of life (meaning at least 80% of the world population) is about 94%.

In terms of the difference that Bruce Willis and his crew could make, we run two scenarios: one where the meteor is exploded and one where it is not. The results of both scenarios are shown in Figure 3.19.

Reading off the values for the probability of "Loss of life" being false we find that we jump from just over 4% (when the meteor is not exploded by Bruce) to 81% (when the meteor is exploded by Bruce). This massive increase in the chance of saving the world clearly explains why it merited an attempt.

The main benefits of this approach are that

- Risk measurement is more meaningful in the context; the BN tells a story that makes sense. This is in stark contrast with the simple "risk equals probability times impact" approach where not one of the concepts has a clear unambiguous interpretation.
- Uncertainty is quantified and at any stage we can simply read off the current probability values associated with any event.
- It provides a visual and formal mechanism for recording and testing subjective probabilities. This is especially important for a risky event that you do not have much or any relevant data

about (in the *Armageddon* example this was, after all, mankind's first mission to land on a meteorite).

Although the approach does not explicitly provide an overall risk score and prioritization these can be grafted on in ways that are much more meaningful and rigorous. For example, we could

- Simply read off the probability values for each risk event given our current state of knowledge. This will rank the risks in order of probability of occurrence (this tells you which are most likely to happen given your state of knowledge of controls and triggers).
- Set the value of each risk event in turn to be fixed and read off the resulting probability values of appropriate consequence nodes. This will provide the probability of the consequence given that each individual risk definitively occurs. The risk prioritization can then be based on the probability values of consequence nodes.

Above all else the approach explains why Bruce Willis's mission really was viable.

3.7 Decisions and Utilities

Although all the nodes in the Armageddon example were treated as uncertain Boolean variables, there are clear differences between them from a "decision theory" perspective. Both the control node ("Explode meteor") and the mitigant node ("Move people underground") really represent *decisions* that we may choose to perform or not. In contrast, the other nodes really are *chance* nodes. There is also something very important missing from the model, namely explicit *utilities*. In general any decision (such as exploding the meteor or moving people underground) will have a cost (which we can think of as a negative utility) and every consequence node (such as "loss of life") will have either a cost or benefit (negative or positive utility). The "correct" model therefore is the one shown in Figure 3.20, where decision nodes are represented as rectangles and utility nodes as diamonds (such a diagrammatic model is called an influence diagram).

In practice if we are choosing which of a set of decisions is optimal we may need to be explicit about the utility of each decision and outcome. Once we do this the optimal risk strategy is the one that involves the set of decisions with the maximum overall utility.

Models that include explicit decision nodes and utility nodes (influence diagrams) can be considered as a special type of BN model, called an influence diagram. Whereas in a normal BN we are interested in seeing how each uncertain node is updated when we observe evidence, in an influence diagram we are also interested in "solving" the problem of determining which decisions optimise the overall utility. Chapter 11 deals with such influence diagrams in depth and describes how AgenaRisk is used to calculate the necessary solutions.

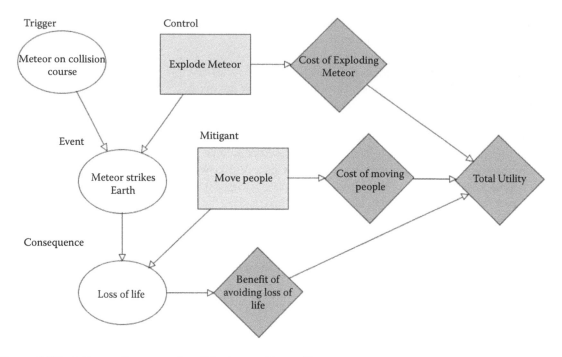

Figure 3.20 Influence diagram version of the Armageddon problem.

3.8 Summary

The aim of this chapter was to show that we can address some of the core limitations of the traditional statistical approaches to risk assessment by using causal or explanatory models. Hopefully, the examples helped convince you that identifying, understanding, and quantifying the complex interrelationships underlying even seemingly simple situations can help us make sense of how risks emerge, are connected, and how we might represent our control and mitigation of them.

We have shown how the popular alternative approaches to risk measurement are, at worst, fundamentally flawed or, at least, limiting. By thinking about the hypothetical causal relations between events we can investigate alternative explanations, weigh the consequences of our actions, and identify unintended or (un)desirable side effects.

Of course it takes mental effort to make the problem tractable: care has to be taken to identify cause and effect, the states of variables need to be carefully defined, and probabilities need to be assigned that reflect our best knowledge. Likewise, the approach requires an analytical mindset to decompose the problem into classes of events and relationships that are granular enough to be meaningful but not too detailed that they are overwhelming. If we were omniscient we would have no need of probabilities; the fact that we are not gives rise to our need to model uncertainty at a level of detail that we can grasp, that is useful, and that is accurate enough for the purpose required. This is why causal modeling is as much an art (but an art based on insight and analysis) as a science.

Further Reading

Adams, J. (1995). *Risk*. Routledge.

COSO (Committee of Sponsoring Organisations of the Treadway Commission). (2004). Enterprise Risk Management: Integrated Framework, www.coso.org/publications.htm.

Fenton, N. E., and Neil, M. (2011). The use of Bayes and causal modelling in decision making, uncertainty and risk. *UPGRADE*, December 2011.

Hollnagel, E. (2009). *The ETTO Principle: Efficiency-Thoroughness Trade-Off: Why Things That Go Right Sometimes Go Wrong*. Ashgate.

Hubbard, D. W. (2009). *The Failure of Risk Management: Why It's Broken and How to Fix It*, Wiley.

Pearl, J., and Mackenzie, D. (2018). *The book of why: the new science of cause and effect*, Basic Books, New York.

Sowell, T. (1987). *A Conflict of Visions: Ideological Origins of Political Struggles*. William Morrow.

Visit www.bayesianrisk.com for exercises and worked solutions relevant to this chapter.

Measuring Uncertainty: The Inevitability of Subjectivity

4.1 Introduction

Probability is the term used to quantify our uncertainty about some unknown entity—be it a future event, a thing whose properties we are unsure of, or even a past event about which we are unsure. We made numerous references to probability in Chapters 2 and 3 without actually defining it or explaining how to compute it in particular instances. What should be clear from Chapter 2 is that probability reasoning can be extremely confusing and difficult even for highly trained professionals. Moreover, many professional statisticians use a different notion of probability (called *frequentist*) to the one (called *subjective*) that plays a central role in our approach to risk assessment. So, it is clear that we need to get the basic rules of probability right. To prepare for that (which we do in Chapter 5) this chapter lays the necessary groundwork by discussing the different approaches to measuring uncertainty.

Because we are interested in risk, the kinds of uncertain events we have to deal with are very diverse. At one extreme we have events where we apparently have good understanding of the uncertainty, like:

- The next toss on a coin will be a head.
- The next roll of a die will be a 6.

At the other extreme are events where we apparently have a poor understanding of the uncertainty like:

- England will win the next World Cup.
- My bank will suffer a loss (such as a fraud or lost transaction) tomorrow.
- A hurricane will destroy the White House within the next 5 years.

Or even an "unknown" event like

- My bank will be forced out of business in the next two years as a result of a threat that we do not yet know about.

All models in this chapter are available to run in AgenaRisk, downloadable from www.agenarisk.com.

As discussed in Chapter 2, uncertain events are not just those that lie in the future. Because of incomplete information, there is uncertainty about events such as:

- Oliver Cromwell spoke more than 3,000 words on April 23, 1654.
- O.J. Simpson murdered his ex-wife.
- You (the reader) have an as yet undiagnosed form of cancer.

Incomplete or absent information can also take the form of missing variables or causes that are absent from the model. Therefore, subjective probability, as a degree of belief, is conditioned as much by information in the model as information not in the model. Indeed, if all information was present and available our reasoning would be omniscient and Godlike and we would assign deterministic logical rather than probabilities. Box 4.1 provides an example to clarify this.

Box 4.1   Causal Revelation and Absence of Information

Imagine a conversation between a modeller, interested in predicting whether a borrower can repay a financial debt, and an independent observer quizzing the modeller on their probabilities.

Modeller: My model contains two variables "lose job" causes "cannot pay debt." If a borrower loses employment they cannot pay the debt back 90% of the time.

Observer: But by losing income they can still pay debt 10% of the time. Why is that? This looks odd. How can they still have chance of 10% of paying debt without a job?

Modeller: Because they could sell their house and can still pay.

Observer: OK, that isn't in the model, let's add that to model (model now has two causes for "can pay debt": lose job and sell house)

Observer: But if the borrower loses their job but doesn't sell their house what's the chance of paying the debt?

Modeller: Answer is 5%.

Observer: How could someone still pay? There must be some other reason.

Modeller: Perhaps they could sell their grandmother into slavery?

Observer: OK, sounds a bit extreme but let's add that to the model. What's the chance of not paying debt now?

Modeller: If borrower loses job, doesn't sell the house and they don't sell their grandmother into slavery, then the chance is 1%.

Observer: But why 1%?

Modeller: Because they may rob a bank!

Observer: OK, let's add that to the model

...dialogue continues

At some point the modeller reveals all possible causal mechanisms and achieves a zero probability of the borrower not paying their debt in the presence of all possible causes, thus rendering the model deterministic. This is Godlike omniscience (and is, of course, impossible). Einstein said: "God does not play dice with the universe," perhaps meaning that—to God—there are no probabilities only certainties.

What can we learn from this dialogue? That our probabilities represent casual mechanisms that are NOT in the model; they represent our lack of information about possible causes. Also, what is or isn't in the model depends on our cognitive revelation, imagination, experience and availability of information. Hence, different people have different probabilities and build different models.

All probabilities depend on context (model boundary and scope) and at each stage of the conversation the modeller extended the scope realising that other factors may be relevant. Hence, the probabilities changed.

Our approach to quantifying uncertainty has to satisfy the following two requirements:

1. Be able to handle all such uncertain events in a consistent way. We will explain that this inevitably means that we have to be able to accommodate subjective judgments about uncertainty, in other words *personal beliefs*.
2. Enable us to revise our belief when we observe new evidence. So, for example, if we start with a belief that England has a 10% chance of winning the next World Cup, then if we subsequently observe that England wins its first 5 qualifying matches, we have to be able to use this information to revise our belief about England winning the World Cup. We will show that the only rational way to do this updating is by applying Bayes' rule.

These two requirements are problematic for the traditional frequentist approach, which can be very good at handling uncertainty for mechanistic events such as those generated by gambling machines, but offers little help with real life risk assessment problems (which are generally about uncertain events that involve lack of information and uniqueness). This and the next chapter focus on the first requirement. Chapter 6 focuses on the second requirement.

The main objective of this chapter is to explain the two different approaches to uncertainty (*frequentist* and *subjective*) and to convince you that the subjective approach subsumes the frequentist one and must inevitably be used for effective risk assessment. Our approach is to provide a rigorous framework for describing what is meant by uncertain events; we do this using the notion of *experiments* and *outcomes*.

4.2   Experiments, Outcomes, and Events

The act of tossing a coin and observing some *outcome* of interest, such as what side the coin lands on, can be regarded as an *experiment*. An example of an actual observation for the outcome of interest of this experiment would be "head."

To avoid the confusion of different notions of outcome (as explained in the sidebar) we will only ever use the word outcome in the variable sense, that is, when we are talking about the outcome of interest of an experiment. Any actual observed value of an experiment will be referred to as an *event*, never an outcome. For example, in the experiment of rolling a die, where the outcome of interest is "the number showing on top" both "6" and "a number greater than 3" are events.

It is also important to distinguish between the notions of event and *elementary event* (see sidebar). Elementary events may also be called *states*.

It is important to note that an understanding of basic counting rules (including *permutations* and *combinations*) is necessary in what follows. You may think you already know everything about counting from elementary school. But beware: in many situations your intuitive counting skills may not be sufficient. For example, can you work out how many different lottery tickets you would have to buy in order to be certain of winning next Saturday's draw, or how many different possible car number plates can be assigned according to the 2018 Department of Transport rules? If not then you will need to first read Appendix A on this subject.

Two Notions of "Outcome"

- The notion: *Outcome of interest* of an experiment is a variable that can take on a range of values or states. For example, in the coin tossing experiment, the outcome of interest—what side the coin lands on—can be regarded as a variable whose possible values (or states) include head and tail.
- The notion *outcome* is an actual observed value for the experiment. For example, heads.

Elementary and Non-elementary Events

In the die rolling experiment, the event "6" is called an *elementary* event because it refers to something unique, in contrast to the event "a number greater than 3," which can refer to 4, 5, or 6. Identifying the set of all elementary events of an experiment is important because together they compose all the possible events.

In general there may be different outcomes of interest to us for the same experiment. For example, for coin tossing we might also be interested in the outcome "how long coin stays in the air" for which an actual observation might be "1.5 seconds." This is something we shall return to in Chapter 5, but for the time being we will assume that there is a single clear outcome of interest for any given experiment.

All of the examples of uncertain events presented in the Introduction can be regarded as events arising from specific experiments with specific outcomes of interest. Box 4.2 clarifies what the experiment, outcome of interest, and set of elementary events are in each case.

Box 4.2   Experiments, Outcomes of Interest, and Sets of Elementary Events

1. *The next toss on a coin will be a head.* This is an elementary event of the experiment of tossing a coin and observing the outcome "what side the coin lands on next." There are three possible elementary events: "head," "tail," and "neither" (with the latter covering the possibility that the coin lands on its side (quite likely if it is tossed on a muddy field) or is lost (quite likely if you toss it while standing on a rope bridge across a ravine).

2. *The next roll of a die will be a 6.* This is an elementary event of the experiment of rolling a die and observing the outcome "number showing on top." There are seven possible elementary events: 1, 2, 3, 4, 5, 6, and "no number" (with the latter covering the possibility the die lands on its side or is lost).

3. *England will win the next World Cup.* This is an elementary event of the experiment of playing the next World Cup and observing the outcome "winning team." There are several dozen elementary events, namely, all of the national teams who enter for the tournament together with "no winner" (to account for the possibility of the cancellation of the tournament).

4. *My bank will suffer a loss tomorrow.* We can regard this as an event of the experiment of running the bank for a day and observing the outcome "loss type" where the elementary events could be "frauds only," "lost transactions only," "both frauds and lost transactions," "no loss."

5. *A hurricane will destroy the White House within the next 5 years.* We can regard this as an elementary event of the experiment of running the next 5 years and observing the outcome of what happens to the White House. For simplicity we could consider that there are just two elementary events, namely, "a hurricane destroys the White House" or "a hurricane does not destroy the White House." Alternatively, we could also consider it as an experiment with a more detailed set of elementary events, each covering the manner of the White House destruction ("alien attack," "meteor strike," "earthquake," etc., plus the event "not destroyed").

6. *Oliver Cromwell spoke more than 3,000 words on April 23, 1654.* We can regard this as an event of the experiment of Oliver Cromwell living one day where the outcome of interest is the number of words Cromwell speaks. The elementary events are 0, 1, 2, 3, up to some suitable upper limit (200,000 would easily suffice) corresponding to the number of words that were spoken by Oliver Cromwell on that day.

7. *O.J. Simpson murdered his ex-wife.* We can regard this as an elementary event of the experiment of observing the death of O.J. Simpson's ex-wife. There are three elementary events: "O.J. Simpson murdered his ex-wife," "a person other than O.J. Simpson murdered his ex-wife," "O.J. Simpson's ex-wife died but was not murdered."

8. *You (the reader) have an as yet undiagnosed form of cancer.* We can regard this as an elementary event of the experiment of testing you for cancer (assuming a perfectly accurate test). There are two elementary events: "cancer" and "no cancer."

If you have previously had some exposure to traditional probability and statistics then you may feel a little uneasy about the list of elementary events in Box 4.2, especially in the case of the first two. That is because traditional probability and statistics texts normally give these as examples where there are just two and six elementary events, respectively. In other words they ignore the unusual or rare events. In fact, traditional approaches usually also make the following simplifications for these examples:

- The elementary events are all equally likely.
- The experiment can be repeated many times under identical conditions.

These simplifications are certainly unrealistic for examples 3 to 8 (we certainly cannot, for example, repeat the next World Cup). But, it is important to note right from the start that these assumptions are somewhat artificial even for the first two experiments. Even if we could genuinely discount the possibility of a coin landing on its side or being lost, it is impossible to create a coin that would be truly fair (i.e., results in equally likely outcomes) because of minor surface differences at sub atomic level. It is also impossible to recreate the exact conditions under which two different tosses could be made. This is an important point that we shall return to later.

In all of the example cases the set of elementary events is finite. But this is not always the case, as the following examples demonstrate:

- Consider the experiment of testing a software program where the outcome of interest is the number of defects found during testing. Although the actual number found is always finite, the set of elementary events is not, since it consists of the infinite set of numbers 0, 1, 2, 3, ….

  Note that the set 0, 1, 2, 3 although infinite, is *discrete*; this is the technical word used by mathematicians for any set that can be counted.

- Consider the experiment of examining a person where the outcome of interest is the measure of their height in centimeters. The situation here is even more complex than the previous example because not only is the set of possible elementary events infinite, only being bounded by the precision of our measurement instruments, but it is also not discrete. Note that, even though we can impose a practical upper bound on height (for example, we could impose an upper bound of 500 centimeters since no living person has ever or is likely ever to exceed such a height), the set of outcomes is still infinite and continuous.

  An infinite set that is not discrete is called *continuous*.

We have already informally made the distinction between an elementary event and an event. Formally we can state:

An *event* is a collection of one or more elementary events of an experiment.

When we are referring to events and elementary events in abstract terms it is standard practice to use capital letters such as E, E_1, E_2, and so on to refer to events and small letters e, e_1, e_2, and so on to refer to elementary events. So an event that comprises three elementary events might be written as $E = \{e_1, e_2, e_3\}$.

Example 4.1

For the O.J. Simpson experiment the following are all events:

- O.J. Simpson murdered his ex-wife. (This is an elementary event.)
- O.J. Simpson did not murder his ex-wife. (This is an event consisting of the two elementary events {"A person other than O.J. Simpson murdered his wife," "O.J. Simpson's ex-wife died but was not murdered"}.)
- O.J. Simpson's ex-wife was murdered. (This is an event consisting of the collection of two elementary events {"O.J. Simpson murdered his ex-wife," "A person other than O.J. Simpson murdered his ex-wife"}.)
- O.J. Simpson's ex-wife is dead. (This is an event consisting of the collection of all three elementary events {"O.J. Simpson murdered his ex-wife," "A person other than O.J. Simpson murdered his ex-wife," "O.J. Simpson's ex-wife died but was not murdered"}.)

Example 4.2

For the die roll experiment (where the outcome of interest is the number on top) the following are all events:

- Roll a 4. (This is the elementary event 4.)
- Roll an even number. (This is an event consisting of the elementary events 2, 4, and 6.)
- Roll a number larger than 2. (This is an event consisting of the elementary events 3, 4, 5, and 6.)
- Do not roll a 4. (This is an event consisting of the elementary events 1, 2, 3, 5, 6, and "no number.")

Certain relationships between events are sufficiently important to be given special names. The ones we need to know are described in Box 4.3.

Box 4.3 Relationships between Events

The complement of an event—The event "O.J. Simpson did not murder his ex-wife" is called the complement of the event "O.J. Simpson murdered his ex-wife." The event "do not roll a 4" is the complement of the event "roll a 4." In general the complement of an event *E* is the collection of all elementary events that are not part of the event *E*.

The union of two events—The event "O.J. Simpson's ex-wife was murdered" is the union of two events "O.J. Simpson murdered his ex-wife" and "a person other than O.J. Simpson murdered his ex-wife." The event "roll a number bigger than 1" is the union of the two events "roll an even number'" and "roll a number bigger than 2." In general the union of two events is the collection of all elementary events that are in either of the events. This can be extended to more than two events.

The intersection of two events—The event "roll an even number larger than 2" is the intersection of the two events "roll an even number" and "roll a number bigger than 2." In general the intersection of two events is the collection of all elementary events that are in both of the events. This can be extended to more than two events.

Mutually exclusive events—In general two events with no elementary events in common are said to be mutually exclusive. So the two events "O.J. Simpson murdered his ex-wife" and "OJ Simpson's ex-wife died but was not murdered" are mutually exclusive, as are the two events "roll a number bigger than 4" and "roll a number less than 3." This can be extended to more than two events.

Exhaustive events—The union of the two events "O.J. Simpson's ex-wife was murdered" and "O.J. Simpson's ex-wife died but was not murdered" contains every possible elementary event of the experiment. The same is true of the events "do not roll a 4" and "roll a number bigger than 2." In general a collection of one or more events is said to be exhaustive if the union of the collection of events contains every possible elementary event of the experiment. If a set of events is both exhaustive and mutually exclusive then we call it a *mutually exclusive and exhaustive set*. The events "O.J. Simpson's ex-wife was murdered" and "O.J. Simpson's ex-wife died but was not murdered" is a mutually exclusive and exhaustive set but the events "do not roll a 4" and "a number bigger than 2" is not because they are not mutually exclusive (having the outcomes 3, 5, 6 in common).

An important example of a mutually exclusive and exhaustive set of events is any event and its complement because such a union must contain every possible elementary event.

Anybody familiar with basic set theory in mathematics will see that the definitions in Box 4.3 are exactly from that simple theory.

4.2.1 Multiple Experiments

So far we have considered the outcome of a single experiment (also referred to as a *trial*). But we can consider not just the outcome of a single experiment but also the outcome of multiple repeated experiments (or trials).

For example, Figure 4.1 shows all the possible elementary events arising from repeating the experiment of tossing a coin twice. In this tree diagram the edges are labeled by the possible elementary events from the single experiment. Each full path consists of two such edges and hence represents an elementary event of the repeated experiment. So, for example, the top-most path represents the elementary event where the first toss is a head (H) and the second toss is a head (H). As shorthand we write the overall elementary event as (H, H) as shown in the last column. In this example there are nine possible elementary events.

We could do something similar for the bank experiment. Here we would consider successive days (Day 1, Day 2, etc.). On each day the set of possible elementary events are frauds only (F), lost transactions only (L), both frauds and lost transactions (B), and no loss (N). In this case the resulting multiple experiment results in outcomes like (F, L, ...), (F, F, ...), (N, F, ...), and so forth.

To calculate the number of possible elementary events of such repeated experiments we use the counting techniques described in Appendix A. If there are four elementary events of the single experiment, as in the bank case, then the number of elementary events of two experiments is equal to the number of samples of size two (with replacement) from four objects. This number is $4^2 = 16$ (see

Certain and impossible events—We refer to a set of exhaustive events as *certain* since at least one of the elementary events must happen. Moreover, since the complement of an exhaustive event contains no elementary events at all we refer to it as *impossible*.

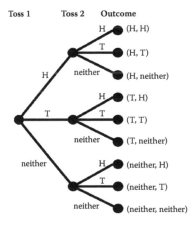

Figure 4.1 Event tree diagram showing outcomes of two successive tosses of a coin.

The notion of multiple experiments applies just as well to the other experiments previously listed. For example, the O.J. Simpson case was tried twice (once in a criminal court and once in a civil court). If we simplify the set of possible elementary events at each trial to have been guilty and not guilty, then the actual outcome of the two trials was the elementary event: (not guilty, guilty). But this was one of only four possible elementary events for the two-trial experiment, the others being (not guilty, not guilty), (guilty, not guilty), and (guilty, guilty). We do not even need two actual trials for the notion of multiple experiments to have value. Think of the 12 jurors at the original trial. Each juror's decision may be regarded as being a single instance of the experiment. The collective view of the jury would then be the experiment repeated 12 times; this involves 2^{12} possible elementary events, of which just one would be a unanimous guilty verdict.

Appendix A). In general, if an experiment has n elementary events and is repeated r times, then the resulting multiple experiment has n^r different elementary events. We call this enumeration. So

- Two successive throws of a die—Assuming the seven outcomes of a single throw, the answer is $7^2 = 49$.
- Five successive throws of a coin—Assuming again the three outcomes of a single throw, the answer is $3^5 = 243$.

4.2.2 Joint Experiments

The idea of multiple experiments is not just restricted to the case where the same experiment is repeated. For example, if we run a gaming club we might be interested in the joint outcome of one coin-tossing game and one die-rolling game. The joint experiment has 21 possible elementary events (with the coin toss listed first in each case):

(H, 1), (H, 2), (H, 3), (H, 4), (H, 5), (H, 6), (H, other)
(T, 1), (T, 2), (T, 3), (T, 4), (T, 5), (T, 6), (T, other)
(other, 1), (other, 2), (other, 3), (other, 4), (other, 5), (other, 6), (other, other)

In general, if A is the set of elementary events of the first experiment and B is the set of elementary events of the second experiment, then the set of elementary events of the joint experiment consists of all pairs of the form (a, b) where a is in A and b is in B.

In this gaming example the two experiments are unrelated, but in many cases we will be interested in joint experiments that are related. For example, the following two experiments:

Readers with a mathematical bent will realize that the joint experiment is simply the *Cartesian product*, written $A \times B$, although curiously probability theorists tend to write it as (A, B) rather than $A \times B$. By the product rule (see Appendix A) the number of such outcomes is simply the size of A times the size of B.

- *Disease X*—This is the theoretical experiment of determining whether a person has disease X. It has two elementary events: yes (when a person has disease X) and no (when a person does not).
- *Test for disease X*—This is the actual experiment of running a diagnostic test for the disease on a person and observing the test result. It has two elementary events: positive and negative.

These are related as shown in Figure 4.2.

The joint experiment (Disease, Test) has four elementary events: (yes, positive), (yes, negative), (no, positive), (no, negative). The outcome (yes, negative) is called a *false negative*, while the outcome (no, positive) is called a *false positive*. Since tests for diseases are rarely 100% accurate, their accuracy is typically measured in terms of the chance of a false negative and the chance of a false positive. What we are especially interested in (and will return to later) is knowing the chance of Disease being yes once we know the outcome of Test.

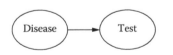

Figure 4.2 The outcome of the test for disease X is clearly influenced by whether the person has disease X.

4.2.3 Joint Events and Marginalization

The notion of an event (being simply a collection of one or more elementary events) extends to joint experiments. So, for example, in the joint experiment of the coin toss and die roll:

- The set of elementary events {(H, 2), (H, 4), (H, 6)} is an event; we can describe this event as "toss a Head and roll an even number."
- The set of elementary events {(H, 1), (H, 2), (H, 3) (H, 4), (H, 5), (H, 6), (H, other), (T, 2), (T, 4), (T, 6), (other, 2), (other, 4), (other, 6)} is an event that we can describe as "toss a head or roll an even number."
- The set of elementary events {(H, 1), (H, 2), (H, 3) (H, 4), (H, 5), (H, 6), (H, other)} is an event that we can describe as "toss a head."

The last of these examples is especially important because it reverts back to an event from the single (coin tossing) experiment. It is as if we are saying: "I know we are considering a joint experiment, but the event I am interested in is one that is relevant only to a single experiment." When we do this we say that we are *marginalizing* the event over the joint experiment.

> The notion of marginalization may seem trivial but it has a profoundly important role to play in the methods used in this book.

Example 4.3

Consider the joint experiment of Disease and Test in Figure 4.2, with elementary events: (yes, positive), (yes, negative), (no, positive), (no, negative).

In this case we may well have information about each of the elementary events such as, for example, from the number of people in a trial as shown in Table 4.1. But we are also interested in

- How many people in the trial have the disease—This is the *marginal event* Disease = yes, and comprises the elementary events (yes, positive), (yes, negative). We can calculate the number by simply adding the number in each of the elementary events (it is 100).
- How many people in the trial test positive—This is the marginal event Test = positive, and comprises the elementary events (yes, positive), (no, positive). Again we can calculate the number by simply adding the number in each of the elementary events (it is 594).

Table 4.1
Trial Results from Test

Disease	No	Yes
Test		
Negative	9,405	1
Positive	495	99

We can generalize Example 4.3 to determine the marginalization of any event as shown in Box 4.4.

Box 4.4   Marginalization of an Event

If we have a joint experiment (A, B) and if E is an event in A, then the marginal event E in (A, B) is simply the set of all elementary events (e, b) where e is in E and b is in B.

So, suppose A has the outcomes a1, a2, a3, a4, a5, a6; and B has the outcomes b1, b2, b3, b4. Suppose E is the event {a2, a3, a5}. Then the marginal event E in the joint experiment (A, B) is the set of elementary events that include any of {a2, a3, a5}, which is {(a2, b1), (a2, b2), (a2, b3), (a2, b4), (a3, b1), (a3, b2), (a3, b3), (a3, b4), (a5, b1), (a5, b2), (a5, b3), (a5, b4)}.

Table 4.2
Information about the Balls in the Urn

		Color	
		Black	White
Texture	Gloss	60	50
	Matte	40	20
	Vinyl	20	10

Consider, for example, an experiment of drawing balls from an urn. Each ball has both a color (black or white) and a texture (gloss, matte, or vinyl). We can think of the experiment of drawing balls as the joint experiment of determining color and determining texture. Then the set of all possible elementary events of the experiment is {(black, gloss), (black, matte), (black, vinyl), (white, gloss), (white, matte), (white, vinyl)}. The marginal event "ball is black" is the set of elementary events {(black, gloss), (black, matte), (black, vinyl)}. The marginal event "ball is gloss" is the set of elementary events {(black, gloss), (white, gloss)}.

Now suppose that we have the information about the number of balls in each outcome category as shown in Table 4.2. Then the marginal event "ball is black" has 120 balls since we can just sum the numbers in the elementary events {(black, gloss), (black, matte), (black, vinyl)}. Similarly the marginal event "ball is gloss" has 110 balls since we can just sum the number in the elementary events {(black, gloss), (white, gloss)}. An easy way to conceptualize marginalization is to think in terms of different partitions as shown in Figure 4.3.

For the joint experiment the partition has six elementary events. But we can also see the partitions for each of the single experiments: the color experiment (along the top of the diagram) partitions the set in two elementary events, and the texture experiment partitions the set in three elementary events (along the left of the diagram). When we marginalize an event like "ball is black" we simply restrict the joint experiment elementary events to those that lie in the outcome of the black part of the color partition. Similarly, when we marginalize an event like "ball is gloss" we simply restrict the joint experiment elementary events to those that lie in the outcome of the gloss part of the texture partition.

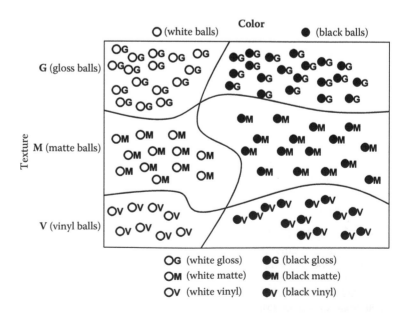

Figure 4.3 Marginalization viewed as different partitions of the set of balls in the urn.

4.3 Frequentist versus Subjective View of Uncertainty

When we consider statements about uncertain events like

- The next toss on a coin will be a head.
- A hurricane will destroy the White House within the next 5 years.

what we really want to do is *measure* the uncertainty of such events. In other words we want to be able to make statements like

- There is a 1 in 2 (or equivalently 50%) chance that the next toss on a coin will be a head.
- There is a 1 in 10 million (or equivalently 0.000001%) chance that a hurricane will destroy the White House within the next 5 years.

Although these statements are superficially similar, there are fundamental differences between them, which come down to the nature of the experiments that give rise to these outcomes. Specifically, whether the following assumptions are reasonable:

- Assumption 1 (repeatability of experiment)—The experiment is repeatable many times under identical conditions.
- Assumption 2 (independence of experiments)—Assuming the experiment is repeatable then the outcome of one experiment does not influence the result of any subsequent experiment.

For the coin-toss statement, the assumptions seem reasonable, both in practice and in theory. If we really have no idea whether the coin is fair, then we could actually carry out the coin tossing experiment many times and simply record the frequency of heads. We would see something like the graph of Figure 4.4. As the number of throws increases, the graph converges toward a number, in this case 0.5 (it is

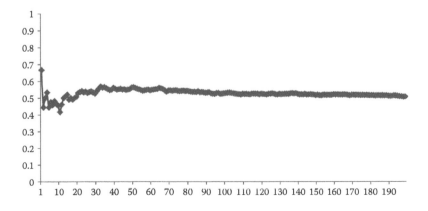

Figure 4.4 Graph plotting average relative frequency of heads.

a fair coin), and we define that empirical percentage to be the chance of a head.

If we wished to assume that there are just two elementary events (H and T), with both equally likely, and that the experiment can be repeated many times under identical conditions, then we do not even need to carry out any experiment, since theoretically we know that the proportion of heads must converge to 50%. The same principle occurs in the next two examples:

Example 4.4

If we assumed that a fair die genuinely had exactly six equally likely elementary events (1, 2, 3, 4, 5, 6), then after many throws the proportion of times we throw a 4 will get increasingly close to 1/6. Hence, we can reasonably state that the chance of getting a 4 in a throw of a fair die is 1 in 6.

Example 4.5   Urn

Consider an urn containing 50 equally sized and weighted balls of which 30 are known to be red, 15 are known to be white, and 5 are known to be black. Assume a ball is drawn randomly from the urn and then replaced. The event "next ball drawn from the urn" has three elementary events whose respective chances are

If the frequency cannot be determined theoretically (as in the preceding examples), then it is necessary to actually undertake as large a number of repeated experiments as possible. In that case the chance of an elementary event is defined as the frequency with which the event occurs in the actual experiments.

Red—(30 in 50) or equivalently 60%
White—(15 in 50) or equivalently 30%
Black—(5 in 50) or equivalently 10%

With the assumptions of repeatability and independence, we can therefore state the *frequentist* definition of chance as follows:

Frequentist definition of chance of an elementary event—
The chance of an elementary event is the frequency with which that event would be observed over an infinite number of repeated experiments.

In the special case where all of the elementary events are equally likely (such as in the roll of a fair die) the frequentist definition of chance of an outcome is especially simple:

Frequentist definition of chance of an elementary event (where elementary events are equally likely)
If an experiment has n equally likely elementary events, then the chance of each particular elementary event is $1/n$.

Example 4.6

The rule for combinations (see Appendix A) explains why there are 13,983,816 possible elementary events in a lottery that involves selecting 6 numbers from 49 (until 2016 the UK National Lottery had this format). Since it is reasonable to assume that each of these combinations of 6 numbers is equally likely to be chosen as the jackpot winner in any one draw, it follows that the chance of any single ticket winning the jackpot is

$$\frac{1}{13,983,816}.$$

Box 4.5 looks at a similar example with a pack of cards.

Box 4.5   Why Every Shuffled Pack of Cards Is a Miracle

If you need to be convinced that truly incredible events happen all the time, then just take a pack of 52 cards. Shuffle the pack thoroughly and then turn the cards over one by one. Imagine if the cards came out in the perfect sequence of Figure 4.5.

You would regard this as nothing short of a miracle. In fact, you would probably regard it as a life-changing moment that you would recount to your grandchildren.

Indeed, as is shown in Appendix A, the total number of possible sequences of 52 playing cards is 52! (that is 52 times 51 times 50 times ... times 2 times 1). So the probability that such an event would happen is 1 divided by 52!

Now 52! is a very large number. It is the number:

8065817517094387857166063685640376697528950544088832778240
00000000000

In fact it is such a big number that it is actually a much bigger number than the number of atoms in our galaxy (see http://pages.prodigy.net/jhonig/bignum/indx.html for further information about big numbers).

So the probability of turning over exactly this perfect sequence of cards is less than the probability of finding one specific atom in our galaxy. You would indeed be justified, therefore, in regarding it as a life-changing event if it happened.

Yet, the probability of getting the (apparently random) sequence of cards that really is revealed is exactly the same as any other sequence and is also the same as this tiny probability. So every sequence really is a miracle.

Figure   4.5   A   miraculous sequence?

The frequentist definition of chance extends easily to events rather than just elementary events. The event "roll a number bigger than 4" comprises the elementary events 5 and 6. So if we repeated the die-rolling experiment many times, then the frequency of the event would be equal to the frequency of getting a 5 (1 in 6) plus the frequency of getting a 6 (1 in 6). In other words, the chance of this event is 2/6 (i.e., 1/3).

Similarly, in the case of the urn in Example 4.5 the chance of drawing a red or a white ball is equal to the frequency of red balls (30/50) plus the frequency of white balls (15/50), which is 45 in 50, or 90%.

We can generalize this to the following frequentist definition of chance:

Frequentist definition of chance of an event—The chance of an event is simply the sum of the frequencies of the elementary outcomes of the event.

For experiments where each elementary event is equally likely, this amounts to:

Frequentist definition of chance of an event (equally likely elementary events)—If an experiment has n equally likely elementary events, then the chance of any event is m/n, where m is the number of elementary events for the event.

Example 4.7

What is the chance that the jackpot winner in a 6 from 49 ball lottery consists entirely of even numbers?

Solution: As we already saw there are 13,983,816 equally likely elementary events for the lottery. We have to calculate the number of these in which all numbers are even. There are 24 different even numbers (namely, the numbers 2, 4, 6, ..., 46, 48) in the draw. So the number of different elementary events in which all numbers are even is equal to the number of combinations of 6 objects from 24. That is

$$\frac{24!}{6! \times 18!} = \frac{24 \times 23 \times 22 \times 21 \times 20 \times 19}{6 \times 5 \times 4 \times 3 \times 2 \times 1} = 134,596$$

So the chance of the jackpot winner consisting entirely of even numbers is

$$\frac{134,596}{13,983,816} = 0.0096$$

that is, just below 1%.

Unfortunately, there is no reasonable frequentist interpretation of the statement: "There is a 1 in 10 million (or equivalent 0.000001%) chance that a hurricane will destroy the White House within the next 5 years." Unlike a lottery, die throw, or set of cards, we cannot play the next five years over and over again counting the number of times in which the White House is destroyed. We can only provide a *subjective measure of uncertainty* based on our current state of knowledge. Hence, this is what we call the subjective definition of chance.

Some people (including even clever ones) feel comfortable with the frequentist approach but so uncomfortable with the subjective approach that they reject it as invalid. In the law there is even a quasi-legal distinction made between them: frequentist measures of uncertainty are considered scientific and hence suitable to be presented in court; subjective measures are considered nonscientific and hence not suitable for presentation in court. Think of it another way: objective means it is a material property that belongs to the object and is therefore independent of any observer.

However, it turns out that the objections to subjective measures of uncertainty are irrational. Specifically, the primary objections are

1. The subjective measure cannot be validated.
2. Different experts will give different subjective measures.

They are irrational because, if you scratch the surface just a little, you will find that in most situations *these objections are equally applicable to frequentist measures.*

Remember that in the frequentist approach, we either make some assumptions about the elementary events (such as each is equally likely) or we have to conduct real repeated experiments. In the former case, there are inevitably some purely subjective assumptions made, for example,

- Just to assume that there are only two elementary events for the coin tossing experiment means we have to rule out many environments in which the experiment could theoretically be carried out.
- It is impossible to create a coin that is genuinely fair, so in practice a convergence to a perfect 50% relative frequency for heads would never actually be observed. This means we can never truly validate the 50% figure by material observation.

In the latter case (where we actually conduct the experiments), there is no true test to validate the figure and there are two reasons for this:

1. Even if you toss a perfectly fair coin 100,000 times, it is almost certain that heads will not come up on exactly 50,000 occasions. So, no matter how many tosses you make, you are unlikely to arrive at an exact figure of 50%.
2. Even if one expert did establish the true convergence frequency as 50%, it is almost certain that a different expert running different sequences of 100,000 tosses would arrive at a different convergence frequency because of the impossibility of perfectly replicating the experiments.

In the extreme case of 2, it has been demonstrated that it is possible to toss a fair coin in a particular way to ensure that it does not flip over at all. This means you can ensure that the toss is always a head. This

The notion that an objective property is one that is independent of any observer is rooted in what philosophers of science call *naturalism*, a scientific methodology that seeks truth from observing events in nature independently of metaphysical interventions or personal prejudice. Any discussion of naturalism or indeed competing philosophies of science are, in general, outside the scope of this book. We consider such a constraint unrealistic when dealing with complex phenomena.

poses another fundamental problem for the frequentist approach that is highlighted in Box 4.6.

What is clear is that in practice even the most definitive frequentist example (tossing a fair coin) inevitably involves a range of assumptions and subjective judgments.

Box 4.6   What If You Toss 100 Consecutive Heads?

Your job (as a risk expert) is not to calculate the chance of tossing a head. Rather, your job is to calculate the chance that the *next* toss of the coin is a head (just look back at the original problems posed at the start of this chapter). So, if you observe what is known to be a fair coin being tossed 100 times and each time the result is heads, what do you believe are the chances of the next coin being heads?

A frequentist, given the fair coin assumption, would insist the answer is still 50%. This is because the frequentist, with these assumptions, does not actually require any coin tosses to take place in practice. To the frequentist, the fair coin assumption means that the chance is always 50% on each throw. In other words, in making a prediction the frequentist must ignore the actual data that has been seen. The 100 consecutive heads would simply be considered a freak coincidence, that is, no more or less likely than any other random sequence of heads and tails. But then, the frequentist must ignore, for example,

1. The possibility that a fair coin can be tossed in such a way that makes heads more likely
2. That the coin tossed was not actually the fair coin assumed

In fact, we will see that such assumptions are irrational given the type of actual data observed. Only the subjective approach coupled with Bayesian reasoning will work effectively in such cases.

If you need further convincing that the frequentist approach is not somehow pure, superior, and clearly distinctive from the subjective approach then you should read Box 4.7.

Box 4.7   When Frequentist and Subjective Approaches Merge

Consider the following two statements:

1. There is a 50.9% chance that a new born baby in the United Kingdom is a girl.
2. There is a 5% chance of the Spurs winning the FA Cup next year.

On the surface there seems to be no doubt that statement 1 is explained by a frequentist argument: Over the last 100 years 50.9% of all births recorded in the United Kingdom have been girls.

There is also no doubt that statement 2 has no such frequentist explanation (and hence must be subjective) since there is only one FA Cup next year, and we cannot somehow play the tournament many times in the same year and count the number of occasions on which the Spurs win.

But if we dig a little deeper here, things get rather murky. The 50.9% figure in statement 1 is actually based on many years of data that may disguise crucial trend information.

Suppose we discover that the percentage of girls born is increasing; say a hundred years ago 48.5% of babies were girls compared with 51.2% last year. Then surely the probability of a randomly selected newborn being a girl now is higher than 50.9% (and higher than 51.2% if the figures have been steadily increasing). And what exactly do we mean by a "randomly" selected baby. Surely, what we are interested in are specific babies such as "the next baby born to Mrs. Roberts of 213 White Hart Lane, London N17." In that case the frequency data may need to be adjusted to take account of specific factors relevant to Mrs. Roberts. Both the general trend adjustments and the case specific adjustments here clearly require the subjective judgment of relevant experts. But that means, according to the frequentists, that their own approach is no longer valid since, as we saw earlier:

- The measure cannot be validated
- Different experts will give different subjective measures

Now look at statement 2 about the FA Cup in comparison. Although it is true that we cannot play the FA Cup more than once next year, we can nevertheless consider the number of times the Spurs won the FA Cup in the last 100 years as a key factor informing our subjective judgment. Of course, past form (especially of the distant past) is not a strong indicator of current form, but can we say with conviction that the situation was much different for the past "form" of babies born (is it not feasible that drastic changes in national figures could result from sudden environmental changes). And just as the Spurs might invest in the world's greatest players to increase their chances of winning the FA Cup next year, so a particular mother might apply a range of techniques to dramatically increase or decrease the chances of having a baby girl.

So, although the frequentist approach (especially in the case where we can assume equally likely outcomes and repeatable, independent experiments) is extremely useful, it should be clear that we cannot generally rely on it to measure uncertainty. This means that subjective measures, like it or not, are here to stay. They are already used so extensively that the fabric of modern society would break down without them. Hence, bookies will provide odds on events (such as England winning the World Cup) based on subjective measures, and insurance companies will do the same in determining policy premiums and governments when determining economic policies.

The subjective approach accepts unreservedly that different people (even experts) may have vastly different beliefs about the uncertainty of the same event. Hence, Norman's belief about the chances of the Spurs winning the FA Cup next year may be very different from Daniel's. Norman, using only his knowledge of the current team and past achievements, may rate the chances at 10%. Daniel, on the other hand, may rate the chances as 2% based on some inside knowledge he has about key players having to be sold in the next 2 months.

A further example is provided in Box 4.8.

Any attempt to measure uncertainty inevitably involves some subjective judgment. There is no truly objective frequentist approach.

Box 4.8 Honest Joe's and Shady Sam's

You visit a reputable casino, called Honest Joe's, in a good neighborhood in a city you know well. When there you see various civic dignitaries, such as the chief of police, the mayor, and local politicians. You decide to play a dice game where you win if the die comes up 6. What is the chance of the die coming up a 6?

Honest Joe's casino closes at midnight (well, it is for clean living types) forcing you to gamble elsewhere. Shady Sam's is recommended by a cab driver. Upon entry there the doormen give you a hard time, there are prostitutes at the bar, and hustlers all around. The same dice game as at Honest Joe's is on offer. What is the chance of the die coming up a 6?

Clearly, the long run property of the die is no longer a reliable estimate and no longer the sole factor governing any decision you might make on whether to gamble at Shady Sam's. Your uncertainty is related to your trust in the honesty of the casino and you typically use signifiers of vice or virtue as surrogates or complements for statistical uncertainty. By including these factors you are acting in a subjectively rational way. Ignoring them would be irrationally objective: you prized the theory over experience or insight.

Figure 4.6 shows the causal Bayesian network structure that reflects the uncertainty in this problem and indicates the inferences being made about whether the die is fair or not (do not worry about the direction of the arrows at this point or the distinction between causal direction and inference). The interesting thing about this is that our belief in the fairness of the die is a subjective probability about a chance event: we term this a *second-order probability* and will revisit the idea in later chapters.

A frequentist statistician might claim that we could determine objective probabilities even for this problem by visiting a sufficiently large number of casinos and determining the frequency with which these indicators of vice or virtue were present and whether they were associated with fair dice. Carrying out such research would certainly be fun, despite its obvious impracticability.

Figure 4.6 Bayesian network for Shady Sam's example.

4.4 Summary

The aim of this chapter was to demonstrate that subjectivity is an inevitable component in any measure of uncertainty. We explained two different approaches to uncertainty, the frequentist and the subjective, and showed that the frequentist offers little help in quantifying risk and that, in any case, the subjective approach subsumes it. However, whichever approach is used, it will be necessary (as in any type of modeling) to continuously monitor and assess the validity of your assumptions, use the most reliable sources for your data, and be cautious about overstating the extent to which your model generalizes.

We introduced the notions of experiments, outcomes, and events to help structure uncertain statements and have used counting rules to help count the number and frequency of events as the basic building blocks of calculation.

For easy reference Table 4.3 provides a simple list of the differences between frequentist and subjective approaches.

Table 4.3

Frequentist versus Subjective Views about Measuring Uncertainty

Frequentist	Subjective
• Can be legitimately applied only to repeatable problems.	• Is an expression of a rational agent's degrees of belief about uncertain propositions.
• Must be a unique number capturing an objective property in the real world.	• Rational agents may disagree. There is no "one correct" measure.
• Only applies to events generated by a random process.	• Can be assigned to unique events.
• Can never be considered for unique events.	

Further Reading

Gigerenzer, G. (2002). *Reckoning with Risk: Learning to Live with Uncertainty*. Penguin Books.
Haigh, J. (2003). *Taking Chances: Winning with Probability*. Oxford University Press.

The Basics of Probability

5.1 Introduction

In discussing the difference between the frequentist and subjective approaches to measuring uncertainty, we were careful in Chapter 4 not to mention the word *probability*. That is because we want to define probability in such a way that it makes sense for whatever reasonable approach to measuring uncertainty we choose, be it frequentist, subjective, or even an approach that nobody has yet thought of. To do this in Section 5.2 we describe some properties (called axioms) that any reasonable measure of uncertainty should satisfy; then we define probability as any measure that satisfies those properties. The nice thing about this way of defining probability is that not only does it avoid the problem of vagueness, but it also means that we can have more than one measure of probability. In particular, we will see that both the frequentist and subjective approaches satisfy the axioms, and hence both are valid ways of defining probability.

In Section 5.3 we introduce the crucial notion of probability distributions. In Section 5.4 we use the axioms to define the crucial issue of *independence* of events. An especially important probability distribution—the Binomial distribution—which is based on the idea of independent events, is described in Section 5.5. Finally in Section 5.6 we will apply the lessons learned in the chapter to solve some of the problems we set in Chapter 2 and debunk a number of other probability fallacies.

5.2 Some Observations Leading to Axioms and Theorems of Probability

Before stating the axioms of probability we are going to list some points that seem to be reasonable and intuitive for both the frequentist and subjective definitions of chance. So, consider again statements like the following:

- There is a 50% chance of tossing a head on a fair coin.
- There is a 0.00001% chance of a hurricane destroying the White House in the next 5 years.

All models in this chapter are available to run in AgenaRisk, downloadable from www. agenarisk.com.

Despite the very different nature of these two statements it is reasonable to conclude the following point for both frequentist and subjective approaches.

Point 5.1

Statements expressing our uncertainty about some event can always be expressed as a percentage chance that the event will happen.

It is important to note that a very common way of expressing uncertainty, especially by bookmakers, is the use of "odds." Box 5.1 describes this approach (and the main national variations). Although this approach does not express uncertainty in terms of percentage chance, Box 5.1 explains how it can easily be translated into a percentage chance.

Box 5.1   Bookmakers' "Odds" and Percentage Uncertainty

Bookmakers usually express uncertainty about an event using what is called an "odds" expression. In the United Kingdom this most commonly has the following form: "The odds of England winning the next World Cup are 4 to 1." This is written 4/1 and verbally is sometimes stated as "4 to 1 against." If you bet and win at these odds then for every pound you bet you will win 4 pounds (plus your stake).

When bookies express their uncertainty in this way what they are actually expressing is the following *odds ratio*:

$$\frac{\text{The chance of the event not happening}}{\text{The chance of the event happening}}.$$

So odds of 4/1 means that it is 4 times more likely that the event will not happen than it will happen. But that is the same as saying there is 1 in 5 chance or 20% chance it will happen.

If the right-hand number in the odds is not equal to 1, the same principle still applies. So odds of 7 to 2 (written 7/2) implies that for every 7 times that the event would not happen there are 2 times that the event would happen. Thus, the event will happen on 2 out of 9 times. This is the same as saying there is a 2 in 9 chance or equivalently a 22.22% chance.

So, in general any odds will be of the form "n to m" (written n/m). This translates to an "m in $(n+m)$ chance," which as a percentage, is

$$100 \times \left(\frac{m}{n+m} \right)$$

Note that there is no reason why m cannot be bigger than n. For example, a strong favorite in a two-horse race might have odds of 1/4 meaning there is a 4 in 5 (i.e., 80%) chance that the favorite wins. In this case bookies often say "4 to 1 on" (in contrast to "4 to 1 against" in the case of 4/1 odds).

One other special case is the odds 1:1 (which converts to a 50% chance). Bookies refer to this as "evens," meaning the chance of it not happening is the same as the chance of it happening.

In addition to the way odds are presented in the United Kingdom, there are the other main variations shown in Table 5.1.

Table 5.1
Converting Odds

Chance of event happening	20%	80%	22.22%	50%	66.66%
Odds ratio	80/20 = 4	20/80 = 0.25	77.7/22.22 = 3.5	50/50 = 1	33.33/66.66 = 0.5
UK odds	4/1	1/4	7/2	1/1	1/2
Decimal (European odds)	5	1.25	4.5	2	1.5
US odds	+400	−400	+350	100	−200

In particular note that

- *Decimal (also called European) odds* is simply one plus the odds ratio stated earlier. The resulting number is precisely the number of dollars you will receive back (including your stake) if you bet one dollar and win.
- *U.S. odds* is 100 times the odds ratio when the odds ratio is at least one, and is −100 times the inverse odds ratio otherwise. So, a positive number, such as +400, which corresponds to 4/1 odds means you will win $400 for each $100 you bet. A negative number such as −200, which corresponds to odds of 1/2, means you will have to bet $200 to win $100.

Finally, it is important to note that, because bookmakers want to make a profit, the odds they offer will generally not reflect the actual chance of the event but rather will be an overestimate (to enable them to make a profit). So, for example, if they were offering odds on the toss of a fair coin, instead of offering 1/1 on both heads and tails they might offer something like 4/5 on both. This corresponds to 5 in 9 chance, which is higher than 50%. A winning punter will win $4 for each $5 bet. Assuming that punters will win 50% of the time, the bookmaker will only have to pay out on average $90 for each $100 gambled, yielding a profit margin of 10% (profit margin is defined as: profit as percentage of selling price or revenue, i.e. 10/100)

Now, thinking of the frequentist view of probability the following point is clear from the definition:

Point 5.2

The percentage that expresses our uncertainty about an event can never be more than 100.

But this is also true for the subjective view. Why? Because the most certain we can ever be about an event happening is 100%. And events that have a 100% chance of happening are certain events. For example, the chance that a man will either live or die tomorrow is 100% because we are actually certain that either outcome will happen and we cannot be more certain than that. It might not be certain that it will rain in Manchester on at least one day next year, but it is surely extremely close to 100%; what it cannot be is any more than 100%.

Although people do say things like "I am 110% sure of this," it is nothing more than a figure of speech and it is no more meaningful than a football coach who says he expects 110% effort from each of his players.

In much the same way as the uncertainty about an event can never be more than 100%, we also have the following point.

Point 5.3

The percentage that expresses our uncertainty about an event can never be less than zero.

Again this is clear from the frequentist definition, but it is also true for the subjective view. Why? Because the least certain we can ever be about an event is 0%; such events, like rolling the number 7 on a normal six-sided die are certain not to happen, so the chance they happen is 0%. The chances of English-speaking martians landing on Earth next year might not be 0% but must be very close to 0%; what it cannot be is any less than 0%.

So, for both the frequentist and subjective approaches, the following point is valid:

Point 5.4

If you divide the percentage uncertainty of an event by 100 you must end up with a number that lies between zero and one.

We can now state the following axiom:

Axiom 5.1

The probability of any event is a number between zero and one.

When mathematicians can prove something they call it a theorem. If they cannot prove it, but need to assume it, they call it an axiom.

Because of Point 5.4 we know that both the frequentist and subjective approaches to uncertainty satisfy Axiom 5.1.

We know that there are different ways of measuring uncertainty (frequentist and subjective) and that in both approaches you might end up with different numbers for the same event. The nice thing about an axiom like Axiom 5.1 is that what it really means is that no matter how you define the probability, this statement should be true. It is true for existing measures and needs be true for other measures of probability that have not yet been invented (otherwise we will not call such measures probability measures).

Mathematicians denote the probability of an event E as $P(E)$.

Example 5.1

Consider again the event "next toss of coin is a head," which we will write as E. Then:

- A person who assumes that this coin is truly fair might reasonably assume, using frequentist arguments, that $P(E) = 0.5$.

- A person who assumes the coin is fair but also assumes there is a small chance it will land on its side, might reasonably assume using subjective arguments that $P(E) = 0.499$.
- A person who has seen the coin tossed 100 times and observed 20 heads might assume, using a frequentist argument, that $P(E) = 0.2$.
- A person who has inspected a coin and can see that both sides of the coin are heads might reasonably assume using subjective arguments that $P(E) = 1$.

Although all four probability values for the same event are different we cannot say that any one is wrong or right. They are all valid according to Axiom 5.1. However, if a person insisted that $P(E)$ was 1.3, then that would not be a valid probability value since it does not satisfy Axiom 5.1.

Recall from Chapter 4 that, in any given experiment, an event of special interest is the *exhaustive* event that consists of every possible elementary event that can result from the experiment. For example:

- For the coin toss experiment, where the outcome of interest is "what side the coin lands on next" the event {head, tail, other} (or {head, tail} if we can genuinely discount any elementary event other than head or tail) is the exhaustive event.
- For the six-sided die throw experiment, where the outcome of interest is "number showing on top" the set of elementary events: {1, 2, 3, 4, 5, 6, other} is the exhaustive event. So is the set of events {1, not 1}.
- For the World Cup experiment, where the outcome of interest is "next winner" the event {England wins, England does not win} is the exhaustive event.

Mathematicians refer to the exhaustive event as the *whole sample space*.

As we already saw from our discussion around Point 5.2, for both the frequentist and subjective approaches the chance of the exhaustive event happening is 100%. So, the following axiom should come as no surprise:

Axiom 5.2

The probability of the exhaustive event is one.

Whereas the first two axioms impose constraints on individual events, the next axiom provides both a mechanism and a constraint on how we reason about probabilities involving more than one event.

In Chapter 4 we introduced the notion of mutually exclusive events. For example, for the die roll experiment, consider these two events:

- E_1—Roll a number bigger than 4 (i.e., the set of elementary events {5, 6})
- E_2—Roll a number less than 4 (i.e., the set of elementary events {1, 2, 3})

These two events are mutually exclusive because they have no elementary events in common. We also introduced the notion of the union of two events. The union of the two events is the event "either E_1 or E_2," that is, the union of the set of elementary events of the two events (meaning in this case the event "roll any number except 4"), which is the event consisting of the elementary events $\{1, 2, 3, 5, 6\}$. Calculating the chance of each of the events separately we get:

■ The chance of the event $\{5, 6\}$ (i.e. E_1) is equal to 2/6, or 1/3.
■ The chance of the event $\{1, 2, 3\}$ (i.e. E_2) is equal to 3/6, or 1/2.
■ The chance of the event $\{1, 2, 3, 5, 6\}$ (i.e. E_1 or E_2) is equal to 5/6.

It follows that, in this and all other frequentist cases, the chance of the union of two mutually exclusive events E_1 and E_2 is the sum of the chances of the two separate events E_1 and E_2. The same conclusion seems reasonable in the subjective case. Not surprisingly, therefore, we have the following axiom:

Mathematicians write Axiom 5.3 as $P(E_1 \cup E_2) = P(E_1) + P(E_2)$, where E_1 and E_2 are mutually exclusive events.

Axiom 5.3

For mutually exclusive events, the probability of either event happening is the sum of the probabilities of the individual events.

Axiom 5.3 extends to any number of mutually exclusive events: the probability of the union is the sum of the probabilities of the individual events. To write this mathematically it is most convenient to use the sigma notation described in Box 5.2 (we will use this notation in various places subsequently in the mathematical explanations).

Box 5.2   The Sigma and Pi Notation

The sigma notation is mathematical shorthand for writing sums of similar looking expressions. For example, suppose we have the six expressions

$$\frac{1}{100}, \frac{2}{100}, \frac{3}{100}, \frac{4}{100}, \frac{5}{100}, \frac{6}{100}$$

then instead of writing the full expression

$$\frac{1}{100} + \frac{2}{100} + \frac{3}{100} + \frac{4}{100} + \frac{5}{100} + \frac{6}{100}$$

we write

$$\sum_{i=1}^{6} \frac{i}{100}$$

More generally if we have expressions of the form

$$f(1), f(2), f(3), ..., f(n),$$

then instead of writing the full expression

$$f(1) + f(2) + f(3) + \cdots + f(n)$$

we write

$$\sum_{i=1}^{n} f(i)$$

or, if it is clear from the context what the range of values for i is, we would simply write

$$\sum_{i} f(i)$$

The sigma notation can also be used to sum over the elements in an infinite or indeterminate set. So, for example,

$$\sum_{i=1}^{\infty} f(i)$$

is shorthand for the infinite sum $f(1) + f(2) + f(3) + f(4) + f(5) + ...$ and

$$\sum_{e \in S} f(e)$$

is shorthand for the sum of all terms of the form $f(e)$, where e is a member of the set S.

The pi notation is analogous to the sigma notation but is for products instead of sums. Hence

$$\prod_{i=1}^{6} f(i)$$

is shorthand for the product $f(1) \times f(2) \times f(3) \times \cdots \times f(n)$ and

$$\prod_{e \in S} f(e)$$

is shorthand for the product of the terms of the form $f(e)$, where e is a member of the set S.

Similarly we also use the union and intersection symbols:

$$\bigcup_{e \in S} f(e)$$

is shorthand for the union of all terms of the form $f(e)$, where e is a member of the set S.

$$\bigcap_{e \in S} f(e)$$

is shorthand for the intersection of all terms of the form $f(e)$, where e is a member of the set S.

Bringing this all together we can now state the general form of Axiom 5.3: For any set of mutually exclusive events $A_1, A_2, ..., A_n$

$$P\left(\bigcup_{i-1}^{n} A_i\right) = \sum_{i=1}^{n} P(A_i)$$

Taken together the three axioms introduced so far impose considerable restrictions on what constitutes a valid measure of probability.

Example 5.2

In the coin toss experiment with set of elementary outcomes {H, T, and "other"}, suppose we assign a subjective probability of 0.001 to "other." It would then be invalid to define $P(H) = 0.5$ and $P(T) = 0.5$. To see why, note that the elementary outcomes can be considered as three mutually exclusive events. Hence, by Axiom 5.3 the probability of the union of the three events would be equal to

$$P(H) + P(T) + P(\text{other}) = 1.001$$

But this contradicts both Axiom 5.1 (the probability is bigger than 1) and Axiom 5.2 (since the union of the three events is the exhaustive event, which must have probability equal to exactly 1).

If we insist that $P(H) = 0.5$ and $P(\text{other}) = 0.001$ then it follows by the axioms that $P(T)$ **must** be assigned the value 0.499 because this is the only value that will satisfy Axiom 5.3.

Using just the axioms we can derive useful theorems that follow logically from them. The first good example of such a theorem is one that tells us how to calculate the probability of the complement of an event. In frequentist experiments like dice throwing it is easy to see that the probability of the complement of an event is one minus the probability of the event.

Example 5.3

Take the event E, "roll the number 1," for the experiment of rolling a die. If we assume that this is a fair six-sided die where the only elementary outcomes are {1, 2, 3, 4, 5, 6}, then the complement of E (roll a number not equal to 1) is the event {2, 3, 4, 5, 6}. With these assumptions we calculate:

$$P(E) = 1/6$$

$$P(\text{not } E) = 5/6$$

But suppose we cannot make the fair die assumption of Example 5.3. Suppose instead that, based either on subjective judgment or actual experiments, we assign a probability of 0.15 to E. Then what can we say about P(not E) in this case? It turns out that, to be consistent with the probability axioms, P(not E) must be equal to 1—$P(E)$, that is, 0.85. This is because we can prove the following theorem from the axioms.

Theorem 5.1   A Probability Theorem (Probability of the Complement of an Event)

The probability of the complement of an event is equal to one minus the probability of the event.

Theorem 5.1 can be proved from the axioms as follows:

1. Any event E and its complement "not E" are mutually exclusive.
2. From Axiom 5.3 it therefore follows that the probability of the combined event "E or not E" is equal to the probability of E plus the probability of not E.
3. But the event "E or not E" is exhaustive, therefore its probability (from Axiom 5.2) is equal to one.
4. From 2 and 3 we conclude that the probability of E plus the probability of not E is equal to one.
5. From 4 we conclude that the probability of not E is equal to one minus the probability of E.

Mathematicians would write this proof more concisely as shown in Box 5.3.

Note that mathematicians write the expression "not E" as "$\neg E$".

Box 5.3   Example of a Mathematical Proof

Theorem: For any event E, $P(\neg E) = 1 - P(E)$
   Proof:

   1. For any E, events E and $\neg E$ are mutually exclusive
   2. The events E and $\neg E$ are exhaustive
   3. $P(E \cup \neg E) = P(E) + P(\neg E)$ (from 1 and Axiom 5.3)
   4. $P(E \cup \neg E) = 1$ (from 2 and Axiom 5.2)
   5. $P(E) + P(\neg E) = 1$ (from 3 and 4 above)
   6. $P(\neg E) = 1 - P(E)$ (from 5 above)

It follows from Theorem 5.1 that if we used either judgment or experimentation to determine that $P(E) = 0.3$ and P(not E) = 0.6, then P is an invalid measure of probability.

It follows from Theorem 5.1 that for any experiment where the outcome of interest has just two elementary events, once we assign a probability to one of the elementary events, the probability of the other is predetermined (as one minus the probability assigned to the first event).

Example 5.4

A bookmaker who offers odds of 4/5 on both heads and tails in the toss of a fair coin is failing to define a valid probability measure because the odds 4/5 equate to a probability of 0.555 and the two probabilities therefore sum to more than 1.

More generally, we can conclude the following:

For any experiment where the outcome of interest has n elementary events, once we assign probability values to $(n-1)$ of the elementary events, the probability of the other elementary event is predetermined (as one minus the sum of the other assigned probabilities).

Another provable consequence (which mathematicians call a *corollary*) of Theorem 5.1 is the following:

Corollary 5.1

The probability of any impossible event must be equal to zero. This is because the impossible event is the complement of the exhaustive event whose probability is 1.

The next useful theorem we can prove is one that tells us how to calculate the probability of the union of two events that are not necessarily mutually exclusive (Axiom 5.3 only dealt with mutually exclusive events).

Theorem 5.2

For any two events, the probability of either event happening (i.e., their union) is the sum of the probabilities of the two events minus the probability of both events happening (i.e., the intersection). Mathematicians write this as
$$P(A \cup B) = P(A) + P(B) - P(A \cap B)$$

The proof of Theorem 5.2 is shown in Box 5.4. It is based on the simple idea of breaking the union of the two events down into events that are mutually exclusive.

Box 5.4   Probability of the Union of Two Events

Proof that for any two events A and B, then $P(A \cup B) = P(A) + P(B) - P(A \cap B)$.
From Figure 5.1 it is clear that

1. $A \cup B$ is the union of the mutually exclusive events $(A \cap B)$, $(A \cap \neg B)$, and $(B \cap \neg A)$.
2. A is the union of mutually exclusive events $(A \cap \neg B)$ and $(A \cap B)$.
3. B is the union of mutually exclusive events $(B \cap \neg A)$ and $(A \cap B)$.

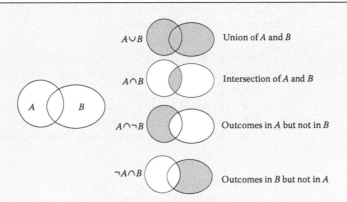

Figure 5.1 Events represented by shaded areas.

Therefore:

4. $P(A \cup B) = P(A \cap B) + P(A \cap \neg B) + P(B \cap \neg A)$ by Axiom 5.3 applied to 1.
5. $P(A) = P(A \cap \neg B) + P(A \cap B)$ by Axiom 5.3 applied to 2.
6. So $P(A \cap \neg B) = P(A) - P(A \cap B)$ (rearranging 5).
7. $P(B) = P(B \cap \neg A) + P(A \cap B)$ by Axiom 5.3 applied to 3,
8. So $P(B \cap \neg A) = P(B) - P(A \cap B)$ (rearranging 7).
9. $P(A \cup B) = P(A \cap B) + P(A) - P(A \cap B) + P(B) - P(A \cap B)$ by substituting 6 and 8 into 4.
10. $P(A \cup B) = P(A) + P(B) - P(A \cap B)$ by simplifying 9.

Example 5.5

Consider the experiment of drawing one card from a standard pack of 52. What is the probability of drawing an ace or a red card, assuming that all 52 cards are equally likely to be drawn?

Solution: Let A be the event "card is an ace" and let B be the event "card is red." The event whose probability we wish to calculate is "card is either an ace or red." This is the event $A \cup B$. Now

- $P(A) = 4/52$ (there are four aces)
- $P(B) = 26/52$ (half of the cards are red)
- $P(A \cap B) = 2/52$ (comprises two elementary events, namely $A\heartsuit$ and $A\diamondsuit$).

Hence, by Theorem 5.2

$$P(A \cup B) = \frac{4}{52} + \frac{26}{52} - \frac{2}{52}$$

$$= \frac{28}{52}$$

$$= \frac{7}{13}$$

In Example 5.5 we could have calculated $P(A \cup B)$ directly by counting the number of elementary events in $A \cup B$ (the 26 red cards plus the

2 red aces). However, in many cases this will not be possible as the next important example demonstrates.

Example 5.6

Suppose that on any given workday at a particular bank:

- The probability of (at least one) fraudulent transaction is 0.6.
- The probability of (at least one) lost check is 0.3.
- The probability of both a fraudulent transaction and lost check is 0.15.

We wish to calculate the probability that in any given day there is neither a fraudulent transaction nor a lost check.

This is typical of many probability problems in the sense that the experiment, the outcome of interest, and the set of elementary events are implied but not made explicit. The first step in the solution is to make these explicit. We are then able to calculate the required probability from the axioms and theorems.

In this case we can think of the experiment as observing a day in the life of a bank where the outcome of interest is "unfortunate incidents occurring." We are assuming that the only unfortunate incidents are fraudulent transactions and lost checks. Then the set of elementary events is:

- e1—Fraud only
- e2—Lost check only
- e3—Both fraud and lost check
- e4—No incidents

The difficulty in this case is that there is only one elementary event for which we are given the probability (namely, e3, whose probability is 0.15); in all our previous examples we were given the probability of each elementary event and used this information to calculate the probability of derived events.

However, in this case we are given the probability of two of the events. Specifically:

- The event A "fraudulent transaction," which is {e1, e3}, has probability 0.6
- The event B "lost check," which is {e2, e3}, has probability 0.3

We also know that $A \cap B$ is the elementary outcome e3 and so $P(A \cap B) = 0.15$.

What we have to calculate is the probability of the elementary outcome e4. Now e4 is the complement of the event {e1, e2, e3}, which is the event $A \cup B$. So $P(e4) = 1 - (A \cup B)$.

But, by Theorem 5.2, $P(A \cup B) = P(A) + P(B) - P(A \cap B) = 0.75$. So $P(e4) = 0.25$.

In all the discussion so far in the chapter we have assumed that there is an experiment with one particular outcome of interest. However, we made it clear at the start of Chapter 4 that the same experiment may have different outcomes of interest. In such cases, we have to be very careful with the probability notation to avoid genuine ambiguity. This is explained in Box 5.5, which formally introduces the important notion of *variable* outcomes.

**Box 5.5   Probability Notation Where There Are Different
Outcomes of Interest for the Same Experiment**

Consider the experiment of rolling two fair dice. There are many different outcomes of interest for this experiment including the following:

- The sum of the two dice rolled (let's call this outcome X).
- The highest number die rolled (let's call this outcome Y).

These two different outcomes of interest have different sets of elementary events.

- Outcome X has eleven elementary events: 2, 3, 4, 5, 6, 7, 8, 9, 10, 11, 12.
- Outcome Y has six elementary events: 1, 2, 3, 4, 5, 6.

If we are not careful about specifying the particular outcome of interest for the experiment, then there is the potential to introduce genuine ambiguity when calculating probabilities.

For example, consider the elementary event "2." What is the probability of observing this event for this experiment? In other words what is $P(2)$? The answer depends on whether we are considering outcome X or outcome Y:

- For outcome X, the probability $P(2)$ is 1/36 because there are 36 different ways to roll two dice and only one of these, the roll (1, 1), results in the sum of the dice being 2.
- For outcome Y, the probability $P(2)$ is 1/12 because of the 36 different ways to roll two dice there are three ways, the rolls (1, 2), (2, 1) and (2, 2), that result in the highest number rolled being 2.

Because of this ambiguity it is common practice, when there are different outcomes of interest for the same experiment to include some notation that identifies the particular outcome of interest when writing down probabilities. Typically, we would write $P(X = 2)$ or $P(Y = 2)$ instead of just $P(2)$.

The notation extends to events that comprise more than one elementary event. For example, consider the event E defined as "greater than 3":

- For outcome X, the event is E is equal to {4, 5, 6, 7, 8, 9, 10, 11, 12}.
- For outcome Y, the event is E is equal to {4, 5, 6}.

We calculate the probabilities as

- For event X, $P(E) = 11/12$.
- For event Y, $P(E) = 3/4$.

Typically we would write $P(X = E)$ or $P(X \geq 3)$ for the former and $P(Y = E)$ or $P(Y \geq 3)$ for the latter.

In this example the outcomes X and Y can be considered as variables whose possible values are their respective set of elementary events. In general, if there is not an obviously unique outcome of interest for an experiment, then we need to specify each outcome of interest as a named variable and include this name in any relevant probability statement.

5.3 Probability Distributions

Consider the experiment of selecting a contractor to complete a piece of work for you. We are interested in the outcome "quality of the contractor." Since, as discussed in Box 5.5, this is just one of many possible outcomes of interest for this experiment (others might be price of contractor, experience of contractor, etc.) it is safest to associate a variable name, say Q, with the outcome "quality of the contractor." Let us assume that the set of elementary events for Q is {very poor, poor, average, good, very good}.

On the basis of our previous experience with contractors, or purely based on subjective judgment, we might assign the probabilities to these elementary events for Q as shown in the table of Figure 5.2(a). Since the

Variables and Their States

Since the actual observed outcome for Q may vary every time we run the experiment, we can also say that Q is a variable with states {very poor, poor, average, good, very good}. It follows that the set of states is exactly the same as the set of elementary events. Henceforth, we shall therefore use the word *state* instead of *elementary event*.

Should We Use Horizontal or Vertical Bars for Probability Distributions Displayed as Graphs?

Within the Bayesian network community it is common practice to use horizontal bar graphs for probability distributions. However, as the number of states increases, the probability distribution is much more naturally displayed with vertical bars, as in the example of Figure 5.3, which shows the distribution of a variable "defects" whose possible states are 0, 1, 2, ..., 30.

As a rule of thumb, if the set of states is a numeric range, then the number of states tends to be large or even infinite, so it is more natural to use the vertical bar graph; if the set of states is nonnumeric it is more natural to use the horizontal bar graph. That is the convention we will adopt throughout this book (it is also the default setting in AgenaRisk, although users may choose to switch).

Very poor	0.4
Poor	0.3
Average	0.15
Good	0.1
Very good	0.05

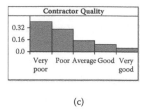

(a)   (b)   (c)

Figure 5.2 Probability distribution. (a) Probability table for Q. (b) Distribution of Q shown as graph (horizontal bars). Horizontal scale is the probability value expressed as a percentage. (c) Distribution of Q shown as graph (vertical bars). Vertical scale is the probability value.

numbers are all between 0 and 1, and since they sum to 1, this assignment is a valid probability measure for Q (i.e., for the experiment with outcome Q) because it satisfies the axioms.

A table like the one in Figure 5.2(a), or equivalent graphical representations like the ones in Figure 5.2(b) and Figure 5.2(c), is called a *probability distribution*. In general, for experiments with a discrete set of elementary events:

A probability distribution for an experiment is the assignment of probability values to each of the possible states (elementary events).

There is a very common but somewhat unfortunate notation for probability distributions. The probability distribution for an outcome such as Q of an experiment is often written in shorthand as simply: $P(Q)$. If there was an event referred to as Q then the expression $P(Q)$ is ambiguous since it refers to two very different concepts. Generally it will be clear from the context whether $P(Q)$ refers to the probability distribution of an outcome Q or whether it refers to the probability of an event Q. However, there is one situation where there is genuine ambiguity, namely, when the

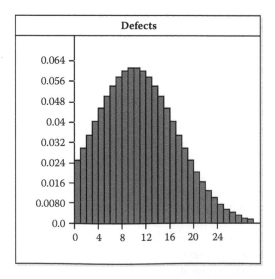

Figure 5.3 Probability distribution with many outcomes.

outcome of interest for an experiment is Boolean, meaning it has just two states (true and false, or yes and no) - see the sidebar.

Once we have the probability distribution defined we can easily calculate the probability of any event:

The *probability of an event E* is simply the sum of the probabilities of the elementary events that make up the event E. This follows directly from Axiom 5.3 since the elementary events are mutually exclusive.

Example 5.7

In the earlier experiment of selecting a contractor where the outcome of interest is quality Q, suppose E is the event "quality is at least average," then, using the notation of Box 5.5, $E = \{Q = \text{average}, Q = \text{good}, Q = \text{very good}\}$ and so, from the table in Figure 5.2 and the probability axioms: $P(E) = P(Q = \text{average}) + P(Q = \text{good}) + P(Q = \text{very good}) = 0.15 + 0.1 + 0.05 = 0.3$

In many situations assumptions about the states of the experiment enable us to automatically define the whole probability distribution rather than having to assign individual probabilities to each state. The simplest example of this is the case of experiments like the fair coin toss and the fair die roll whereby the relevant probability distribution is called the Uniform distribution (see sidebar).

We shall see more interesting examples of such automatically defined distributions throughout the book.

5.3.1 Probability Distributions with Infinite Outcomes

For experiments with a continuous set of states, such as measuring the height of a person in centimeters, it is hopeless to assign probability values to each possible state because there are an infinite number of them. For example, should you assign a different probability to 2.3126 compared to 2.3125? Instead, we do one of the following:

1. Divide the continuous range of outcomes into a set of discrete intervals (this process is called *discretization*) and assign probability values to each interval. For example, if the continuous range is 0 to infinity we might define the discrete set as: [0, 100), [100, 110), [110, 120), [120, 130), [130, 140), [140, 150), [150, infinity).
2. Use a continuous function whose area under the curve for the range is 1. One common example of this that we have already seen in Chapter 2 is the *Normal distribution*. For example, we could use a Normal distribution such as shown in Figure 5.4(a) for our height experiment (in a tool like AgenaRisk you simply enter the function as shown in Figure 5.4(b) where we simply have to specify the mean and variance).

When $P(X)$ Really Has Two Different Meanings

Consider the experiment of a legal trial when one (but by no means the only) outcome of interest is whether the defendant is guilty. This experiment is Boolean in the sense that it has just two elementary events: *true* and *false*. But *true* is the same as saying *guilty*, and false is the same as saying *not guilty*. So guilty can refer to the (variable) outcome of interest, but it can also refer to one of two specific elementary events. Hence, if we write $P(guilty)$ we could mean either:

The probability distribution for guilty, that is, an assignment like $P(guilty = true) = 0.9$ and $P(guilty = false) = 0.1$

or

$P(guilty = true)$.

In Chapter 6 we shall use both of these interpretations, but it should still be clear from the context which one is relevant.

Uniform Distribution (Discrete)

Where we can assume that each of the states of an experiment is equally likely, it follows directly from Chapter 4 that the probability of each state is simply $1/n$ where n is the number of states. Such a probability distribution is called a *Uniform* distribution.

The notation $[a, b)$ means all numbers in the range a to b including a but not including b.

Variance and Standard Deviation

The variance and standard deviation (which we introduced in Chapter 2) are both statistics that measure how much variation there is from the "average." The variance is simply the square of the standard deviation. It is more usual to specify the variance than the standard deviation.

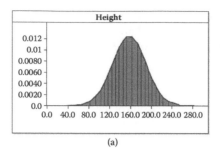
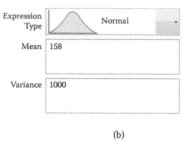

(a)

(b)

Figure 5.4 A continuous probability distribution. (a) Normal distribution (with mean 158 and variance 1000) displayed as a graph. (b) Defining the distribution in AgenaRisk.

However, note that although the graph plotted here is shown only on the range 0 to 300, the Normal distribution extends from minus infinity to plus infinity. As such there is a (very small in this case) "probability" of negative values meaning that the Normal distribution can only ever be a crude approximation to the "true" distribution.

If you are using a continuous function as a probability distribution you need to understand that, unlike discrete probability distributions, it is not meaningful to talk about the probability of a (point) value. So, we cannot talk about the probability that a person's height is 158 cm. Instead we always need to talk in terms of a range, like the probability a person's height is between 157.9 and 158.1 cm. The probability is then the proportion of the curve lying between those two values. This is shown in Box 5.6, which also describes the important example of the continuous Uniform distribution.

Box 5.6   Calculating Probabilities That Are Defined with Continuous Functions

Suppose that we have modeled the outcomes of measuring people's height using the Normal distribution shown in Figure 5.5. As was explained in Chapter 2, the model is an idealized distribution for the underlying population. We cannot use the model to determine the probability that a person has some exact (point value), but we can use it to calculate the probability that a person's height is within any nonzero length range. We do this by computing the area under the curve between the endpoints of the range. So in the figure the shaded area is the probability that a person is between 178 and 190 cm tall. In general the area under a curve is calculated by integrating the mathematical function that describes the distribution. Even for experienced mathematicians this can be very difficult, so they use statistical tables or tools like MS Excel or AgenaRisk to do it for them.

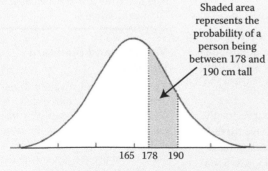
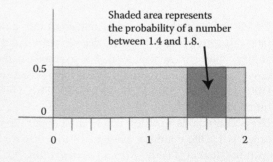

Figure 5.5   Probabilities as area under the curve.

Figure 5.6   Uniform distribution over the range 0 to 2 (written Uniform[0,2]).

> However, one important continuous distribution for which it is easy to compute relevant probabilities is the *Uniform distribution*. We have already seen the discrete Uniform distribution. The continuous Uniform distribution is relevant when an experiment of interest results in numerical outcomes within some finite range such that an outcome within any subrange is just as likely as an outcome in any other subrange of the same size. For example, suppose we use a computer to generate random real numbers between 0 and 2 for assigning security codes to a group of personnel. Then we would expect that a number in the range 0.2 to 0.4 would be just as likely as a number in the range 1.4 to 1.8. In such situations we would use the continuous Uniform distribution shown in Figure 5.6.
>
> The area under the Uniform distribution is simply a rectangle whose length is the length of the range and whose height is 1 divided by the length of the range. Hence, to calculate the probability a value lies between any particular range (such as 1.4 and 1.8 in Figure 5.6) we simply multiply the length of the particular range (0.4 in this case) by the height (0.5). In this case the probability is 0.2.

5.3.2 Joint Probability Distributions and Probability of Marginalized Events

In Chapter 4 we defined the notion of joint experiments and events. Recall that if we have an experiment where the outcome of interest A comprises the set of elementary events $\{a1, a2, a3\}$ and if we have a second experiment where the outcome of interest B comprises the set of elementary events $\{b1, b2, b3, b4\}$ then the set of elementary events of the joint experiment is just the set (A, B), that is, the collection of all combinations of pairs from A and B:

$$\{(a1,b1), (a1,b2), (a1,b3), (a1,b4), (a2,b1),$$
$$(a2,b2),(a2,b3), (a2,b4), (a3,b1), (a3,b2),(a3,b3), (a3,b4)\}$$

Recall also that an event in the joint experiment is simply a subset of (A, B).

From now on, instead of referring explicitly to an experiment with an outcome A of interest for which there is a set of elementary events, it is more convenient (see sidebar) simply to refer to A as a *variable* whose *states* are just the set of elementary events.

It therefore follows that:

The joint probability distribution of A and B is simply the probability distribution for the joint event (A,B).

Example 5.8

With the fair coin and fair die assumptions the joint probability distribution for (coin toss, die roll) might reasonably be asserted as Table 5.2.

Example 5.9

To assess the suitability of potential subcontractors, each organization is rated for both "quality of previous work" (with possible outcomes: poor, average, good) and whether they are "certified" (with possible outcomes:

As we discussed at the start of this section, if an experiment has an outcome of interest A for which the set of elementary events is $\{a1, a2, ..., an\}$, then this is equivalent to characterizing the experiment as a variable A with states $\{a1, a2, ..., an\}$.

Table 5.2
Joint Probability Distribution for (Coin Toss, Die Roll)

(H, 1)	1/12
(H, 2)	1/12
(H, 3)	1/12
(H, 4)	1/12
(H, 5)	1/12
(H, 6)	1/12
(T, 1)	1/12
(T, 2)	1/12
(T, 3)	1/12
(T, 4)	1/12
(T, 5)	1/12
(T, 6)	1/12

Table 5.3
Joint Probability
Distribution for (Quality of
Previous Work, Certified)

(poor, no)	0.3
(poor, yes)	0.2
(average, no)	0.25
(average, yes)	0.15
(good, no)	0.02
(good, yes)	0.08

yes or no). Regarding quality and certification as experiments on organizations we can consider the joint experiment and, based on our subjective knowledge of subcontractors we have worked with, we might define the joint probability distribution as in Table 5.3.

In Chapter 4, we defined the notion of marginalization for events in joint experiments. We explained how in many situations we were interested in information about the marginal event. What we did not say then, but should be obvious now, is that we are especially interested in the *probability* of the marginal event. Returning to the examples we used earlier we will now see how easy it is to calculate the marginal probabilities.

Table 5.4
Trial Results from Test (with
Frequencies Added)

Disease	No	Yes
Test		
Negative	9405	1 (0.01 %)
	(94.05%)	
Positive	495	99
	(4.95%)	(0.99%)

Example 5.10

For the joint experiment Disease and Test in Chapter 4 (Example 4.3) we have the results shown in Table 5.4. Then the joint probability distribution for the experiment is shown in Table 5.5.

But we are interested in the probability of the marginal event "Disease is yes." This event is simply the collection of states (yes, negative), (yes, positive). From the probability axioms the probability of the marginal event is just the sum of the probabilities of the individual states that make up the marginal event, so that

$$P(Disease = Yes)$$

$$= P(Disease = Yes, \ Test = Negative)$$

$$+ \ P(Disease = Yes, \ Test = Positive)$$

$$= 0.0001 \ + \ 0.0099$$

$$= 0.01$$

Table 5.5
Joint Probability Distribution
for (Disease, Test)

(No, Negative)	0.9405
(No, Positive)	0.0495
(Yes, Negative)	0.0001
(Yes, Positive)	0.0099

Similarly, for the marginal event "Test is positive" the probability is

$$P(Test \ = \ Positive)$$

$$= P(Disease = No, \ Test = Positive)$$

$$+ \ P(Disease = Yes, \ Test = Positive)$$

$$= 0.0495 \ + \ 0.0099$$

$$= 0.0594$$

Example 5.11

For the balls in urn joint experiment (color and texture) we might have the results shown in Table 5.6. Then the probability distribution for (color, texture) is as shown in Table 5.7.

Table 5.6
Information about the Balls in the Urn (with Frequencies Added)

		Color	
		Black	**White**
Texture	Gloss	60 (30%)	50 (25%)
	Matte	40 (20%)	20 (10%)
	Vinyl	20 (10%)	10 (5%)

Table 5.7
Joint Probability Distribution for (Color, Texture)

(black, gloss)	0.3
(black, matte)	0.2
(black, vinyl)	0.1
(white, gloss)	0.25
(white, matte)	0.1
(white, vinyl)	0.05

But we are interested in the probability of the marginal event "ball is black." Since this marginal event is simply the collection of states (black, gloss), (black, matte), (black, vinyl) it follows from the probability axioms that

$$P(Color = black)$$

$$= P(Color = black, Texture = gloss)$$

$$+ P(Color = black, Texture = matte)$$

$$+ P(Color = black, Texture = vinyl)$$

$$= 0.3 + 0.2 + 0.1$$

$$= 0.6$$

Similarly, for the marginal event "ball is gloss" the probability is

$$P(Texture = gloss)$$

$$= P(Color = black, Texture = gloss)$$

$$+ P(Color = white, Texture = gloss)$$

$$= 0.3 + 0.25$$

$$= 0.55$$

When there are many states it is more convenient to use the sigma notation for marginalization as shown in Box 5.7.

Box 5.7 Sigma Notation for Marginalization

In Example 5.11 the notation got a little longwinded with a long sequence of plus (+) operators. So to make things a little neater and to save space we use the Σ symbol introduced in Box 5.2. For this example we get:

$$P(color = black) = \sum_{texture} P(color = black, texture)$$

If we are interested in marginalizing out texture for all states of the variable color, in other words getting the full probability distribution for color, we write this as

$$P(color) = \sum_{texture} P(color, texture)$$

In general, if $A = \{a_1, a_2, ..., a_n\}$ and $B = \{b_1, b_2, ..., b_m\}$ then the $P(A = a_i)$ is obtained by marginalizing over B in (A, B):

$$P(A = a_i) = \sum_{j=1}^{m} P(a_i, b_j)$$

The full probability distribution of A is written as

$$P(A) = \sum_{j=1}^{m} P(A, b_j)$$

or more simply as just

$$P(A) = \sum_{B} P(A, B)$$

The notion of joint probability distributions and their marginalization extends to more than two variables. We deal with this next.

5.3.3 Dealing with More than Two Variables

Suppose that in our search for a suitable contractor (as in Example 5.7) instead of just "quality of previous work," A, and "certification," B, we are also interested in "experience of staff," C, "number of previous similar projects," D, and "credit rating," E.

The kind of problems that occur in critical risk assessment turn out to be problems that can be considered exactly as predicting the outcome of a large multiple or joint experiment. So, instead of just two variables of interest, there may in general be many (as in the sidebar example).

Generalizing the approach of the previous section, the joint probability distribution for variables in the sidebar example

$$P(A, B, C, D, E)$$

is simply the set of all probability tuples:

$P(a, b, c, d, e)$ for every a in A, b in B, c in C, d in D, and e in E.

Moreover the approach to marginalization also generalizes. So if, for example, having obtained the joint distribution of $P(A, B, C, D, E)$ we wish to focus only on the joint distribution $P(B, C)$ then we simply marginalize over the variables A, D, E:

$$P(B,C) = \sum_{A,D,E} P(A,B,C,D,E)$$

So, in general, if $A_1, A_2, ..., A_n$ are uncertain variables relevant for a particular problem then what we are interested in is knowing the joint probability distribution

$$P(A_1, A_2, ..., A_n),$$

that is, the probability of each possible combination of states in $A_1 = \{a_{11}, a_{12}, ...\}$, $A_2 = \{a_{21}, a_{22}, ...\}$, and so on.

Once we know each of these individual probability values, then we have seen that we can calculate the probability of any event by simply summing up the relevant probabilities. In principle, this gives us a complete and quite simple solution to all our risk prediction problems. But, of course there is a bit of a snag. The total number of combinations of individual probability values that we need to know is

$$\text{(Number of states in } A_1) \times$$
$$\text{(Number of states in } A_2) \times$$
$$\text{(Number of states in } A_3) \times \dots \text{ etc.}$$

In the disease diagnosis problem of Example 5.10 the number of these state combinations is just 4. But, suppose that in the previous example A has 5 states, B has 2 states, C has 5 states, D has 1000 states, and E has 20 states. Then the number of state combinations is 1 million.

If there are many variables relevant for a problem then it will generally be impossible to define the joint probability distribution in full. The same is true even if there are just a small number of variables that include some that have a large number of states.

> Much of what follows in this book addresses this issue in one way or another. Specifically, the challenge is to calculate probabilities efficiently, and in most realistic situations this means finding ways of avoiding having to define joint probability distributions in full.

We next make a start on this crucial challenge.

5.4 Independent Events and Conditional Probability

So, where we have joint events, we can regard the outcomes that result as one big event and we can work out all the probabilities we need from the joint probability distribution. We have just seen some examples of this. It is possible to go a long way with this approach. In fact every problem of uncertainty can ultimately be reduced to a single experiment, but as just discussed, the resulting joint probability distributions may be too large to be practicable. In fact, even in relatively simple situations it turns out that using the joint probability distribution may be unnecessarily complex as the following examples illustrate.

Example 5.12

Suppose we get one friend, Amy, to toss a fair coin and her brother Alexander to roll a fair die. We are interested in the probability that Amy tosses a Head and Alexander rolls a 5 or 6. From Section 4.2.2 we know that the joint experiment has the following pairs of outcomes (ignore the "other" outcomes discussed there):

(1, H), (2, H), (3, H), (4, H), (5, H), (6, H), (1, T), (2, T), (3, T), (4, T), (5, T), (6, T)

Using the frequentist assumptions we assume are all equally likely, so we can assume that each has probability 1/12. The event "Amy tosses a head and Alexander rolls a 5 or 6," consists of the two outcomes (H, 5) and (H, 6). So its probability is 2/12, that is, 1/6.

Although the calculation is correct there is something unsatisfactory about it. Amy tossing a coin and Alexander rolling a die appear to be independent events in the sense that the outcome of one has no influence on the outcome of the other. It should be possible to calculate the probability of the combined event simply in terms of the probabilities of the two separate events. The probability of Amy tossing a head is 1/2. The probability of Alexander rolling a 5 or 6 is 1/3. Now, if we simply multiply the two separate probabilities together we get:

$$\frac{1}{2} \times \frac{1}{3} = \frac{1}{6}$$

which is exactly the probability we calculated for the joint event.

Example 5.13

From a shuffled pack of 52 cards, the top card is drawn. The card is replaced in the pack, which is again shuffled and the top card is again drawn. We want to calculate the probability that both cards drawn were kings.

The joint experiment here is drawing a card (with replacement) twice. From Chapter 4 we know that the number of possible outcomes for this joint experiment is equal to 52^2. Therefore, the number of such outcomes is 2704. Assuming each is equally likely this means each outcome has probability 1/2704.

Next we have to establish how many of these outcomes consist of two kings. This is asking how many ways can we arrange four kings in two positions with replacement. This is equal to $4^2 = 16$. It therefore follows that the probability of drawing two kings is 16/2704.

But again this seems unnecessarily complex. We know that the probability of drawing a king as the first card is 4/52 (this is an experiment with 52 equally likely outcomes). But drawing the second card is intuitively independent of what happened on the first draw. The probability of drawing a king is again 4/52. Now, if you simply multiply the two separate probabilities together we get:

$$\frac{4}{52} \times \frac{4}{52} = \frac{16}{2704}$$

which is exactly the probability we calculated for drawing two kings.

In both of these examples we can therefore conclude that:

Of course we cannot conclude that Observation 5.1 is generally true and the observation also relies on a notion of independence that we have not yet formally defined. However, it is clearly true for all similar examples that you may care to try out.

Observation 5.1   Probability of Independent Events

If A and B are independent events, then the probability that both A and B happen is equal to the probability of A times the probability of B. In other words

$$P(A \cap B) = P(A) \times P(B)$$

$P(A \cap B)$ is often also written as $P(A,B)$, that is, the joint event.

It turns out that, even if the result of one experiment does impact on another experiment, it may still be unnecessarily complex to consider the set of all outcomes of the single joint experiment, as the following example demonstrates.

Example 5.14

Suppose that as in Example 5.13 we draw the top card from a shuffled pack of cards. However, this time instead of replacing the card and shuffling the pack we draw the next top card. Again we want to calculate the probability that we draw a pair of kings.

To cast this as a single experiment we have to consider the set of all possible outcomes that result from drawing two cards without replacement. From Appendix A we know that this is the set of all permutations of size 2 from the 52 cards. Therefore the number of such outcomes is $52 \times 51 = 2652$. Assuming each is equally likely this means each outcome has probability 1/2652.

How many of these outcomes consist of two kings? This is asking how many ways can we arrange four kings in two positions. This is the number of permutations of two from four, which is equal to $4!/2! = 12$. It therefore follows that the probability of drawing two kings in this case is 12/2652.

However, once again although the calculation is correct, it seems to be unnecessarily complex. Again we know that the probability of drawing a king as the first card is 4/52. In this case the drawing of the second card is not independent of the first. However, if we know that the first card is a king then that leaves 51 cards of which only 3 are kings. So the probability of drawing a king on the second card once the first card was a king is 3/51 (since this is an experiment with 51 equally likely outcomes). Now if you simply multiply the two probabilities together you get

$$\frac{4}{52} \times \frac{3}{51} = \frac{12}{2652}$$

which is exactly the probability of drawing two kings without replacement.

In the particular case of Example 5.10 we can therefore conclude Observation 5.2:

Observation 5.2   Probability of Events That Are Not Necessarily Independent

The probability that both A and B happen is equal to the probability of A times the probability of B given A. Writing the event "B given A" as $B \mid A$ this means:

$$P(A \cap B) = P(A) \times P(B \mid A)$$

The expression $P(B \mid A)$ is stated as "the probability of B given A." This conditional probability is very important later on when we introduce Bayes' theorem.

---

and hence:

Observation 5.3

Providing that $P(A)$ is not equal to zero it follows from Observation 5.2 that

$$P(B \mid A) = \frac{P(A \cap B)}{P(A)}$$

Again we cannot conclude that Observation 5.2 (and by consequence Observation 5.3) is generally true even though it clearly is for the examples seen so far. But nothing can stop us defining the notion of *conditional probability* and *independence* in terms of the aforementioned observations. Specifically we define:

Axiom 5.4   Fundamental Rule (of Conditional Probability)

Let A and B be any event with $P(A)$ not equal to zero. The event "B occurs given that A has already occurred" is written as $B \mid A$ and its probability is defined as

$$P(B \mid A) = \frac{P(A \cap B)}{P(A)} = \frac{P(A,B)}{P(A)}$$

Example 5.15

In any year the probability that it snows in Minneapolis on 1 January is 0.6. The probability that it snows in Minneapolis on 1 January and on 2 January is 0.35. If we know that it has snowed in Minneapolis on 1 January then what is the probability it will snow the following day?

Solution: Let A be the event "snows on 1 January" and B the event "snows on 2 January." We have to calculate $P(B \mid A)$. Since we know $P(A)$ = 0.6 and $P(A \cap B)$ = 0.35 it follows from Axiom 5.4 that

$$P(B \mid A) = \frac{0.35}{0.60} = 0.583$$

Box 5.8 provides a demonstration of why joint probabilities and Axiom 5.4 make sense for frequentist arguments.

Box 5.8   Demonstrating Joint Probabilities and Axiom 5.4 Using Frequencies

Consider a class of 100 students where 40 are Chinese and 60 are non-Chinese. There are 10 female Chinese students and 20 female non-Chinese students. By elimination we know there must be 70 male students (of whom 30 are Chinese and 40 are non-Chinese).

We want to use this frequency information to show that the probability formulas for joint distributions, marginals, and conditional probability match the frequency counts. So, let C represent the variable "nationality." Then C has two states $\{c, n\}$ where c represents Chinese and n represents non-Chinese.

Let S represent the variable sex, so that S has two states $\{f, m\}$ corresponding to female and male.

The joint probabilities are easily calculated from the frequencies of each joint event. This provides us with everything we need to know about the population.

$$P(C = c, S = f) = \frac{10}{100} = 0.10 \quad \text{(10 of the 100 students are Chinese and female)}$$

$$P(C = c, S = m) = \frac{30}{100} = 0.30 \quad \text{(30 of the 100 students are Chinese and male)}$$

$$P(C = n, S = f) = \frac{20}{100} = 0.20 \quad \text{(20 of the 100 students are non-Chinese and female)}$$

$$P(C = n, S = m) = \frac{40}{100} = 0.40 \quad \text{(40 of the 100 students are non-Chinese and male)}$$

The marginal probabilities are therefore

$$P(C = c) = P(C = c, S = f) + P(C = c, S = m) = \frac{10}{100} + \frac{30}{100} = 0.40$$

$$P(C = n) = P(C = n, S = f) + P(C = n, S = m) = \frac{20}{100} + \frac{40}{100} = 0.60$$

$$P(S = f) = P(C = c, S = f) + P(C = n, S = f) = \frac{10}{100} + \frac{20}{100} = 0.30$$

$$P(S = m) = P(C = c, S = m) + P(C = n, S = m) = \frac{30}{100} + \frac{40}{100} = 0.70$$

Notice that these derived marginal probabilities, when multiplied by 100, correspond precisely to the frequencies we started with. Hence the method for calculating marginals is justified.

Next we see if Axiom 5.4 is justified in terms of the frequency counts.

First, we want to know the probability a student is Chinese given that the student is female. For Axiom 5.4 we get

$$P(C = c \mid S = f) = \frac{P(C = c, S = f)}{P(S = f)} = \frac{10}{100} / \frac{30}{100} = \frac{10}{30}$$

and indeed this corresponds to the 10 out of 30 from the frequency count.

Similarly, the probability a student is not Chinese given that the student is female is

$$P(C = n \mid S = f) = \frac{P(C = n, S = f)}{P(S = f)} = \frac{20}{100} / \frac{30}{100} = \frac{20}{30}$$

which corresponds to the 20 out of 30 from the frequency count.

Note also that $P(C = c \mid S = f) + P(C = n \mid S = f) = 1$, which confirms Theorem 5.1.

The probability a student is Chinese given that the student is male is

$$P(C = c \mid S = m) = \frac{P(C = c, S = m)}{P(S = m)} = \frac{30}{100} / \frac{70}{100} = \frac{30}{70}$$

and this corresponds to the 30 out of 70 from the frequency count.

Finally, the probability a student is not Chinese given that the student is male is

$$P(C = n \mid S = m) = \frac{P(C = n, S = m)}{P(S = m)} = \frac{40}{100} / \frac{70}{100} = \frac{40}{70}$$

which corresponds to the 40 out of 70 from the frequency count.

So, using frequency counts we have been able to justify the various probability formulas and axioms.

Note that, from Axiom 5.4, we can immediately deduce a theorem that gives us the easier intuitive way to calculate the probability of $A \cap B$ as suggested in the discussion following Example 5.10:

Theorem 5.3

For any two events A and B:

$$P(A \cap B) = P(A)P(B \mid A)$$

Note that if $P(A) = 0$ then A is an event with no possible outcomes. It follows that $A \cap B$ also contains no possible outcomes and so $P(A \cap B) = 0$.

Sidebar 5.1

It is well known that the bookies' profit margin on "combined event" bets—such as "Spurs score first and win" or "Harry Kane scores and Spurs win" are significantly higher than for normal single event bets (typically 6% compared to 4%). Yet these types of bets are among the most popular. Why? Because football fans like to bet on their team winning but are generally reluctant to place a bet for which the profit is less than the stake (i.e. for which the odds offered are less than "evens"). Only rarely will the "Spurs win" bet be offered at odds of better than evens. However, the odds offered for a bet like "Harry Kane scores and Spurs win" will always be better than evens and, hence, looks relatively attractive. In the long run punters would be far better off sticking to "single event" bets.

Example 5.16

A very common type of bet that bookmakers offer in English football is for a team to score first and win. Suppose that Spurs score the first goal in 40% of their matches and that they win in 70% of the matches in which they score first. If the bookmaker offers odds of 2 to 1 for Spurs to score first and win in their next match should you take the bet?

Solution: Let A be the event "Spurs score the first goal" and B the event "Spurs win." Making the frequentist assumptions, including the assumption that the next match will be much like all previous matches, we can assume that $P(A) = 0.4$ and $P(B \mid A) = 0.7$. Since we want to know $P(A \cap B)$, it follows from Theorem 5.3 that $P(A \cap B) = 0.28$. Recall from Box 5.1 that the bookmaker's odds of 2 to 1 are equivalent to a probability of 1/3, i.e. 0.33. The bookmaker is therefore overstating the true probability, although the "mark-up" is fairly typical to ensure his profit in the long run. If you have some inside knowledge that the next match is likely to be easier than normal for Spurs (and, hence, that the true probability is higher even than what the bookmaker is offering) then you should take the bet. See sidebar 5.1 for how Bookies exploit punters on such popular "combined" event bets.

A very important result that we will need for Bayesian networks is the generalization of Theorem 5.3 to more than two events. This generalization is called the *chain rule* and it is explained in Box 5.9.

Box 5.9   Chain Rule

In Example 5.16 we were interested in calculating the probability of the joint event $A \cap B$ where:

  A is the event "Spurs win"
  B is the event "Spurs score first"

By Theorem 5.3 we know that $P(A \cap B) = P(A \mid B) \times P(B)$ and we used this in our calculation.
  Now suppose that we introduce a third event C where

  C is the event "Spurs score the second goal"
  Again we are interested in the probability of the joint event $A \cap B \cap C$. But we can apply Theorem 5.2 to help us as follows:

$$P(A \cap B \cap C) = P(A \cap (B \cap C))$$

$$= P(A \mid (B \cap C)) \times P(B \cap C) \text{ (substituting } (B \cap C) \text{ for } B \text{ in Theorem 5.3)}$$

$$= P(A \mid (B \cap C)) \times P(B \mid C) \times P(C) \text{ (applying Theorem 5.3 to } (B \cap C))$$

Here the notation $P(A)$ actually refers to the probability that A is true. But, as we noted in Section 5.2 the notation $P(A)$ can also refer to the full probability distribution of A. So, in this case $P(A)$ could also mean the pair of values:

$$P(A = \text{True})$$

$$P(A = \text{False})$$

The same applies to $P(B)$ and $P(C)$. So, instead of just thinking of the above formula as an expression for calculating the probability of the joint event "A is true and B is true and C is true" we can also think of it an expression for the full joint probability distribution of A, B, and C. So, in general we can write

$$P(A,B,C) = P(A \mid B,C) \times P(B \mid C) \times P(C)$$

and we can use this expression for any combination of outcomes of A, B, and C. So suppose, for example, we wished to calculate the joint probability that A is false, B is true, and C is false. Then (writing "not A" for "A is false," B for "B is true," and "not C" for "C is false") we have

$$P(\text{not } A, B, \text{not } C) = P(\text{not } A \mid B, \text{not } C) \times P(B \mid \text{not } C) \times P(\text{not } C)$$

We can extend the approach to any number of events. In general, if we have n events (or uncertain variables) $A_1, A_2, \ldots, A_n$ then the full joint probability distribution can be expressed as

$$P(A_1, A_2, A_3, \ldots, A_n) = P(A_1 \mid A_2, A_3, \ldots, A_n)$$

$$\times P(A_2 \mid A_3, \ldots, A_n) \times \ldots \times P(A_{n-1} \mid A_n) \times P(A_n)$$

In general we refer to this as the <u>chain rule</u>.

Using the product notation introduced in Box 5.2 we can also write the chain rule more compactly as

$$P(A_1, A_2, A_3, \ldots, A_n) = \prod_{i=1}^{n} P(A_i \mid A_{i+1}, \ldots, A_n)$$

Having defined the notion of conditional probability axiomatically we can now provide a formal definition of the notion of independence of events:

Definition: Independent Events

Let A and B be any events with $P(A)$ not equal to zero. Then we define A and B as being independent if

$$P(B) = P(B \mid A)$$

It follows from Theorem 5.3 that if A and B are independent events then

$$P(A \cap B) = P(A) \times P(B)$$

Example 5.17

Using the assumptions of Example 5.12 in the joint experiment of tossing a coin and rolling a die the event A "toss a head" and the event B "roll a 6" are independent. To see this, we note that the marginal probability $P(A) = 1/2$ and the marginal probability $P(B) = 1/6$. So $P(A)P(B) = 1/12$. But $A \cap B$ consists of just one outcome (H, 6), so $P(A \cap B) = 1/12$.

Example 5.18

Sometimes it can be quite surprising to find that certain events are independent. In the experiment of rolling a fair die, consider the two events

- A—Roll an even number: {2, 4, 6}
- B—Roll a multiple of 3: {3, 6}

These two events are independent because $A \cap B$ has a single outcome, 6, so $P(A \cap B) = 1/6$ while $P(A) \times P(B) = (1/2) \times (1/3) = 1/6$.

Thus, knowing that we have rolled an even number has no influence on whether or not the number is a multiple of three:

$P(B) = P(B \mid A) = P(\text{multiple of six} \mid \text{even}) = 1/3$

5.5   Binomial Distribution

Whenever possible we would like to be able to automatically define probability distributions rather than have to define the probability of each outcome separately. Whether we can do this of course depends on the nature of the underlying experiment and the assumptions that we can make about it. So far we have seen the Uniform distribution as an example of a probability distribution that could automatically be assigned given certain assumptions about the underlying experiment. Now that we have defined the notion of independent events we can describe a more interesting case (and one that we will use extensively), namely the *Binomial distribution.*

Suppose that a factory mass produces a particular electronic component. Previous tests have shown that the failure rate of these components is very consistent: 20% of them will fail within one year under standard conditions of use. A company buys five of these components. They are interested in predicting the number that will fail within one year of standard use.

Clearly this is an experiment with six possible elementary events: {0, 1, 2, 3, 4, 5}. Equivalently, as discussed, this means that the number of failures is a variable with states {0, 1, 2, 3, 4, 5}.

We assume that the components are completely independent and that they therefore fail independently of each other. Therefore, it is reasonable to think of the experiment as being made up of five independent experiments (or trials), one for each component. In each of the independent experiments the component has a 20% chance (i.e., 0.2 probability) of failing within one year and an 80% chance (i.e., probability 0.8) of not failing.

So, if we let F represent "a component fails," and "not F" represent "component does not fail," it follows from the definition of independence that $P(0)$, the probability of zero components failing, is equal to

$$P(\text{not } F) \times P(\text{not } F) \times P(\text{not } F) \times P(\text{not } F) \times P(\text{not } F)$$

$$= 0.8 \times 0.8 \times 0.8 \times 0.8 \times 0.8 = 0.8^5 = 0.32768$$

It is just as easy to calculate $P(5)$, the probability that all of the five components fails as this is equal to

$$P(F) \times P(F) \times P(F) \times P(F) \times P(F)$$

and this is equal to

$$0.2 \times 0.2 \times 0.2 \times 0.2 \times 0.2 = 0.2^5 = 0.00032$$

But what about, say $P(2)$ the probability of exactly two components failing? To calculate this we need to use the definition of independence as well as our knowledge of combinations (Appendix A). First think about a particular case where the first two components fail and the last three do not. This event has probability

$$P(F) \times P(F) \times P(\text{not } F) \times P(\text{not } F) \times P(\text{not } F)$$

and this is equal to

$$0.2 \times 0.2 \times 0.8 \times 0.8 \times 0.8 = 0.2^2 \times 0.8^3 = 0.02048$$

But the number of ways we can have two components fail in five positions is equal to the number of combinations of size 2 from 5 and this is equal to

$$\frac{5!}{2! \times 3!} = \frac{5 \times 4}{2 \times 1} = 10$$

Since each of these combinations has the same probability, namely, 0.02048, it follows that

$$P(2) = 10 \times 0.02048 = 0.2048$$

We can use the same method to calculate $P(3)$, the probability of exactly three failures. We consider the particular case where the first three components fail and the last two do not. This event has probability

$$P(F) \times P(F) \times P(F) \times P(\text{not } F) \times P(\text{not } F)$$

and this is equal to

$$0.2 \times 0.2 \times 0.2 \times 0.8 \times 0.8 = 0.2^3 \times 0.8^2 = 0.00512$$

But the number of ways we can have three components fail in five positions is equal to the number of combinations of size 3 from 5 and this is equal to:

$$\frac{5!}{3! \times 2!} = \frac{5 \times 4}{2 \times 1} = 10$$

Since each of these combinations has the same probability, namely, 0.00512, it follows that

$$P(3) = 10 \times 0.00512 = 0.0512$$

We use the same approach to calculate $P(1)$ and $P(4)$ so that we get the distribution shown in Table 5.8. You can check that these probability values sum to exactly 1.

If you found these calculations a little tricky, you do not need to worry. Box 5.10 provides a general formula for the probability distribution (based on a generalization of the previous arguments).

Table 5.8
Full Probability Distribution for Number of Failures of Five Components

$P(0)$	0.32768
$P(1)$	0.4096
$P(2)$	0.2048
$P(3)$	0.0512
$P(4)$	0.0064
$P(5)$	0.00032

Box 5.10   General Formula for the Binomial Distribution

In general, suppose there are n independent trials and suppose that on each trial there is the same probability p of a success. (The terms here are used quite generally. A trial could refer to a component over one year as in the earlier example and a success in this case could actually mean either the component does not fail or we could equally well define success as the component fails. A trial could also refer to the experiment of selecting a person at random, with success being defined as the "person has gray hair.") The probability of exactly r successes in n trials is equal to

$$\frac{n!}{r! \times (n-r)!} \times p^r (1-p)^{n-r}$$

So, for example, if $n = 10$, then

$$P(0) = \frac{n!}{0! \times n!} \times p^0 (1-p)^n = (1-p)^n$$

$$P(1) = \frac{n!}{1! \times (n-1)!} \times p^1 (1-p)^{n-1} = n \times p \times (1-p)^{n-1}$$

$$P(2) = \frac{n!}{2! \times (n-2)!} \times p^2 (1-p)^{n-2} = \frac{n \times (n-1)}{2!} \times p^2 \times (1-p)^{n-2}$$

...

$$P(n) = \frac{n!}{n! \times 0!} \times p^n (1-p)^0 = p^n$$

The most important thing to remember about the Binomial distribution is that it is completely defined by two parameters n (the number of trials) and p (the success probability). In AgenaRisk this is all you need to know to generate the whole distribution. Figure 5.7 shows the dialogue for defining the distribution (assuming you have created an integer node) and Figure 5.8 shows the resulting distribution.

Example 5.19

In a family with eight children, what is the probability that every child is a boy if we assume that boy and girl births are equally likely, and that the sex of successive children is independent? What is the probability that exactly two of the children are boys?

Solution: To solve this problem we can assume a Binomial distribution where the number of trials is eight and the probability of success is the probability of a child being a boy (namely, 1/2). We have been asked to calculate two particular outcomes $P(8)$ (probability of eight boys) and $P(2)$ (probability of two boys). Using the formula

$$P(8) = \left(\frac{1}{2}\right)^8 = \frac{1}{256} = 0.00390625$$

$$P(2) = \frac{8!}{2! \times 6!} \times \left(\frac{1}{2}\right)^2 \times \left(\frac{1}{2}\right)^6 = \frac{8 \times 7}{2!} \times \left(\frac{1}{2}\right)^8 = \frac{28}{256} = 0.109375$$

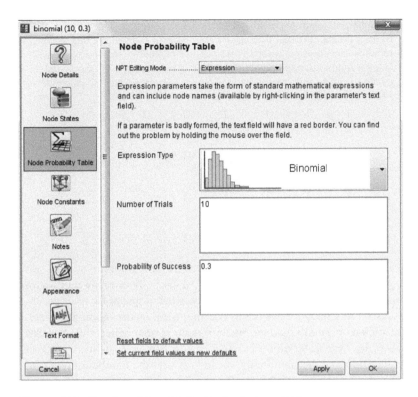

Figure 5.7   Creating a Binomial distribution in AgenaRisk.

Example 5.20

A software program is estimated to contain 20 hard-to-find bugs. During final testing it is assumed that each bug has a 90% chance of being found and fixed. Assuming that the bugs are independent, what is the probability that at most two bugs will remain after testing?

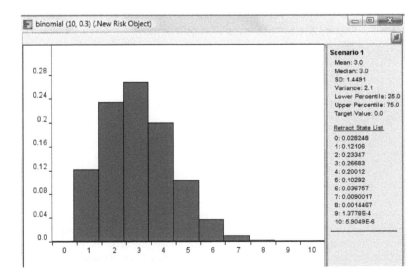

Figure 5.8   Binomial distribution displayed.

Solution: Again we assume a Binomial distribution. The number of trials is 20 and it is best to consider success here as meaning that a bug remains after testing. Since the probability of finding and fixing a bug is 0.9 it follows that the probability of a bug remaining is 0.1. So, to calculate the probability that at most two bugs remain we have to calculate $P(0) + P(1) + P(2)$:

$$P(0) = (0.9)^{20} = 0.1216$$

$$P(1) = 20 \times 0.1 \times (0.9)^{19} = 0.27017$$

$$P(2) = \frac{20 \times 19}{2!} \times (0.1)^2 \times (0.9)^{18} = 0.28518$$

so $P(0) + P(1) + P(2) = 0.67695$

In the preceding problem we calculated the probability of at most two bugs remaining. But what if we wanted to know the average number of bugs that remain? We first need to agree what is meant by average. As discussed in Chapter 2 we could use the mean or the median. In fact, statisticians normally use the mean (which is also called the *expected value*). Its formal definition is explained in Box 5.11. It turns out that for the Binomial distribution the mean is always the number of trials (n) times the probability of success (p) so in this case the mean number of bugs found is 18, and the mean number of bugs remaining to 2.

Box 5.11   Expected Value (Mean) of a Discrete Probability Distribution

For distributions like the Binomial in which the states are all numeric, it makes sense to ask what is the expected value or equivalently the most likely value. For example, if we roll a fair die 120 times we want to know what is the expected value for the number of 6s rolled? Intuitively in this case the number is 20, since the probability of a 6 is 1/6.

In general, if $P(X)$ is a probability distribution where the variable X has numeric states $\{x_1, x_2, ..., x_k\}$ then, using the sigma notation explained in Box 5.2, the expected value (mean) of X is defined as

$$\sum_{i=1}^{k} x_i P(x_i)$$

For example, if X is the number of heads tossed in two tosses of a fair coin then we know that X has three states—0, 1, 2—and that the probability distribution of X is

$$P(0) = 1/4; \ P(1) = 1/2; \ P(2) = 1/4$$

Hence the mean of X is

$$0 \times \frac{1}{4} + 1 \times \frac{1}{2} + 2 \times \frac{1}{4} = 1$$

It can be shown that for a Binomial distribution (n,p) (meaning a Binomial distribution where the number of trials is n and the probability of success is p) the mean is equal to np. You can work this out by substituting the general expression presented in Box 5.10 into the formula for expected value above (the proof is left as an exercise). Hence, if we roll a fair die 120 times the expected number of 6s rolled is $120 \times 1/6 = 20$, since this is a Binomial (120, 1/6) distribution.

5.6 Using Simple Probability Theory to Solve Earlier Problems and Explain Widespread Misunderstandings

We can now use what we have learned to solve some of the problems we posed in Chapter 2 and explain a number of other widespread misunderstandings and fallacies.

5.6.1 The Birthday Problem

Recall that we have 23 children and wish to calculate the probability that at least 2 children share the same birthday.

The easiest way to solve this problem is to calculate the probability of the complement event, that is, the probability that none of the children share the same birthday. If we know the probability p of the complementary event, then the probability of the event we are interested in is just $1 - p$.

We provide two different approaches to finding p. One, based purely on the counting rules and frequentist definition of probability, is shown in Box 5.12. But there is an easier way using independence. Let's first suppose there are just four children. Imagine we line the children up and label them 1, 2, 3, and 4. Then the probability that the birthdays of children 1, 2, 3, and 4 are all different is the probability that

(2 is different to 1) and (3 is different to 2 and 1)
and (4 is different to 1, 2, and 3)

Box 5.12   Calculation of Birthdays Probability Using Counting Rules

Assuming that there are 365 distinct birthdays, the total number of different ways we can assign birthdays to 23 children is the number of samples of size 23 from 365 with replacement. That is equal to 365^{23}. We assume that any particular assignment of birthdays to the 23 children is just as likely as any other. So any specific assignment of birthdays has a probability of $1/365^{23}$. That means, for example, that the probability that every child was born on January 1 is $1/365^{23}$ and the probability that 10 were born on May 18 and 13 were born on September 30 is also $1/365^{23}$.

What we need to calculate is the number of assignments, N, in which no two birthdays are equal. Then, according to Appendix A the probability that 23 children all have different birthdays is

$$N\left(\frac{1}{365}\right)^{23}$$

But N is equal to the number of samples of size 23 from 365 objects with replacement (to ensure all are different) but where order is important. This is equal to the number of permutations of 23 objects from 365 and is

$$\frac{365!}{342!} = 365 \times 364 \times 363 \times 362 \times \cdots \times 344 \times 343$$

It follows that p is

$$\frac{365 \times 364 \times 363 \times 362 \times \cdots \times 344 \times 343}{365^{23}} \approx 0.49$$

But since these events are independent, the probability that all four are different is

$$P(2 \text{ is different to } 1) \times P(3 \text{ is different to } 2 \text{ and } 1)$$
$$\times P(4 \text{ is different to } 1, 2, \text{ and } 3)$$

So let's calculate each of these probabilities in turn:

$$P(2 \text{ is different to } 1) = 364/365$$

because once we know the first child's birthday there are 364 "unused" other days for the second child's birthday to be different.

$$P(3 \text{ is different to } 2 \text{ and } 1) = 363/365$$

because for the third child's birthday to be different than both the first and second child it must fall on one of the 363 unused birthdays.

$$P(4 \text{ is different to } 1, 2, \text{ and } 3) = 362/365$$

because for the fourth child's birthday to be different to each of the first three children it must fall on one of the 362 unused birthdays.

Hence the probability that all four birthdays are different is

$$\frac{364}{365} \times \frac{363}{365} \times \frac{362}{365}$$

Carrying on this argument for 23 children we find that the probability that none of the children share the same birthday is:

$$\frac{364}{365} \times \frac{363}{365} \times \frac{362}{365} \times \frac{361}{365} \times \cdots \times \frac{344}{365} \times \frac{343}{365} \approx 0.49$$

This is exactly the same result as produced in Box 5.12.

So the probability that at least two children share the same birthday is 1 minus 0.49, which is equal to 0.51. This is just better than a one in two chance.

We can use the same method to calculate the probability that at least two children share the same birthday in a class of size n (where n is any number). For example, in a class size of 40 (which was the typical class size when we were growing up) the probability is about 0.9. It is therefore not surprising that there were invariably kids with the same birthday in almost every class of which we were members.

The fact that people underestimate this kind of probability so badly is often used by people to trick you in various ways. Bookmakers use it to trick you out of money by offering odds that look more generous than they really are. Magicians and self-proclaimed psychics use it to fool you into thinking they can predict incredible events, which actually turn out to be very likely indeed.

Why are people so surprised by this result? Probably because they think in terms of someone else sharing the same birthday as themselves. That probability is fairly low as explained in Box 5.13.

We have made some simplifying assumptions in the birthday calculations. We have ignored leap-year birthdays (which makes the probability lower by a tiny fraction) but set against this we have ignored the fact that in practice birthdays are not spread evenly throughout the year; in fact if you take into account this nonuniform distribution of birthdays (far more birthdays occur in May in England than in October even though both months have 31 days), it turns out that you need just 20 children in a class for the probability of finding a duplicate birthday to be about 0.5, in other words about 50% of any classes of 20 children will have children with duplicate birthdays.

Box 5.13   The Probability That Another Child Shares Your Birthday

If you are one of 23 people, what is the probability that at least one of the other people shares your birthday? Again it is easier to calculate the probability of the complement of the event, namely, the probability that no other person shares your birthday. For each of the other people the probability that they have a different birthday to you is 364/365. So there are 22 events each with a probability of 364/365. Moreover, each of these events is independent. So the probability that they all occur is simply the product of each of the probabilities, that is,

$$\left[\frac{364}{365}\right]^{22}$$

which is approximately 0.94. That means the probability at least one of the people shares your birthday is 0.06 or about a 1 in 17 chance.

Note that you need to pay careful attention to the wording used in these examples to describe the event of interest. In Box 5.13 you were asked to answer a question involving "at least" one child in addition to yourself, whereas in Box 5.12 you were asked about the probability of "any two" children having the same birthday. Seemingly trivial differences in problem wording like this imply completely different event definitions and ultimately very different answers.

5.6.2   The Monty Hall Problem

Recall that we have three doors, behind which one has a valuable prize and two have something worthless. After the contestant chooses one of the three doors Monty Hall (who knows which door has the prize behind it) always reveals a door (other than the one chosen) that has a worthless item behind it. The contestant can now choose to switch doors or stick to his or her original choice. The sensible decision is always to switch because doing so increases your probability of winning from 1/3 to 2/3 compared to sticking with the original choice. There are many explanations of this but we feel the following (shown graphically in Figure 5.9) is the easiest to understand.

The key thing to note is that Monty Hall will always choose a door without the prize, irrespective of what your choice is. So consequently the event "door first chosen wins" is independent of Monty Hall's choice. This means that the probability of the event "door first chosen wins" is the same as the probability of this event conditional on Monty Hall's choice. Clearly, when you first choose, the probability of choosing the winning door is 1/3. Since Monty Hall's choice of doors has no impact on this probability, it remains 1/3 after Monty Hall makes his choice. So, if you stick to your first choice the probability of winning stays at 1/3; nothing that Monty Hall does can change this probability.

Since there is a 1/3 probability that your first choice is the prize-winning door, it follows that the probability that your first choice is not the prize-winning door is 2/3. But if your first choice is not the prize-winning door then you are guaranteed to win by switching doors. That is because Monty Hall always reveals a door without the prize (so, for example, if you chose door 1 and the prize is behind door

Figure 5.9 All three cases once you have chosen a particular door (assumed to be 1). In 2 out of 3 cases you are certain to win by switching, in 1 you are certain to lose by switching.

Another reason why people over-estimate the probability of winning if they stick is that when Monty Hall reveals a door that they did not choose has no prize it tends to confirm in their mind that their first choice was correct. This is an example of what is commonly referred to as *confirmation bias*. It is quite irrational of course because Monty Hall must always reveal a door without a prize, irrespective of your choice. We shall return yet again to this problem in Chapter 7 by showing a solution using a Bayesian network.

2 Monty Hall would have to reveal door 3, which has no prize). So, since you always win by switching in the case where your first choice was not the prize-winning door, there is a 2/3 probability of winning by switching.

It may be easier to appreciate the solution by considering the same problem with 10 doors instead of just 3. In this case there are 9 doors without the prize behind them and one door with the prize. The contestant picks a door. Monty Hall then has to reveal 8 of the other doors without the prize leaving just the contestant's door and one other door. Unless the contestant chooses the one door with the prize (a 1/10 probability) the only door left unopened by Monty must be the one with the prize behind it. So there is a 9/10 probability that the other door will contain the prize, as 9 out of 10 times the player first picked a door without the prize. The contestant's probability of winning is therefore 9/10 if he switches compared to 1/10 if he does not.

The trick about the three-door game is that it creates a misleading impression; the contestant is always presented with 1/3 proportions. There is a 1/3 chance of winning, the host reveals 1/3 of the mystery, and the player is allowed to switch to the other 1/3 option. All options seem equal, yet they are not.

5.6.3 When Incredible Events Are Really Mundane

The birthdays problem is a classic example of a very common fallacy of probabilistic reasoning, namely, to underestimate the true probability of an event that seems to be unlikely. More dramatic examples appear in the media all the time. You will all have come across newspaper articles and television reports of events that are reported to be one in a million, one in a billion, or maybe even one in a trillion. But it is usually the case that the probability of such events is much higher than stated. In fact, it is often the case that such events are so common that it would be more newsworthy if it did *not* happen.

Example 5.21

A few years ago the *Sun* newspaper carried a story about a woman who had just given birth to her eighth child, all of whom were boys. It said that the probability of this happening was less than one in a billion.

The fallacy here, as in all such stories, is to confuse the specific with the general. The probability of a specific mother (for example, your mother) giving birth this year to her eighth child, all of which are boys, is indeed very low (as we will explain later). But the probability of this happening to at least one mother in the United Kingdom during a one-year period is almost a certainty. Why?

In any given year there are approximately 700,000 births in the United Kingdom. Among these approximately 1000 are to mothers having their eighth child. Now, in a family of eight children, the probability that all eight are boys is 1 in 256; this was explained in Example 5.19 using the Binomial distribution. So there is a probability of 1/256 that a mother having her eighth child will have all boys. So, out of 1000 such mothers how many will have all boys? This is another case of the Binomial distribution, this time with $n = 1000$ and $p = 1/256$. The probability that none of the 1000 mothers have all boys is

$$\left(\frac{255}{256}\right)^{1000} = 0.01996$$

That is about a 1 in 50 chance. So there is about a 98% chance that in any given year at least one mother will give birth to her eighth child, all boys. This is, therefore, hardly a newsworthy event. In fact it would be far more newsworthy if, in any given year, there was no case of a mother giving birth to her eighth child all of which are boys.

So where does the one in a million or one in billion come from? Well first let's just restrict ourselves to one of the specific 700,000 mothers giving birth this year. For any such mother chosen at random, the probability that she will be having her eighth child is just 1 in 700. Now, in a family of eight children, the probability that all eight are boys is 1 in 256. So the probability that the chosen mother gives birth to eight boys is 1 divided by 700 divided by 256. This is about 1 in 180,000. But the probability of any specific mother in the United Kingdom (of whom there are about 15 million) giving birth at all this year is about 1 in 20, so you need to divide the 1 in 180,000 probability by 20 to arrive at a figure of about 1 in 4 million. If, in addition, you decide to narrow the focus to, say, the probability that any specific mother gives birth to her eighth boy in this particular week or even day, then you can see how easy it is to argue that such a probability is less than one in a billion.

There are many similar examples of supposedly impossibly rare events that the media loves to report. All of the following come from real news items.

In Chapter 4, Example 4.7, we saw that the probability of any single ticket winning the jackpot (assuming 6 out of 49 numbers are selected) is 1/13,983,816. Applying the rule for independent events we just multiply 1/13,983,816 by itself to get the probability that exactly two lottery tickets for separate lottery draws both win the jackpot. The resulting number is 1 divided by approximately 200 billion.

Example 5.22 "American Woman Wins Lottery Jackpot Twice"

This was cited in a media article as being a 1 in 200,000,000,000,000 chance. In fact the chance of any specific person winning the jackpot on two specific separate occasions with exactly two tickets is indeed that low, as explained in the sidebar.

But in, say, a 20-year period in a country as large as the United States, the probability that at least one person wins the jackpot twice is actually quite high. In the United States around 120 million lottery tickets are bought every week. Let's say, typically, this means 60 million people buying two tickets each week. For simplicity let's also assume just one lottery a week so the probability of a typical player winning in any week is about 1 in 7 million. In a 20-year period each of those 60 million people will play in 1000 lotteries. What is the probability that a given person will win two lottery games? We can use the Binomial distribution to get the solution:

$$P(\text{one player wins twice})$$
$$= Binomial(n = 1000, r = 2, p = 1/7 \text{ million})$$
$$= 499500(1/7 \text{ million})^2(1-1/7 \text{ million})^{998}$$
$$= 1.02 \times 10^{-8}$$

Sidebar 5.2

Question: If we run n trials where the probability of success is p, what is the probability that we achieve at least one success?

Solution: By the Binomial theorem the probability that we achieve no successes is equal to $(1-p)^n$ *. So the answer we need is* $1-(1-p)^n$ *.*

This is a very small number indeed, but this is the probability of a given individual player winning twice over a 20-year period not the probability of any player winning. What we really want to calculate is the probability that at least one player, from the playing population of 60 million regular players, wins twice in a 20-year period. This is an example of the problem shown in sidebar 5.2; where $p = 1.02 \times 10^{-8}$ and $n = 60,000,000$. So the answer is

$$= 1-(1-1.02\times10^{-8})^{60000000}$$
$$= 1-0.54$$
$$= 0.46$$

Therefore the probability of at least one person winning twice in a 20-year period is roughly 46%. This is much higher than people expect. However, even this may be pessimistic given that the initial win is likely to lead to the winner funding higher stake investments in future lottery tickets with a corresponding increase in the chance of winning.

Example 5.23 "Same Six Numbers Come Up in Bulgarian Lottery Two Weeks in Row"

On September 6, 2009, the winning Bulgarian lottery numbers were 4, 15, 23, 24, 35 and 42. Amazing? Hardly. But on the very next play of the lottery the winning numbers were 4, 15, 23, 24, 35 and 42. In fact, even this apparently amazing event is not so unusal when we consider, say, a 20-year period and the fact that there are about 300 different national lotteries (some countries have several).

Here is why: The Bulgarian lottery involves drawing 6 balls from 42 without replacement. Using the same method as Chapter 4, Example 4.6, the probability of any specific set of 6 numbers being drawn is 1/5245786 (about 1 in 5 million). So this is the probability that on any two successive draws the numbers on the second draw are exactly the same as on the previous one. But over a 20-year period there are 1040 weeks, which means there are 1039 pairs of consecutive draws.

The probability that none of these 1039 consecutive draws result in the same set of numbers is

$$\left(1 - \frac{1}{5245786}\right)^{1039} = 0.9998$$

Therefore the probability that at least one of the consecutive draws over 20 years yields identical results is $1 - 0.9998 = 0.00019785$.

However, if there are 300 lotteries of a similar design running worldwide, then the probability that in at least one of them the same numbers turn up in consecutive weeks is (using the same formula as in the previous example):

$$= 1 - (1 - 0.00019785)^{300}$$

$$= 1 - 0.9423 = 0.0577$$

So, the probability of this happening is just under 6%.

Example 5.24 "Long-Lost Brothers Die on the Same Day"

No probability figure was stated but the implication was this was an extremely rare event. But, in the United Kingdom alone there are probably 500,000 pairs of brothers who never see each other. Any pair of brothers is likely to be close in age and so is likely to die within, say, 10 years of each other, about 4000 days. So the probability of any specific pair of brothers dying on the same day is about 1 divided by 4000. But the probability of at least one pair dying on the same day is one minus the probability that none of the 500,000 pairs of brothers dies on the same day. This latter probability is 3999/4000 to the power of 500,000. This number is so small that it is more likely that Martians will be represented in the next Olympic games.

The fallacy of regarding highly probable events as highly improbable is especially common and concerning within the legal system. From our own experience as expert witnesses we have seen numerous instances of prosecutors exaggerating the relevance of evidence because of this fallacy. It is usually associated with a claim like "there is no such thing as coincidence." The example in Box 5.14 is based on claims made in a real case.

Box 5.14   When Highly Probable Events Are Assumed to be Highly Improbable: A Fallacy of Legal Reasoning

Imagine that a defendant is charged with five similar crimes. One of the prosecution's first key strategies is to demonstrate that the crimes are related and must have been committed by the same person. That way, if they can provide hard evidence linking the defendant to any one of the five crimes the assumption is that they must also be guilty of the other four. Consider the following prosecution argument:

"The following facts about these five crimes are indisputable:

- *All 5 crimes were robberies.*
- *All 5 crimes took place in the same district (within an area of approximately one square mile).*
- *All 5 crimes took place between the hours of 7 p.m. and 11:30 p.m.*
- *All 5 crimes took place within 100 meters of a theater.*
- *All 5 crimes took place within 100 meters of a bus stop.*
- *In all 5 crimes the attacker did not know the victim.*
- *In all 5 crimes the attacker stole belongings from the victim.*
- *In all 5 crimes the attacker struck from behind to avoid being seen.*

The sheer number of similarities between these crimes is such that we can disregard the probability that they are unrelated. There is no such thing as coincidence, certainly not on this kind of scale."

In fact the prosecutor's argument is fundamentally flawed. Let us suppose that the crimes all took place within a 3-year period in the midtown area of New York. It is known that there were several thousands of robberies that took place in that area during that period. Being the central theater district, the vast majority of these muggings happened during the hours of 7 and 11:30 p.m. when the streets were filled with theatergoers and other tourists. In every recorded case the attacker did not know the victim and in almost every case the attacker struck from behind and stole something. Within the downtown area almost everywhere that people congregate is within 100 meters of both a bus stop and a theater. It therefore follows that almost all of the thousands of robberies that took place downtown during that period have exactly the same set of similarities listed.

For simplicity (and to help the prosecution case) let's give the prosecutor the benefit of the doubt and assume there were just 1000 such crimes. Nobody is claiming that the same person committed all of these crimes. What is the maximum proportion of those crimes that could have been committed by the same person? Let's suppose there is a truly super-robber who committed as many as 200 of the crimes. Then, if we randomly select any one of the 1000 crimes, there is a 20% chance (0.2 probability) that the super-robber was involved.

Now pick out any five of these crimes randomly. What is the probability that the super-robber committed them all? The answer, assuming the Binomial distribution (where $n = 5$ and $p = 0.2$), is

$$0.2^5 = 0.00032$$

In other words, here are five crimes that satisfy the exact similarities to those claimed by the prosecutor. But, contrary to the claim of near certainty that these were committed by the same person, the actual chance of them all being committed by the same person are about 1 in 300. Hence, in the absence of any other evidence, the claim that the crimes are all related is essentially bogus.

At the heart of the prosecutor's erroneous reasoning lie two key issues:

1. The commonly observed fallacy of *base rate neglect* whereby people reasoning in probabilistic terms fail to take account of the true size of the underlying population. In this case the population is the number of all muggings with the same characteristics in the area during the period in question.
2. The fallacy of treating dependent events as if they were independent. Although the prosecutor did not make this explicit, it was implicit in the claim about "no such thing as coincidence" and that we could "disregard the probability" that the crimes were unrelated. Consider, for example, the following events.

- *A*—Crime is a mugging in the main tourist area.
- *B*—Crime takes place within 100 meters of a theater.

- C—Crime takes place within 100 meters of a bus stop.
- D—Crime in which the attacker does not know the victim.

On the basis of data for all crimes in the United States the prior probabilities for these events might be something like

$$P(A) = 0.001; P(B) = 0.0001; P(C)=0.01; P(D) = 0.4$$

If these events were all independent the probability of them all occurring would indeed be incredibly low since it would be the product of the individual probabilities. But of course they are not. The probabilities $P(B \mid A)$, $p(C \mid A)$, and $p(D \mid A)$ are all close to 1. And in this case we are restricted to cases where A is true.

What the prosecutor does further to confuse the jury is to compound the fallacy by pointing out that the defendant was known to be "close to the crime scene" on the occasions of each of the five crimes. But again the value of such evidence is minimal. The defendant has, in fact, lived and worked in the area for many years. As such, even if he were innocent, there is a high probability that he will be in the area at any given time. Suppose, in fact, that a typical person living and working in this area leaves the area once every 7 days. Then, on any given evening, there is a probability of 6/7 that such a person will be in the area. That means that the probability such a person will be in the area on the night of each of the five crimes is, using the Binomial distribution,

$$(6/7)^5 = 0.46$$

But there are thousands of people (say 50,000) who live and work in that area. For *each* of them there is a 0.46 probability that they were in the area on the night of each of the five crimes. Using the formula for the mean of the Binomial distribution (Box 5.11), this implies there are on average 23,000 other such people. In the absence of any other evidence against the defendant, he is no more likely to be guilty than any of them. And this would be true even if the five crimes were genuinely known to be committed by the same person.

5.6.4 When Mundane Events Really Are Quite Incredible

The flipside of the fallacy of underestimating the probability of an event that appears to be very unlikely is to overestimate the probability of an event that seems to be likely. In Chapter 4, Box 4.4, we explained, for example, why every ordering of a pack of cards was a truly incredible event. As another more concrete example, suppose a fair coin is tossed 10 consecutive times. Which of the following outcomes is most likely?

- H H T H T H T T H T
- T T T T T T T T T T

Most people believe that the first is more likely than the second. But this is a fallacy. They are equally likely. People are confusing randomness (the first outcome seems much more random than the second) with probability. The number of different outcomes for 10 tosses of a coin is, as shown in Appendix A, equal to 2^{10}. That is 1024. So each particular outcome has probability 1/1024. Since the two sequences are just two of these particular outcomes, each has the same probability 1/1024.

The appearance of randomness in the first sequence compared to the second (which is the reason why people think the first sequence is more likely) arises because it looks like lots of other possible sequences. Specifically, this sequence involves five heads and five tails, and there are many sequences involving five heads and five tails. In fact, as explained in Appendix A (or by using the Binomial distribution), 252 of the possible 1024 sequences contain exactly five heads and five tails. This contrasts with the second sequence; it really looks unique because there is only one sequence (this one) involving 10 tails. The probability of getting a sequence that contains five heads and five tails is 252/1024, which is, of course, much higher than the probability of getting a sequence with 10 tails. But the probability of getting the particular sequence of five heads and five tails presented earlier is still only 1/1024.

An example of how this kind of fallacy can lead to irrational recommendations in practice is shown in Box 5.15.

Box 5.15   Irrational Recommendations for Playing the National Lottery

We have seen that in a lottery consisting of 49 numbers (1 to 49) any sequence of 6 numbers is just as likely to be drawn as any other sequence. For example, the set of numbers 1, 2, 3, 4, 5, 6, is just as likely to be drawn as, say, 5, 19, 32, 37, 40, 43.

A common misconception is to assume that certain sequences (like the latter) are somehow more likely because they seem more random and more similar to the kind of sequences that we have observed as winning. The fact that each sequence has the same probability is enough to be able to debunk a lot of popular myths and shams relating to the lottery. For example, a very popular site offering hints and tips for winning the lottery is www.smartluck.com/index.html. An example of one of many tips offered there that you can now easily debunk as nonsense is the following:

"Try to have a relatively even mix of odd and even numbers. All odd numbers or all even numbers are rarely drawn, occurring less than 1 percent of the time. The best mix is to have 2/4, 4/2, or 3/3, which means two odd and four even, or four odd and two even, or three odd and three even. One of these three patterns will occur in 81 percent of the drawings."

There are however some tips that are worth pursuing should you decide to play the lottery. These exploit known cognitive biases in how people choose numbers. One strategy involves exploiting the knowledge that most people are generally less likely to choose two consecutive numbers simply because they look less random and therefore less likely to come up. Choosing consecutive numbers will not increase your probability of winning the lottery but it will reduce the chance of having to share the winnings with other players who had chosen the same numbers.

Similarly, the numbers 32 to 49 are chosen far less than other numbers for the simple reason that many players choose numbers based on the birthdays of their friends and family. So, again, choosing numbers between 32 and 49 will generally result in a higher payout in the unlikely event of you winning.

Finally there is one set of numbers you should definitely avoid: the set 1, 2, 3, 4, 5, 6. Although this has the same chance as any other sequence of winning, in the unlikely event that it does win, the lottery jackpot will need to be shared between the 5000 or so people who, in a typical week, select this particular set.

5.7   Summary

In this chapter we have introduced the basic machinery of probability calculations in the form of the axioms and theorems that are central to probability theory. We have then expanded this to include probability distributions for discrete and continuous variables and examined how we can express probabilities as conditional events, where one uncertain statement about some event might depend on some other event, and joint events where events may be considered together as a set or group. Both joint and conditional probability rely on the idea of independence, which involves us assessing whether two or more variables are logically or causally linked to each other. Indeed this notion of independence is shown to be a powerful concept that governs not only how we can carry out probabilistic calculations in a feasible way but also helps us express properties of the real world in the causal models we might use to predict risk and make better decisions.

Along the way we have visited some "toy" problems such as the Monty Hall game-show problem and demonstrated how probability theory can help us avoid the mistakes people typically make when faced with situations, such as lotteries, where the odds may seem more attractive than they really are (and vice versa).

Further Reading

Cosmides, L., and Tooby, J. (1996). Are humans good intuitive statisticians after all? Rethinking some conclusions from the literature on judgment under uncertainty. *Cognition*, 58, 1–73.
Kahneman, D., Slovic, P., and Tversky, A. (1982). *Judgement Under Uncertainty: Heuristics and Biases*, Cambridge University Press.
Mosteller, F. (1987). *Fifty Challenging Problems in Probability with Solutions*, Dover.

6

Bayes' Theorem and Conditional Probability

6.1 Introduction

From Chapter 5 it should be clear that the only thing that truly distinguishes frequentist and subjective approaches to measuring uncertainty is how we assign the initial probabilities to elementary events. We argued that, since there must always be some subjective assumptions in the initial assignment of probabilities, it was inevitable that all probability was subjective. What this means is that the probability assigned to an uncertain event A is always conditional on a context K, which you can think of as some set of knowledge and assumptions. It is this central role of conditional probability that lies at the heart of the Bayesian approach described in this chapter. Fortunately, there is actually very little extra to introduce by way of theory.

We will present Bayes' theorem and show it is simply an alternative method of defining conditional probability. The theorem is easily derived from Axiom 5.4 (see Chapter 5; true Bayesians would actually use Bayes' theorem as the axiom). The key benefit of Bayes' theorem is that, in most situations, it provides a much more natural way to compute conditional probabilities.

We will show how Bayes' theorem enables us to debunk more of the fallacies introduced in Chapter 2 and show how it can be used to make inferences from prior beliefs by taking account of data to arrive at new, posterior beliefs in a rational way.

6.2 All Probabilities Are Conditional

Even when frequentists assign a probability value of 1/6 to the event "rolling a 4 on a die," they are conditioning this on assumptions of some physical properties of both the die and the way it can be rolled.

It follows therefore that any initial assignment of probability to an event A is actually a statement about conditional probability. Hence, although we have used the terminology $P(A)$ it would be more accurate to write it as $P(A \mid K)$, where K is the background knowledge or context.

All models in this chapter are available to run in AgenaRisk, downloadable from www.agenarisk.com.

Example 6.1

In the experiment of rolling a die let A be the event "4." Then the following are all reasonable:

- $P(A \mid K) = 1/6$, where K is the assumption that the die is genuinely fair and there can be no outcome other than 1, 2, 3, 4, 5, 6.
- $P(A \mid K) = 1/6$, where K is the assumption that there can be no outcome other than 1, 2, 3, 4, 5, 6 and I have no reason to suspect that any one is more likely than any other.
- $P(A \mid K) = 1/8$, where K is the assumption that the only outcomes are 1, 2, 3, 4, 5, 6, lost, lands on edge, and all are equally likely.
- $P(A \mid K) = 1/10$, where K is the assumption that the results of an experiment, in which we rolled the die 200 times and observed the outcome "4" 20 times, is representative of the frequency of "4" that would be obtained in any number of rolls.

In practice, if the same context K is assumed throughout an analysis, then it makes sense to simply write $P(A)$ rather than $P(A \mid K)$.

Since the focus of this book is on risk in practice, rather than in casinos or theoretical games of chance, most of the events for which we have to assign an initial probability have no reasonable frequentist approach for doing so. We are therefore forced to use at least some subjective judgment. This is why a subjective probability $P(A)$ is often referred to as a "degree of belief." However, we are driven by a quest for improvement. Having started with some initial probability assignment for an event A, we look for, or observe, evidence that can help us revise the probability (our degree of belief in A). In fact we all do this every day of our lives in both mundane and routine, as well as important, decision making.

Cromwell's Rule

It is dangerous to assign a prior probability of zero (one) to any event unless it is logically impossible (certain). For example, if you set the prior probability of zero to the hypothesis that intelligent life exists elsewhere in the universe, then even an alien invasion will fail to convince you.

In this and all subsequent examples where a variable X is Boolean (i.e., represents an uncertain event having states true and false) we will use the shorthand (but ambiguous) notation described in Chapter 5 whereby $P(X)$ means the event $P(X = \text{true})$. It should be clear from the context when $P(X)$ means the probability distribution of X.

Example 6.2

As I do not like having to take an umbrella to work, every day I have to make a decision based on my degree of belief that it will rain. When I wake in the morning I have some prior belief that it will rain. Specifically, if I did not hear any weather forecast the previous evening I assign a probability value of 0.1 to rain based on my informal judgment that it seems to rain about 1 day in 10; so $P(\text{rain}) = 0.1$ conditioned on K, where K is a combination of informal frequency data and purely personal belief. Between waking up and actually leaving for work I can gather or observe evidence that will change my belief about $P(\text{rain})$. For example, if I look out of the window and see very dark clouds my belief about $P(\text{rain})$ is increased, as it will be even more so if I hear a local weather forecast predicting heavy rain. However, it will decrease if the weather forecast confidently states that the cloud will lift within 30 minutes to leave a bright sunny day. So every time I get some evidence, E, I revise my belief about the $P(\text{rain})$. In other words I calculate $P(\text{rain} \mid E)$. Whatever the evidence, E, at the time I have to leave the house, if my revised belief, namely, $P(\text{rain} \mid E)$, is greater than 0.3, then I will reluctantly take the umbrella.

Example 6.3

The entire legal process (innocent until proven guilty) is based on the idea of using evidence to revise our belief about the uncertain event "Defendant is innocent" (let's call this variable H).

Now, everyone will have a belief about whether H is true or not. If you know absolutely nothing about the defendant or the crime, your belief about H will be different to that of, say, the police officer who first charged the defendant or the prosecuting lawyer who has seen all of the evidence. In other words different people will have a different context, K. Although such people might be reluctant to state their belief as a probability value, they will nevertheless have such a value in their mind at least informally. The police officer who first charged the defendant might believe that H is true with a probability of 0.1 (and hence false with a probability of 0.9), whereas you may have no reason to believe that the defendant is any more likely to be guilty of the particular crime than any other able-bodied person in the country. So, if there are say 10 million such people, your belief will be that H is false with a probability of 1 in 10 million) and hence that H is true with a probability of 9,999,999 in 10 million).

If there were no witnesses to the crime, then the only person whose belief about H is not uncertain is the defendant himself (notwithstanding esoteric cases such as the defendant suffering amnesia or being under a hypnotic trance). If he really is innocent, then his belief about H being true is 1 and if he is not innocent his belief about H being true is 0. Whatever your prior belief, as the case proceeds you will be presented with pieces of evidence, E. You can think of evidence as being causally related to the hypothesis, H, as shown in Figure 6.1.

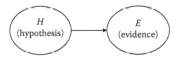

Figure 6.1   Causal view of evidence.

If the hypothesis is false (i.e., defendant is guilty), then you will be more likely to receive evidence that links the defendant to the crime (such as witness statements or blood samples), whereas if the hypothesis is true you will be more likely to receive evidence that distances the defendant from the crime (such as an alibi). In general, some of the evidence may favor the hypothesis being true and some may favor it being false.

In all cases you are informally calculating a revised belief about H given E, in other words $P(H \mid E)$. If E is the set of all evidence, then, if you were a member of the jury, you would be expected to find the defendant guilty if $P(H \mid E)$ was close to zero. What we mean by "close" is itself subjective since, for criminal cases in many jurisdictions it is assumed that you should find the defendant guilty only if the probability is "beyond reasonable doubt." Ideally the probability value corresponding to beyond reasonable doubt should be agreed in advance and should always be independent of the particular case (for example, if you thought "beyond reasonable doubt" meant that a wrong conviction can happen less than 1 out of 100 times, then the threshold probability value is 0.01). Unfortunately, in practice this never happens. However, in civil cases the law is clear about the value of $P(H \mid E)$ needed to find the defendant guilty. It must be less than 0.5, since the law specifies "on the balance of probabilities" meaning that you only need to conclude that $P(H \mid E)$ is smaller than $P(\text{not } H \mid E)$. By Axiom 5.2 this will be true whenever $P(H \mid E) < 0.5$.

The examples make clear that much probabilistic reasoning involves us:

1. Starting with a hypothesis, H, for which we have a belief $P(H)$ called our *prior belief about H*.
2. Using evidence, E, about H to revise our belief about H in the light of E; in other words, to calculate $P(H \mid E)$, which we call the *posterior belief about H*.

$P(H{\cap}E)$ is also written simply as $P(H,E)$.

Axiom 5.4 already gives us a method to calculate $P(H \mid E)$ in terms of $P(H)$. However, it involves us having to know the joint probability $P(H \cap E)$. In many situations we will not know $P(H \cap E)$. However, what we often know is $P(E \mid H)$, which we call the *likelihood of the evidence E*, that is, how likely we are to see this evidence given the hypothesis.

Example 6.4

In a particular chest clinic 5% of all patients who have been to the clinic are ultimately diagnosed as having lung cancer, while 50% of patients are smokers. By considering the records of all patients previously diagnosed with lung cancer, we know that 80% were smokers. A new patient comes into the clinic. We discover this patient is a smoker. What we want to know is the probability that this patient will be diagnosed as having lung cancer.

In this problem the hypothesis, H, is "patient has lung cancer" and the evidence, E, is "patient is a smoker."

- Our prior belief in H is 5%, so $P(H) = 0.05$.
- Our prior belief in E is 50%, so $P(E) = 0.5$.

What we want to do is update our belief in H given evidence, E, that the patient is a smoker. In other words we want to calculate $P(H \mid E)$. Axiom 5.4 gives us a means to calculate $P(H \mid E)$:

$$P(H \mid E) = \frac{P(H,E)}{P(E)}$$

Unfortunately, we do not have direct information about $P(H, E)$. But we do know the likelihood of the evidence $P(E \mid H) = 0.8$.

If we know the likelihood of the evidence, E, then we can compute the posterior probability of the hypothesis, H, using Bayes' theorem, which we introduce next.

6.3   Bayes' Theorem

Bayes' theorem gives us a simple method to calculate $P(H \mid E)$ in terms of $P(E \mid H)$ rather than $P(H, E)$:

Bayes' Theorem

$$P(H \mid E) = \frac{P(E \mid H) \times P(H)}{P(E)}$$

So, in Example 6.4 we can compute

$$P(H \mid E) = \frac{0.8 \times 0.05}{0.5} = 0.08$$

This means our belief in the patient having lung cancer increases from a prior of 5% to a posterior of 8%. This makes us more concerned that the patient will have lung cancer, but this prospect is still very unlikely.

Bayes' theorem is easily derived from Axiom 5.4 (the fundamental rule), as shown in Box 6.1.

Box 6.1   Proof of Bayes' Theorem

By Axiom 5.4 (the fundamental rule) we know that

$$P(H \mid E) = \frac{P(H,E)}{P(E)}$$

But, interchanging H and E, we can use the fundamental rule again to give us

$$P(E \mid H) = \frac{P(H,E)}{P(H)}$$

And rearranging this equation gives us

$$P(H , E) = P(E \mid H)P(H)$$

Substituting this value of $P(H, E)$ into the first equation gives us

$$P(H \mid E) = \frac{P(E \mid H)P(H)}{P(E)}$$

which is Bayes' theorem.

In fact, the pure Bayesian approach to probability actually uses Bayes' theorem as the fourth axiom. This is how conditional probability is defined rather than being defined in terms of the joint probability $P(H, E)$.

Another immediate result of Bayes' theorem is that it demonstrates the difference between $P(H \mid E)$ (which is 0.08 in the earlier example) and $P(E \mid H)$ (which is 0.8 in the example). Many common fallacies in probabilistic reasoning arise from mistakenly assuming that $P(H \mid E)$ and $P(E \mid H)$ are equal. We shall see further important examples of this.

Box 6.2 provides a justification of why Bayes' theorem produces the "right" answer by using frequency counts for the same example.

Box 6.2   Justification of Bayes' Theorem Using Frequencies

Suppose the chest clinic has 1000 patients. Of these

- 50 have cancer (i.e., 5%)
- 500 are smokers (i.e., 50%)
- 40 of the 50 cancer patients are smokers (i.e., 80%)

When we asked the question (in Bayes' theorem) "What is the probability a patient has cancer given that they smoke?" this is the same, using frequency counts, as asking "What proportion of the smokers have cancer?"

Solution: Clearly 40 of the 500 smokers have cancer. That is 8%, which is exactly the same result we got when applying Bayes' theorem.

We will see further intuitive justifications of Bayes' theorem later in the chapter. But it is also worth noting that, using frequencies, Bayes is equivalent to the following formula in this case:

$$= \frac{\dfrac{40}{50} \times \dfrac{50}{1000}}{\dfrac{500}{1000}} = \frac{40}{500}$$

The reason for using Bayes' theorem, as opposed to Axiom 5.4, as the basis for conditional probability was that in many situations it is more natural to have information about $P(E \mid H)$ (the likelihood of the evidence) rather than information about $P(H,E)$ (the joint probability). However, in Bayes (as well as Axiom 5.4) we still need to have information about $P(E)$. This raises a key question.

What happens if information about $P(E)$ is not immediately available?

It turns out that in many practical situations we can use marginalization to determine $P(E)$.

Example 6.5

In Example 2.4 of Chapter 2 we have the following information (where we will assume H represents the Boolean variable "Disease" and E represents the Boolean variable "Test positive for the disease"):

$$P(H) = 0.001, \text{ so } P(\text{not } H) = 0.999$$
$$P(E \mid \text{not } H) = 0.05$$
$$P(E \mid H) = 1$$

By marginalization we know that

$$P(E) = P(E, H) + P(E, \text{not } H)$$

But by Axiom 5.4 this is the same as

$$P(E) = P(E \mid H)P(H) + P(E \mid \text{not } H)\, P(\text{not } H)$$

We know each of the values on the right-hand side here, so

$$P(E) = 1 \times 0.001 + 0.05 \times 0.999 = 0.05095$$

We can now use Bayes' theorem to answer the problem that so confounded the staff and students at Harvard. What we need to know is the probability that a person actually has the disease if we know that they test positive. In other words we want to know $P(H \mid E)$. By Bayes' theorem:

$$P(H \mid E) = \frac{P(E \mid H)P(H)}{P(E)} = \frac{1 \times 0.001}{0.05095} = 0.01963$$

In other words there is a less than 2% chance that a person testing positive actually has the disease. This is very different from the 95% that most people in the study calculated.

The marginalization in Example 6.5 provides us with the more general version of Bayes' theorem shown in Box 6.3.

Box 6.3   Bayes' Theorem (General Versions)

If H is a Boolean variable then

$$P(H \mid E) = \frac{P(E \mid H) \times P(H)}{P(E \mid H) \times P(H) + P(E \mid not\ H) \times P(not\ H)}$$

In fact an even more general version of Bayes' theorem applies when H has more than two states. For example, there may be three possible diseases for which the same test may provide information. In that case the variable H has four possible states: $h1$, $h2$, $h3$, and $h4$, where $h1$, $h2$, and $h3$ represent the three different diseases and $h4$ represents the "no disease" outcome. Assuming the four outcomes are mutually exhaustive and exclusive we know that by marginalization

$$P(E) = P(E \mid h1)P(h1) + P(E \mid h2)P(h2) + P(E \mid h3)P(h3) + P(E \mid h4)P(h4)$$

Hence, for any i from 1 to 4, we have

$$P(h_i \mid E) = \frac{P(E \mid h_i)P(h_i)}{P(E \mid h_1)P(h_1) + P(E \mid h_2)P(h_2) + P(E \mid h_3)P(h_3) + P(E \mid h_4)P(h_4)}$$

Using the sigma notation we write this as

$$P(h_i \mid E) = \frac{P(E \mid h_i)P(h_i)}{\sum_{i=1}^{4} P(E \mid h_i)P(h_i)}$$

Obviously the formula extends to the case where H has any number of states.

Although the Bayesian argument presented in Example 6.5 clearly demonstrates that the doctors in the original study were wrong, there is still a problem. Most lay people (including doctors and lawyers) cannot follow such an argument because they are intimidated by the mathematical formula. Readers of this book may also have found it hard to follow. Now one of the objectives of this book is to show you how complex Bayesian arguments can be performed without ever having to do a Bayesian calculation by hand or even understanding every step of such a calculation (see Box 6.4). But such calculations will only be accepted by people who at least believe that the results of a simple Bayesian calculation are rational and correct. If a doctor or lawyer cannot see why the result in the simple case of Example 6.5 is correct, then we still have a major problem, because such people would never accept the results of a more complex Bayesian argument. It is possible to overcome this major problem by presenting visual explanations that are equivalent to Bayes. We already presented an informal visual explanation for this problem in Chapter 2 (Figure 2.22). An equivalent (but more repeatable and standard) visual argument is to use an *event tree* as shown in Figure 6.2.

Indeed, as our risk assessment problems become more complex (with more variables, more states, and more dependencies between variables) event tree representations become intractable. That is why we will need to turn to BNs as an alternative visual representation of Bayesian arguments.

Note that, in the event tree (and in contrast to the formal Bayesian argument), we start with a hypothetical large number of people to be tested. This simple enhancement has a profound improvement on comprehensibility. But the event tree is essentially an explanation of Bayes' theorem and if people can understand and accept it, then they are also more likely to accept the results of more complex Bayesian arguments (even if they do not understand them).

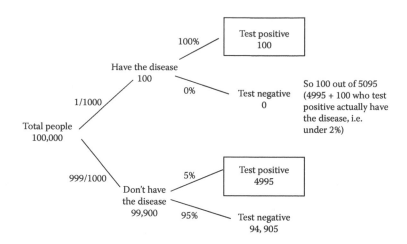

Figure 6.2   Event/decision tree explanation.

Box 6.4   Why You Never Need to Worry About Those Nasty Bayesian Calculations

You are reading this book because you have an interest in Bayes and risk. Moreover, because you have access to a Bayesian network software application, such as AgenaRisk, you never need to have to do any Bayesian calculation yourself. In fact, you do not even have to understand Bayes' theorem. This will become crystal clear from Chapter 7 onward, but in the case of Example 6.3 we can easily convince you right away.

Open AgenaRisk and create two nodes H and E with the link between them shown in Figure 6.3.

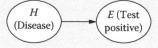

Figure 6.3   BN with two linked nodes.

By default the nodes will each have two values: true and false. Open the node probability table for H and enter the following values (these are simply the assumptions we have already made), as shown in Table 6.1. Do the same for the node probability table of E (which is conditional on H), as shown in Table 6.2.

Table 6.1
Prior Probabilities for Node H

False	0.999
True	0.001

Table 6.2
Conditional Probabilities for Node E

H (Disease)	False	True
False	0.95	0.0
True	0.05	1.0

Now simply run the model by pressing the calculate button. The result you will see first is simply the marginal probabilities for *H* and *E*, as shown in Figure 6.4.

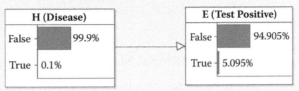

Figure 6.4 BN with two linked nodes with probabilities shown.

But now enter the observation "True" for *E* and calculate again. You will get the result shown in Figure 6.5. This is exactly the result of the Bayes' theorem calculation.

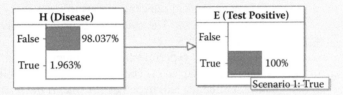

Figure 6.5 BN with two linked nodes with probabilities shown and observations made.

6.4   Using Bayes' Theorem to Debunk Some Probability Fallacies

If you know that all horses are four-legged animals, can you conclude that all four-legged animals are horses?

Of course not. Yet people, including genuine experts in critical situations, make an equally nonsensical deduction when they conclude that:

> Since we know the probability of some evidence given the hypothesis is very low, we can conclude that the probability of the hypothesis given the evidence must also be very low.

The problem with this type of fallacious probabilistic reasoning is especially common in the legal profession. We saw examples already in Chapter 5 and will see more in Chapter 15.

In other words people are assuming that $P(H \mid E) = P(E \mid H)$. This is called the *fallacy of the transposed conditional*. If you simply look at the formula for Bayes' theorem it is clear that the only circumstances under which $P(H \mid E) = P(E \mid H)$ are when the priors $P(H)$ and $P(E)$ are themselves equal.

We saw two very common examples of this fallacy in Chapter 2:

■ In traditional statistical hypothesis testing
■ The prosecutor's fallacy

Whereas we only provided informal explanations in Chapter 2, we can now provide rigorous explanations of what the proper conclusions should have been using Bayes' theorem. In discussing the prosecutor's fallacy we also expose the defendant's fallacy and explain why an alternative form of Bayes' theorem (called the "odds" form) using the notion of "likelihood ratio" is more natural in legal arguments.

6.4.1 Traditional Statistical Hypothesis Testing

For any uncertain hypothesis the *null hypothesis* is the assertion that the hypothesis is false. For example, if we are interested in the hypothesis "variable X and variable Y are related" then the null hypothesis is the assertion "there is no relationship between variable X and variable Y."

When there is a claim that some hypothesis of interest is true, the standard approach statisticians use is to assume the opposite (i.e., the null hypothesis) and run experiments to see if the evidence "contradicts" the null hypothesis. The evidence, E, will typically be the kind of data shown in Table 6.3.

In such cases experimenters will compute the p-value for the data (as explained in Chapter 2). The p-value is the probability of observing the data assuming the null hypothesis is true. If the p-value is low, for example 0.01, statisticians would regard this as highly significant (specifically, significant at the 1% level) and would reject the null hypothesis. In many cases they wrongly assume that the result demonstrates that the probability of the null hypothesis is 0.01. But this is an example of the fallacy of the transposed conditional.

Denoting the null hypothesis as H, what the data shows is that $P(E \mid H) = 0.01$. To assume that $P(H \mid E) = 0.01$ is to wrongly assume that $P(E \mid H) = P(H \mid E)$. In fact, $P(H \mid E)$ will, as always, be heavily influenced by our prior belief about H. By ignoring this prior, experimenters

> Some examples of null hypotheses we might wish to test:
>
> - Smoking does not cause cancer.
> - Social class has no bearing on salary earned.
> - Drug Oomph does not lead to weight loss.

Table 6.3
Typical Data for Hypothesis Testing

Patient	Smoker	Cancer	Customer	Class	Income	Subject	Weight Change (lb) Placebo	Oomph
Patient 1	y	y	Customer 1	A	10,000	Subject 1	−1	
Patient 2	n	y	Customer 2	B	15,000	Subject 2		10
Patient 3	n	n	Customer 3	A	9,000	Subject 3		25
Patient 4	y	n	Customer 4	C	25,000	Subject 4	5	
Patient 5	y	y	Customer 5	C	22,000	Subject 5		13
Patient 6	n	n	Customer 6	A	7,000	Subject 6		12
Patient 7	n	y	Customer 7	B	12,000	Subject 7	6	
Patient 8	n	n	Customer 8	A	12,000	Subject 8	0	
Patient 9	n	n	Customer 9	A	8,000	Subject 9	−2	
Patient 10	y	n	Customer 10	B	23,000	Subject 10		20
Patient 11	n	y	Customer 11	C	29,000	...	...	...
...	...	...	...	...	...			
(a) Smoking/Cancer Hypothesis Data			(b) Social Class/Earnings Data			(c) Drug Oomph Weight Loss Data		

cannot conclude anything about $P(H \mid E)$. To make this absolutely clear, consider Example 6.6.

Example 6.6

Suppose that we have 999 fair coins and one coin that is known to be biased toward "heads." We assume that the probability of tossing a "head" is 0.5 for a fair coin and 0.9 for a biased coin. We select a coin randomly. We now wish to test the null hypothesis that the coin is not biased. To do this we toss the coin 100 times and record the number of heads, X. As good experimenters we set a significance level in advance at a value of 0.01. What this means is that if we observe a p-value less than or equal to 0.01 (see sidebar) then we will reject the null hypothesis if we see at least 63 heads (in fact 62 would be very close to rejection also).

So, assuming that H is the null hypothesis (fair coin) and E is the observed event "at least 63 heads in 100 throws" then since $P(E \mid H) <$ 0.01 (in fact $P(E|H)=0.006$) the null hypothesis is rejected and statisticians conclude that the coin is biased. It is typical to claim in such situations something like "there is only a 1% chance that the coin is fair, so there is a 99% chance that the coin is biased." But such a conclusion assumes that $P(E|H) = P(H|E)$.

The coin is, in fact, still very likely to be a fair coin despite the evidence. The key point is that the correct prior probability for the null hypothesis, H, is 0.999, because only one of the 1000 coins is biased. Also, $P(E \mid$ not $H)$ is clearly very close to 1 (it is 0.999999999999899). Hence, we have:

$$P(H \mid E) = \frac{P(E \mid H) \times P(H)}{P(E \mid H) \times P(H) + P(E \mid not\ H) \times P(not\ H)}$$

$$= \frac{0.006 \times 0.999}{0.006 \times 0.999 + 0.999999999999899 \times 0.001} = 0.86$$

So, given the evidence E, the probability that the coin is fair has come down from 0.999 to 0.86. In other words, there is still an 86% chance the coin is fair.

The conclusions drawn by people using the p-value are actually even more misleading than what we have already revealed. That is because the event they actually observe is not "at least 63 heads" but a *specific* number of heads. So suppose they actually observe the number 63. In this case (using the Binomial distribution) it turns out that $P(E|H) = 0.0027$ and $P(E|$ not $H)$ is 0.0000000000000448.

This means that, from Bayes:

$$P(H \mid E) = \frac{P(E \mid H) \times P(H)}{P(E \mid H) \times P(H) + P(E \mid not\ H) \times P(not\ H)}$$

$$= \frac{0.0027 \times 0.999}{0.0027 \times 0.999 + 0.0000000000000448 \times 0.001} \approx 1$$

In other words, far from the evidence proving that the coin is biased, the evidence actually proves the exact opposite: the coin is almost certainly unbiased. The statistical hypothesis test gets it completely wrong.

We would need to see a much higher number of heads in 100 tosses (than suggested by the predefined p-value) to in any way justify rejecting the null hypothesis.

The scenario in Example 6.6 is similar to what lies behind the prosecutor's fallacy.

For Example 6.6, the Binomial distribution (explained in Chapter 5) with $n = 100$ and $p = 0.5$ provides a formula for the probability of tossing X number of heads from 100 tosses of a fair coin. But what we want to know is the 99th percentile of the distribution, that is, the number X for which 99% of the probability distribution lies to the left of X. Or, equivalently, 1% of the distribution lies to the right of X. It turns out that the 99th percentile in this case is between 62 and 63. The probability of at least 62 heads in 100 tosses of a fair coin is approximately 0.0105 and the probability of at least 63 heads is 0.006.

The key in Example 6.6 is that we started with a very strong prior belief that the null hypothesis is true. The evidence alone is nowhere near sufficient to shift the prior enough for us to believe it is false. The Bayesian approach to hypothesis testing is covered in Chapter 12.

6.4.2　The Prosecutor Fallacy Revisited

The main evidence, E, against defendant Fred is that his blood matches that left at the scene by the person who committed the crime. About 1 in a 1000 people have this blood type. So if H is the hypothesis "Fred is innocent" (which in this case we take as equivalent to "not at the scene of the crime") then what we actually know here is that $P(E \mid H)$ is 1/1000.

But of course what we really want to know, and what the prosecutor mistakenly confuses for $P(E \mid H)$, is the probability that Fred is innocent given the evidence of the blood match, that is, $P(H \mid E)$. This probability will, of course, depend on the prior probability $P(H)$. In the absence of any other evidence against Fred, let us suppose that any adult male in the town where the crime was committed can be considered a suspect. Then, if there are 10,000 such adults, we can state the probability assumptions as

$$P(H) = 9{,}999/10{,}000, \text{ so } P(\text{not } H) = 1/10{,}000$$

$$P(E \mid H) = 1/1000$$

We can also assume that if it was Fred at the scene, then the blood will certainly match, so

$$P(E \mid \text{not } H) = 1$$

One of the reasons why the fallacy is especially confusing for lawyers is that if you assume that $P(H) = 0.5$ (in other words your prior assumption is that the defendant is just as likely to be guilty as innocent), then in this case it turns out that $P(H \mid E) = 0.001001$, which is indeed very close to the value for $P(E \mid H)$.

Now applying Bayes' theorem we get

$$P(H \mid E) = \frac{\frac{1}{1000} \times \frac{9{,}999}{10{,}000}}{\frac{1}{1000} \times \frac{9{,}999}{10{,}000} + 1 \times \frac{1}{10{,}000}} = \frac{9{,}999}{10{,}999} \approx 0.91$$

So, even given the evidence of the blood match, the probability that Fred is innocent is still very high (contrary to the prosecutor's claim).

6.4.3　The Defendant's Fallacy

The flip side of the prosecutor's fallacy is a fallacy known as the *defendant's fallacy*. An example of the defendant's fallacy, using exactly the same assumptions as earlier, would be an argument presented by the defense lawyer along the following lines:

> The evidence presented by the prosecution leads us to conclude that there is actually a very high probability that the defendant is innocent. Therefore this evidence is worthless even for the prosecutor's argument and so can safely be ignored.

This argument is wrong because it fails to recognize that the evidence has still genuinely led to a significant change from our prior belief in guilt. The prior probability was 1 in 10,000. But after we get the evidence the posterior probability of guilt is 9/100. That is still unlikely but by no means as unlikely as before. In fact, the evidence has decreased the belief in innocence by a factor of 1000. Such evidence can certainly not be discounted, especially if it is just one of several pieces of evidence against the defendant. This type of reasoning leads to a different formulation of Bayes' theorem, which, in situations like this, can enable us to determine the probative value of the evidence without even having to worry about the prior values. This formulation is called the "odds form of Bayes" and involves the important notion of the *likelihood ratio*.

6.4.4 Odds Form of Bayes and the Likelihood Ratio

In situations like the legal ones we have discussed the variable, H, we are interested in has just two states (true and false). For any piece of evidence, E, we can consider two different likelihoods:

$P(E \mid H)$—The probability of the evidence E if H is true
$P(E \mid \text{not } H)$—The probability of the evidence E if H is false

In the prosecutor fallacy example the evidence, E, was that of a blood match. We know that

$P(E \mid H) = 1/1000$ (the probability of the evidence if the defendant is innocent)

We also assumed that

$P(E \mid \text{not } H) = 1$ (the probability of the evidence if the defendant is guilty)

So we are 1000 times more likely to see this particular evidence if the defendant is guilty. Intuitively this means that, whatever our prior belief was about the odds on innocence, we must revise it accordingly in favor of guilty. This suggests that the ratio of the likelihoods

$$\frac{P(E \mid H)}{P(E \mid \text{not } H)}$$

(which is called the *likelihood ratio (LR)*) can have a key role in determining the impact of the evidence, E.

In fact, it follows from the formal explanation in Box 6.5 that the likelihood ratio has the following multiplicative effect on the odds of H once we observe the evidence, E:

$$\text{Posterior odds} = \text{Likelihood ratio} \times \text{Prior odds}$$

Box 6.5 Odds Form of Bayes' Theorem

Bayes' theorem tells us that

$$P(H \mid E) = \frac{P(E \mid H) \times P(H)}{P(E)}$$

But equally it tells us that

$$P(\text{not } H \mid E) = \frac{P(E \mid \text{not } H) \times P(\text{not } H)}{P(E)}$$

If we simply divide the first equation by the second we get the following

$$\frac{P(H \mid E)}{P(\text{not } H \mid E)} = \frac{P(E \mid H) \times P(H)}{P(E \mid \text{not } H) \times P(\text{not } H)}$$

$$= \frac{P(E \mid H)}{P(E \mid \text{not } H)} \times \frac{P(H)}{P(\text{not } H)}$$

But recall from Chapter 4 that the probability of an event divided by its complement is simply the odds of the event. So the preceding equation is the same as saying

Odds of H given E = Likelihood ratio × Odds of H

Example 6.7

In a variation of the disease and test example previously discussed, suppose that a test for a disease has a 0.1 probability of a false-positive result and a 0.8 probability of a true-positive result. In other words

$$P(E \mid H) = 0.8$$
$$P(E \mid \text{not } H) = 0.1$$

where E is "Test positive" and H is "Has disease."

Suppose we know that the disease is prevalent in 1 in a 1000 people. Then

$$P(H) = 1/1000$$

and so

$$P(\text{not } H) = 999/1000$$

The prior odds of a person having the disease are therefore $P(H)/P(\text{not } H)$, which is 1/999.

Suppose we know that a person tests positive. Then, by the odds version of Bayes' theorem the posterior odds of the person having the disease are 1/999 times the likelihood ratio, which is 0.8 divided by 0.1, that is, 8. So the posterior odds are 8/999. The person is still unlikely to have the disease, despite the positive test, but the chances have increased by a factor of 8.

6.5   Likelihood Ratio as a Measure of Probative Value of Evidence: Benefits and Dangers

What is clear from the odds version of Bayes—as was demonstrated in Example 6.7—is that the LR is valuable for evaluating the importance of the evidence. In particular, it suggests we can evaluate the importance of the evidence without even having to consider prior probabilities. So, in the example, irrespective of what the prior odds were for disease, we know that a positive test result increases the odds by a factor of 8. Similarly, in the prosecutor fallacy example we now know that, irrespective of the prior odds of innocence, the blood match evidence decreases the odds of innocence by a factor of 1000.

This suggests that the LR can be considered as some kind of measure of the "probative value" of evidence. Indeed, as we shall see in Chapter 15, it is especially prominent in the use of Bayesian reasoning in legal arguments. The special attraction is because it enables us to evaluate the impact of evidence **without having to consider the prior odds of a hypothesis** and that is especially important in legal situations, since lawyers are reluctant to quantify the prior probability of innocence or guilt.

However, there is much confusion and misunderstanding surrounding its use. In particular, the popular notion that the LR can **define** the "probative value" turns out to be, in general, only valid under certain constrained assumptions. In this section, we provide a natural formal definition of probative value of evidence and explain why it only makes sense for where we have a well-defined hypothesis and where the alternative hypotheses is its well defined "negation." In other words, the two hypotheses must be mutually exclusive and exhaustive. In such situations, we will show the LR is essentially equivalent to the natural formal definition of probative value. However, when the LR is applied to two "alternative" hypotheses that are not mutually exclusive and exhaustive great care must be taken in the way the LR is interpreted: in the mutually exclusive and exhaustive case the LR tells us directly something about the change in the posterior probability of the hypothesis. However, when the hypotheses are not mutually exclusive and exhaustive the LR can only tell us about the extent to which the evidence supports one hypothesis over the other.

6.5.1 The (Only) Natural Definition of Probative Value of Evidence

The reason for our interest in Bayes theorem is that we are generally interested in the extent to which we update our prior belief in a hypothesis when we observe evidence. If the evidence makes no change to our belief in a hypothesis then clearly it is not probative with respect to that hypothesis. Hence, the only natural way in which we would regard evidence as being probative is if it actually *changes* our belief about the hypothesis. In other words:

Evidence That Is Probative

Evidence E is **probative** with respect to a hypothesis H if the posterior belief in H is different from the prior belief in H. In other words, E is probative if $P(H|E) \neq P(H)$ and not probative if $P(H|E) = P(H)$

When $P(H|E) > P(H)$, the evidence E results in an increased belief in H. This means we can say that the evidence E *is probative in support of the hypothesis H*.

Therefore, to determine if the evidence E is probative, we must use Bayes theorem to calculate $P(H|E)$. Suppose, for example, our hypothesis H is that a coin (which we cannot inspect) is biased. For simplicity, suppose that there are only two types of coins in circulation: fair coins for which the probability of tossing a head is exactly 0.5 and biased coins for which the probability of tossing a head is exactly 0.6. Our hypothesis H is that the coin is biased. Suppose we have evidence E that the coin was tossed twice and that both times a head was observed. For simplicity, we will rule out "land on edge" and "lost" as possible outcomes of tossing the coin. Under this hypothesis H, we know that $P(E|H)$ is the probability of two consecutive heads when on each toss the probability of a head is 0.6. This is $0.6 \times 0.6 = 0.36$. Similarly, $P(E|\text{not } H)$ is the probability of two consecutive heads when on each toss the probability of a head is 0.5, which is 0.25. Hence, if our prior belief in H being true was 50% (i.e. $P(H) = 0.5$) then, by Bayes theorem $P(H|E)$

$$P(H|E) = \frac{P(E|H)P(H)}{P(E|H)P(H) + P(E|\text{not } H)P(\text{not } H)}$$

$$= \frac{0.36 \times 0.5}{(0.36 \times 0.5) + (0.25 \times 0.5)} = 0.59$$

Therefore, the posterior probability of H being true has increased from 50% to 59% as a result of observing the evidence. This means that the evidence is probative in support of H.

If the evidence E had been one tail and one head then we would have $P(E|H) = 0.48$ and $P(E|\text{not } H) = 0.5$. Hence in this case $P(H|E) = 0.49$, which is a decrease from the prior probability. In this case, the evidence E is not probative in support of H. However, because the hypotheses H and not H are mutually exclusive and exhaustive, we know that

$$P(\text{not } H|E) = 1 - P(H|E) = 0.51$$

Therefore, since the probability of the hypothesis not H being true has increased from 50% to 51% and so it is probative in favour of not H.

6.5.2 A Measure of Probative Value of Evidence

Consider the ratio R of the posterior probability of H divided by the prior probability:

$$R = \frac{P(H|E)}{P(H)}$$

It follows that, for mutually exclusive and exhaustive hypotheses H and not H the ratio has the following properties:

- $R > 1$ is the same as saying that $P(H|E) > P(H)$
- $R < 1$ means $P(H|E) < P(H)$ and hence that:

$$1 - P(\text{not } H|E) < 1 - P(\text{not } H)$$

which means

$$P(\text{not } H|E) > P(\text{not } H)$$

- $R = 1$ means that $P(H|E) = P(H)$ and also $P(\text{not } H|E) = P(\text{not } H)$ meaning that the evidence does not change our prior belief in either hypothesis.

Hence R is a measure of the probative value of evidence in the following natural sense:

- When $R > 1$ the evidence is probative in favour of H (and the higher R is the more probative the evidence).
- When $R < 1$ the evidence is probative in favour of the alternative hypothesis not H (and the closer R is to zero the more probative the evidence).
- When $R = 1$ the evidence has no probative value

The reason R is almost never used, however, as a measure of probative value of evidence is that it requires us to specify prior probabilities. The LR on the other hand, as we have seen, does not make this requirement.

Box 6.6 explains that the LR is genuinely "valid" as a measure of probative value of evidence only when the two hypotheses are mutually exclusive and exhaustive.

Box 6.6   When the LR Is Genuinely "Valid" As a Measure of Probative Value of Evidence

In the case where we are considering two alternative hypotheses H and H', the LR is the ratio of $P(E|H)$ divided by $P(E|H')$. It follows that

- When LR > 1 the evidence E is more likely under H than H'.
- When LR < 1 the evidence E is more likely under H' than H.
- When LR = 1 the evidence E is no more likely under H than H' and vice versa.

By the odds form of Bayes we also know that, in all cases, the posterior odds of H against H' increase by a factor of LR times the prior odds. Therefore, if LR > 1, then the posterior odds of H against H' increase, so in this case the evidence "supports" H over H'. However, unlike the measure R none of these statements are direct statements about the posterior probability of H.

It turns out that, when H and H' are mutually exclusive and exhaustive (and *only* when that is true)—in other words when H' is the hypothesis not H—we *can* conclude the following:

- When LR > 1 the evidence is probative in favour of H (and the higher LR is the more probative the evidence).
- When LR < 1 the evidence is probative in favour of the alternative hypothesis not H (and the closer LR is to zero the more probative the evidence).
- When LR = 1 the evidence has no probative value.

To prove this we have to show that, when H' is the same as not H:

Property 1: LR > 1 implies that $P(H|E) > P(H)$
Property 2: LR < 1 implies that $P(H'|E) > P(H')$
Property 3: LR = 1 implies that $P(H|E) = P(H)$

From the odds version of Bayes' theorem we know

$$\frac{P(H|E)}{P(H'|E)} = \frac{P(E|H)}{P(E|H')} \times \frac{P(H)}{P(H')} = LR \times \frac{P(H)}{P(H')}$$

Therefore, when LR > 1 it follows that:

$$\frac{P(H|E)}{P(H'|E)} > \frac{P(H)}{P(H')} \tag{6.1}$$

However, because H' is equal to not H (meaning H and H' are mutually exclusive and exhaustive) we know that

$$P(H') = 1 - P(H) \quad \text{and} \quad P(H'|E) = 1 - P(H|E)$$

Hence, substituting these into the above inequality equation (6.1) we get

$$\frac{P(H|E)}{1 - P(H|E)} > \frac{P(H)}{1 - P(H)}$$

Hence,

$$P(H|E)(1 - P(H)) > P(H)(1 - P(H|E))$$

Hence,

$$P(H|E) - P(H|E)P(H) > P(H) - P(H|E)P(H)$$

Hence,

$$P(H|E) > P(H)$$

This proves Property 1.

The proof of Property 2 follows by symmetry, while the proof of Property 3 follows by substituting "=" for ">" in the above proof.

So, for mutually exclusive and exhaustive hypotheses H and H' the LR is a valid and meaningful measure of the probative value of evidence with respect to H.

6.5.3 Problem with the LR

In Chapter 15 we will see that the LR is used widely in law and forensics as a measure of probative value of evidence. Box 6.6 shows that this requires us to assume that the two hypotheses are mutually exclusive and exhaustive. Unfortunately, as we will show in Chapter 15, this is commonly ignored and the ramifications can be that the value of evidence is seriously misjudged.

Even when we have the mutually exclusive and exhaustive hypotheses H and not H the meaning of both H and not H will generally be ambiguous. For example, if a defendant is charged with murdering her husband

and we assume that H means "Guilty" what exactly is "Not guilty?" It could reasonably refer to any of the following hypotheses:

- Defendant is completely innocent of the crime
- Defendant killed her husband completely by accident
- Defendant killed her husband, but only in self-defence
- Defendant killed her husband as an act of mercy at the husband's request
- Defendant's lover killed her husband
- Husband is missing and not dead
- Husband committed suicide
- etc.

In most situations, even if we have a well-defined hypothesis H it will generally be difficult or even impossible to define an alternative hypothesis that is the negation of H and for which we can compute the necessary likelihood $P(E|\text{not } H)$. What happens therefore in practice is that we are forced to consider an "alternative" hypothesis H' that is definitely not the negation of H. In such situations for any evidence E the LR may not be a measure of probative value with respect to H and so we have to be extremely careful how we interpret the LR. The following coin example (which we use simply because it avoids us getting diverted into arguments about the particular likelihood values) proves the point:

In Chapter 12 we will consider Bayesian hypothesis testing in much greater detail

Example 6.8

Suppose it is known that a coin is either biased in favour of heads or biased in favour of tails. Suppose we know that the coins biased in favour of heads have a probability $P = 0.6$ of tossing a head. The hypothesis H is that the coin is biased in favour of heads. What is the alternative hypothesis in this case? There are many possibilities. It could be that the coin is a double-sided tail. It could be that it has a probability 0.49 of tossing a head. For convenience, we choose the alternative hypothesis H' corresponding to probability 0.4 of tossing a head. The evidence E consists of two consecutive coin tosses in which we observe a head followed by a tail. This means that:

$$P(E|H) = 0.6 \times 0.4 = 0.24$$

$$P(E|H') = 0.4 \times 0.6 = 0.24$$

It follows that the LR = 1 suggesting that the evidence has no probative value. In fact, while it may have no probative value in terms of distinguishing between which of H and H' is more likely to be true it may well be probative in support of H. Suppose, for example, that a small number (1%) of the tails-biased coins are double-tailed. Then "not H" must include both H' and the additional hypothesis H'': "coin is double-tailed." Then

$$P(E|\text{not } H) = 0.01 \times P(E|H'') + 0.99 \times P(E|H) = 0 + 0.99 \times 0.24 = 0.238$$

In this case, it follows from Box 6.6, that the evidence E is probative in favour of H since $P(H | E) > P(H)$.

Example 6.9

Although this is another coin-tossing example it is more representative of the kind of problem that arises in legal arguments. Fred claims to be able to toss a fair coin in such a way that about 90% of the time it comes up heads. The hypothesis we are interested in testing is

H: Fred has genuine skill (90% heads)

To test the hypothesis, we observe him toss a coin 10 times. It comes out heads each time. So, our evidence E is 10 out of 10 heads. Our alternative hypothesis is

H': Fred is just lucky (50% heads)

Using the binomial distribution assumptions discussed in Section 5.5 (assuming $n = 10$ and $p = 0.5$ and 0.9, respectively) we compute:

$P(E \mid H)$ is approximately 0.35
$P(E \mid H')$ is approximately 0.001

So the LR is about 350, strongly in favour of H *over* H'. It seems reasonable to conclude that the evidence is highly probative in favour of H. However, again it is likely that H and H' are not exhaustive. There could be another hypothesis

H'': "Fred is cheating by using a double-headed coin" (100% heads) Now, clearly $P(E \mid H'') = 1$.

If we assume that H, H' and H'' are the only possible hypotheses (i.e. they are exhaustive) and that the priors are equally likely, that is each is equal to 1/3 then the posteriors after observing the evidence E are

H: 0.25907
H': 0.00074
H'': 0.74019

Therefore, although as we saw above the LR of H with respect to H' was about 350, meaning that the evidence strongly favoured H over H', once we take account of the possibility of H'', it turns out that, despite the high LR the posterior probability for H has actually *decreased* after observing the evidence meaning that E does not support H.

Real examples—good and bad—of the use of the LR in criminal trials are presented in Box 6.7.

Box 6.7   Using the Likelihood Ratio to Properly Evaluate the Impact of Evidence

The Barry George case

The evidence, E, of gunpowder residue on Barry George's coat was key to the jury convicting him in 2001 of the murder of TV presenter Jill Dando. The forensic expert told the court that the probability of seeing such evidence in the absence of actual gunpowder was low (although he did not state an actual number in court, he subsequently provided a value of 0.01). This low probability of $P(E \mid \text{not guilty})$ clearly had an impact on the jury. However, the court never heard any testimony about $P(E \mid \text{guilty})$. The subsequent appeal in 2008 presented expert testimony that $P(E \mid \text{guilty})$ was also approximately 0.01. Since the likelihood ratio was close to one the value of the gunpowder evidence was deemed to have no probative value. The appeal court accepted that presenting only the original gunpowder evidence may have biased the jury toward the guilty verdict. A more detailed analysis of this evidence explains why the likelihood ratio argument was over-simplistic. In particular, because of ambiguity about what the hypothesis "not guilty" entailed in the assessment of $P(E \mid \text{not guilty})$ it is possible that the chosen hypotheses were not mutually exclusive and exhaustive. Consequently although the LR with respect to the chosen hypotheses was 1, it is possible that the evidence was probative in support of the guilty hypothesis.

Birmingham Six case

The main evidence E against those convicted was a positive test result for traces of nitro-glycerine on their hands; the forensic expert claimed this gave overwhelming support for the hypothesis H that they had handled high explosives. The expert testimony amounted to a statement that $P(H \mid E)$ was close to one. This information, presented on its own, would have had a profound impact on the jury. In fact all that could be concluded was that $P(E \mid H)$ was close to one.

The defendants' successful appeal was on the basis of considering $P(E \mid \text{not } H)$—which had not been done at the original trial. Specifically, the Griess test that was used to test traces of nitro-glycerine can also produce positive results in the presence of many other commonly deposited substances, including cigarettes and playing cards, and the kind of paint found on the surfaces of trains. Since the defendants (some of whom smoked) were known to have been playing cards on a train before being arrested it was demonstrated that $P(E \mid \text{not } H)$ was, like $P(E \mid H)$, quite high and hence the value of the evidence was not as great as implied at the original trial. Although explicit probabilities were not provided, if we assume $P(E \mid \text{not } H)$ was 0.5 and that $P(E \mid H) = 0.99$ then the likelihood ratio shows that the evidence E increases the prior odds of guilt by a factor of about 2. In the absence of any other compelling evidence the conviction was considered unsafe.

6.5.4 Calculating the LR When We Know $P(H \mid E)$ Rather Than $P(E \mid H)$

While we have indicated above some common errors in using the LR, there is especially great confusion in its use where we have data for $P(H \mid E)$ rather than $P(E \mid H)$. For example, look at the argument in Figure 6.6

Some of the calculations get lost in the bar room hubbub, but they are as follows: Muslims make up around 90% of offenders from the lists of convictions, where guilt and innocence had been tested in a courtroom. But Muslims make up 5% of the population. Their calculations were:

- The likelihood of individual Muslims to commit this offence is: $90 \div 5\% = 1,800$
- The likelihood of individual non-Muslims to commit this offence is: $10 \div 95\% = 10.53$
- So the likelihood of individual Muslims to commit this offence, compared to individual non-Muslims, is: $1,800 \div 10.53 = 170.94$

In other words, a Muslim man is 170 times more likely than a non-Muslim man to commit this crime. There is mathematical squabbling, and swearing about whether the figure can really be so steep. But that is the figure settled on.

Figure 6.6 Use of likelihoods and the LR. In McLoughlin, P. "Easy Meat: Inside Britain's Grooming Gang Scandal." (2016).

Although it is unclear, the authors are actually attempting to determine the LR of the evidence E with respect to the hypothesis H where:

$H:$ "Offence is committed by a Muslim" (so not H means "Offence is committed by a non-Muslim")
$E:$ "Offence is a child rape"

In this case, the population data provide our priors $P(H) = 0.05$ and, hence, $P(\text{not } H) = 0.95$. However, we also have the data on child rape convictions that gives us $P(H|E) = 0.9$ and, hence, $P(\text{not } H|E) = 0.1$.

Although we do not have data on either $P(E|H)$ or $P(E|\text{not } H)$, we can still use Bayes theorem to calculate the LR since:

$$LR = \frac{P(E|H)}{P(E|\text{not } H)} = \frac{\dfrac{P(H|E) \times P(E)}{P(H)}}{\dfrac{P(\text{not } H|E) \times P(E)}{P(\text{not } H)}} = \frac{\dfrac{P(H|E)}{P(H)}}{\dfrac{P(\text{not } H|E)}{P(\text{not } H)}}$$

$$= \frac{P(H|E) \times P(\text{not } H)}{P(H) \times (P(\text{not } H|E))}$$

Therefore, in the example we get

$$LR = \frac{0.9 \times 0.95}{0.05 \times 0.1} = 171$$

Hence, while the method described in Figure 6.6 is unclear, the conclusion arrived at is (almost) correct (the slight error in the result, namely 170.94 instead of 171, is caused by the authors rounding $10 \div 95\%$ to 10.53).

6.6 Second-Order Probability

Recall the Honest Joe's and Shady Sam's example we encountered in Chapter 4 (Box 4.7). In that example we expressed a belief in the chance of the die being fair or not, as a probability, while being aware that the fairness is also expressed as a probability. At first glance this looks very odd indeed since it suggests we are measuring a "probability about a probability." We call this a *second-order probability*. If we think of the fairness of the die, and its chance of landing face up on a 6, as a property of the die and how it is thrown, then we are expressing a degree of belief in the chance of the die having a given value. This is no different from expressing a degree of belief in a child reaching a given height when they mature in the sense that height and chance are both unknown properties of a thing that is of interest to us.

Example 6.10 shows how we might model such second-order probabilities in practice.

Example 6.10   Honest Joe's and Shady Sam's Revisited

Let us assume someone has smuggled a die out of either Shady Sam's or Honest Joe's, but we do not know which casino it has come from. We wish to determine the source of the die from (a) a prior belief about where the die is from and (b) data gained from rolling the die a number of times.

We have two alternative hypotheses we wish to test: Joe ("die comes from Honest Joe's") and Sam ("die comes from Shady Sam's"). The respective prior probabilities for these hypotheses are:

$$P(\text{Joe}) = 0.7 \qquad P(\text{Sam}) = 0.3$$

This is justified by the suspicion that our smuggler may be deterred by the extra personal risk in smuggling a die from Shady Sam's compared with Honest Joe's.

The data consists of 20 rolls of the die, observing there was one "6" and nineteen "not 6" results. So, we need to compute the likelihoods P(Joe | data) and P(Sam | data) and combine these (by Bayes theorem) with our prior beliefs about the hypotheses to get our posterior beliefs. To compute the likelihoods, recall that we believed that a die from Honest Joe's was fair, with a chance of a 6, $p = 1/6$, and from Shady Sam's it was unfair, say, $p = 1/12$. We can use these assumptions with the Binomial distribution to generate the likelihoods we need for the data, X successes in 20 trials, given each of our hypotheses:

$$P(X = x \mid p, 20) = \binom{20}{x} p^x (1 - p)^{20-x}$$

The results are shown in Table 6.4. Notice that we are expressing beliefs in hypotheses that are equivalent to beliefs about probabilities, in this case

$$P(\text{Joe}) \equiv P(p = 1/6)$$

and

$$P(\text{Sam}) \equiv P(p = 1/12)$$

From Table 6.4 we can see that we now should favor the hypothesis that the die was sourced from Shady Sam's rather than Honest Joe's. This conclusion reverses our prior assumption (which had favored Honest Joe's).

Table 6.4

Shady Sam's versus Honest Joe's

Prior	Likelihood	Posterior
$P(\text{Joe})$	$P(data \mid \text{Joe})$	$P(\text{Joe} \mid data)$
$\equiv P\left(p = \dfrac{1}{6}\right)$	$\equiv \dbinom{20}{1}\left(\dfrac{1}{6}\right)^1 \left(\dfrac{5}{6}\right)^{19} = 0.10434$	$= \dfrac{\text{likelihood} \times \text{prior}}{P(data)}$
(this is given as 0.7)		$= \dfrac{0.10434 \times 0.7}{0.08257} = 0.432$
$P(\text{Sam})$	$P(data \mid \text{Sam})$	$P(\text{Sam} \mid data)$
$\equiv P\left(p = \dfrac{1}{12}\right)$	$\equiv \dbinom{20}{1}\left(\dfrac{1}{12}\right)^1 \left(\dfrac{11}{12}\right)^{19} = 0.31906$	$= \dfrac{0.31906 \times 0.3}{0.08257} = 0.568$
(this is given as 0.3)		

Where $P(data) = [P(data \mid \text{Joe}) \times P(\text{Joe})] + [P(data \mid \text{Sam}) \times P(\text{Sam})] = 0.08257$

Note that our uncertainty about p was expressed as two alternative single point values (1/6 and 1/12 respectively under hypotheses Joe and Sam). In fact, in practice this may be unrealistic. In Chapter 10 we will show that our uncertainty may be expressed as a continuous distribution.

6.7   Summary

Bayes' theorem is adaptive and flexible because it allows us to revise and change our predictions and diagnoses in light of new data and information. In this way if we hold a strong prior belief that some hypothesis is true and then accumulate a sufficient amount of empirical data that contradicts or fails to support this, then Bayes' theorem will favor the alternative hypothesis that better explains the data. In this sense Bayes is said to be scientific and rational since it forces our model to "change its mind."

We argued that, since there must always be some subjective assumptions in the initial assignment of probabilities, it is inevitable that all probability is subjective. We have presented Bayes' theorem and shown that it is simply a method for defining and manipulating conditional probabilities and that it provides a natural way to compute probabilities. We have demonstrated the power of Bayes by showing how it helps avoid common fallacies of probabilistic reasoning, with a range of examples including those in legal reasoning. However, it should be clear to you that the kinds of fallacies we have exposed are widespread in many domains; indeed sloppy reasoning that contradicts Bayes' theorem is the norm rather than the exception.

Further Reading

Blair, P. J., & Rossmo, D. K. (2010). Evidence in Context: Bayes' Theorem and Investigations. Police Quarterly, 13(2), 123–135.

Fenton, N. E., D. Berger, D. Lagnado, M. Neil and A. Hsu. (2013). When 'neutral' evidence still has probative value (with implications from the Barry George Case), *Science and Justice*, http://dx.doLorg/10.1016/j.scijus.2013.07.002
   Article explains the role of the likelihood ratio in the Barry George case.

Fenton, N. E., and Neil, M. (2011). Avoiding legal fallacies in practice using Bayesian networks. *Australian Journal of Legal Philosophy* vol. 36, 114–150.
   This paper provides an extensive discussion of the probabilistic fallacies of reasoning in the law.

Greer, S. (1994). Miscarriages of criminal justice reconsidered. *The Modern Law Review* 57(1), 58–74.
   This paper discusses the Birmingham six and other cases.

Lindley, D. (1985). *Making Decisions*, 2nd ed., John Wiley & Sons.

McGrayne, S. B. (2011). *The theory that would not die: How Bayes; Rule cracked the Enigma Code, hunted down Russian submarines, and emerged triumphant from two centuries of controversy*. Yale University Press.
   Excellent story of the history of Bayes' Theorem.

7

From Bayes' Theorem
to Bayesian Networks

7.1 Introduction

We have now seen how Bayes' theorem enables us to correctly update a prior probability for some unknown event when we see evidence about the event. But in any real-world risk assessment problem there will be many unknown events and many different pieces of evidence, some of which may be related. We already saw in Chapter 3 examples of such risk assessment problems. When we represent such problems graphically (with all uncertain variables being represented as nodes and an edge between two nodes representing a relationship) we have a Bayesian Network (BN). It turns out that Bayes' theorem can be applied to correctly update evidence in these more general complex problems, and the purpose of this chapter is to explain exactly how this is done and under what constraints.

We start in Section 7.2 by considering the simplest type of risk assessment problem with just two variables. We can think of this as a two-node BN. In this case if we enter evidence on one variable we can use Bayes' theorem exactly as described in Chapter 6 to update the probability of the other variable. So essentially Section 7.2 explains fully how to do computations in a two-node BN. In Section 7.3 we show how Bayes' theorem can be extended to perform the necessary probability updating when we consider an extended version of the problem (with four nodes). The calculations quickly become quite tricky, but in Section 7.4 we demonstrate how the underlying idea of Bayesian "propagation" through the network provides some very powerful types of reasoning. This leads to a formal definition of BNs in Section 7.5, which is based on the crucial notion of what is meant when two variables are not directly linked (the so-called independence assumptions). In Section 7.6 we describe the basic structural properties of BNs, and it is a recognition of these properties that form the basis for the general propagation algorithm that is described in Section 7.7. In Section 7.9 we show how BNs can be used to "solve" two of the paradoxes discussed in earlier chapters (Monty Hall and Simpson). Finally, in Section 7.8 we provide guidelines for building and running BN models.

Readers who do not wish to understand any of the mathematical underpinnings of BNs could go straight to Section 7.8 for a practical description of how to build and run BNs.

All models in this chapter are available to run in AgenaRisk, downloadable from www.agenarisk.com.

This chapter contains more mathematical material than in previous chapters, but the most demanding material appears in boxes (which can be avoided by those readers who do not need to understand how the BN models work), and the full description of the propagation algorithm appears in Appendices B and C.

7.2 A Very Simple Risk Assessment Problem

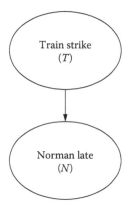

Figure 7.1 Will Norman be late?

In real-world examples a variable such as "Norman late" might be continuous on a scale of 0 to, say, 200, representing the number of minutes late.

Since it is important for Norman to arrive on time for work, a number of people (including Norman himself) are interested in the probability that he will be late. Since Norman usually travels to work by train, one of the possible causes for Norman being late is a train strike. Because it is quite natural to reason from cause to effect we examine the relationship between a possible train strike and Norman being late. This relationship is represented by the causal model shown in Figure 7.1 where the edge connects two nodes representing the variables "Train strike" (T) to "Norman late" (N).

It is obviously important that there is an edge between these variables since T and N are not independent (using the language of Chapter 5); common sense dictates that if there is a train strike then, assuming we know Norman travels by train, this will affect his ability to arrive on time. Common sense also determines the direction of the link since train strike causes Norman's lateness rather than vice versa.

To ensure the example is as simple as possible we assume that both variables are discrete, having just two possible states: true and false.

Let us assume the following prior probability information:

1. The probability of a train strike is 0.1 (and therefore the probability of no train strike is 0.9). This information might be based on some subjective judgment given the most recent news or it might be based on the recent frequency of train strikes (i.e. one occurring about every 10 days). So the prior probability distribution for the variable "Train strike" is as shown in Table 7.1.

2. The probability Norman is late given that there is a train strike is 0.8 (and therefore the probability Norman is not late given that there is a train strike is 0.2). The probability Norman is late given that there is not a train strike is 0.1 (and therefore the probability Norman is not late given that there is not a train strike is 0.9). So, the (conditional) probability distribution for "Norman late" given "Train strike" is as shown in Table 7.2.

Table 7.1

Probability Table for "Train Strike"

False	0.9
True	0.1

Table 7.2

Probability Table for "Norman Late" Given "Train Strike"

Train strike	False	True
False	0.9	0.2
True	0.1	0.8

The first thing we want to calculate from Table 7.1 and Table 7.2 is the *marginal* (i.e. unconditional) probability that Norman is late, $P(N)$. As we saw in Chapter 5 this is the probability that Norman is late when we do not know any specific information about the events (in this case the train strike) that influence it. The probability tables do not provide this value directly, but as we saw in Chapter 5, they do provide it indirectly by virtue of the following equation for marginalization:

$$P(N = True) = P(N = True \mid T = True)P(T = True)$$

$$+P(N = True \mid T = False)P(T = False)$$

$$= 0.8 \times 0.1 + 0.1 \times 0.9$$

$$= 0.17$$

Notice that we are presenting the probabilities as tables here rather than lists of numbers. This is something we will do more here and in subsequent chapters. We call these tables node probability tables (NPTs) or conditional probability tables (CPTs).

So there is a 17% chance Norman is late. Now suppose that we actually observe Norman arriving late. In the light of such an observation we wish to revise our belief about the probability of a train strike. There are numerous reasons why we might wish to do this including, most obviously, that such information will have an impact on whether other people who take the train will also be late.

To calculate this posterior probability of a train strike we use Bayes' theorem:

$$P(T = True \mid N = True) = \frac{p(N = True \mid T = True)p(T = True)}{p(N = True)}$$

$$= \frac{0.8 \times 0.1}{0.17} \tag{7.1}$$

$$= 0.47059$$

So knowing that Norman is late leads to an increased probability of a train strike (from 0.1 to 0.47050), that is, from 10% to 47%.

We stress again that you should never have to do the Bayesian calculations manually. Simply run AgenaRisk and create two nodes with an edge link between them as shown in Figure 7.1. Then complete the probability tables for the two nodes with the values given in Table 7.1 and Table 7.2. When you run the resulting model you will see the marginal distributions for the two nodes (so you will see that the probability N is true equals 0.17). This is shown in Figure 7.2(a). When you enter the observation N is true and run the model again you will see that the revised posterior probability of T being true is 0.47059. This is shown in Figure 7.2(b).

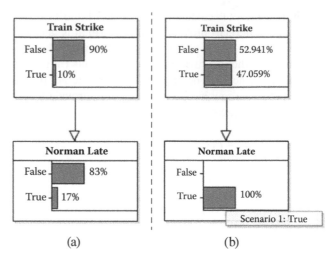

Figure 7.2   (a) Initial (prior) marginal state of model; (b) Revised (posterior) state of model after observation is entered.

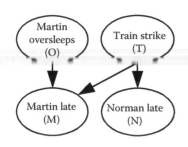

Figure 7.3   Revised causal model.

7.3   Accounting for Multiple Causes (and Effects)

Norman is not the only person whose chances of being late increase when there is a train strike. Martin is also more likely to be late, but Martin depends less on trains than Norman and he is often late simply as a result of oversleeping. These additional factors can be modeled as shown in Figure 7.3.

You should add the new nodes and edges using AgenaRisk. We also need the probability tables for each of the nodes *Martin oversleeps* (Table 7.3) and *Martin late* (Table 7.4).

The table for node *Martin late* is more complicated than the table for *Norman late* because *Martin late* is conditioned on *two* nodes rather than one. Since each of the parent nodes has two states, true and false (we are still keeping the example as simple as possible), the number of combinations of parent states is four rather than two.

If you now run the model and display the probability graphs you should get the marginal probability values shown Figure 7.4(a). In particular, note that the marginal probability that Martin is late is equal to 0.446 (i.e. 44.6%). Box 7.1 explains the underlying calculations involved in this.

But if we know that Norman is late, then the probability that Martin is late increases from the prior 0.446 to 0.542 as shown in Figure 7.4(b). Box 7.1 explains the underlying calculations involved.

Table 7.3
Probability Table for "Martin Oversleeps"

False	0.6
True	0.4

Table 7.4
Probability Table for "Martin Late"

| Martin oversleeps (O) | | False | | True | |
Train strike (T)		False	True	False	True
Martin late	False	0.7	0.4	0.4	0.2
(M)	True	0.3	0.6	0.6	0.8

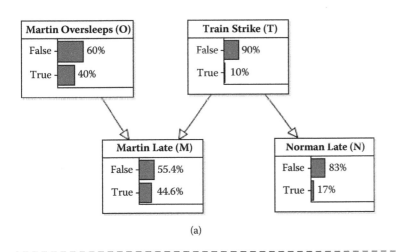

(a)

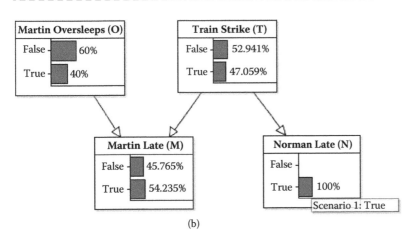

(b)

Figure 7.4 Running the model in BN tool. (a) Initial marginal state of model. (b) Revised state of model after observation is entered.

Box 7.1   The Underlying Bayesian Network (BN) Calculations Involved in Figure 7.4

1. Calculating the prior marginal probability that "Martin late" is true.

    Martin late (M) is conditioned on *Martin oversleeps* (O) and *Train strike* (T). The marginal probability for $M = True$ is therefore

$$P(M = True) = \sum_{O,T} P(M = True \mid O,T)P(O)P(T)$$

$$= P(M = True \mid O = True, T = True)P(O = True)P(T = True)$$

$$+ P(M = True \mid O = True, T = False)P(O = True)P(T = False)$$

$$+ P(M = True \mid O = False, T = True)P(O = False)P(T = True)$$

$$+ P(M = True \mid O = False, T = False)P(O = False)P(T = False)$$

$$= (0.8 \times 0.4 \times 0.1) + (0.6 \times 0.4 \times 0.9) + (0.6 \times 0.6 \times 0.1) + (0.3 \times 0.6 \times 0.9)$$

$$= 0.446$$

where we simply have substituted in the probability values from Table 7.1, Table 7.3, and Table 7.4.

2. Revising the probability that Martin late is true once we know Norman late is true.

    In Section 7.2 we already showed how to calculate the revised probability the train is on strike given that Norman is late using Bayes' theorem. Specifically, from Equation (7.1), $P(T = True \mid N = True) = 0.47059$. By Theorem 4.1, we also deduce that $P(T = False \mid N = True) = 0.52941$.

    By marginalization we know that

$$P(M = True \mid N = True)$$

$$= \sum_{O,T} P(M = True \mid O,T)P(O)P(T \mid N = True)$$

Using the preceding results for $P(T = True \mid N = True)$ and $P(T = False \mid N = True)$ along with Table 7.4 we get

$$\sum_{O,T} P(M = True \mid O, T)P(O)P(T \mid N = True)$$

$$= P(M = True \mid O = True, T = True)P(O = True)P(T = True \mid N = True)$$

$$+ P(M = True \mid O = True, T = False)P(O = True)P(T = False \mid N = True)$$

$$+ P(M = True \mid O = False, T = True)P(O = False)P(T = True \mid N = True)$$

$$+ P(M = True \mid O = False, T = False)P(O = False)P(T = False \mid N = True)$$

$$= (0.8 \times 0.4 \times 0.47059) + (0.6 \times 0.4 \times 0.52941) + (0.6 \times 0.6 \times 0.47059) + (0.3 \times 0.6 \times 0.52941)$$

$$= 0.542353$$

Hence the revised marginal probability $P(M = True)$ given Norman late is 54.235%, and by Theorem 4.1, the revised marginal probability $P(M = False) = 45.765\%$.

In Box 7.1 we have made use of the marginalization notation introduced in Chapter 5. Box 7.2 highlights the need to be careful when using this notation.

Box 7.2   Taking Care with Marginalization Notation

Each time we marginalize the result is dependent on the current state of the joint probability model and as we find out new information, as evidence or data, the state of the probability model changes. Some books on probability and Bayesian networks do not make this change clear in their notation.

For example, we might marginalize to get $P(B)$ in the model $P(A,B) = P(B\,|\,A)P(A)$:

$$P(B) = \sum_A P(B\,|\,A)P(A) \qquad (1)$$

And then, given we know $A = \textit{True,}$ we might marginalize again to get:

$$P(B) = \sum_A P(B\,|\,A = \textit{True})P(A = \textit{True}) \qquad (2)$$

However, we do not know whether $P(B)$ from Equation (1) is the same as $P(B)$ from Equation (2) since the notation is ambiguous and in Equation (2) we have evidence. Some books clarify this by expressing the new marginal in Equation (2) as $P(B\,|\,A = \textit{True})$, rather than $P(B)$, to show that this marginal is dependent on the evidence, which in this case is $A = \textit{True}$, since

$$P(B\,|\,A = \textit{True}) = \sum_A \frac{P(B, A = \textit{True})}{P(A = \textit{True})}$$

$$= \sum_A \frac{P(B\,|\,A = \textit{True})P(A = \textit{True})}{P(A = \textit{True})} = P(B)$$

This is fine and in this case is equivalent to simply looking up $P(B\,|\,A = \textit{True})$ in the NPT, but in more complex cases the probability calculus will be needed.

Another way to make the change in marginal probability clear, due to conditioning on new evidence, is to use an asterisk (*) as a superscript or subscripts to show that the marginal has changed in some way, such as this:

$$P(B\,|\,A = \textit{True}) \equiv P_{A=\textit{True}}(B) \equiv P^*(B)$$

7.4   Using Propagation to Make Special Types of Reasoning Possible

When we enter evidence and use it to update the probabilities in the way we have seen so far we call it *propagation*. In principle we can enter any number of observations anywhere in the BN model and use propagation to update the marginal probabilities of all the unobserved variables.

This can yield some exceptionally powerful types of analysis. For example, without showing the computational steps involved, if we first enter the observation that Martin is late we get the revised probabilities shown in Figure 7.5(a).

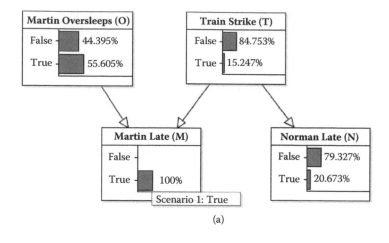

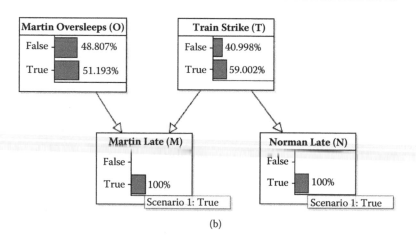

(b)

Figure 7.5 The notion of "explaining away" evidence. (a) When we observe Martin is late the most likely cause is oversleeping. (b) When we also observe Norman is late the most likely cause now is train strike.

What the model is telling us here is that the most likely explanation for Martin's lateness is Martin oversleeping; the revised probability of a train strike is still low. However, if we now discover that Norman is also late (Figure 7.5(b)) then *Train strike* (rather than *Martin oversleeps*) becomes the most likely explanation for Martin being late. This particular type of (backward) inference is called *explaining away* (or sometimes called nonmonotonic reasoning). Classical statistical tools alone do not enable this type of reasoning and what-if analysis.

In fact, as even the earlier simple example shows, BNs offer the following benefits:

■ *Explicitly model causal factors*—It is important to understand that this key benefit is in stark contrast to classical statistics whereby prediction models are normally developed by purely data-driven approaches. For example, the regression models introduced in Chapter 2 use historical data alone to produce equations relating dependent and independent variables. Such approaches not only fail to incorporate expert judgment in scenarios where there is insufficient data, but also fail to accommodate causal explanations. We will explore this further in Chapter 9.

- *Reason from effect to cause and vice versa*—A BN will update the probability distributions for every unknown variable whenever an observation is entered into any node. So entering an observation in an "effect" node will result in back propagation, that is, revised probability distributions for the "cause" nodes and vice versa. Such backward reasoning of uncertainty is not possible in other approaches.

- *Reduce the burden of parameter acquisition*—A BN will require fewer probability values and parameters than a full joint probability model. This modularity and compactness means that elicitation of probabilities is easier and explaining model results is made simpler.

- *Overturn previous beliefs in the light of new evidence*—The notion of explaining away evidence is one example of this.

- *Make predictions with incomplete data*—There is no need to enter observations about all the inputs, as is expected in most traditional modeling techniques. The model produces revised probability distributions for all the unknown variables when any new observations (as few or as many as you have) are entered. If no observation is entered then the model simply assumes the prior distribution.

- *Combine diverse types of evidence including both subjective beliefs and objective data*—A BN is agnostic about the type of data in any variable and about the way the probability tables are defined.

- *Arrive at decisions based on visible, auditable reasoning*—Unlike black-box modeling techniques (including classical regression models and neural networks) there are no "hidden" variables and the inference mechanism is based on Bayes' theorem.

This range of benefits, together with the explicit quantification of uncertainty and ability to communicate arguments easily and effectively, makes BNs a powerful solution for many types of risk assessment problems.

7.5 The Crucial Independence Assumptions

Take a look again at the BN model of Figure 7.3 and the subsequent calculations we used. Using the terminology of Chapter 5 what we have actually done is use some crucial simplifying assumptions in order to avoid having to work out the full joint probability distribution of:

(*Norman late, Martin late, Martin oversleeps, Train strike*)

We will write this simply as (*N, M, O, T*)

For example, in calculating the marginal probability of *Martin late* (*M*) we assumed that *M* was dependent only on *Martin oversleeps* (*O*) and

Train strike (*T*). The variable *Norman late* (*N*) simply did not appear in the equation because we assume that none of these variables are directly dependent on *N*. Similarly, although *M* depends on both *O* and *T*, the variables *O* and *T* are independent of each other.

These kind of assumptions are called *conditional independence* assumptions (we will provide a more formal definition of this later). If we were unable to make any such assumptions then the full joint probability distribution of (*N*, *M*, *O*, *T*) is (by the chain rule of Chapter 5)

$$P(N, M, O, T) = P(N \mid M, O, T)P(M \mid O, T)P(O \mid T)P(T)$$

However, because *N* directly depends only on *T* the expression $P(N \mid M, O, T)$ is equal to $P(N \mid T)$, and because *O* is independent of *T* the expression $P(O \mid T)$ is equal to $P(O)$.

Hence, the full joint probability distribution can be simplified as:

$$P(N, M, O, T) = P(N \mid T) \, P(M \mid O, T) \, P(O) \, P(T)$$

and this is exactly what we used in the computations. Indeed, the four expressions on the right-hand side correspond precisely to the four probability tables that we used:

- $P(N \mid T)$ is Table 7.2 (probability Norman late given Train strike)
- $P(M \mid O, T)$ is Table 7.4 (probability Martin late given Martin oversleeps and Train strike)
- $P(O)$ is Table 7.3 (probability Martin oversleeps)
- $P(T)$ is Table 7.1 (probability of Train strike)

This leads to the following simple but profound observation:

The Crucial Graphical Feature of a BN

The crucial graphical feature of a BN is that it tells us which variables are *not* linked, and hence it captures our assumptions about which pairs of variables are not directly dependent.

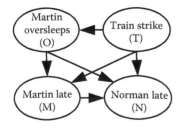

Figure 7.6 BN without any assumptions about independence between variables.

If we were unable to make any assumptions about independence for a particular risk problem, then there would be little point in using a graphical model at all. Indeed, without independence assumptions every variable must be linked directly to every other variable. So our example model would look like Figure 7.6 (it is what mathematicians call a *complete graph*). In this case there would be no simplifications possible in calculating the full joint probability distribution.

We are now in a position to define a BN:

Definition of a Bayesian Network

A Bayesian Network (BN) is an explicit description of the direct dependencies between a set of variables. This description is in the form of a directed graph and a set of node probability tables (NPTs):

- *Directed graph*—The directed graph (also called the *topology* or *structure* of the BN) consists of a set of nodes and arcs. The nodes correspond to the variables and the arcs link directly dependent variables. An arc from A to B encodes an assumption that there is a direct causal or influential dependence of A on B; the node A is then said to be a *parent* of B. We also insist that there are no cycles in the graph (so, for example, if we have an arc from A to B and from B to C then we cannot have an arc from C to A). This avoids circular reasoning.
- *NPT*—Each node A has an associated probability table, called the node probability table (NPT) of A. This is the probability distribution of A given the set of parents of A. For a node A without parents (also called a *root node*) the NPT of A is simply the probability distribution of A.

As we have seen, the existence of unlinked nodes in a BN reduces the complexity of the full joint probability distribution. A completely general expression, derived from the chain rule for Bayesian networks, is shown in Box 7.3.

Box 7.3 General Expression for Full Joint Probability Distribution of a BN

Suppose a BN consists of n variables $A_1, A_2, ..., A_n$. The chain rules tells us that the full joint probability distribution is

$$P(A, A_2, A_3, ..., A_n) = P(A_1 \mid A_2, A_3, ..., A_n)$$

$$P(A_2 \mid A_3, ..., A_n)...P(A_{n-1} \mid A_n)P(A_n)$$

which, as we saw in Chapter 5 is written using the product symbol as

$$P(A_1, A_2, A_3, ..., A_n) = \prod_{i=1}^{n} P(A_i \mid A_{i+1}, ..., A_n)$$

However, we can simplify this by using our knowledge of what the parents of each node are. For example, if node A_1 has exactly two parents, A_3 and A_5, then the following part of the joint probability distribution:

$$P(A_1 \mid A_2, ..., A_n)$$

is equivalent to:

$$P(A_1 \mid A_3, A_5)$$

So, in general let *Parents*(A_i) denote the set of parent nodes of the node A_i. Then the full joint probability distribution of the BN can be simplified as

$$P(A_1, A_2, A_3, ..., A_n) = \prod_{i=1}^{n} P(A_i \mid \text{Parents}(A_i))$$

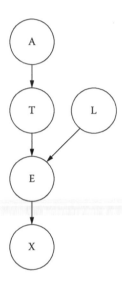

Figure 7.7   BN with five nodes.

Example 7.1

Consider the BN in Figure 7.7. The full joint probability distribution of this BN can be simplified as

$$P(A, T, L, E, X) = P(A)\, P(T \mid A)\, P(L)\, P(E \mid T,L)\, P(X \mid E)$$

You can think of this expression as an equivalent representation of the topology of the BN. Like the graphical model itself, it encodes our assumptions about direct dependencies between the variables and it tells us exactly which NPTs we need to define; in this case we need the five NPTS for

$$P(A),\ P(T \mid A),\ P(L),\ P(E \mid T, L),\ P(X \mid E)$$

Using the BN in Example 7.1, Box 7.4 describes how we can compute the marginal distribution for any variable (node) in a BN by the process of *marginalization by variable elimination*.

Box 7.4   Marginalization by Variable Elimination

In explaining both how and why marginalization of variables in a BN works it is important to remember the following two key facts (for any variables A and B)

$$\text{Fact 1: } \sum_{B} P(A \mid B)P(B) = P(A)$$

This is simply how marginalization of A over B was defined in Chapter 5.

$$\text{Fact 2: } \sum_{A} P(A \mid B) = 1$$

This follows from Bayes' theorem and Fact 1, since

$$\sum_{A} P(A \mid B) = \sum_{A} \left(\frac{P(B \mid A)P(A)}{P(B)} \right) = \frac{\sum_{A} P(B \mid A)P(A)}{P(B)} = \frac{P(B)}{P(B)} = 1$$

Using these two facts we show how marginalization works in the BN of Figure 7.7. We have already noted that the joint probability distribution can be written as

$$P(A, T, L, E, X) = P(A) \, P(T \mid A) \, P(L) \, P(E \mid T, L) \, P(X \mid E)$$

Suppose we want to use marginalization to get $P(T)$. We know that the formula for this is

$$P(T) = \sum_{A,L,E,X} P(A)P(T \mid A)P(L)P(E \mid T,L)P(X \mid E)$$

But to both explain why this works and also how to actually do the necessary calculations, we note that marginalization is a distributive operation over combinations. This means that we can marginalize the global joint probability by marginalizing local NPTs, which in this example consists of the following factors:

$$P(T) = \left(\sum_{A} P(A)P(T \mid A) \left(\sum_{E} \left(\sum_{L} P(E \mid T,L)P(L) \left(\sum_{X} P(X \mid E) \right) \right) \right) \right) \tag{1}$$

The variable elimination computation has two interleaving steps: in one step we multiply some factors together. In the next step we marginalize the results of the previous step to smaller factors, which we later use again. The basic idea is to compute the intermediate factors that we created in the elimination procedure while working right to left through the summation, and then reuse them as we work left to right.

So, working from the right we note that the term $\sum_{X} P(X \mid E) = 1$ by Fact 2. Hence, we have eliminated X from the equation for $P(T)$:

$$P(T) = \left(\sum_{A} P(A)P(T \mid A) \left(\sum_{E} \left(\sum_{L} P(E \mid T,L)P(L) \right) \right) \right)$$

We next want now to simplify the expression $\sum_{L} P(E \mid T,L)P(L)$. From Fact 1 we know

$$\sum_{L} P(E \mid T,L)P(L) = P(E \mid T)$$

Hence, we have eliminated L from the equation for $P(T)$:

$$P(T) = \left(\sum_{A} P(A)P(T \mid A) \left(\sum_{E} P(E \mid T) \right) \right)$$

But from Fact 2 we know that $\sum_{E} P(E \mid T) = 1$. Hence we have eliminated E from the equation for $P(T)$:

$$P(T) = \left(\sum_{A} P(A)P(T \mid A) \right)$$

The final elimination of A is simply another instance of Fact 1.
Note that there is no unique elimination order when models get larger.

A number of algorithms use variable elimination, or variants thereof, for computation in BNs. This is where the term "propagation" comes from to describe the process because it is suggestive of the effects of evidence propagating through the BN graph, which is easier to visualize than variable elimination through a joint probability model.

In Chapter 8 we discuss practical and theoretical issues we might encounter when building the structure of a BN. Here we introduce three fragments from which every BN model might be constructed. Understanding and manipulating these fragments lie at the heart of the computational power of the algorithms we use to calculate probabilities on BNs so it's worthwhile understanding how they operate.

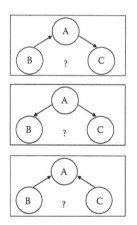

Figure 7.8 Three related variables.

If (as in the example here) the relationships in a serial connection are causal, then it is called a *causal trail*. If we are interested in reasoning from *C* to *B* (as in the case here when we wish to use information about whether Norman is late to determine whether there is a signal failure) then we call this an *evidential trail*.

Although the independence assumptions generally lead to major simplifications in the Bayesian calculations, it is clear that even in the simple example BN we have been using, the propagation calculations are still extremely complicated to work out manually. Fortunately, using a BN tool means you never actually have to do these calculations manually in practice.

7.6 Structural Properties of BNs

In BNs the process of determining what evidence will update which node is determined by the conditional dependency structure. The main formal area of guidance for building sensible BN structures therefore requires some understanding of different types of relationships between variables and the different ways these relationships are structured.

Generally we are interested in the following problem. Suppose that variable *A* is linked to both variables *B* and *C*. There are three different ways the links can be directed as shown in Figure 7.8. Although *B* and *C* are not directly linked, under what conditions in each case are *B* and *C* independent of *A*?

Knowing the answer to this question enables us to determine how to construct appropriate links, and it also enables us to formalize the different notions of conditional independence that we introduced informally in Chapter 6.

The three cases in Figure 7.8 are called, respectively, *serial, diverging*, and *converging* connections. We next discuss each in turn.

7.6.1 Serial Connection: Causal and Evidential Trails

Consider the example of a serial connection as shown in Figure 7.9. Suppose we have some evidence that a signal failure has occurred (*B*). Then clearly this knowledge increases our belief that the train is delayed (*A*), which in turn increases our belief that Norman is late (*C*). Thus, evidence about *B* is transmitted through *A* to *C* as is shown in Figure 7.10.

However, now suppose that we know the true status of *A*; for example, suppose we know that the train is delayed. Then this means we have hard evidence for A (see Box 7.5 for an explanation of what hard and uncertain evidence are and how they differ).

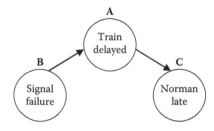

Figure 7.9 Serial connection: causal and evidential trail.

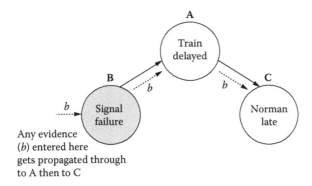

Figure 7.10 Evidence in *B* propagates through to *C*.

Box 7.5 Hard Evidence and Uncertain Evidence

If we know for certain that Norman is late on a particular day then the state of the node "Norman late" is true. In all subsequent calculations this means that *P(Norman late = true)* = 1. This is an example of *hard evidence* (also called *instantiation*). In this case we would say that *the node is instantiated* as "true."

But, suppose that on a different day we check Norman's office at 10:15 a.m. and find he is not there. We are not certain he is late because there is a small chance he might have popped out for a coffee, but this is quite rare. We cannot say that *P(Norman late = true)* = 1, but we might say it is 0.9. This kind of evidence is called *uncertain evidence*. However, there is much confusion in the BN literature about the notion of uncertain evidence because there are at least two possible meanings:

1. It could mean the result of propagating the BN is such that the marginal distribution of the node *Norman late* is exactly (0.9, 0.1). In other words, it is a statement about the posterior value; in this case the posterior for true must be 0.9.
2. It could mean that the probability of true being 0.9 is simply a likelihood. So, if the prior probability for true is, say 0.2, the posterior probability will be

$$\frac{0.9 \times 0.2}{0.9 \times 0.2 + 0.1 \times 0.8} = 0.69$$

To distinguish these two types of uncertain evidence, researchers now generally refer to the first as *soft evidence* and the second as *virtual evidence*. The confusion is compounded by the fact that most BN tools (including AgenaRisk) implement only the second type (virtual evidence), but call it soft evidence. Soft evidence is generally much harder to implement.

In this case any knowledge about *B* is irrelevant to *C* because our knowledge of *A* essentially overwrites it. Any new information about the likelihood of a signal failure (*B*) is not going to change our belief about Norman being late once we know that the train is delayed. In other words the evidence from *B* cannot be transmitted to *C* because *A blocks* the channel. This is shown in Figure 7.11.

In summary, in a serial connection evidence can be transmitted from *B* to *C* unless *A* is instantiated. Formally, we say:

In a serial connection *B* and *C* are conditionally independent given *A* (or equivalently *B* and *C* are *d*-separated given *A*).

Box 7.6 provides a proof of this assertion.

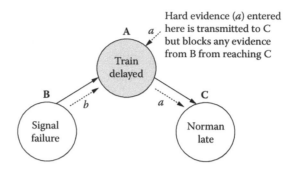

Figure 7.11 Hard evidence for *A* blocks any evidence from *B* reaching *C*.

Box 7.6   Proof of *d*-Separation for Serial Connection

In our serial example shown in Figure 7.9, the joint probability given by the BN structure is $P(A, B, C) = P(C \mid A)$ $P(A \mid B) P(B)$. We are interested in demonstrating that given $A = a$, then any evidence on B will make no difference to the marginal distribution for C.

Let's first add $A = a$ to the BN:

$$P(C \mid B, A = a) = \sum_{B,A} P(C \mid A = a)P(A = a \mid B)P(B)$$

$$= \sum_{A}\left(P(C \mid A = a)\left(\sum_{B} P(A = a \mid B)P(B) \right)\right)$$

$$= \sum_{A} P(C \mid A = a)P(A = a) \tag{1}$$

Now compare this to when we also add $B = b$ to the BN:

$$P(C \mid B = b, A = a) = \sum_{B,A} P(C \mid A = a)P(A = a \mid B = b)P(B = b)$$

$$= \sum_{A}\left(P(C \mid A = a)\left(\sum_{B} P(A = a \mid B = b)P(B = b) \right)\right)$$

$$= \sum_{A} P(C \mid A = a)P(A = a) \tag{2}$$

Since Equations (1) and (2) are equal we can deduce that C is independent of B given evidence at A.

From a practical model-building viewpoint what you need to know is that, if you model the relationship between A, B, and C as a serial connection, then the behavior of the model with respect to evidence propagation is exactly as defined here. Suppose, in contrast to the assumptions made earlier, that Norman might decide to drive to work if he hears that there is a signal failure. Then the signal failure evidence can influence whether Norman is late even after we discover that the train is delayed. Or, to put it another way, signal failure is a better reason for train delay than any other. In general, if evidence about *B* can influence *C* even after hard evidence is entered at *A* then the serial connection is *not* the correct representation of the relationship. You may need a direct link from *B* to *C* or a different type of connection.

7.6.2   Diverging Connection: Common Cause

Consider the example of a diverging connection shown in Figure 7.12. Any evidence about the train being delayed, *A*, is transmitted to both Martin late, *B*, and Norman late, *C*.

For example, if we have hard evidence that increases our belief the train is delayed, then this in turn will increase our belief in both Martin being late and Norman being late. This is shown in Figure 7.13.

But we want to know whether information about *B* can be transmitted to *C* (and vice versa). Suppose we have no hard evidence about *A* (that is, we do not know for certain whether the train is delayed). If we have some evidence that Martin is late, *B*, then this increases our belief that the train is delayed *A* (here we reason in the opposite direction of the causal arrow using Bayes' theorem). But the increased belief in the train being delayed in turn increases our belief in Norman being late. In other words, evidence about *B* (Martin late) is transmitted through to *C* (Norman late). This is shown in Figure 7.14. Of course, by symmetry, we can similarly conclude that evidence entered at *C* is transmitted to *B* through *A*.

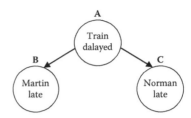

Figure 7.12   Diverging connection: common cause.

If, as in the example here, the relationships are causal, then we call *A* a *common cause* since it is a cause of more than one effect variables.

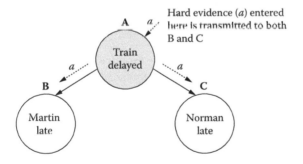

Figure 7.13   Evidence at *A* is transmitted to both *B* and *C*.

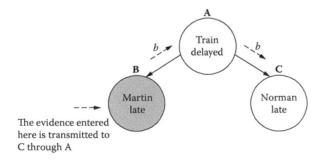

Figure 7.14   Evidence at *B* is transmitted to *C* through *A*.

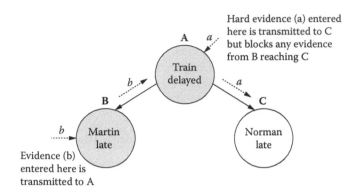

Figure 7.15 Hard evidence at A blocks evidence from B from reaching C.

However, suppose now that we have hard evidence about A; for example, suppose we know for certain the train is delayed as shown in Figure 7.15. This evidence will increase our belief about Norman being late, C. But now any evidence about Martin being late will not change our belief about Norman being late; the certainty of A blocks the evidence from being transmitted (it becomes irrelevant to C once we know A for certain).

Thus, when A is known for certain, B and C become independent. Because the independence of B and C is conditional on the certainty of A, we say formally that:

In a diverging connection B and C are conditionally independent (given A) (or equivalently B and C are d-separated given A).

In summary: Evidence can be transmitted from B to C through a diverging connection A unless A is instantiated. Box 7.7 provides a proof of this assertion.

Box 7.7   Proof of d-Separation for Diverging Connection

In our diverging example shown in Figure 7.12, the joint probability given by the BN structure is $P(A, B, C) = P(C \mid A)\, P(B \mid A)\, P(A)$. We are interested in demonstrating that given $A = a$ then any evidence on B will make no difference to the marginal distribution for C.

Let's first add $A = a$ to the BN since this is the common cause:

$$P(C \mid A = a, B) = \sum_{B, A} P(C \mid A = a) P(B \mid A = a) P(A = a)$$

$$= \sum_{A} \left(P(C \mid A = a) \left(\sum_{B} P(B \mid A = a) \right) P(A = a) \right)$$

$$= \sum_{A} P(C \mid A = a) P(A = a) \tag{1}$$

Now compare this to when we also add $B = b$ to the BN:

$$P(C \mid A = a, B = b) = \sum_{B,A} P(C \mid A = a)P(B = b \mid A = a)P(A = a)$$

$$= \sum_{A} \left(P(C \mid A = a) \left(\sum_{B} P(B = b) \mid A = a \right) P(A = a) \right)$$

$$= \sum_{A} P(C \mid A = a)P(A = a) \qquad (2)$$

Since Equations (1) and (2) are equal we can deduce that C is independent of B given evidence at A.

So, from a practical model-building viewpoint what you need to know is that, if you model the relationship between A, B, and C as a diverging connection, then the behavior of the model with respect to evidence propagation is exactly as defined here. So if, for example, you feel that evidence about B can influence C even after hard evidence is entered at A (as would be the case if both Martin and Norman also sometimes traveled by a different mode of transport like a bus), then the diverging connection is not the correct representation of the relationship. You may need a direct link from B to C (or vice versa) or a different type of connection.

7.6.3 Converging Connection: Common Effect

Consider the example of a converging connection shown in Figure 7.16.

Clearly any evidence about either the train being delayed (B) or Martin oversleeping (C) will lead us to revise our belief about Martin being late (A). So evidence entered at either or both B and C is transmitted to A as shown in Figure 7.17.

However, we are concerned about whether evidence can be transmitted between B and C. If we have no information about whether Martin is late, then clearly whether Martin oversleeps and whether there are train delays are independent. In other words if nothing is known about A then A's parents (B and C) are independent, so no evidence is transmitted between them.

However, suppose we have some information about A (even just uncertain evidence) as shown in Figure 7.18. For example, suppose that Martin usually hangs up his coat in the hall as soon as he gets in to work. Then if we observe that Martin's coat is not hung up after 9:00 a.m. our belief that he is late increases (note that we do not know for certain that he is late; we have uncertain evidence as opposed to hard evidence because on some days Martin does not wear a coat). Even this uncertain evidence about Martin being late increases our belief in both Martin oversleeping, B, and train

If, as in the example here, the relationships are causal then we call A a *common effect* since it is shared by more than one cause.

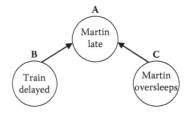

Figure 7.16 Converging connection.

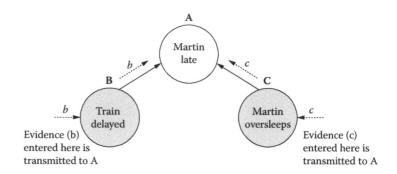

Figure 7.17   Evidence at *B* is transmitted to *A* but not to *C*.

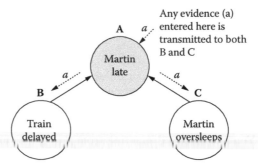

Figure 7.18   Once any evidence is entered for *A*, the nodes *B* and *C* become dependent.

delay, *C*. If we find out that one of these is false the other is more likely to be true. Under these circumstances we say, therefore, that *B and C are conditionally dependent on A* (or, equivalently, that they are *d-connected given A*).

It follows that in a converging connection evidence can only be transmitted between the parents *B* and *C* when the converging node *A* has received some evidence (which can be hard or uncertain). A proof is shown in Box 7.8.

So, from a practical model-building viewpoint what you need to know is that, if you model the relationship between *A*, *B*, and *C* as a converging

---

Box 7.8   Proof of *d*-Connection for Converging Connection

In our converging connection example shown in Figure 7.16, the joint probability given by the BN structure is $P(A, B, C) = P(A \mid B, C) \, P(B) \, P(C)$. We are interested in demonstrating that given $A = a$ then any evidence on *B* will update the marginal distribution for *C*.

Let's first add $A = a$ to the BN since this is the common cause:

$$P(C \mid A = a, B) = P(C) \sum_{B,A} P(A = a \mid B, C) P(B) \qquad (1)$$

Now let's also add $B = b$ to the BN:

$$P(C \mid A = a, B = b) = P(C) \sum_{B,A} P(A = a \mid B = b, C) P(B = b) \qquad (2)$$

Equation (1) $\neq$ (Equation (2) so C is dependent on B given evidence at A.

connection, then the behavior of the model with respect to evidence propagation is exactly as defined here. So if, for example, you feel that evidence about B can influence C even without any evidence entered at A (as would be the case if Martin generally listens to the radio and goes back to sleep if he hears there might be train delays), then the converging connection is not the correct representation of the relationship. You may need a direct link from B to C (or vice versa) or a different type of connection.

7.6.4 Determining Whether Any Two Nodes in a BN Are Dependent

In each of the three types of connection (serial, divergent, and convergent) we defined the conditions under which pairs of nodes were dependent (formally this was the notion of d-connected). By analyzing paths in a BN in terms of the three types of connection, it is possible to determine in general whether any two nodes are dependent in the sense of being d-connected. Box 7.9 explains how this is done in general. Although you will not need to determine such general dependencies in any practical model building, this notion is important if you wish to understand the details of BN algorithms. Box 7.10 describes the formal notation used in some books to represent the various notions of independence.

Box 7.9 Determining d-Connectedness in Large BNs

The following lists what we have seen from the discussion on the three different types of connections:

1. *Causal and evidential trails*—In a serial connection from B to C via A, evidence from B to C is blocked only when we have hard evidence about A.
2. *Common cause*—In a diverging connection where B and C have the common parent A evidence from B to C is blocked only when we have hard evidence about A.
3. *Common effect*—In a converging connection where A has parents B and C any evidence about A results in evidence transmitted between B and C.

In cases 1 and 2 we say that the nodes B and C are d-separated when there is hard evidence of A. In case 3 B and C are only d-separated when there is no evidence about A. In general two nodes that are not d-separated are said to be d-connected.

These three cases enable us to determine in general whether any two nodes in a given BN are dependent (d-connected) given the evidence entered in the BN. Formally:

Two nodes X and Y in a BN are d-separated if, for all paths between X and Y, there is an intermediate node A for which either:

1. The connection is serial or diverging and the state of A is known for certain; or
2. The connection is converging and neither A (nor any of its descendants) has received any evidence at all.

If X and Y are not d-separated then they are said to be d-connected.

In the BN in Figure 7.19 the evidence entered at nodes B and M represents instantiation, that is, $B = b$ and $M = m$. We start the example by entering evidence at A and seeing how it affects the other nodes in the BN (this is called propagation of effects). We are especially interested in whether it affects node J and G and we can trace the propagation by determining first the d-connections that each node belongs to and the trails through which dependencies can flow.

If evidence is entered at A it affects D and H and K. At B it is blocked by $B = b$.

Since K is part of a converging d-connection triad of nodes with H and I as parents any evidence at M influences K and makes I and H d-connected. Therefore the evidence propagated from A to H influences I.

I is part of a serial connection from E to L, therefore E and L change. E is part of a diverging d-connection triad with C and F, therefore changes at E affect C and F and through to J.

Does node G change? A descendent of node J has been updated, node L, but crucially has not received either hard or uncertain evidence directly itself or on a direct descendant, therefore G and J are d-separated and G remains unchanged. Therefore G is independent of A given $B = b$, $M = m$.

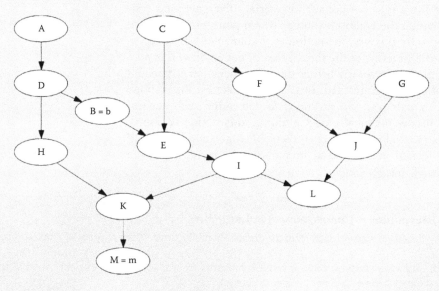

Figure 7.19   Determining whether nodes are d-connected.

Box 7.10   Mathematical Notation for Independence

For compactness, some other books use a special mathematical symbol ⊥. to specify whether variables are conditionally or marginally independent of one another. Specifically:

1. $(X \perp Y) \mid Z = z$ means that X is independent of Y given hard evidence z for Z. For example, in Figure 7.9 and in Figure 7.12 we have $(B \perp C) \mid A=a$ because in the former we have a serial d-connection through A and in the latter we have a diverging d-connection from A.
2. $(X \perp Y) \mid Z$ means that X is independent of Y given *no* evidence in Z. For example, in Figure 7.16 (where there is converging connection on A) we would write $(B \perp C) \mid A$

7.7   Propagation in Bayesian Networks

When there are many variables and links, as in most real-world models, and where the number of states for each variable, is large, calculations become daunting or impossible to do manually. In fact, no computationally efficient solution for BN calculation is known that will work in all cases. This was the reason why, despite the known benefits of BNs over other techniques for reasoning about uncertainty, there was for many years little appetite to use BNs to solve real-world decision and risk problems.

However, dramatic breakthroughs in the late 1980s changed things. Researchers such as Lauritzen, Spiegelhalter, and Pearl published algorithms that provided efficient propagation for a large class of BN models. These algorithms are efficient because they exploit the BN structure, by using a process of *variable elimination*, to carry out modular calculations rather than require calculations on the whole joint probability model.

The most standard algorithm, which is explained informally in this section and formally in Appendix C, is called the *junction tree* algorithm. In this algorithm, a BN is first transformed into an associated tree structure (this is the junction tree) and all of the necessary calculations are then performed on the junction tree using the idea of *message passing*. An example of a BN (the same one we used in Figure 7.7) and its associated junction tree is shown in Figure 7.20. In this case the BN itself was already a tree structure; a more interesting example is shown in Figure 7.21.

Since the first published algorithms there have been many refined versions that use the idea of binary fusion to ensure that the junction tree is binary. In most cases this leads to faster computation times.

The first commercial tool to implement an efficient propagation algorithm was developed in 1992 by Hugin, a Danish company closely associated with Jensen's BN research team at Aalborg University in Denmark. The Hugin tool had a liberating effect, since for the first time it enabled researchers who were not specialists in BNs and statistics to build and run BN models. Since then numerous other BNs have followed, including the state-of-the-art AgenaRisk.

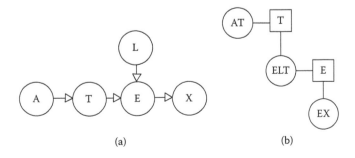

(a)   (b)

Figure 7.20   (a) A BN and (b) its associated junction tree.

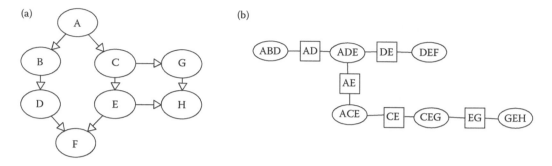

Figure 7.21   (a) A BN and (b) its associated junction tree.

When we introduced variable elimination for the BN of Figure 7.20 we showed (in Box 7.4) that we could marginalize from the inside out and from right to left:

$$P(T) = \left(\sum_A P(A)P(T \mid A) \left(\sum_E \left(\sum_L P(E \mid T, L)P(L) \left(\sum_X P(X \mid E) \right) \right) \right) \right)$$

Recall that we multiplied factors together when performing variable elimination. Here these factors are equivalent to "clusters" of variables, namely, XE, ETL, and AT. In the corresponding junction tree there are round nodes that correspond to each of these *clusters*. The junction tree algorithm is therefore designed to create the tree induced by the variable elimination process where the tree contains clusters (represented as round nodes) and edges connecting the clusters, separated by *separators* (represented as square nodes) that contain those variables that are shared between clusters. Each cluster in the junction tree corresponds to a set of the variables associated with a factor. Each edge in the graph corresponds to a multiplication of factors and connects the clusters together. Shared variables between clusters (factors) form the separators. The edges in the junction tree are not directed.

The algorithm exploits the distributivity property of marginalization (that we discussed in Box 7.4) and allows us to carry out local computations on parts of the tree and "propagate" these calculations to other parts of the tree. So we can get global answers from a BN using local computations. This is neat and is quite ingenious but also pretty complicated.

The variable elimination computation had two interleaving steps: in one step we multiplied some factors together. So here we would multiply clusters together. In the other step we marginalized the results of the previous step to smaller factors, which we later used and this is equivalent to marginalizing to get the separators. Notice that the junction tree contains clusters that are either:

Provided the cluster sizes are not too large the resulting algorithm works very well. Real-world BN models of many hundreds of nodes are more often than not naturally structured in such a way that the cluster sizes are small; hence these propagation algorithms have proven to work on many real-world problems. The methods that we describe in Chapter 8 will help ensure that you only build BN models in which the cluster sizes are small.

- groups of nodes that are parents in a converging *d*-connection, such as *ELT*, or
- part of either serial or diverging connections, such as *AT* and *EX*.

The junction tree algorithm preserves the semantics of d-connectedness discussed earlier since parent nodes in converging d-connections always share membership of at least one common cluster.

For small BNs it is possible to derive the junction tree from the marginalization order by eye but for larger BNs doing so manually is fraught with difficulty. It also does not really matter what variables we wish to marginalize out since the variable elimination order would be much the same and the clusters similar.

As discussed earlier, once the junction tree is constructed we can carry out the local computations on parts of the tree and propagate these calculations to get global answers, to queries asking for the marginal distributions, from a BN. The calculations are performed using the algebra of NPTs for marginalization, division, and multiplication of tables (this is presented in Appendix B).

The way to think about propagation is to visualize the passing of messages in a junction tree. Figure 7.22 shows how messages, in the form of new evidence, are propagated through the junction tree. First NPTs are assigned to clusters to create large tables. Evidence is then *collected* by entering it into the model, updating cluster probability tables, and then sending these updates, as messages, to update the probabilities at the separators. The point of collection is arbitrary: we can select any *root* cluster for this purpose (in Figure 7.22(a) the root cluster is marked with an *X*). The updated separators then send messages to their neighboring clusters, which are then updated in turn, and the process continues until all clusters have received a collect message from one of its neighbors. Once collection has been performed the opposite process is performed and this involves evidence *distribution*. This acts in the opposite direction

The junction tree algorithm requires some knowledge of a branch of mathematics called *graph theory*, which we do not cover in this book, but if you are confident in this area and sufficiently interested in the mechanics of how it works Appendix C provides the detailed explanation.

Large cluster sizes—which make the junction tree algorithm inefficient—are inevitable if there are multiple dependencies between a large set of nodes. For example, in the worst case if there are n nodes in a BN for which each node has a dependency on every other node (as in the BN of Figure 7.6) then there must be a cluster containing all those n nodes. If, for example, $n = 10$ and each of those nodes has six states then there is no avoiding a cluster table with 1 million entries. Consequently, when building BNs it is important to be as sparing as possible in adding dependency links. Complexity can also be managed by decreasing the number of states in key nodes and/ or by introducing synthetic nodes.

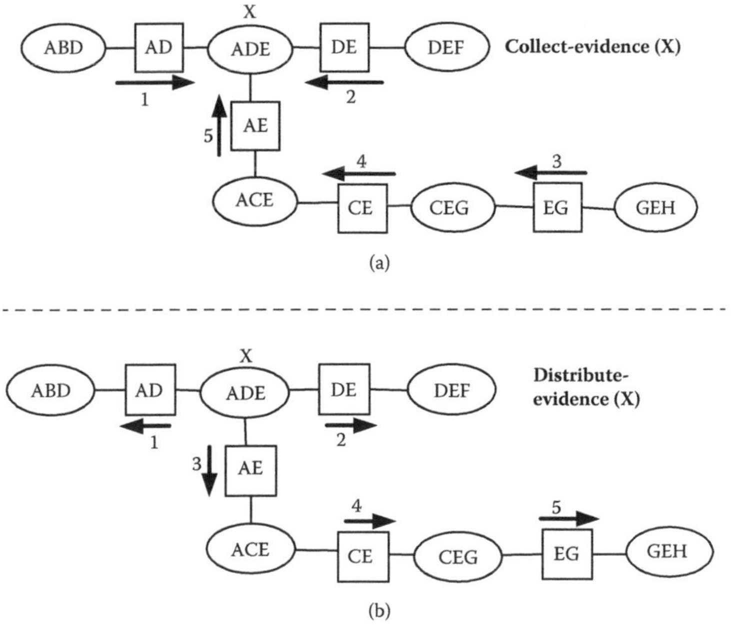

Figure 7.22 Junction tree propagation (for BN of Figure 7.21) using (a) collect and (b) distribute evidence.

and is shown in Figure 7.22(b) where the messages are distributed from the root cluster toward all other clusters. The process terminates once every cluster has been collected from and distributed to, and at this stage the model is consistent in the sense that every cluster has received messages from every other cluster.

In Figure 7.22(a) we might have new evidence entered for node *G* and this updates cluster *GEH* by being multiplied into the cluster. Next we marginalize *EG* from the updated cluster *GEH* and this is said to update the separator *EG*. The new *EG* is then multiplied into cluster *CEG*. The separator *CE* is marginalized from *CEG* and passed as a message to update *CE* and so on. Distribution acts in the same way. Ultimately, at the end of the process the new evidence on node *G* has now been *propagated* through the whole model.

7.8   Steps in Building and Running a BN Model

7.8.1   Building a BN Model

Building a BN to solve a risk assessment problem therefore involves the following steps (much of the rest of this book is concerned with more detailed guidelines on the various steps involved):

1. *Identify the set of variables that are relevant for the problem—* You should always try to keep the number down to the bare minimum necessary. For example, if your interest is in predicting and managing flood risk in a particular city close to a river, then relevant variables might be *River water level, Rainfall, Quality of flood barrier, Flood, Availability of sandbags, Quality of emergency services, People drown,* and *Houses ruined*. Note that your perspective of the problem will determine both the set of variables and the level of granularity. So, if you were a city official trying to assess the general risk to the city then having a single variable *Availability of sandbags* to represent the average availability over all households makes more sense than having a node for each individual household.

> Just as important would be the choices about what variables *not* to model either because they are believed to be irrelevant, uneconomic to address, or require too complex or challenging a model to address at a very granular level. For example, we might decide against modeling every single citizen of the city and their swimming abilities.

2. *Create a node corresponding to each of the variables identified*. This is shown in Figure 7.23. Note that in AgenaRisk you first need to decide if the node is discrete or continuous (continuous nodes are normally defined as simulation nodes—sidebar Figure 7.24 shows how to create discrete and simulation nodes).

3. *Determine the node type*. Depending on whether it is a discrete or continuous simulation node you will have different options for the node types as shown in Figure 7.25. For example: "Flood" could reasonably be a discrete Boolean node; "River Water level" could be a continuous interval simulation node; "Quality of flood barriers" could be a ranked node. The type you choose depends on your perspective (as we shall see later it will also be restricted by issues of complexity).

4. *Specify the states for each non-simulation node*. Having chosen the node type you can specify the names of states and

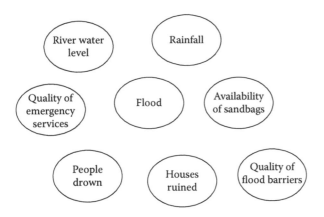

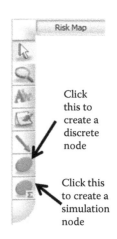

Figure 7.24   Adding a node in AgenaRisk.

Figure 7.23   Create a node for each identified variable.

(for non-Boolean nodes) the number of states. For example, "Quality of flood barriers" could have just three states—high, medium and low—or more as shown in Figure 7.26. Chapters 9 and 10 provide full details about selecting the type of variables and its set of states.

5. *Identify the variables that require direct links.* For example, *River water level* directly affects *Flood* and *Flood* directly affects *Houses ruined* but we do not need a direct link from *River water level* to *Houses ruined* because this link already exists through *Flood.* Earlier sections of this chapter, together with Chapter 8, provide guidance on this step. In AgenaRisk you use the link tool to create the identified links. This is demonstrated in Figure 7.27.

6. *For each node in the BN specify the NPT.* (This is usually the hardest part of the modeling process and that is why much of Chapters 9 and 10 will be devoted to providing detailed guidelines and help on this topic.) Thus, for the NPT for node *People drown* we have to specify *People drown* given *Quality of emergency services* and *Flood.* If all the variables involved have a

(a)   (b)

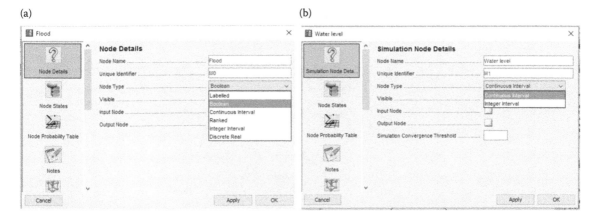

Figure 7.25   Specify the node type and set of states. (a) A Discrete node. (b) A simulation (continuous) node.

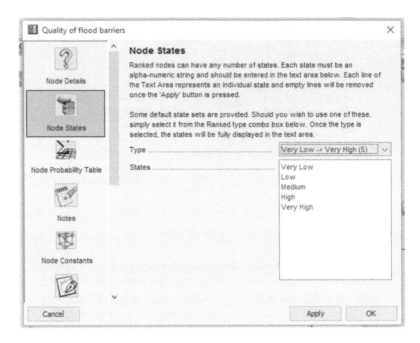

Figure 7.26   Defining node states in AgenaRisk.

finite discrete set of states then the NPT requires us to specify the probability of each state of the node given each combination of states of the parent nodes. Figure 7.28 shows how this is done in AgenaRisk. The case of continuous nodes will be dealt with in Chapter 10.

An obvious and necessary seventh step here would be model validation and testing. This is something that we will return to in later chapters.

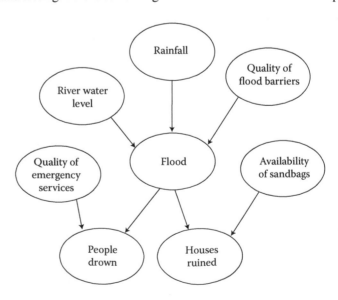

Figure 7.27   Adding relevant arcs in the BN.

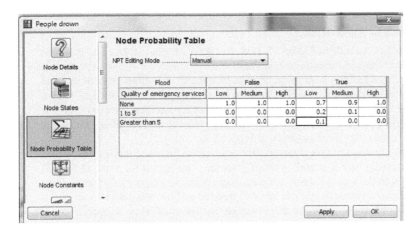

Figure 7.28 Specifying the NPT for the node *People drown*.

7.8.2 Running a BN Model

To execute a model in AgenaRisk you click on the appropriate toolbar button (Figure 7.29). This will result in all the marginal probability values being calculated.

But what we are most interested in is entering observations and recalculating the model to see the updated probability values. Entering an observation (also called evidence) for a variable means that you specify a particular state value for that variable (see Figure 7.30). As discussed in Box 7.5, this is called hard evidence, in contrast to uncertain evidence.

In AgenaRisk uncertain evidence is called soft evidence (even though, as explained in Box 7.6 it as actually virtual evidence that is implemented, not soft evidence). The way soft evidence is entered is explained in Box 7.11.

Figure 7.29 Running a model in AgenaRisk.

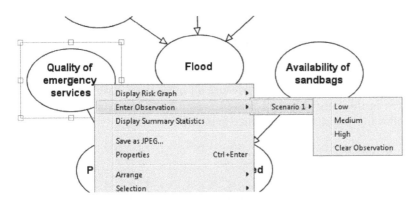

Figure 7.30 Entering evidence on a node.

Box 7.11   Entering Soft Evidence in AgenaRisk

Consider the node *Norman late* in the BN example of Figure 7.3. If we know for certain that Norman is late on a particular day we can enter the value "true" in the node *Norman late*.

But, as discussed in Section 7.6.1, whereas this is an example of hard evidence, we might sometimes wish to enter soft evidence, such as would be the case if we observe Norman is not in his office at 10:00 a.m. but are not certain he is late because there is a small chance he might have popped out for a coffee. We might say that the evidence Norman is late is something like 0.9. To enter this kind of soft evidence in AgenaRisk you simply assign the appropriate values to each state of the node as shown in Figure 7.31.

Any resulting calculations will take account of this evidence. But you need to be aware that—as explained in Box 7.5—the soft evidence implemented here is actually *virtual evidence*. This means the value of the soft evidence you enter does not stay fixed in the way that hard evidence does. When you enter hard evidence like setting *Norman late* to be true, the probability that Norman is late will be fixed as 1 no matter what other observations are entered into the model; but when you enter the soft evidence that the probability Norman is late is 0.9, this value will generally change as the model gets updated.

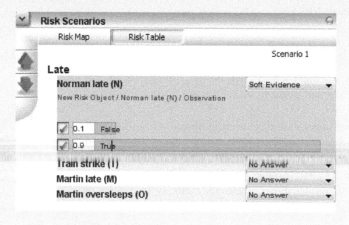

Figure 7.31   Entering soft evidence.

In most BN tools there is an option that determines whether the model is automatically recalculated when you enter new evidence. If the model is large it is always best to have this option switched off, in which case the recalculation will only take place when you press the button.

Once any type of evidence is entered the model needs to be recalculated so that you can see the revised probability values.

7.8.3   Inconsistent Evidence

It is important to understand that sometimes evidence you enter in a BN will be impossible. For example, look at the NPT for the node *People drown* in Figure 7.28. If the node *Flood* is false then, no matter how good or bad the *Quality of the emergency services*, the state of the node *People drown* must be *None* because all other states have a probability zero in this case. There is, however, nothing to stop you entering the following two pieces of evidence in the BN:

- *Flood* is "false."
- *People drown* is "1 to 5."

If you attempt to run the model with these two observations you will (quite rightly) get a message telling you that you entered *inconsistent evidence.* What happens is that the underlying inference algorithm first takes one of the observations, say *Flood* is false and recomputes the other node probability values using Bayesian propagation. At this point the probability of *People drown* being 1 to 5 is calculated as zero. When the algorithm then tries to enter the observation *People drown* is "1 to 5" it therefore detects that this is impossible.

Entering inconsistent evidence will not always be as obvious as in the preceding example. One of the most common confusions when using BNs in practice occurs when entering evidence in large complex BNs. It is often the case that entering particular combinations of evidence will have an extensive ripple effect on nodes throughout the model. In some cases this will result in states of particular nodes having zero probability, in other words they are now impossible. These nodes might not be directly connected to the nodes where evidence was entered. If, in these circumstances, the user subsequently enters evidence that one of the impossible states is "true" the model, when computed, will produce an inconsistent evidence message that may surprise the user (who may wrongly assume that the algorithm has failed). In such circumstances the user also has the tricky task of identifying which particular observation caused the inconsistent evidence.

7.9   Using BNs to Explain Apparent Paradoxes

To provide examples of how BN models are constructed and used we present BN solutions that explain two of the apparently paradoxical problems introduced in Chapter 2: the Monty Hall problem and Simpson's paradox. We will also show how a simple BN model refutes the commonly held belief that "If there is no correlation then there cannot be causation."

7.9.1   Revisiting the Monty Hall Problem

We provide two BN solutions: one with a simple structure and one more complex. They both give the same results but the complex one has a purer causal structure.

7.9.1.1   Simple Solution

The simple model (in its initial state) is shown in Figure 7.32. The NPT for the node *Door shown empty* is given in Table 7.5. Suppose that the contestant picks the red door. Then, as shown in Figure 7.33, at this point each door is still equally likely to win. However Figure 7.34 shows the result of Monty Hall revealing a door (blue in this case). All of the probability that was previously associated with the two doors not chosen by the contestant is loaded into the green door. There is therefore a 2/3 probability that the contestant will win by switching doors.

7.9.1.2   Complex Solution

In this solution we explicitly model the three events *Switch or Stick*, *Doors After Choice*, and *Win Prize* as shown in Figure 7.35.

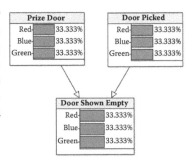

Figure 7.32 Structure of model and its initial state.

Table 7.5

NPT for the Node *Door Shown Empty*

Prize Door	Red			Blue			Green		
Door Picked	Red	Blue	Green	Red	Blue	Green	Red	Blue	Green
Red	0.0	0.0	0.0	0.0	0.5	1.0	0.0	1.0	0.5
Blue	0.5	0.0	1.0	0.0	0.0	0.0	1.0	0.0	0.5
Green	0.5	1.0	0.0	1.0	0.5	0.0	0.0	0.0	0.0

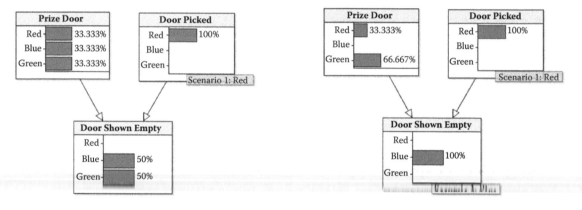

Figure 7.33   Contestant chooses red door.

Figure 7.34   Monty Hall shows blue door empty.

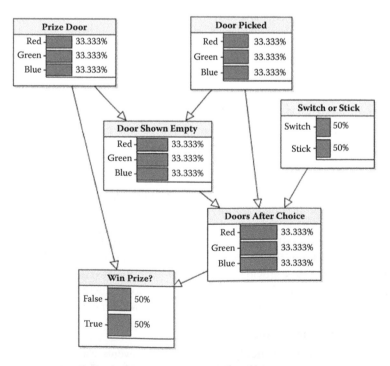

Figure 7.35   Complex model structure and initial state.

Table 7.6
NPT for the Node *Win Prize*

Doors After Choice	Red			Green			Blue		
Prize Door	Red	Green	Blue	Red	Green	Blue	Red	Green	Blue
False	0.0	1.0	1.0	1.0	0.0	1.0	1.0	1.0	0.0
True	1.0	0.0	0.0	0.0	1.0	0.0	0.0	0.0	1.0

The NPT for *Door Shown Empty* is exactly the same as in the simple model (Table 7.5). The NPT for the node *Win Prize* is shown in Table 7.6. The NPT for the node *Door After Choice* is shown in Table 7.7.

Notice that some of the probabilities we have to assign correspond to columns representing *impossible events* (see sidebar), such as the combination "Door picked = red" and "Door shown empty = red." The fact that this combination is impossible is already encoded elsewhere in the model in Table 7.5 for node *Door Shown Empty*.

Suppose that the contestant picks the red door. Then, as shown in Figure 7.36, at this point each door is still equally likely to win, but note that (since we assume switch and stick are equally likely) the model at this stage also reveals that there is 50% chance of winning.

When Monty Hall reveals a door (Figure 7.37) the only thing that changes is that the other door now has a 2/3 chance of being the winning door.

If the contestant switches (Figure 7.38) it is now clear that the chances of winning the prize are 2/3.

Node Probability Table (NPT) Entries for Impossible Events

Although the logical structure of a BN may mean that certain columns of an NPT correspond to impossible events (i.e. combinations of states that can never occur) we still have to assign some values to these columns in the NPT. It turns out that we can assign any probabilities we like for these impossible events, but the standard approach (as adopted in Table 7.7) is to assign equal probability to each column entry (hence the assignment of 1/3's in the table). We will discuss this issue further in Chapter 8.

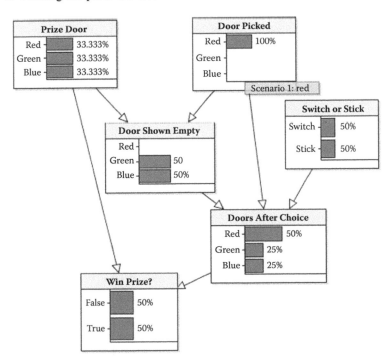

Figure 7.36 Contestant picks red door.

Table 7.7
NPT for the Node *Door after Choice*

Door Picked	Red						Green						Blue					
Door Shown Empty	Red		Green		Blue		Red		Green		Blue		Red		Green		Blue	
Switch or Stick	Switch	Stick	Switch	Stick	Switch	Stick	Switch	Stick	Switch	Stick	Switch	Stick	Switch	Stick	Switch	Stick	Switch	Stick
Red	0.33	0.33	0.0	1.0	0.0	1.0	0.0	0.0	0.33	0.33	1.0	0.0	0.0	0.0	1.0	0.0	0.33	0.33
Green	0.33	0.33	0.0	0.0	1.0	0.0	0.0	1.0	0.33	0.33	0.0	1.0	1.0	0.0	0.0	0.0	0.33	0.33
Blue	0.33	0.33	1.0	0.0	0.0	0.0	1.0	0.0	0.33	0.33	0.0	0.0	0.0	1.0	0.0	1.0	0.33	0.33

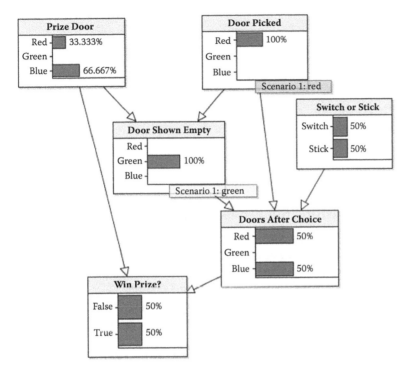

Figure 7.37   Monty Hall reveals an empty door.

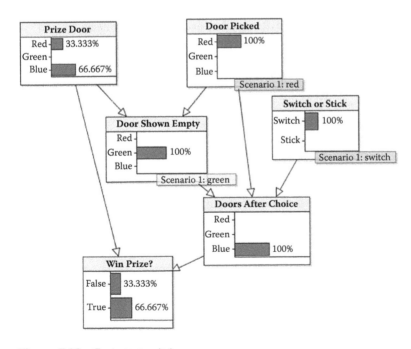

Figure 7.38   Contestant switches.

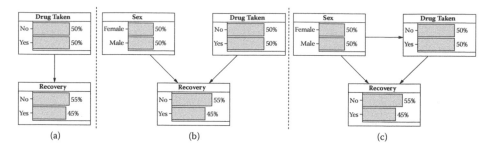

Figure 7.39 All models in initial state.

7.9.2 Revisiting Simpson's Paradox

In Chapter 2 we saw the example of Simpson's paradox in respect of drug trials, whereby the effect of the drug was positive when we considered all subjects, yet was negative when we considered men and women separately. The full data is shown again in Table 7.8.

In Chapter 2 we stated that there were three possible causal models that could be used. These, with their marginal probabilities, are shown in Figure 7.39.

So how are these models constructed and how can we use them to explain and avoid the paradox? We consider each in turn:

- *Model a*—This has only two nodes (*Drug taken* and *Recovery*), and so we need only define two probability tables.
  - *Drug taken*—To determine this table we simply need to know the number of subjects who did and did not take the drugs. From Table 7.8 we can see that 40 took the drug (30 males and 10 females) and 40 did not (10 males and 30 females). So the probability table is simply that shown in Table 7.9.
  - *Recovery | Drug taken*—To determine this table we first calculate (from Table 7.8) the numbers in Table 7.10. This results in the NPT for *Recovery | Drug taken* shown in Table 7.11.
- *Model b*—This has three nodes (*Drug taken*, *Sex*, and *Recovery*), and so we need the three probability tables:
  - *Drug taken*—This is exactly the same as in model *a*.
  - *Sex*—From Table 7.8 there are 40 males and 40 females, so entering these values results in Table 7.12.

Table 7.8

Drug Trial Results with Sex of Patient Included

Sex	Female		Male	
Drug taken	No	Yes	No	Yes
Recovered				
No	21	8	3	12
Yes	9	2	7	18

Table 7.9

NPT for Drug Taken

No	0.5
Yes	0.5

It is useful to note that, in AgenaRisk, you can simply enter absolute values (in this case 40 and 40) into an NPT and it will automatically scale the numbers in each column to sum to 1.

Table 7.10

Numbers for Recovery| Drug Taken

Drug taken	No	Yes
Recovered		
No	24	20
Yes	16	20

Table 7.11

NPT for Recovery|Drug Taken

Drug taken	No	Yes
No	0.6	0.5
Yes	0.4	0.5

Table 7.12

NPT for Sex

Female	0.5
Male	0.5

- *Recovery | Drug taken, Sex*—For this table we simply enter the exact same numbers as in Table 7.8, which results in Table 7.13.
- *Model c*—This is the same as model b, but with the additional arc from *Sex* to *Drug Taken*. The only change therefore is to the node *Drug taken*. From Table 7.8 we get Table 7.14. Entering these values into the NPT for *Drug taken* results in Table 7.15.

Having constructed all of the models, when you run them without observations they all produce the same recovery rate of 45% (Figure 7.39).

However, compare Figure 7.40 and Figure 7.41 where, respectively, we have entered as an observation in all models *Drug taken* as yes and no. Models a and c both produce the same predictions for recovery rate (50% and 40%, respectively in Figures 7.40 and 7.41) and so both seem to agree that the drug is effective since the recovery rate is higher when it is taken. But model b produces the opposite prediction: a 40% recovery rate when the drug is taken and a 50% recovery rate when it is not. So which model is correct?

The key clue is to note that in model c the observation changes the belief about Sex. Specifically, there is a 75% chance that sex is male once we know the drug is taken and a 75% chance that sex is female once we know the drug is not taken.

So, let us see what happens when we fix the sex. Figure 7.42 and Figure 7.43 show the results of entering Sex = Male in the respective cases of drug taken and drug not taken.

Both models b and c agree in both cases: the recovery rate is 60% when the drug is taken and 70% when it is not. So, for males, the drug is not effective. But, similarly we can enter the observation Sex = Female as shown in Figure 7.44 and Figure 7.45. Again, models b and c agree that the drug is not effective because the recovery rate is 20% when taken and 30% when not taken.

Since model a is clearly misleading (as it ignores the crucial impact of sex) we can discount it. But which model, b or c, is correct? The answer is that ideally model b ought to be correct, because ideally in such trials the drug taken should not be dependent on sex. In fact, if this

Table 7.13
NPT for Recovery|Drug Taken, Sex

Sex	Female		Male	
Drug taken	No	Yes	No	Yes
No	0.7	0.8	0.3	0.4
Yes	0.3	0.2	0.7	0.6

Table 7.14
Numbers for Drug Taken/Sex

Sex	Female	Male
Drug taken		
No	30	10
Yes	10	30

Table 7.15
NPT for Drug Taken/Sex

Sex	Female	Male
No	0.75	0.25
Yes	0.25	0.75

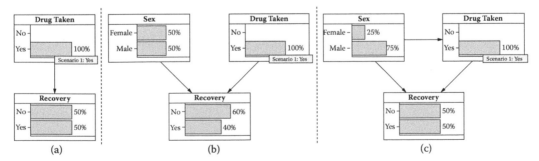

Figure 7.40 Drug taken = yes.

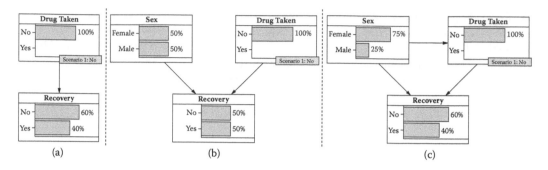

Figure 7.41   Drug taken = no.

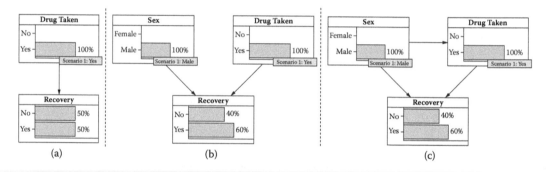

Figure 7.42   Drug taken – yes and Sex – male.

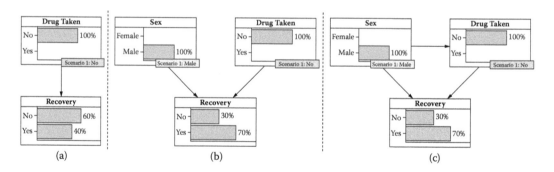

Figure 7.43   Drug taken = no and Sex = male.

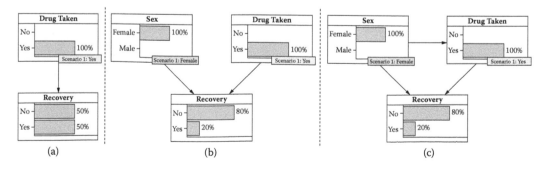

Figure 7.44   Drug taken = and Sex = female.

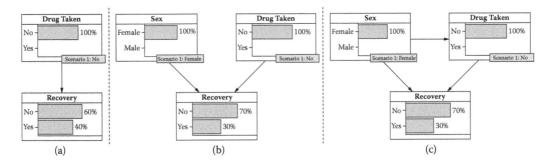

Figure 7.45 Drug taken = no and Sex = female.

had been the case as assumed by model b (i.e. if the same number of males as females had been given the drug), then we would not have had to go further than entering the observations for drug taken (yes and no) to discover that the drug was not effective, in contradiction to model a. However, because in this particular trial there really was a dependence of drug taken on sex, model c is the correct model in this case. However, in situations like this, if we want to use the BN model to make inferences about the effect of an intervention (in this case "taking the drug") then we need to perform some temporary "surgery" on the model as explained in the Section 7.10.

7.9.3 Refuting the Assertion "If There Is No Correlation Then There Cannot be Causation"

One of the key lessons in Chapter 2 was that "correlation implies causation" is false. While this is widely known and accepted, many people do not realise that the converse implication

"Causation implies correlation"

is also false.

Indeed that, and the following equivalent assertions, are often wrongly assumed to be true:

"If there is no association (correlation) then there cannot be causation."
"If there is causation there must be association (correlation)."

We can disprove these (equivalent) assertions with a simple counter-example using two Boolean variables a, and b, that is whose states are true or false. We do this by introducing a third, latent, unobserved Boolean variable c. Specifically, we define the relationship between a,b and c via the Bayesian network in Figure 7.46.

By definition b is completely causally dependent on a. This is because, when c is true the state of b will be the same as the state of a, and when c is false the state of b will be the opposite of the state of a.

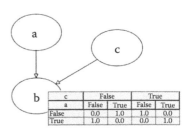

c	False		True	
a	False	True	False	True
False	0.0	1.0	1.0	0.0
True	1.0	0.0	0.0	1.0

Figure 7.46 BN that disproves the assertion "causation implies correlation."

However, suppose—as in many real-world situations—that c is both hidden and unobserved (i.e. a typical confounding variable as described in Chapter 2). Also, assume that the priors for the variables a and c are uniform (i.e. 50% of the time they are false and 50% of the time they are true).

Then when a is false there is a 50% chance b is false and a 50% chance b is true. Similarly, when a is true there is a 50% chance b is false and a 50% chance b is true. In other words, what we actually observe is zero association (correlation) despite the underlying mechanism being completely (causally) deterministic.

7.10 Modelling Interventions and Counterfactual Reasoning in BNs

7.10.1 Interventions

While the model (c) in Figures 7.40–7.45 is the one that most accurately reflects all the data available, it actually gives the "wrong" result to the following basic question:

Is it beneficial to take the drug?

This is because there is a crucial difference between **entering an observation** (which is what we did in Figure 7.41(c)) and **performing an intervention** (which is what we need to do to answer the question). When we enter an observation that the drug is taken, the model takes account of the dependence on sex of the subjects in the study. It determines that, having observed that the drug was taken, it is far more likely that the subject was male—the probability the sex is male increases from 50% to 75% after the observation is entered. As a result the probability of recovery *increases* from 45% to 50%. While the result correctly reflects the overall impact of taking the drug on the study's full cohort of subjects, it is not the correct result to the intervention question.

Clearly, if a person takes the drug it does not change their sex. So, if we want to determine the effect of taking the drug as an *intervention* "Sex" should not be updated. Hence, we have to break the link from "Sex" to "Drug Taken" in the model. In other words we need to convert from the model (c) to model (b) in order to model the effect of an intervention. In model (b) entering "Drug taken" = true is equivalent to modelling the intervention, since "Sex" is unchanged. In this case, the probability of recovery *decreases* from 45% to 40%.

To differentiate between an observation and an intervention in a BN model Pearl introduced the "do-calculus." While this calculus is defined in mathematical terms, it can be summarised as performing the following:

If you wish to model an intervention on a node X in AgenaRisk then you should make a copy of the model and perform the "graph surgery" (i.e. remove all arcs entering node X) in the copy. This way you will not "lose" the NPT information for node X in the original model.

Graph surgery

To model an intervention on node X, remove all arcs entering X and enter the required "intervention" as an observation on X.

7.10.2 Counterfactuals

Continuing with the simple drug trial example, suppose we know that a person who took the drug did not recover. Without knowing the sex of the person, what is the probability that the person would have recovered *if they had not taken* the drug? This is an example of what is referred to as a counterfactual question. We are asking a question about a hypothetical alternative world—would the outcome have been different if we had made a different decision/intervention? Contrary to much popular belief, BN models can provide the answer to such counterfactual questions but, once again, they require some simple adjustments to the BN model.

First of all, note that since we are dealing with interventions here (actually taking or not taking the drug) the "correct" model structure is that of Figure 7.39b. What we *know* is that the person took the drug and did not recover. This is represented in Figure 7.47.

Crucially, from this information we know that the person is twice as likely to have been a female as a male (66.67% compared to 33.33%). To answer the counterfactual question we must incorporate this "new" information. One way to do this is to create a "copy" of the model in which the prior probability for sex is 66.67% female and 33.33% male as shown in Figure 7.48(a). The counterfactual question can now be answered by

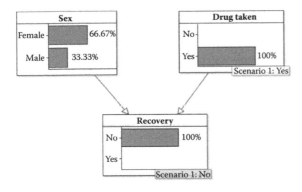

Figure 7.47 We know the person took the drug and did not recover.

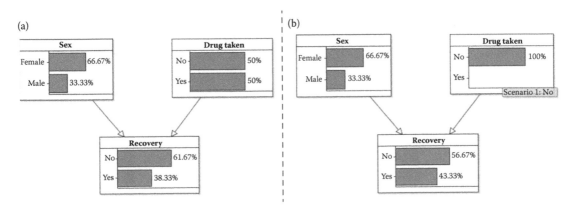

Figure 7.48 Counterfactual model. In (a) the new prior marginal probabilities are shown. In (b) we see the result of not taking the drug.

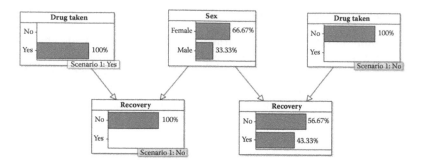

Figure 7.49 Twin network version of the counterfactual argument.

In general, the "twin network" solution involves copying the nodes which have parents and leaving the background nodes (those without parents) shared over the two networks.

observing the intervention "Drug taken" = No as shown in Figure 7.48(b) in this new version of the model. The result shows there is a 43.33% probability the person would have recovered if they had not taken the drug.

Instead of creating a copy of the model that inherits the updated prior probabilities for sex we could also extend the original model with copies of the nodes "Drug take" and "Recovery" as shown in Figure 7.49. Pearl refers to this as the "twin network representation" for counterfactual arguments.

Box 7.12 provides another classic example of a counterfactual problem solved with a BN using the twin networks method.

Box 7.12   Counterfactual Solution to Driving Route Problem

A road commuter has two route options for getting home from work. Route 1 is generally much faster, but there is a slightly higher chance of adverse traffic conditions on this route compared with Route 2. When there are adverse driving conditions, Route 2 is faster. Sometimes adverse driving conditions (such as when there is a rail strike or bad weather) affect both routes.

For simplicity, we categorise time taken for the journey as low, medium and high. One day the driver takes Route 1 and the journey time is high. What is the probability that the journey time would still have been high if he had taken Route 2?

The basic BN model and prior marginal probabilities are shown in Figure 7.50.

The NPT for the node "Time taken" is shown in Table 7.16.

Using the "twin network" approach we get the solution shown in Figure 7.51. There is only a 13% chance the time taken would have been high if the driver had taken Route 2.

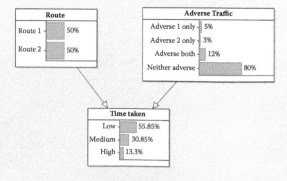

Figure 7.50 Basic BN model with prior marginal probabilities.

Table 7.16
NPT for node "Time Taken"

Route	Route 1				Route 2			
Adverse Traffic	Adverse 1 only	Adverse 2 only	Adverse both	Neither Adverse	Adverse 1 only	Adverse 2 only	Adverse both	Neither Adverse
Low	0.3	0.7	0.3	0.7	0.5	0.4	0.4	0.5
Medium	0.3	0.2	0.3	0.2	0.4	0.4	0.4	0.4
High	0.4	0.1	0.4	0.1	0.1	0.2	0.2	0.1

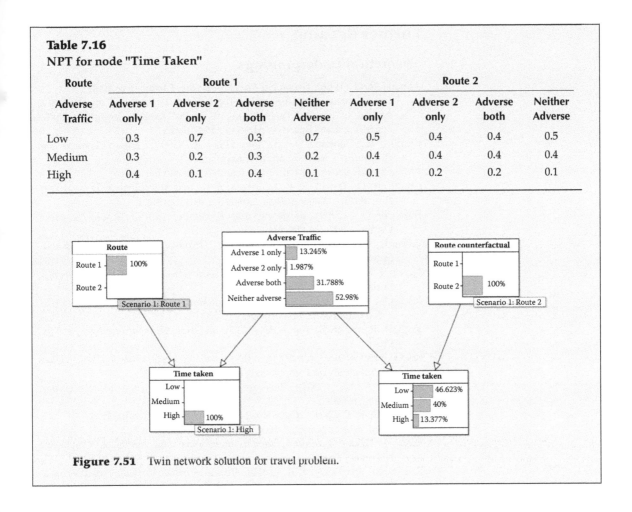

Figure 7.51 Twin network solution for travel problem.

7.11 Summary

In this chapter we have shown how Bayes' theorem, and the other axioms of probability theory, can be represented graphically using Bayesian networks and automatically calculated using the junction tree algorithm. The benefits of BNs are manyfold and allow us to be explicit in our reasoning, reason from cause to effect and vice versa, and also compute answers for large models efficiently.

We have also covered the atomic fragments of BNs, in the form of *d*-connection types. These support modeling multiple causes, multiple consequences, and explaining away, an operation that is central to human reasoning about uncertainty.

We have used BNs to model two well-known probability paradoxes—the Monty Hall game show and Simpson's paradox—and provided some insight into how different BN structures provide a variety of insights into understanding these paradoxes. Also, we have described how to use AgenaRisk to start modeling BNs.

For the motivated reader we have included a sizeable Appendix C covering how the junction tree algorithm operates and have done so in two parts: junction tree construction and evidence propagation. This is advanced material but we show how the roots of the algorithm can be traced to factorization used in variable elimination during the marginalization process.

Further Reading

Theoretical Underpinnings

Almond, R. G. (1996). *Graphical Belief Modeling*, Chapman & Hall.
> Covers many of the issues but the focus is on applying an alternative method of uncertainty reasoning to Bayes, namely, Dempster–Shafer belief functions. Based on Almond's PhD thesis.

Castillo, E., Gutierrez, J. M., and Hadi, A. (1997). *Expert Systems and Probabilistic Network Models*, Springer.
> Extensive coverage but very research based and theoretical.

Cowell, R. G., Dawid, A. P., Lauritzen, S. L., and Spiegelhalter, D. J. (1999). *Probabilistic Networks and Expert Systems*, Springer.

Darwiche, A. (2009). *Modeling and Reasoning with Bayesian Networks*, Cambridge University Press.

Edwards, D. (2000). *Introduction to Graphical Modelling*, 2nd ed., Springer-Verlag.

Jensen, F. V. and Nielsen, T. (2007). *Bayesian Networks and Decision Graphs*, Springer-Verlag.

Jordan, M. I. (1999). *Learning in Graphical Models*, MIT Press.
> An edited collection of papers.

Korb, K. B. and Nicholson, A. E. (2010). *Bayesian Artificial Intelligence*. CRC Press.

Koski, T. and Noble, J. (2009). *Bayesian Networks: An Introduction. Wiley Series in Probability and Statistics.* Wiley.

Madsen, A. L. (2007). *Bayesian Networks and Influence Diagrams,* Springer-Verlag.

Neapolitan, R. E. (2004). *Learning Bayesian Networks*, Prentice Hall.
> Focuses on learning algorithms.

Pearl, J. (2000). *Causality: Models Reasoning and Inference*, Cambridge University Press.

Scutari, M. and Denis, J.-B. (2014). *Bayesian networks: with examples in R.* CRC Press.

BN Applications

Mittal, A. and Kassim, A. (2007). *Bayesian Network Technologies: Applications and Graphical Models*, IGI.
> This is an edited collection of papers.

Pourret, O., Naim, P., and Marcot, B., eds. (2008). *Bayesian Networks: A Practical Guide to Applications* (Statistics in Practice), Wiley.
> This is an edited collection of papers.

Taroni, F., Aitken, C., Garbolino, P., and Biedermann, A. (2006). *Bayesian Networks and Probabilistic Inference in Forensic Science*, Wiley.

Nature and Theory of Causality

Glymour, C. (2001). *The Mind's Arrows: Bayes Nets and Graphical Causal Models in Psychology*, MIT Press.
> This book is essentially a theory of causal reasoning in humans.

Sloman, S. (2005). *Causal Models: How People Think about the World and Its Alternatives*, Oxford University Press.

Uncertain Evidence (Soft and Virtual)

Chan, H. and Darwiche, A. (2005). On the revision of probabilistic beliefs using uncertain evidence. *Artif. Intell.* 163, 67–90.

Fenton, N. et al. (2016). How to model mutually exclusive events based on independent causal pathways in Bayesian network models. *Knowledge-Based Syst.* 113, 39–50.

Defining the Structure of Bayesian Networks

8.1 Introduction

We have found that experts, in a range of application domains, often experience common difficulties and apply common strategies when building Bayesian network (BN) models. The aim of this chapter is to help you avoid the common problems and benefit from the common solutions. One particular difficulty experienced concerns causality and the decisions about direction of edges in a BN; this is described in Section 8.2. In Section 8.3 we focus on exploiting the very similar types of reasoning that experts apply in very different application domains. We have identified a small number of natural and reusable patterns in reasoning to help when building BNs. We call these patterns *idioms*. In Section 8.4 we discuss another aspect of BN modeling, called *asymmetry*, which presents some difficulty. Typically this is the result of conditional causality, where the existence of a variable in the model depends on a state of some other. We look at the most common types of asymmetry and propose pragmatic solutions in each case.

Although idioms are a powerful method for helping to model the structure of a risk assessment problem as an individual BN, we need a different method to help us build models that "scale-up" to address large-scale problems. The most natural way to build realistic large-scale BN models is to link individual models using ideas that originate from object-oriented design. These multiobject BN models are described in Section 8.5.

Finally in Section 8.6 we discuss the missing variable fallacy. This occurs when we neglect to include some crucial variable in the model and can only really reason about its absence by interrogating the model results, in terms of predictions and inferences. The key argument here concerns visibility of assumptions in the model.

Chapter 7 provided an overview of the process involved in building a BN and also provided some important theoretical guidelines to help you build the correct structural relationship between variables in a BN. The objective of this chapter is to provide more practical guidelines to help you build a suitable BN structure.

All models in this chapter are available to run in AgenaRisk, downloadable from www.agenarisk.com.

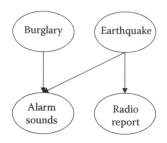

Figure 8.1 A BN model for the earthquake example.

A real difficulty people encounter when building a BN is that, because their thinking is goal oriented, they draw arcs that denote the direction of inference, that is, what they want to know, rather than those that reflect causality. This is a common challenge and real-life problems are rarely as small as this example. We have to provide some answers to the challenge of scaling up what we can learn from small, often fictitious examples, to real world prediction problems.

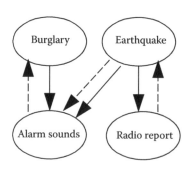

Figure 8.2 A BN model for the earthquake example supplemented with arrows denoting direction of inference.

8.2 Causal Inference and Choosing the Correct Edge Direction

Much of the literature on BNs uses very simple examples to show how to build sensible graphs for particular problems. The standard texts on BNs by Pearl and then Jensen, use examples where Mr. Holmes and Dr. Watson are involved in a series of episodes where they wish to infer the probability of icy roads, burglary and earthquakes from uncertain evidence. One such example is as follows (we have italicized those sections that denote the logical direction of deductions made by Mr. Holmes upon hearing new evidence):

> Mr. Holmes is working at his office when he receives a telephone call from Watson who tells him that Holmes' burglar alarm has gone off. Convinced that a burglar has broken into his house (*alarm sounds → burglary*), Holmes rushes into his car and heads for home. On his way he listens to the radio, and in the news it is reported that there has been a small earthquake in the area (*radio report → earthquake*). Knowing that the earthquake has a tendency to turn the burglar alarm on (*earthquake → alarm sounds*), he returns to his work leaving his neighbors the pleasures of the noise.

Figure 8.1 shows the BN for this example (the nodes here are all Boolean; their states are either true or false). There are two key points to note here:

1. *The example is small enough that the causal directions of the edges are obvious.* A burglary causes the alarm to sound; the earthquake causes the radio station to issue a news report and also causes the alarm to sound.
2. *The actual inferences made can run counter to the causal edge directions.* From the alarm sounding Holmes inferred that a burglary had taken place and from the radio sounding he inferred that an earthquake had occurred. Only when explaining away the burglary hypothesis did Holmes reason along the edge from earthquake to alarm. The directions of deductions are shown in Figure 8.2 as dotted arrows.

Although we propose that edge directions should always be in the direction of cause to effect rather than in the direction implied by the deductions we might wish to make, it is important to note that taking the latter approach does not necessarily lead to an invalid BN. For example, there is nothing to stop us from revising the simple model in Chapter 7 by reversing the direction of the edge as shown in Figure 8.3. The difference is that now:

■ Instead of having to specify an unconditional probability table for node *T*, the probability table we have to specify for *T* is conditioned on *N*.

- We have to specify an unconditional probability table for node *N*, instead of having to specify a probability table for *N* that is conditioned on node *T*.

Mathematically there is actually no reason to choose the original model over the revised one; they are equivalent (this is explained in Box 8.1). However, in practical situations mathematical equivalence is not the sole criterion:

1. If there is a clear cause and effect then it is more natural to use the direction from cause to effect. The train strike causes Norman to be late, not vice versa.
2. In some cases, however, it may make sense to use the direction from effect to cause, simply because the necessary prior probabilities are easier to elicit. For example, although it is generally accepted that smoking may cause cancer (and that cancer does not cause smoking), it might be easier to elicit a prior for smoking given cancer than for cancer given smoking. In the case of the former we can look at records of cancer sufferers and count the proportions who were smokers. In contrast, if we sample smokers, we will not be able to count those who may yet get cancer.
3. There may be other variables and dependencies between them and in such cases the structure of these dependencies will determine which arc directions are feasible.

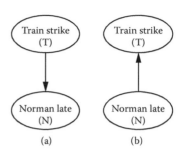

Figure 8.3 Edge direction reversed from Chapter 7 model. (a) Cause to effect. (b) Effect to cause.

Box 8.1 Why Cause to Effect and Effect to Cause Are Mathematically Equivalent

Figure 8.3 shows the two types of model we seek to compare. In Chapter 7 we specified the tables $P(T)$ and $P(N \mid T)$ that were needed for the cause-to-effect model. To specify the probability tables required for the effect-to-cause model (i.e., the tables $P(N)$ and $P(T \mid N)$ rather than $P(T)$ and $P(N \mid T)$ that we used in the original model) we simply use the outputs of the previous Bayesian calculations. Specifically, we computed the marginal probability table for node *N* to be that shown in Table 8.1.

The new table $P(T \mid N)$ is as shown in Table 8.2. We did one of the calculations here, namely, the probability that *Train strike* is true given *Norman is late*, simply applying Bayes' theorem. By Theorem 5.1 the probability that *Train strike* is false given that *Norman late* is false is simply one minus that value; the probability *Train strike* is true given *Norman is not late* is calculated by simply applying Bayes' theorem with the different evidence.

When we run the effect-to-cause model (comparing it directly to the cause-to-effect model) we get the results shown in Figure 8.4.

So the marginal probabilities are identical in both cases. Moreover, if we enter an observation, say *Norman late* is true, we still get results that are identical between the models as shown in Figure 8.5.

Table 8.1
Probability Table for "Norman Late"

False	0.83
True	0.17

Table 8.2
Probability Table for "Train Strike" Given "Norman Late"

Norman Late	False	True
False	0.9759	0.52941
True	0.0241	0.47059

This demonstrates that, using Bayes' theorem, the two models are equivalent even though the edge directions are different.

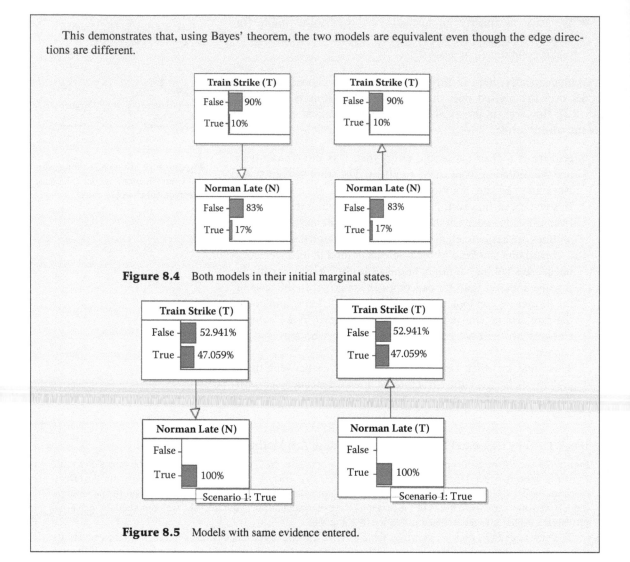

Figure 8.4 Both models in their initial marginal states.

Figure 8.5 Models with same evidence entered.

8.3   The Idioms

An *idiom* is defined in *Webster's Dictionary* (1913) as: "The syntactical or structural form peculiar to any language; the genius or cast of a language." We use the term idiom to refer to specific BN fragments that represent very generic types of uncertain reasoning. For idioms we are interested only in the graphical structure and not in any underlying probabilities (we focus on defining the probabilities in Chapters 9 and 10). For this reason an idiom is not a BN as such but simply the graphical part of one. We have found that using idioms speeds up the BN development process and leads to better quality BNs. We have identified four especially common idioms:

The formal texts on BNs recommend that domain experts should build BNs by explicitly examining different possible *d*-connection structures between nodes (as described in Chapter 7) under different evidence scenarios. This is extremely difficult for most people to do; fortunately, the *d*-connection properties required for particular types of reasoning are preserved by the idioms and emerge through their use.

- *Cause consequence idiom*—Models the uncertainty of a causal process with observable consequences.
- *Measurement idiom*—Models the uncertainty about the accuracy of any type of measurement.

- *Definitional/synthesis idiom*—Models the synthesis or combination of many nodes into one node for the purpose of organizing the BN. Also models the deterministic or uncertain definitions between variables.
- *Induction idiom*—Models the uncertainty related to inductive reasoning based on populations of similar or exchangeable members.

Domain knowledge engineers find it easier to use idioms to help construct their BNs rather than following textbook examples or by using the approach recommended by the more formal texts on BNs.

Also, because each idiom is suited to model particular types of reasoning, it is easier to compartmentalize the BN construction process.

In the remainder of this section we define these idioms in detail. Idioms act as a library of patterns for the BN development process. Experts simply compare their current problem, as described, with the idioms and reuse the appropriate idiom for the job. By reusing the idioms we gain the advantage of being able to identify model components that should be more cohesive and self-contained than those that have been created without any underlying method. Also, the use of idioms encourages reuse and is more productive.

In what follows we distinguish between the notion of idiom and idiom instantiation:

An idiom *instantiation* is an idiom made concrete for a particular problem, by using meaningful node labels.

Like idioms generally, idiom instantiations represent only the graphical part of a BN, that is, they do not require the probability tables.

8.3.1   The Cause–Consequence Idiom

The cause–consequence idiom is used to model a causal process in terms of the relationship between its causes (those events or facts that are inputs to the process) and consequences (those events or factors that are outputs of the process). The basic form of the idiom is shown in Figure 8.6(a), with a simple instantiation in Figure 8.6(b). The direction of the arrow indicates causal direction.

We can identify causal connections by the sequences in which they occur in time: the consequence always follows the cause (so an accident today can cause us to be in hospital tomorrow, but it makes no sense to say that being

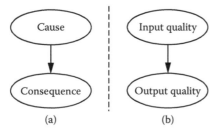

(a)   (b)

Figure 8.6   Cause–consequence idiom (a) with instantiation (b).

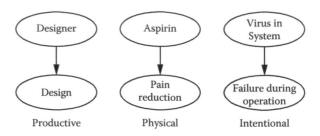

Figure 8.7   More cause–consequence example instantiations.

Neither machines nor statistical models can determine causality because they have no sense of time.

in hospital tomorrow caused the accident today). The task of identifying causal connections often boils down to common sense (see Box 8.2 for a comprehensive example of good and bad uses of causal connections).

Generally there will be a *productive, physical,* or *intentional* relationship between cause and consequence as highlighted by the examples in Figure 8.7. Thus:

■ When the designer (cause) produces a design (consequence) the relationship is productive. Other examples of productive relationships are where we transform an existing input into a changed version of that input (for example, a product before and after testing) or by taking an input to produce a new output (for example, a specification of a product is used to produce the actual product).
■ When the aspirin (cause) leads to a reduction in pain (consequence) the relationship is physical.
■ When a virus inserted by a "hacker" in a computer system (cause) triggers a failure during the operation of the system (consequence) then the relationship is intentional.

A common example of poor BN modeling is to add a node that explicitly represents a process. Using the cause–consequence idiom helps avoid this.

It is important to note that in the cause–consequence idiom the underlying causal process itself is not represented, as a node, in the BN. So, in the examples shown in Figure 8.7, the design process, the process of taking the aspirin, and the process of executing the system are, respectively, not represented as nodes. It is not necessary to do so since the role of the underlying causal process, in the BN model, is represented by the conditional probability table connecting the cause to the consequence. This information tells us everything we need to know (at least probabilistically) about the uncertain relationship between causes and consequences.

Also note that the key issue in representing the parent–child relationship is the chronological ordering of the variables: if one precedes or is contemporaneous with the other, then it is a candidate causal influence on the other and could be represented as a parent node in the BN idiom.

Clearly the examples are very simplistic. Most interesting phenomena will involve many contributory causes and many effects. Joining a number of cause–consequence idioms together can create more realistic models, where the idiom instantiations have a shared output or input node.

Example 8.1

A simple instantiation of two cause–consequence idioms, joined by the common node "failures," is shown in Figure 8.8. Here we are predicting the frequency of software failures based on knowledge about problem difficulty and supplier quality.

This process involves a software supplier producing a product. A good quality supplier will be more likely to produce a failure-free piece of software than a poor quality supplier. However, the more difficult the problem to be solved the more likely it is that faults may be introduced and the software fail.

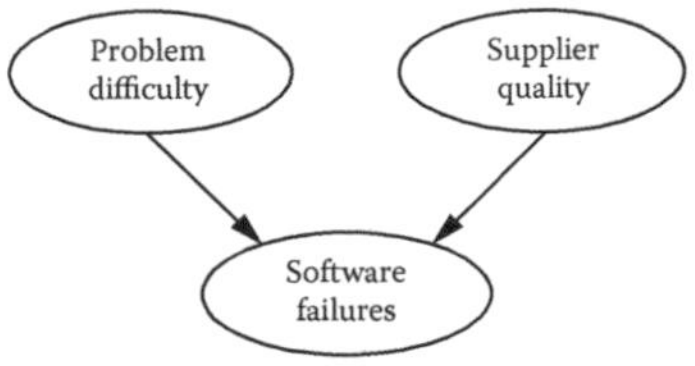

Figure 8.8 Two cause–consequence idiom instantiations joined (software failures).

The models, shown in Figure 8.9, illustrate how cause-consequence idiom instances can be combined when reasoning about risk. This is precisely the generic model (which can also be used for opportunities as well as risks) that we introduced in Chapter 3. In this generic risk modeling, there may be many causes and consequences. There may also be many (or no) controls and many (or no) mitigants. Some mitigants can influence more than one consequence, as in the example of Figure 8.10. So we can see that complex models can be developed very quickly using the cause–consequence idiom.

Note that we can apply the cause–consequence idiom to structure the model regardless of whether we expect to use the model for predictive or diagnostic purposes. The model structure is invariant in both cases; only the direction we may choose to make inferences might differ.

8.3.2 Measurement Idiom

Any node in a BN can represent an uncertain variable, but in many situations there is additional uncertainty arising from the way that we observe or measure the variable. For example, the temperature outside can vary but if, additionally when we measure it, we do not have an accurate thermometer then the temperature that we observe may well be different to the actual temperature. The measurement idiom, shown in Figure 8.11, captures this situation. The edge directions here can be interpreted in a straightforward way. The actual value of the attribute must exist before it is measured by a measurement instrument (person or machine) with a known assessment accuracy,

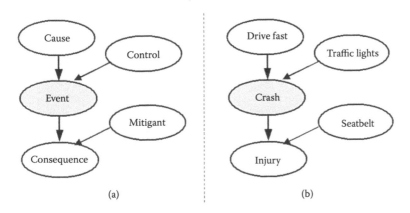

(a)   (b)

Figure 8.9 Risk event modeling as instantiation of joined cause–consequence idiom.

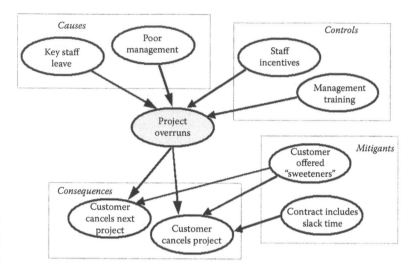

Figure 8.10 Example of cause–consequence idiom instance joined showing multiple causes, consequences, controls, and mitigants.

A key consideration in any measurement process is whether we can actually observe with complete accuracy the thing we wish to measure. In many cases it may be uneconomic, difficult, or even dangerous to observe the thing we wish to measure directly. For example, we could measure the temperature by the expansion of a metal bar but this might not provide timely or convenient results or may be impossible in some environments. Instead we might use a mercury thermometer but again this will have limited accuracy under specific circumstances. Similarly, the absolute safety of a thing or situation may not be directly known but can only be estimated by some process of measurement. The general lesson is that the measurement instrument is estimating something that is not directly observed or completely known, and the measurement idiom clarifies the issues that may affect the accuracy of the inference about the true (but unknown or latent) value from the measured value.

the result of which is an assessed value of the attribute. The higher the assessment accuracy the closer we would expect the assessed value of the attribute to be to the actual value of the attribute. Within the node *Assessment accuracy* we could model different types of inaccuracies including expectation biases and over- and underconfidence biases.

A classic instantiation of the measurement idiom is the testing example shown in Figure 8.12. When we are testing a product to find defects, we use the number of discovered defects as a surrogate for the true measure that we want, namely the number of inserted defects. In fact the measured number is dependent on the node *Testing accuracy*. Positive and encouraging test results could be explained by a combination of two things:

- Low number of inserted defects resulting in low number of discovered defects
- Very poor quality testing resulting in low number of defects detected during testing

Inference in the measurement idiom is diagnostic from effect to cause since our evidence is in the form of some observation and from this we update our estimate on the true, but unknown, value.

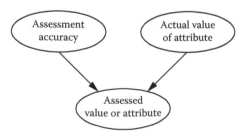

Figure 8.11 Measurement idiom.

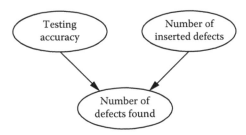

Figure 8.12 Measurement idiom instantiation (testing).

By using the measurement idiom we can therefore explain away (as discussed in Chapter 7) false-positive results, simply because the competing cause—the accuracy node—is explicitly considered.

In two classes of problems the measurement idiom is implemented without the "Assessment accuracy" node, by encoding the "inaccuracy" implicitly into the "Assessed value or attribute" node:

> *Case 1: When the Inaccuracy Is Fixed and Known.* This is the case, for example, where we have a diagnostic test for a specific disease that has known false-positive and false-negative rates (as discussed in Chapter 4). An example is shown in Figure 8.13 in which the standard test is known to have a 1% false-positive rate and 5% false-negative rate. In (a) the full version of the measurement idiom is implemented where *Test accuracy* is classified as either standard or perfect. In (b) we remove the "Test accuracy" node under the assumption that only the standard test is ever applied.
>
> *Case 2 (Indicator Nodes): When It Is Accepted That the Actual Value of the Attribute Cannot be Measured Directly, and Hence That Only Indirect Measures are Possible.* We call

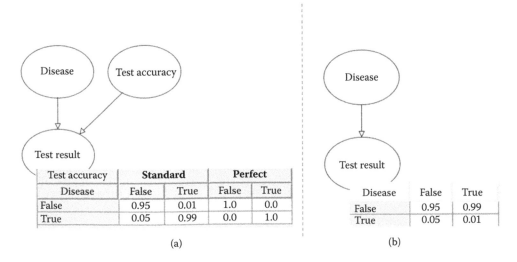

Figure 8.13 Standard measurement idiom (a) and implicit measurement idiom (b).

Figure 8.14 Measurement using indicators.

such indirect measures *indicators*, because it is accepted that they at best provide an indication of the actual attribute, which may itself be latent, difficult to observe or measure. For example, suppose we wish to measure something fairly abstract, like the quality of a manufacturing process. Then we would look for measurable indicators such as defects found, staff turnover and customer satisfaction. The appropriate BN model is then as shown in Figure 8.14. There is clearly a relationship between process quality and the indicators, and the strength of the relationship can be captured directly in the NPT of the indicator node, meaning that there is no need for a separate accuracy node. In Chapter 9 we will show a very simple and efficient method for defining such NPTs.

Box 8.2 provides a real-world example of how instantiations of both the measurement idiom and the cause consequence idiom help to improve a BN structure.

Box 8.2   Commonsense Approach to Causality

Figure 8.15 is a real BN model that was developed by Nikiforidis and Sakellaropoulos (1998) to help doctors determine the severity of a head injury and affect some intervention to aid the patient. The model was produced by data mining to find the correlations between the variables.

The variables are:

- *Age*—This is actually a classification of the age of the patient (infant, child, teenager, young adult, etc.)
- *Brain scan result*—This is a classification of the types of physical damage (such as different types of hemorrhaging) indicated by shadows on the scan.
- *Outcome*—This is classified as death, injury, or life.
- *Arterial pressure*—The loss of pressure may indicate internal bleeding.
- *Pupil dilation and response*—This indicates loss of consciousness or coma state.
- *Delay in arrival*—Time between the accident and the patient being admitted to the hospital.

The arrows are supposed to denote causality. But do the relationships make sense? Let's analyze some actual causal relationships that are relevant here:

- Age may affect the outcome. Older people are less able to recover from serious head injuries and younger people are more likely to suffer serious traumatic injuries as a result of sports participation, fights or motor accidents. But can age cause someone to have a particular brain scan result?
- Does the brain scan result cause the outcome (i.e., can it kill or cure you)? MRI technology is a wonder but it cannot revive Lazarus or kill you.
- The brain scan result is not the same as the actual injury. The scan might be inaccurate because of a faulty machine.
- Other consequences of the injury might be gained from test evidence, such as arterial pressure and pupil dilation.
- Delay in arrival may make moderate injuries more serious. But here delay in arrival looks to be a consequence of the outcome.

Additionally, a key causal influence that is clearly missing from the model is the *treatment*. This is very surprising since the model is being recommended by the authors as a way of identifying treatments. Effective medical treatment may prevent death or injury. However, the database did not contain any data on treatment and

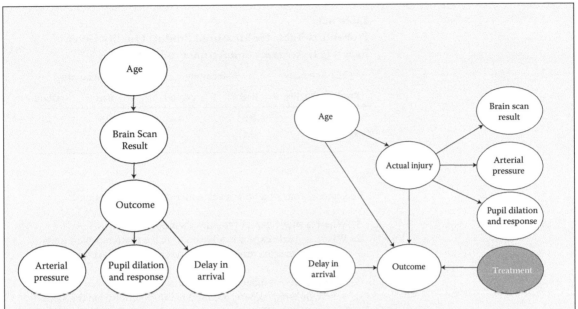

Figure 8.15   Data mined BN model. But does it make sense?   **Figure 8.16**   Revised causal model.

therefore it was missing from the analysis. This is clearly a mistake and shows the dangers of conditioning the analysis on the available data rather than on the problem.

A more rational causal model, therefore, is shown in Figure 8.16.

We have used a number of idiom instantiations here:

- The actual injury is a latent variable and unobservable (this is a topic we will cover in detail in Section 8.6). The doctor runs tests on arterial pressure, brain scan, and pupil dilation to determine what the unknown state of the brain and what the actual injury might be. These are all instantiations of the special case of the measurement idiom indicator nodes (where the accuracy is implicit).
- The outcome is influenced by four causes, including treatment. So this is an instantiation of the cause–consequence idiom. Likewise the actual injury may be a consequence of age since the severity or type of injury suffered by the elderly or infant may differ from that likely to be suffered by others.

We can repeatedly apply the measurement idiom to provide better measures of the truth given a known accuracy of an individual measure from an individual instrument. For example, we could take repeated measures from the same thermometer or from different thermometers or consult multiple experts for their opinions. The accuracy of each thermometer or expert will have a significant effect on our uncertainty about the true value being estimated.

Example 8.2   Cola Test

We wish to estimate the quality of some product, say, the taste of a soft drink product, which is either, and absolutely, good or bad. We might wish to consult multiple drinks experts to do this and can use the measurement idiom as the basis for modeling this in a BN.

Table 8.3

Probability Table for Measured Product Quality Given Expert (E1) Accuracy and Actual Product Quality

E1 Accuracy	Inaccurate		Accurate	
Product Quality	Bad	Good	Bad	Good
Bad	0.6	0.4	0.9	0.1
Good	0.4	0.6	0.1	0.9

Let's model four situations, each with different assumptions:

1. Where a single expert provides their opinion
2. Where a single expert is allowed to make three repeated judgments on the same product (presume the product is disguised in some way)
3. Where different independent experts are used
4. Where different experts are used but they suffer from the same inaccuracy, that is, they are dependent in some significant and important way

We could think of inaccuracy here as bias or imprecision in the judgment of the expert. So in situations 2 and 4 if there was a bias then every measurement taken would suffer the same bias.

Let's assume that each expert has two levels of accuracy: accurate and inaccurate. When they are accurate they get it right 90% of the time, but when they are inaccurate they get it right only 60% of the time (see Table 8.3).

We also assume that the product quality is 50:50 good or bad, and that we are equally ignorant about the accuracy of any expert at the outset and assign them the same 50:50 prior.

Let's look at the results from the BN.

Figure 8.17 Single expert with "Good" opinion.

1. The single expert, E1, pronounces that the drink is of good quality. Figure 8.17 shows the result: there is a 75% probability the product is good, this being simply the average of 0.6 and 0.9 in Table 8.3.
2. The single expert pronounces that the same drink is good, bad, and good on three different taste tests. The BN for this is shown in Figure 8.18. The structure here is different from before because the same accuracy node for the expert, E1, applies to every measurement. Given the inconsistency in the opinions offered, for the same drink, the model revises belief in the accuracy of E1 down from 50% to 27% and estimates the product to have a 68% probability of being of good quality. This is lower than in result 1.
3. Here we have three experts, E1, E2, and E3, each offering opinions on the drink quality, but where each expert is independent of the others. So E1 could be inaccurate, while E2 and E3 might be accurate and vice versa. We therefore model inaccuracy for each expert by a separate node in the BN. The experts pronounce good, bad and good in that order. The results are shown in Figure 8.19, where the model estimates there is a 75%

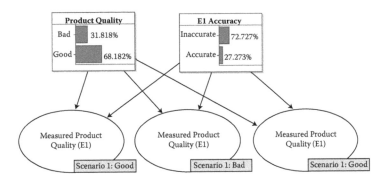

Figure 8.18   Single expert measuring same product with "Good," "Bad," and "Good" opinions.

probability of being of good quality. Notice that the model estimates that expert E2 is the least reliable given the inconsistency
with the other experts. Still, the product is more likely to be
good compared to case 2 when we asked the single expert to
judge the quality three times.

4. Here we have three experts as before but have strong reasons
   to believe they are not independent of each other; they may
   have attended the same university, belong to the same organization, or have the same incentives and preferences. Hence
   the BN structure reflects this and the inaccuracy node in the
   measurement idiom instance applied is shared among all three,
   as shown in Figure 8.20. We extend the BN structure in result 3
   by adding a node to model dependence/independence between
   the experts and also a new node for "expert accuracy," representative of the community of thought to which the experts
   belong. Given the same opinions as before we can see that the
   percentage probability of the product being of good quality is
   only 68%. This is the same result as when we used a single
   expert and they produced just as inconsistent a result. So clearly
   we have not gained anything by consulting three dependent
   experts since they are just as good or poor as a single expert on
   his own. To put it another way they have added no additional
   information to reduce our uncertainty.

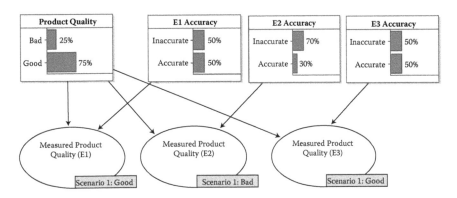

Figure 8.19   Multiple independent experts measuring same product with "Good," "Bad," and "Good" opinions.

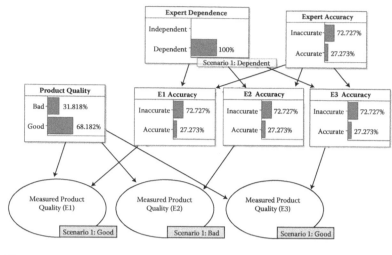

Figure 8.20 Multiple dependent experts measuring same product with "Good," "Bad," and "Good" opinions.

If independent scientists use the same flawed analysis procedure or same data set containing errors as previous scientists, then they will arrive at conclusions that might confirm previous hypotheses. The claim that such "independent confirmation" strengthens the hypothesis is vacuous. Some skeptics in the global warming debate have argued that climate scientists have fallen into this trap. In general, your statistical alarm bells should ring whenever you hear statements like "this hypothesis must be true because the majority of scientists think so."

Even if we have very weak indicators, if we have many of them all pointing the same way, then it will be possible to make strong conclusions about the value of the actual attribute.

Despite its simplicity the lessons from Example 8.2 are quite profound:

- It shows that repeated confirmation of a test result from multiple sources adds little information if those sources are closely dependent on one another. For scientific peer review or even experimental design the implications are that a consensus of expert opinion is meaningless if all the experts suffer from the same errors or biases.
- When designing complex, critical systems it is common to introduce fault tolerance or redundancy to improve reliability. This normally involves secondary backup systems that will operate on stand-by; should the primary system fail the stand-by will step in and maintain operation. However, if the backup and main systems are dependent in some way (for example if they have the same maintenance regime or a shared design flaw) then they might both fail at the same time, thus negating the supposed reliability benefit.

8.3.3   Definitional/Synthesis Idiom

Although BNs are used primarily to model causal relationships between variables, one of the most commonly occurring class of BN fragments is not causal at all. The definitional/synthesis idiom, shown in Figure 8.21, models this class of BN fragments and covers each of the following cases where the synthetic node is determined by the values of its parent nodes using some combination rule.

8.3.3.1   Case 1: Definitional Relationship between Variables

In this case the synthetic node is *defined* in terms of the parent nodes. This does not involve uncertain inference about one thing based on

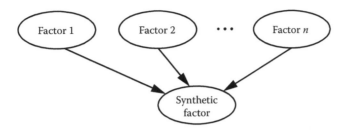

Figure 8.21 Definitional/synthesis idiom.

knowledge of another. For example (as shown Figure 8.22), *velocity* of a moving object is defined in terms of *distance* traveled and time by the functional relationship Velocity = Distance/Time. We might be uncertain about the individual nodes *distance* and *time* in a given situation, but, in contrast, the law that determines *velocity* is not uncertain.

A more typical example of the kind that occurs in systems risk assessment is shown in Figure 8.23. Here we are defining the complexity of a software system in terms of three distinct complexity factors: interface, code, and documentation.

Clearly synthetic nodes, representing definitional relations, could be specified as deterministic functions or axiomatic relations where we are completely certain of the deterministic functional relationship between the concepts. Otherwise we would need to use probabilistic functions to state the degree to which some combination of parent nodes combine to define the child node. These issues are covered in Chapter 9.

8.3.3.2 Case 2: Hierarchical Definitions

One of the first issues we face when building BNs is whether we can combine variables using some sensible hierarchical structure. A hierarchy of concepts is simply modeled as a number of synthesis/definitional idioms joined together. An example, building on Figure 8.23, is shown in Figure 8.24.

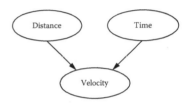

Figure 8.22 Instantiation of definitional/synthesis idiom for velocity example.

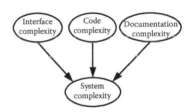

Figure 8.23 Instantiation of definitional/synthesis idiom for system complexity.

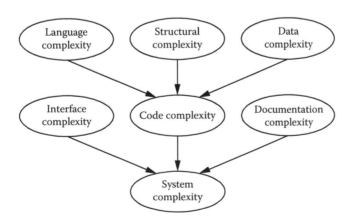

Figure 8.24 Hierarchical definition of system complexity.

In this case, we could have produced a flat hierarchy with system complexity defined in terms of five complexity factors (*interface*, *language*, *structural*, *data*, and *documentation*); however, since *language*, *structural*, and *data complexity* are all components of *code complexity* it makes sense to organize these into the hierarchy shown. This not only improves the readability and understandability of the model, but it also avoids the problem of having a node with too many parents (for reasons that will become clear it is normally advisable to minimize the number of parents where possible).

8.3.3.3 Case 3: Combining Different Nodes Together to Reduce Effects of Combinatorial Explosion ("Divorcing")

The technique of cutting down the combinatorial space using synthetic nodes is called *divorcing*, since a synthetic node divorces sets of parents from each other. It is important to note that divorcing is not something that can be arbitrarily applied in all situations. The conditions under which divorcing can be applied are discussed in Box 8.3 with a detailed example.

In Figure 8.24 the attributes defining *system complexity* were naturally organized into a two-level hierarchy. But in many situations there may be no such natural hierarchical decomposition. For example, if an organization is made up of six separate units, then its overall quality might be defined as shown in Figure 8.25. The problem here is the complexity of the NPT node *Organization quality*.

Although in situations like this there is usually no need to elicit individual NPT values, since we can define some function (deterministic or probabilistic), it is nevertheless extremely inefficient (when it comes to running such models) to include nodes with so many parents because of the large NPT. One commonly used solution is to introduce "synthetic" nodes as shown in Figure 8.26.

If, for each node, *quality* is defined on a 3-point scale (poor, average, good), then the NPT for the node *Organization quality* will have 3^7 (that is 2187) cells, each of which is a value that may need to be elicited. If we were using a 5-point scale then the total number would be 78,125.

Although this introduces additional nodes and hence additional NPTs (four in the example) there is no longer a massive NPT to construct (in the example of Figure 8.26 the largest NPTs have a total of 27 cells). When we restructure a BN in this way, such that each node has no more than two parents, we say that the BN is **binary factorized**.

The edge directions in the synthesis idiom do not indicate causality (causal links can be joined by linking it to other idioms). This would not make sense. Rather, the link indicates the direction in which a subattribute defines an attribute, in combination with other subattributes (or attributes define superattributes, etc.). It is natural to do this in order to handle complexity and we do it routinely when we define concepts using linguistic qualifiers and aggregators.

Figure 8.25 Example of definitional idiom with flat hierarchy.

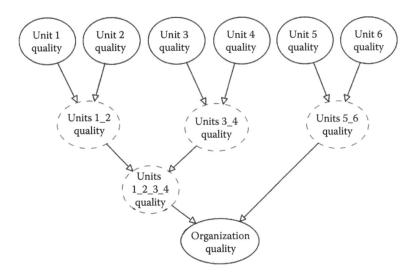

Figure 8.26 Reorganizing a flat hierarchy using synthetic nodes (shown with dotted outlines). Because each node has at most two parents this type of restructuring process is also called ***binary factorisation***.

Box 8.3   Divorcing: When it Does and Does Not Work

Parent nodes can only be divorced from one another when their effects on the child node can be considered separately from the other non-divorced parent node(s). For a synthetic node to be valid some of its parent node state combinations must be exchangeable, and therefore equivalent, in terms of their effect on the child node. These exchangeable state combinations must also be independent of any non-divorcing parents, again in terms of their effects on the child node.

To illustrate this we can consider an example that occurred in a real project concerned with assessing system safety. Figure 8.27 shows the original part of a BN model using the definitional/synthesis idiom. Here *test results* for a system are defined in terms of the *occurrences of failures* (i.e., the number of failures observed during testing), *severity of failures*, *tester experience*, and *testing effort*.

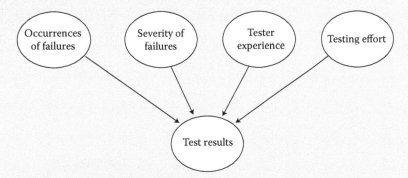

Figure 8.27   Original BN fragment.

Since reliability (or safety) is typically defined in terms of *occurrences of failures* and *severity of failures*, it is uncontroversial to first create the synthetic node *reliability* as shown in Figure 8.28 to divorce the parents' *occurrences of failures* and *severity of failures* from the other two parents.

The next step is to create a synthetic node *test quality* to model the joint effects of *tester experience* and *testing effort* on the *test results* as shown in Figure 8.29.

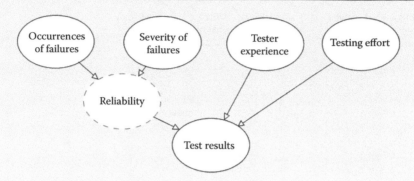

Figure 8.28   Synthetic node *reliability* introduced.

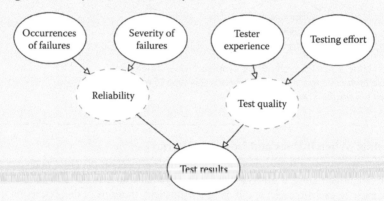

Figure 8.29   Synthetic node *test quality* introduced.

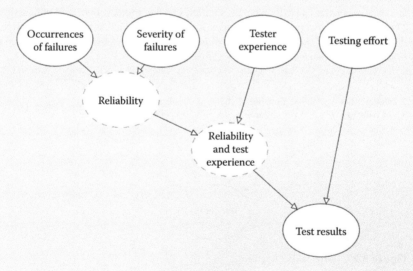

Figure 8.30   Inappropriate use of synthetic node.

Implicit in this model is the assumption that *tester experience* and *testing effort* operate together to define some synthetic notion of *test quality* where, when it comes to eliciting values for *P*(test results | reliability, test quality), it does not matter whether poor quality testing has been caused by inexperienced testers or lack of effort. Now contrast this solution with one shown in Figure 8.30.

> The problem with this solution is that when it comes to eliciting P(test results | reliability and test experience, testing effort), it does matter whether good reliability and test experience has been caused by good reliability or by experienced testers. For example, an experienced tester may be able to discover a lot of severe failures with very little effort at all.

8.3.4 Induction Idiom

The induction idiom (shown in Figure 8.31) is used when we make observations to learn about some population parameter in order to make predictions about it in the future, taking account of differences of context. In fact the induction idiom is a general model for any type of statistical inference that you might wish to represent or calculate using a BN. The key difference here is that we are learning an unknown or partially known parameter about some population of interest from some known data.

The induction idiom is simply a model of statistical induction to learn some parameter that might then be used in some other BN idiom. For example, we might learn the accuracy of a thermometer and then use the mean accuracy as a probability value in some NPT using the measurement idiom.

Example 8.3

Suppose that a manager of a call center is concerned about the number of lost calls that are occurring. A lost call is when a call center employee fails to complete the call to the satisfaction of the caller. In the example instantiation in Figure 8.31(b) the manager wishes to find out the true (population) mean number of lost calls per day and so decides to observe the actual number of lost calls on 10 successive days. These 10 observations are used to update any prior knowledge about the population mean and this feeds into a forecast of the number of lost calls on the 11th day. In this case the contextual differences that impact the forecast revolve around the fact that staff knew they were being monitored on the previous 10 days and therefore may have behaved differently than normal.

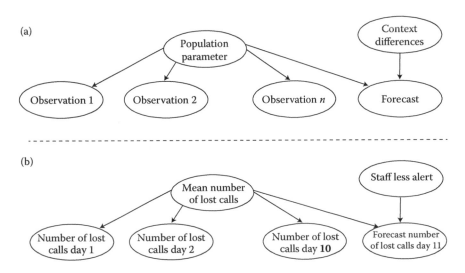

Figure 8.31   Induction idiom. (a) Generic. (b) Instantiation.

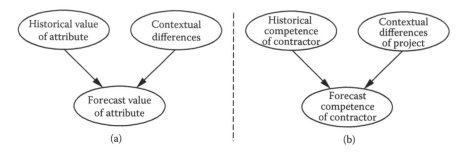

Figure 8.32 Simplified version of induction idiom. (a) Generic. (b) Instantiation.

None of the reasoning in the induction idiom is explicitly causal. Specifically, the idiom has two components:

1. It models Bayesian updating to infer the relevant parameter of the population where the entities from this population are assumed to be exchangeable.
2. It allows the expert to adjust the estimates produced if the entity under consideration is expected to differ from the population, that is, if it is not exchangeable because of changes in context.

In Figure 8.31 each "observation" is used to estimate the "population parameter" used to characterize the population. This then forms the prior for the next observation. This can be repeated recursively to provide more accurate estimates for the population. Finally, we can use the population parameter distribution to forecast the attribute of the entity under consideration. Essentially, we use the induction idiom to learn the probability distribution for any node in instantiations of the measurement or cause–consequence idioms. We shall see concrete examples in Chapter 10.

In some situations it may not be possible to set up a sequence of observations in order to learn the distribution of the population parameter. However, there may be other ways to elicit this distribution (such as from the memory/experience of an expert). In these cases the induction idiom simply involves three nodes as shown in Figure 8.32(a), where the node *Historical value of attribute* is the elicited distribution.

In the example instantiation in Figure 8.32(b) we might wish to evaluate the competence of a contractor that we have used on previous projects. The contractor's competence on previous projects is used to determine the historical competence. If the next project is similar for the contractor then we would expect their competence on it to be similar. However, if it is very different to any project we have previously seen, then the historical competence will not provide great insight.

8.4 The Problems of Asymmetry and How to Tackle Them

There are a number of situations where an apparently natural BN structure gives rise to a very unnatural problem of being forced to consider variables and states that are logically impossible.

We already saw an example of this in Chapter 7, Section 7.9.1 (in the BN for the Monty Hall problem, in which an NPT contained combinations of parent states that were impossible). It felt unsatisfactory to have to provide probability values for the states associated with an impossible event. It turns out that this is an example of a very common problem, which we call the *asymmetry problem*.

Most current algorithms and implementations of BN technology assume that all parent node states can causally interact with each other; the result is that we are forced to consider all combinations of parent states, even when certain combinations are logically impossible. When the nodes are discrete the NPTs represent a symmetric unfolding of all possible outcomes and events. In other words, we are forced into a symmetric solution of what is fundamentally an asymmetric problem.

In this section we consider the most common situations that lead to asymmetry problems, and propose practical solutions for them. These are, respectively, where we have impossible paths, mutually exclusive paths, distinct causal pathways, and mutual exclusivity from taxonomic classifications.

8.4.1 Impossible Paths

Consider the simple example of a person descending a staircase where there is a risk they will slip. If they slip they fall either forward or backward. What appears to be a simple and natural BN structure for modeling this risk problem is shown in Figure 8.33. But there is a major problem with this model when we come to define the NPT for the node *Falls* (Table 8.4). How should we complete the entries for the column when *Slips* = No?

If a person does not slip, then the states for *Falls* are irrelevant— neither forward nor backward can occur and therefore have zero probability of existence. The only way out of this "conditional causal existence" quandary would be to break the axioms of probability or to dynamically remove or add variables to the BN to accommodate the causal unfolding of new events. The former option is hardly attractive and the latter is challenging using the current generation of algorithms. In the Monty Hall example we solved the problem by assigning equal values to the entries for the impossible column. In fact we could have assigned any values. But the solution is unsatisfactory because we should not be forced to consider such states.

The pragmatic solution is to add explicit "don't care" or "not applicable" states to nodes where there are impossible state combinations, as shown in Table 8.5. Specifically, we assign the value 0 to the NA state when the parent states are impossible and 1 otherwise.

The forced symmetry imposed on NPTs in BN algorithms creates unnecessary complexity and effort, both in defining NPTs and running the resulting models. Our objective is to show that careful modeling can minimize some of the worst effects of this problem.

We shall see other situations where asymmetry is a dominant feature later in the book. For example, in Chapter 9 we shall see that there is asymmetry resulting from the use of the *NoisyOR* function, which assumes causal independence between the parents.

Table 8.4
NPT for Node
(*Falls* | *Slips*)

Slips	Yes	No
Forward	0.1	??
Backward	0.9	??

Table 8.5
Revised NPT for Node
(*Falls* | *Slips*)

Slips	Yes	No
Forward	0.1	0.0
Backward	0.9	0.0
NA	0.0	1.0

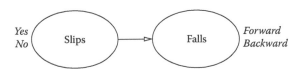

Figure 8.33 Simple BN model for "Slips" showing node states.

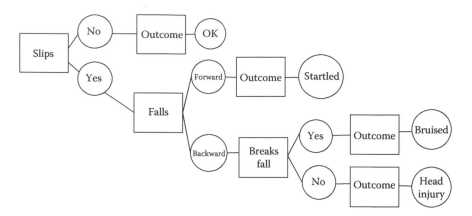

Figure 8.34 Event tree for slip example (circles are states and rectangles are variables).

Although this solution increases the size of the NPT it is conceptually simpler and makes sense. Unfortunately, when we extend this simple example we run into a different asymmetry problem. Suppose that:

- If the person falls forward they will be startled (but otherwise unharmed).
- If the person falls backward they might break their fall or not.
- If they break their fall they will be bruised, but if not they will suffer a head injury.

Before considering a BN solution, we can model this problem by an alternative graphical model, namely, an *event tree*, as shown in Figure 8.34. Notice that the event tree is asymmetric and contains four paths in total, leading respectively to the four different outcomes.

An apparently natural BN solution is shown in Figure 8.35. However, the NPT needed for the node *Outcome* is far from natural. It has 72 entries (since it has 4 states and 18 combinations of parent states). Moreover, most of these entries correspond to impossible state combinations.

Compared to there being just four paths in the event tree, the total number of NPT entries required for this BN is 89 (the node *Slips* has 2 entries, *Falls* has 6, *Breaks fall* has 9, and *Outcome* 72).

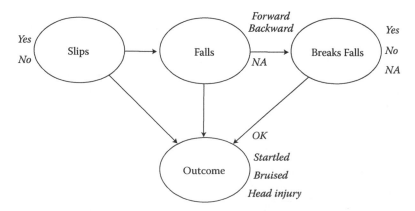

Figure 8.35 Natural BN solution for extended problem?

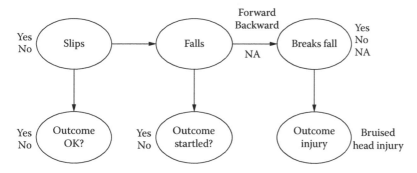

Figure 8.36 Improved BN solution for "Slips" problem.

The pragmatic solution to this problem is to create the BN as shown in Figure 8.36, which minimizes the number of impossible state combinations by recognizing that the different outcomes are best modeled as distinct variables rather than as states of a single outcome variable.

Even allowing for NA states, this BN has no single NPT with more than 9 entries and has a much-reduced total of 33 entries in all. In this example the pragmatic BN essentially mimics the structure of the event tree model. However, for problems involving more variables event tree solutions are computationally intractable.

8.4.2 Mutually Exclusive Paths

Consider the following "mountain pass" problem:

- We want to arrive at an appointment to visit a friend in the next town.
- We can either take a car or go by train.
- The car journey can be affected by bad weather, which might close the mountain pass through which the car must travel.
- The only events that might affect the train journey are whether the train is running on schedule or not; bad weather at the pass is irrelevant to the train journey.

The event tree model for this problem is shown in Figure 8.37. Notice that the event tree is asymmetric in the sense that only some of the variables are causally dependent. The weather and pass variables are irrelevant conditional on the train being taken and the train running late is irrelevant conditional on the car being taken.

Most common attempts to model this problem with a BN fail. For example, the natural structure in Figure 8.38(a) is unsatisfactory because, not only does the *Make appointment?* node have many impossible states, but the model fails to enforce the crucial assumption that *Take car* and *Take train* are mutually exclusive alternatives with corresponding mutually exclusive paths. Although it is possible to ease the problems associated with the NPT for *Make appointment?* by introducing synthetic nodes as shown in Figure 8.38(b), the problem of mutual exclusivity is still not addressed in this version.

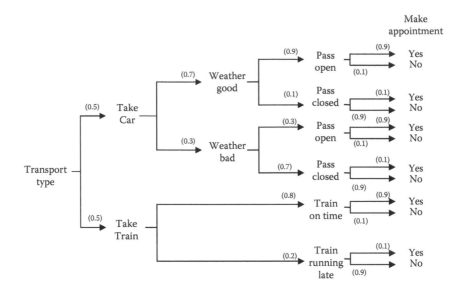

Figure 8.37   Event tree for mountain pass problem.

The mutually exclusive paths problems is solved using the BN shown in Figure 8.39. On the left-hand side we have one causal path containing the nodes *Weather* and *Pass open?*. On the right-hand side we have *Train late?*. Clearly the only variable shared by the different causal paths is *Make appointment?*. We introduce a conditioning "switch" node into the BN to separate the causal paths. This is the node called *Mode of transport*. When *Mode of transport* = *car* the left-hand causal path is activated and when *Mode of transport* = *train* the right-hand side is activated.

This switching strategy, to cope with mutually exclusive paths, is sometimes called *partitioning*. We can then think of the switch as partitioning the model so that it can switch between different causal paths, models, or assumptions, each of which is mutually exclusive.

Switch activation of left and right is governed by the NPT for the node *Make appointment* in Table 8.6.

Notice that in the NPT we are forced to declare and consider all state combinations. This gives rise to the unfortunate but necessary duplication of probabilities to ensure consistency. For example, we need to consider *Pass open?* states for *Mode of transport = train* and *Train late?* states for *Mode of transport = car* and so simply repeat the probabilities to show they are independent of the irrelevant causal path in each case (i.e., these irrelevant states do not affect the probabilities).

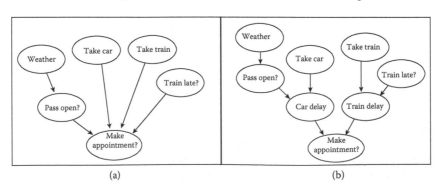

Figure 8.38   Unsatisfactory BN models of mountain pass problem. (a) Simple model. (b) Revised model.

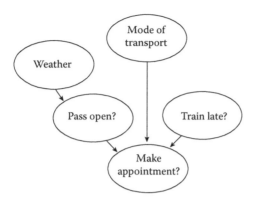

Figure 8.39 BN for mountain pass example.

Table 8.6

NPT for *Make Appointment?*

Mode of Transport	Train				Car			
Train Late?	False		True		False		True	
Pass Open?	Open	Closed	Open	Closed	Open	Closed	Open	Closed
False	0.1	0.1	0.9	0.9	0.1	0.9	0.1	0.9
True	0.9	0.9	0.1	0.1	0.9	0.1	0.9	0.1

8.4.3 Mutually Exclusive Events and Pathways

In the mountain pass example traveling by car and traveling by train were mutually exclusive options. Only by modeling these options as mutually exclusive states within the same node (*mode of transport*) were we able to "solve" the mutually exclusive paths problems. However, in certain situations this neat solution will not work (we actually already saw one such situation in the "slips" problem whereby it was infeasible to capture mutually exclusive outcomes simply by declaring them as states of a single outcome node).

In particular, we are concerned with the situation where there are two or more mutually exclusive states, which each belong to a separate causal pathway. Merging the causal pathways into a single node may detract from the semantics of the model and make elicitation and communication difficult.

Consider, for example, an inquest into the death of Joe Smith; there are three possible mutually exclusive causes: "natural," "unlawful" and "suicide." There is evidence Joe may have had a serious illness and there is also evidence he may have suffered depression which could have caused suicide. A suicide note was found by Joe's body, but it is not known if it was written by Joe and there are possible suspicious wounds found on his body.

In the BN model of Figure 8.40a we have used a single node whose states correspond to the possible causes of death to ensure mutual exclusivity. However, this solution requires us to complete NPTs which (in

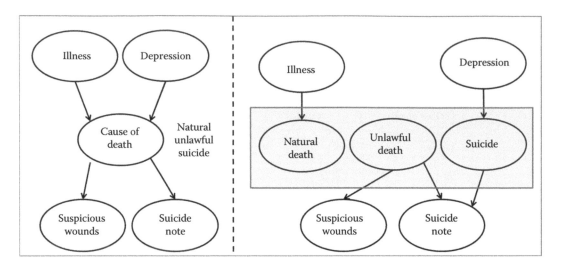

Figure 8.40   BN structural solutions for mutually exclusive events and pathways. (a) Single node whose states represent the mutually exclusive events. (b) Separate Boolean nodes for each mutually exclusive event.

realistic examples with multiple causes and alternatives) are infeasibly large and for which the vast majority of entries are either redundant or meaningless. For example:

- *The NPT for the cause of death node*: Although each parent node influences only one possible outcome they are forced to give (redundant) separate probabilities conditioned on every combination of all the other causal factor states. For example, although "Illness" only influences whether or not Joe died of natural causes, we have to specify the probability of death from natural causes conditioned on illness and every possible combination of values for the other parent states—none of which is relevant.
- *The NPT for child nodes of the cause of death node*: Since some of these are also only relevant for at most a single cause of death, we again have to unnecessarily specify separate probabilities conditioned on each of the different alternative causes of death.

Only by modelling the alternatives for cause of death as distinct nodes can we hope to separate out the different causal pathways. However, if we separate them as in Figure 8.40(b) we run into the problem of not being able to enforce mutual exclusivity between the alternatives. No matter how we define the NPTs in this model, if we know that one of the alternatives, say "suicide" is true, the model does not exclude the possibility that the others are true.

Jensen and Nielsen (2007) proposed a solution by introducing a Boolean constraint node (as shown in Figure 8.41 for the simplest case where there are two alternatives) and setting it to be *true*. The NPT for the constraint node is defined as *true* when exactly one of the parents is *true* and *false* otherwise (so this easily generalises to arbitrary *n* nodes).

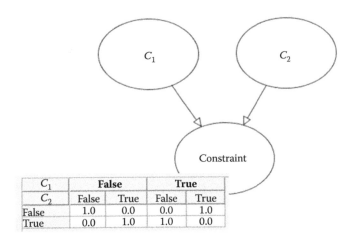

C_1	**False**		**True**	
C_2	False	True	False	True
False	1.0	0.0	0.0	1.0
True	0.0	1.0	1.0	0.0

Figure 8.41 Enforcing mutual exclusivity by introducing simple Boolean constraint node.

Providing the constraint is always set to be *true* when the model is run, mutual exclusivity is satisfied.

However, there is a fundamental problem with this proposed solution. The constraint node has to be set to true the prior probabilities of C_1 and C_2 are not preserved except in the case where they are equal. For example, suppose $P(C_1 = \text{true}) = 0.7$ and $P(C_2 = \text{true}) = 0.3$. Then:

$$P(C_1 = \text{true} \mid \text{Constraint} = \text{true}) = 0.8448$$
$$P(C_1 = \text{true} \mid \text{Constraint} = \text{true}) = 0.1552$$

Therefore, when the constraint is set to true (as is necessary) the priors for the cause nodes change even though no actual evidence has been entered. In other words the proposed solution changes the semantics of the model.

The solution (see Figure 8.42 for the structure) is to add both a constraint node and an auxiliary classification node, which is a common

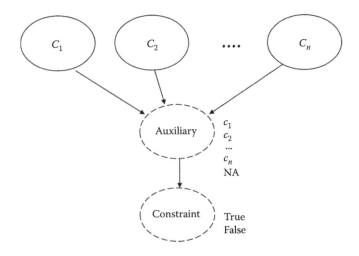

Figure 8.42   Structure for general solution.

Table 8.7
NPT for Auxiliary Cause Node

	C_1		False				True	
	C_2	False		True		False		True
	...							
	C_n	False	True	False	True	False	True	False	True
c_1		0	0	0	0	1	0	0	0
c_2		0	0	1	0	0	0	0	0
...									
c_n		0	1	0	0	0	0	0	0
NA		0	0	0	1	0	1	1	1

Specifically: $P(c_i = 1)$ when $C_i =$ true and $C_j =$ false for each $i \neq j$; otherwise $P(c_i = 0)$; $P(NA) = 0$ if exactly one C_i is true and 1 otherwise.

Table 8.8
NPT for Constraint Node

Auxiliary	c_1	c_2	...	c_n	NA
False	x_1	x_2	...	x_n	1
True	$1 - x_1$	$1 - x_2$	...	$1 - x_n$	0

child of the C_i nodes. Unlike the proposed solution described in Section 4.3, setting the constraint node to true does **not** change the prior probabilities for the C_i nodes.

The auxiliary node has $n + 1$ states, namely the n original states c_i (for $i = 1 - n$) of the node S plus a special NA state standing for "Not Applicable" and representing impossible combinations. Assuming that the prior probability for c_i being true is x_i for $i = 1 - n$ it turns out that if we define the NPTs for the auxiliary node and constraint node as shown in Tables 8.7 and 8.8, respectively, then when we set the constraint node to true not only is the mutual exclusivity between the c_i preserved but so are the priors x_i.

8.4.4 Taxonomic Classification

The mutual exclusivity problem is especially pertinent when we need to introduce classifications into a model.

With a taxonomic hierarchy we aim to classify hidden attributes of an object using direct or inferred knowledge. There are many reasons why we might wish to model such a thing using a BN. The key feature of a taxonomic hierarchy is that mutual exclusivity expresses the logical existence constraints that exist at a given level in a taxonomy; so an object can be classified as {Mammal, Dog, Alsatian} but not as a {Mammal, Cat, Abyssinian} at the same time.

The taxonomic hierarchy is therefore governed by values denoting membership rather than probability assignments. Taxonomic relations are examples of the definitional/synthesis idiom.

Figure 8.43 shows a taxonomic hierarchy for military asset types. Here we present a simple classification hierarchy where the more abstract class is refined into more particular classes. This could continue to any particular depth as determined by the problem domain. A class is denoted by a rectangle and an edge denotes class membership between parent class and child class.

At the top level we have three mutually exclusive classes: {Land, Sea, and Air}. If an asset is a Land unit it cannot be a Sea or Air unit or vice versa. We then further refine the classes of Land unit into {Armored, Armored Reconnaissance, Mechanized Infantry, Infantry, and Supply}. We can have similar refinements for Air and Sea units and the key thing in the taxonomy is that the mutual exclusivity is maintained as we progress through the levels in the hierarchy.

We can use this information to help build a BN and then supplement this BN with additional nodes that reflect measurement idioms and so on. Each value in the parent class becomes a child node in the BN and this is further refined as necessary. A taxonomic decomposition like this is very simple. However, when translated into a BN, as the size and depth of the classification grows the need to manage the complexity increases. The problem is that, because of the mutually exclusive states, we find that very quickly we have a model with a very large state space; meaning large, complex NPTs.

We can recast a classification model by treating each subclass as a separate "variable," but to ensure that the subclasses remain mutually exclusive we must introduce a new state—NA, for not applicable—to each child class. Whenever the parent class takes a value inconsistent with the child class under consideration NA is set to true. These NA states must then be maintained through all child nodes of each class node to ensure that the mutual exclusivity constraints are respected. The necessary NPTs are then set up as shown by the example in Table 8.9.

Note that:

- We use a uniform prior to model the membership of the subclasses with the parent class. So each probability (except NA) in the first column is 0.2 (because we have five subclasses).
- We assign zero probability where subclasses do not belong to the parent class, and one where the complement, NA, applies.

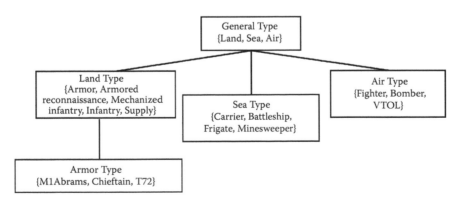

Figure 8.43 Classification hierarchy for military asset types.

Table 8.9

NPT for *Land Type* Conditioned on *General Type*

General Type	Land	Sea	Air
Armor	0.2	0.0	0.0
Armored reconnaissance	0.2	0.0	0.0
Mechanized infantry	0.2	0.0	0.0
Infantry	0.2	0.0	0.0
Supply	0.2	0.0	0.0
NA	0.0	1.0	1.0

Figure 8.44 shows the resulting BN for this classification problem. Figure 8.44(a) shows the marginal probabilities for each of the classes and subclasses. Note that the sum of the subclass probabilities is equal to the parent class probability.

When we actually observe that an entity belongs to a subclass (or class) we enter an observation on the model, as shown in Figure 8.44(b) where *Land Type = Infantry*. We can see that the parent class is now *Land Type*, as expected, and the other classes have zero probability and the subclasses are assigned NA. Thus mutual exclusivity is enforced.

In this example we might wish to use intelligence and sensor sources to identify an unknown enemy asset type from observations made and data received. Therefore, we might want to also encode the following information in the BN:

- Prior knowledge about possible attribute values (probabilistic)
- Effects on other objects (causal)
- Signals received by sensors (infrared, radar, etc.)
- Indirect measures from tracking and filtering (max speed, location)
- Relationship with other objects (proximity to valued asset)

We can include, within the classification hierarchy, distinguishing attributes that identify (perhaps with uncertainty) membership of the parent class in the taxonomy. For example, we might know that some classes of military asset carry a particular type of radar and that this is shared across mutually exclusive classes or subclasses. So, observing a signal from that radar might lead us to believe that the enemy asset identified is one of those classes of assets.

We can define the observable attributes as leaf classes on any node. Here we will denote them using ellipses, as shown in Figure 8.45. We have three types of attributes: *Physical Location* = {*Land, Sea, Air*}, *Armor Strength* = {*Hard, Soft*}, *Emitter* = {*Radar, Radio*}, and *Speed* = {*Slow, Fast*}. Emitters are shared over all classes (all assets have one) but Armor strength is only relevant to Armored land assets; similarly while physical location is relevant to all classes, *Speed* is only relevant to aircraft. The key thing about attributes is that they are defined at whatever level in the model where the discrimination between attributes matters (this might be determined by the sensor or measurement capability).

Attributes could themselves be decomposed into mutually exclusive classes. You could also deploy different hierarchies dependent on the features of interest.

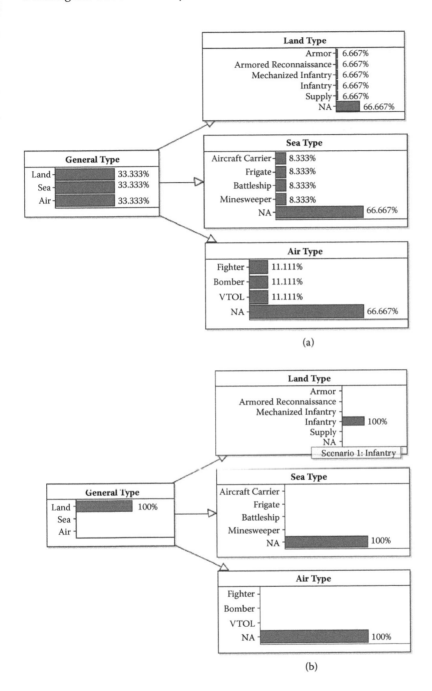

Figure 8.44 Classification hierarchy, represented by a BN, for military assets supplemented by marginal probabilities in (a) initial state and (b) mutual exclusivity enforced when subclass observed (note the *General type* node does not require an NA state value).

Figure 8.46 shows a BN with nodes for each of the attributes of interest. Note that, for convenience, we have decomposed the signal source (emitter type) for land, sea, and air.

Table 8.10 shows the NPT for the *Land Signal Source* attribute node. The probabilities allocated represent our beliefs about the relative

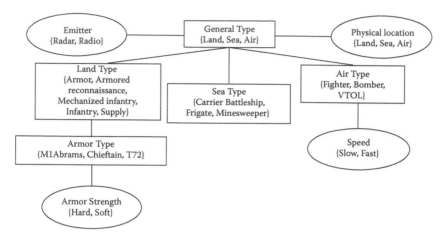

Figure 8.45 Classification hierarchy for military asset types with attributes added.

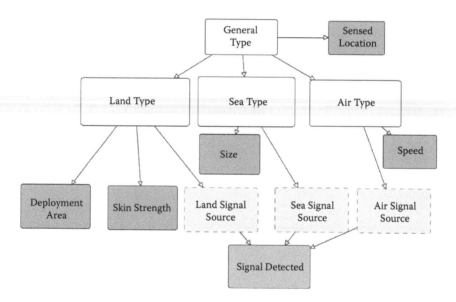

Figure 8.46 Expanded BN version of classification for military asset types with attributes added.

Table 8.10

NPT for *Land Signal Source* Attribute Node

Land Type	Armor	Armored Reconnaissance	Mechanized Infantry	Infantry	Supply	NA
Radio type A	0.9	0.0	0.333	0.0	0.0	0.0
Radio type B	0.0	0.83	0.333	1.0	1.0	0.0
Radar type X	0.09	0.16	0.333	0.0	0.0	0.0
Radar type Y	0.001	0.01	0.0	0.0	0.0	0.0
NA	0.0	0.0	0.0	0.0	0.0	1.0

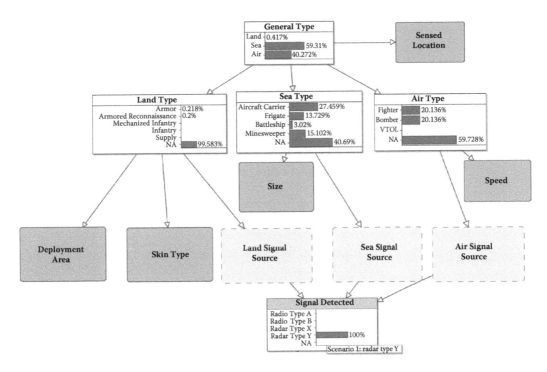

Figure 8.47 Posterior classification of military asset from sensor signal detected "Radar Type Y."

probability of each land unit having a particular radar or radio type. So observing radar type Y will mean that only an armored or armored reconnaissance unit is the possible asset: the others are ruled out. Obviously if *Land Type* = NA then the signal source must also be NA.

The NPT for the *Signal Detected* node needs a little explanation as it is fairly complex (you can see the full table in the AgenaRisk model). The sensor can detect a signal from each of the emitter types but we must be careful to set the signal detected state to be true only when one and only one of the parent state results equals the same state. This is equivalent to an exclusive OR function except applied to a labeled node with a large number of states rather than a Boolean.

Exclusive OR functions are covered in Chapter 9.

Now that we have the complete model we can use it to help identify an unknown enemy military asset from the data observed and our given taxonomy. Figure 8.47 shows the posterior classification probabilities given we have observed that a military unit, of unknown type, is using a radar of type Y. We are approximately 60% sure the asset is a sea asset and 40% sure it is an air asset. Of the sea assets, aircraft carrier is the more likely. Notice how all of the classes in the taxonomy are updated appropriately.

8.5 Multiobject Bayesian Network Models

There are two types of situations in which it becomes inefficient or impractical to model a problem using a single BN:

1. *When the model contains so many nodes that it becomes conceptually too difficult to understand.* One of the key benefits of BNs is that they are such a powerful visual aid. But, unless there is some especially simple structure, as soon as a model contains more than, say, 30 nodes its visual representation becomes hard to follow. It needs to be broken up into smaller understandable chunks.

2. *When the model contains many similar repeated fragments,* such as the case when there are repeated instances of variable names that differ only because they represent a different point in time. For example, the model in Figure 8.48 (which represents a sequence of days in which we assess the risk of flood) contains several such repeated variables that differ only by the day they are recorded.

In both cases we seek to decompose the model into smaller component models. For example, the large model shown in Figure 8.49 ought to be decomposable somehow into three smaller models as indicated by the grouping of the nodes, whereas the flood model ought to be decomposable into a sequence of identical models of the type shown in Figure 8.50.

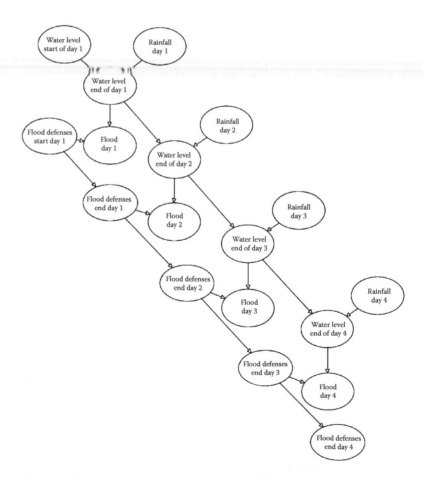

Figure 8.48 A model with repeated nodes that differ only in time.

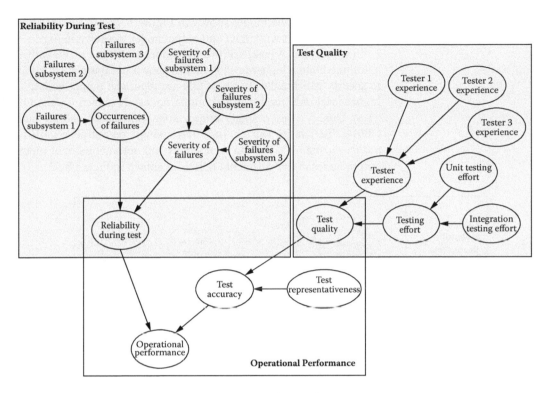

Figure 8.49 A large model with a possible decomposition indicated.

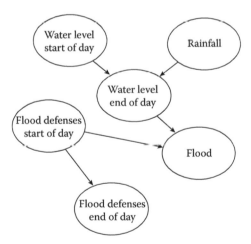

Figure 8.50 Simple flood model.

The component models we need are called *object-oriented BNs* (OOBNs), because they have some of the properties associated with object-oriented modeling.

An OOBN is simply a BN with certain additional features that make it reusable as part of a larger BN model. The most important feature of an OOBN is that it will generally have input and/or output nodes. These represent the "external interface" of the BN and enable us to link OOBNs in a well-defined way.

Key Principles of Object-Oriented Modeling

- *Abstraction* allows the construction of classes of objects that are potentially more reusable and internally cohesive.
- *Inheritance* means that objects can inherit attributes and operations of parent classes.
- *Encapsulation* ensures that the methods and attributes naturally belonging to objects are self-contained and can only be accessed via their public interfaces.

For example, the model shown in Figure 8.49 can be decomposed into three OOBNs with input and output nodes as shown in Figure 8.51.

In a tool like AgenaRisk the individual OOBNs can be embedded and linked into a higher-level model as shown in Figure 8.52. You need to specify which nodes in an OOBN are input and output nodes; at the higher level only the input and output nodes are shown (so the OOBN "Test Quality" has a single output node called *test quality* and the OOBN "Reliability During Test" has a single output node called *reliability during test*). We can then simply link relevant pairs of input and output nodes in two different OOBNs as shown in Figure 8.52.

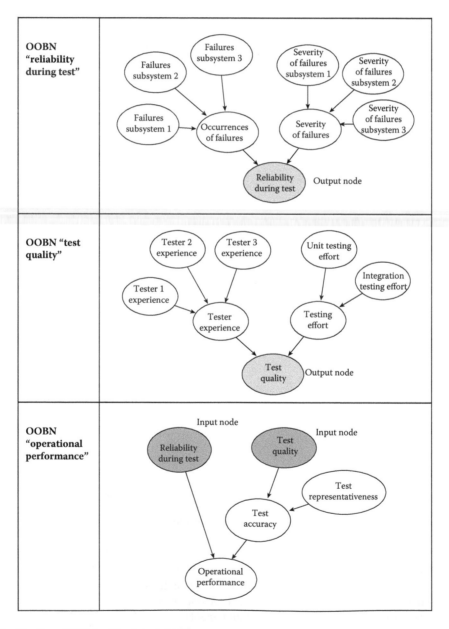

Figure 8.51   The three OOBNs of the larger model.

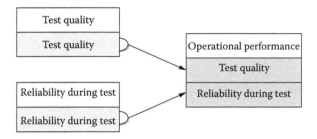

Figure 8.52   Top-level view of model.

In this example any updates to the OOBNs "Test Quality" and "Reliability During Test" will be reflected in the OOBN "Operational Performance" via the nodes *test quality* and *reliability during test*, respectively.

In the case of the flood model the one-day model can be thought of as a single object BN as shown in Figure 8.53.

To create an OOBN version of the 4-day model that we saw in Figure 8.49 we simply create four instances of the 1-day model and link them as shown in Figure 8.54.

All of these figures are screenshots from the AgenaRisk toolset. To create an OOBN in AgenaRisk you simply:

1. Create individual OOBNs in the same way you create any BN model. The only difference is that you will additionally declare (in the node properties) whether a node is an input/output node.
2. Import or copy previously saved OOBNs into a new model (in the model view).
3. Link appropriate input and output nodes.

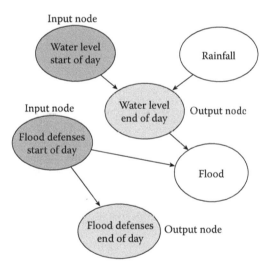

Figure 8.53   Single object BN for flood (with inputs and outputs shown).

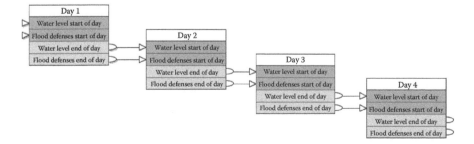

Figure 8.54   OOBN version of the flood model (4 days).

The various types of link connections that are allowed in an OOBN in AgenaRisk are shown in Box 8.4.

Box 8.4   Types of Link Connections That Are Allowed in an OOBN (AgenaRisk)

1. *Where the input and output nodes are exactly the same type with exactly the same number of state values* (this is the default connection). The result of this linking is to pass the entire set of probability values from the input node to the output node.
2. *From a continuous node to a continuous node.* In this case you can either pass the full set of marginals (as in the default case) or the value of a summary statistic as a constant. So, for example, the output node might represent a variable "height" and the input node might represent a variable "mean height." In this case the link type you would select would be the summary statistic "Mean."
3. *From noncontinuous node to a continuous node.* In this case you can pass the value of a single state as a constant. For example, the node *Flood* in the previous example is a Boolean node. We could link this to a continuous node (with a range 0 to 1) called *Flood probability* in another risk object and specify that the value passed is the value of true. If the value of the state true is 0.6 in the node *Flood*, then the node *Flood probability* will have the value 0.6.

The various constraints on OOBNs (including which nodes are allowed to be input and output nodes) are described in Box 8.5.

Box 8.5   Constraints on OOBNs

1. *Only nodes without parents can be input nodes.* The reason for this is that an input node should only ever be dependent on the output node of some other OOBN. It would make no sense for it to be an input node if it was dependent on some other internal parent.
2. *It is not possible to connect more than one output node to a single input node.* For the same reason as in 1, this makes no sense. If, for example, you wanted to aggregate the values of two output nodes and use this aggregated value as the input to a new risk object then the way to do this would be to ensure that the aggregation is done in the new risk object as shown in Figure 8.55 (generally the new risk object would have an aggregation node whose two parents are the two input nodes). That way you can achieve the desired effect by connecting the two outputs to the two inputs nodes of the new risk object. Note that an output node can be connected to input nodes in more than one risk object.
3. *Loops are not allowed between risk objects.* Thus, risk object *A* cannot link to object *B* if *B* is already linked to *A*.
4. *Between risk objects only forward propagation is performed in AgenaRisk.* Predictive calculations are performed by sequentially passing the resulting marginal distributions from one (ancestor) risk object to a connected descendant risk object. You must therefore be careful to only split a model into component risk objects according to the purpose of the model. In purely diagnostic models ancestor objects will be completely unaffected by observations entered in any or all ancestor risk objects. The flow of calculation is illustrated by means of the river flooding example again as shown in Figure 8.56. Also note that only marginal probability distributions, or summary statistics, are passed between risk objects and this results in loss of accuracy given that these may not be independent.

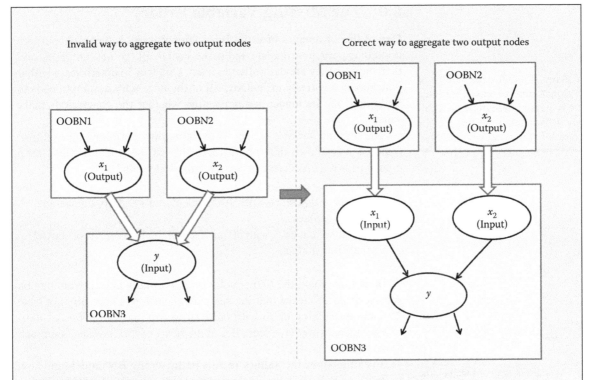

Figure 8.55   Aggregating output nodes.

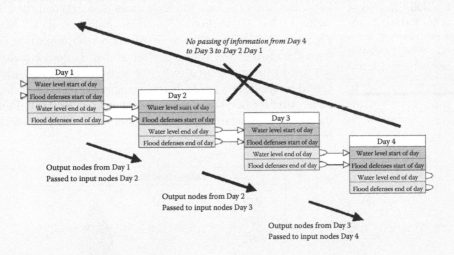

Figure 8.56   Flow of information in OOBN.

8.6   The Missing Variable Fallacy

One of the advantages of using Bayesian networks is that it encourages domain experts to articulate and make visible all their assumptions and then ensures they are carefully modeled. One way to determine whether your model has really articulated all of the necessary assumptions is to build and run the model and determine whether the conclusions make sense.

In many situations the failure of the conclusions to make sense in a BN is wrongly regarded as a fallacy. It is much more likely to be the result of missing key variables.

The missing variable fallacy is explained by the following example. Suppose it is known that, on average, 50% of the students who start a course pass it. Is it correct to conclude the following?

  a. A course that starts with 100 students will end up, on average, with 50 passes.
  b. A course that ends with 50 passes will, on average, have started with 100 students.

In fact, although the first conclusion is normally correct you may be very surprised to learn that the second conclusion is normally not true. This has everything to do with the way we reason with *prior assumptions*, which, as we have seen, lies at the heart of the Bayesian approach to probability.

In AgenaRisk you specify the variance, rather than the standard deviation, for the Normal distribution. So in this case the variance (which is the standard deviation squared) is set as 400. Since about 95% of the data lies within ±2 standard deviations of the mean, it follows that there is a 95% chance that a course will have between 130 and 230 students.

To explain how this fallacy results in the wrong BN model (and how to fix it) we will add some more detail to the example. The crucial prior assumption in this case is the probability distribution of student numbers who start courses. Let's suppose that these are courses in a particular college where the average number of students per course is 180. We know that some courses will have more than 180 and some less. Let's suppose the distribution of student numbers looks like that in Figure 8.57. This is a Normal distribution whose mean is 180. As we have already seen in Chapter 2, Normal distributions are characterized not just by the mean but also by the standard deviation, which is how spread out the distribution is. In this example the standard deviation is 20.

Because the number of students who pass is obviously influenced by the number who start, we represent this relationship by the BN shown in Figure 8.58(a).

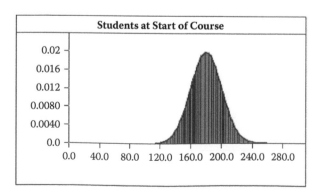

Figure 8.57 Prior distribution of student numbers on courses.

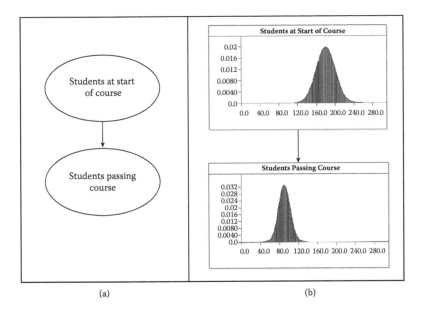

Figure 8.58 Number of students passing the course given number starting the course. (a) Underlying causal model. (b) Resulting marginal distributions.

To run this BN we need first to define the NPT for students passing course, that is, we need to define the conditional probability of students passing course given students at start of course. We know that on average the number of students who pass is 0.5 multiplied by the number of students who start the course. It follows, from Chapter 5, that the appropriate probability distribution is the Binomial distribution with $p = 0.5$ and n = number of students at start of course. The marginal distributions shown in Figure 8.58(b) are the result of running the Bayesian calculations in AgenaRisk with these prior distributions.

What we can now do is enter some specific values. Suppose we know that 100 students start the course. Entering this value, as an observation, results in the (predicted) distribution shown in Figure 8.59(a) for the number who will pass. As you would expect, the mean of the predicted distribution is 50.

However, suppose we do not know the number who start but we know that 50 passed a particular course. In this case (using Bayes' theorem) the model reasons backward to give the result shown in Figure 8.59(b). The mean of the predicted distribution for the number who started this course is not 100 but is higher; it is about 120. This seems to be wrong. But in fact, the fallacy is to assume that the mean should have been 100.

What is happening here is that the model is reasoning about our uncertainty in such a way that our prior assumptions are taken into consideration. Not only do we know that the average number of people who start a course is 180, but we also know that it is very unlikely that fewer than 120 people start a course (there are some such courses but they are rare). On the other hand, although a course with, say, 150 starters

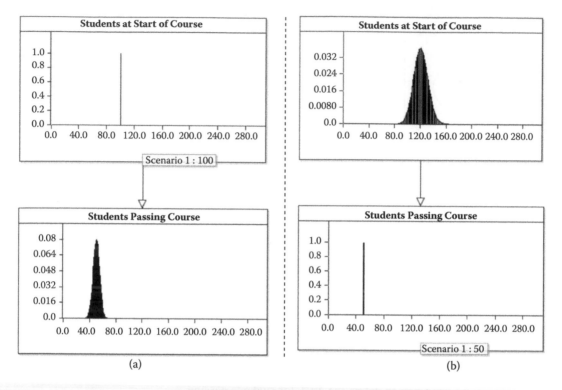

(a)   (b)

Figure 6.60   Entering observations: (a) Students at start of course (b) Students passing course

will on average result in 75 passes, about 5% of the time the number of passes will be 50 or lower.

Hence, if we know that there is a very low number, 50, of passes on a course there are two possible explanations.

1. There might have been far fewer students start the course than the 180 we expected.
2. The pass rate, that is, the probability of passing, for the course might have been far lower than the 50% we expected.

What the model does is shift more of the explanation on the latter than the former, so it says

> I am prepared to believe that there were fewer students starting this course than expected but the stronger explanation for the low number of passes is that the pass rate was lower than expected.

This reasoning uncovers the root of the fallacy in that we are given *some* information rather than *all* of the information we need about the problem and are then surprised that the model behaves irrationally. This is because we tend to focus on the information at hand and invest it with high relevance, but often fail to think it through to identify all of the parameters that are actually important.

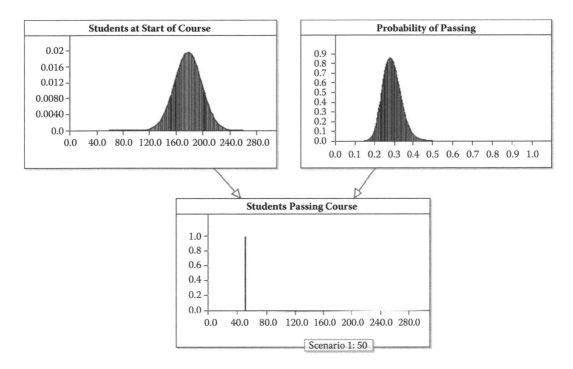

Figure 8.60   Incorporating the missing variable into the model.

When we incorporate uncertainty about the pass rate into the Bayesian model it provides a more coherent explanation of the data.

So, what has happened here is that the original model failed to incorporate our uncertainty about the pass rate. The correct model is the one shown in Figure 8.60 where we have added a node representing the variable "probability of passing."

Let's assume we are ignorant about the pass rate and hence specify probability of passing as *Uniform*[0,1].

Suppose we discover 50 students pass a course. Then in this case, as can be seen in Figure 8.60, the posterior marginal distributions in the model show that:

1. We are now much more uncertain about the number of students in the course.
2. We are very confident that the pass rate is less than 50%.

This type of problem is extremely common. Box 8.6 describes an analogous example drawn from our experience of working with military vehicle reliability experts.

Box 8.6 Forecasting Necessary Vehicle Fleet Sizes

This was a project tackling the problem of attrition rates for classes of military vehicles in combat. On the one hand we needed to know the likely number of vehicles left operational at the end of combat, given certain combat scenarios. On the other hand we also wanted to know: given a requirement for a minimum number of vehicles to be operational at the end of combat, what is the minimum number of vehicles to start with?

Although the model we needed involved many variables you can think of it conceptually in terms exactly like the student pass rate model where

- Vehicles at start of combat replaces students starting course
- Operational vehicles at end of combat replaces students passing course

As in the student example users of the model could not understand why it predicted, say, 50 vehicles at the end of combat, given 100 vehicles at the start, yet predicted over 120 vehicles at the start given 50 at the end. Since the prior distribution for vehicles had been provided by the domain experts themselves, the model was, in fact, working correctly and rationally even if it did not produce the results that the users thought were sensible. In this case it was the strength of the prior distribution that the users had to review for correctness, resulting in an improved model where previously missing variables were added.

8.7 Conclusions

Building the structure of a large-scale BN leads to similar challenges faced by designers of complex systems in many engineering disciplines. Just as in any other form of complex design, we need

- Methods for identifying known solutions and mapping these to the problem
- Ways of building the complex system by combining smaller components.

Specific types of BN modeling problems encountered in practice involve difficulties in

- Determining sensible edge directions in the BN given that the direction of inference may run counter to causal direction
- Applying notions of conditional and unconditional dependence to specify dependencies between nodes
- Building the BN using a "divide and conquer" approach to manage complexity

To address these problems we have described a set of generally applicable building blocks, called idioms, to capture the experience embodied in previously encountered BN patterns. Idioms can be combined together into objects. These can then in turn be combined into larger BNs, using the method of multiobject BNs.

The method and idioms described have been used in a wide range of real applications. This experience has demonstrated that relative BN novices can build realistic BN graphs using idioms and multiobject BNs. Although we do not claim that the set of idioms is complete for all application domains they have proven remarkably robust for a wide range of risk assessment problems.

Another major issue faced by BN model designers is how best to tackle complex mutual exclusivity constraints and relationships between nodes in the model. We have shown how we can do this in a variety of situations where we may have mutually exclusive causal paths in a model, states, or where we have a rich taxonomic hierarchy.

Of course, in this chapter we have only provided a partial solution to the modeling challenge. We have explained how to build appropriate BN graphs but have provided few guidelines on how to define the crucial NPTs associated with the nodes of these graphs. We address this in the next chapter.

Further Reading

Fenton, N. E., Neil, M., Lagnado, D., Marsh, W., Yet, B., Constantinou, A. (2016). How to model mutually exclusive events based on independent causal pathways in Bayesian network models. *Knowledge-Based Systems*, 113, 39–50. http://dx.doi.org/10.1016/j.knosys.2016.09.012
This paper gives the general formula for setting the soft evidence values that solve the mutual exclusivity modeling problem.

Jensen, F. (1996). *Introduction to Bayesian Networks*, Springer-Verlag.

Jensen, F. V., and Nielsen, T. (2007). *Bayesian Networks and Decision Graphs*, Springer-Verlag.

Koller, D., and Pfeffer, A. (1997). Object-Oriented Bayesian Networks, in *Proceedings of Thirteenth Annual Conference on Uncertainty in Artificial Intelligence*, Morgan Kaufmann.

Laskey, K. B., and Mahoney, S. M. (1997). Network fragments: Representing knowledge for constructing probabilistic models, in *Proceedings of Thirteenth Annual Conference on Uncertainty in Artificial Intelligence*, D. Geiger and P. Shenoy, eds., Morgan Kaufmann.

Lauritzen, S. L., and Spiegelhalter, D. J. (1988). Local computations with probabilities on graphical structures and their application to expert systems (with discussion). *Journal of the Royal Statistical Society, Series B* 50(2), 157–224.

Neil, M., Fenton, N. E., and Nielsen, L. (2000). Building large-scale Bayesian Networks. *The Knowledge Engineering Review* 15(3), 257–284.

Nikiforidis, G., and Sakellaropoulos, G. (1998). Expert system support using Bayesian belief networks in the prognosis of head-injured patients of the ICU. *Medical Informatics* (London) 23(1), 1–18.
This is the paper (referred to in Box 8.2) in which a BN model for head injury prognosis was learned from data.

Pearl, J. (2000). *Causality: Models Reasoning and Inference*, Cambridge University Press.

Visit www.bayesianrisk.com for exercises and worked solutions relevant to this chapter.

Building and Eliciting Node Probability Tables

9.1 Introduction

This chapter discusses one of the main challenges encountered when building a Bayesian network (BN) and attempting to complete the node probability tables (NPTs) in the BN. This is that the number of probability values needed from experts can be unfeasibly large, despite the best efforts to structure the model properly (Section 9.2). We attempt to address this problem by introducing methods to complete NPTs that avoid us having to define all the entries manually. Specifically, we use functions of varying sorts and give guidance on which functions work best in different situations. The particular functions and operators that we can use depend on the type of node: In this chapter we focus on discrete node types: labeled, Boolean, discrete real, and ranked. In Section 9.3 we show that for labeled nodes comparative expressions (e.g., IF, THEN, ELSE) can be used, and in Section 9.4 for Boolean nodes we show that we can also use a range of Boolean functions including OR, AND, and special functions such as NoisyOR. We also show how to implement the notion of a weighted average. In Section 9.5 we describe a range of functions that can be used for ranked nodes. Finally Section 9.6 describes the challenges of eliciting from experts probabilities and functions that generate NPTs. This draws on practical lessons and knowledge of cognitive psychology.

The previous two chapters provided concrete guidelines on how to build the structure, that is, the graphical part, of a BN model. Although this is a crucial first step, we still have to add the node probability tables (NPTs) before the model can be used.

The focus of this chapter is on discrete variables. We will deal with probabilities for continuous variables in Chapter 10.

9.2 Factorial Growth in the Size of Probability Tables

Recall that the NPT for any node of a BN (except for nodes without parents) is intended to capture the strength of the relationship between the node and its parents. This means we have to define the probability of the node, given every possible state of the parent nodes. In principle this looks like a daunting problem as explained in Box 9.1.

All models in this chapter are available to run in AgenaRisk, downloadable from www.agenarisk.com.

Box 9.1   Factorial Growth in Probability Tables

Consider a simple BN shown in Figure 9.1. The NPT for the node *Water level at end of day* has 3 states and its parent nodes have 3 and 4 states, respectively, making 36 possible state combinations to consider. When you create this BN model in a tool like AgenaRisk and examine the NPT for the node *Water level at end of day* you will see a table with all cells equal:

Water Level at Start of Day	Low				Medium				High			
Rainfall	None	<2 mm	2–5 mm	>5 mm	None	<2 mm	2–5 mm	>5 mm	None	<2 mm	2–5 mm	>5 mm
Low	1.0	1.0	1.0	1.0	1.0	1.0	1.0	1.0	1.0	1.0	1.0	1.0
Medium	1.0	1.0	1.0	1.0	1.0	1.0	1.0	1.0	1.0	1.0	1.0	1.0
High	1.0	1.0	1.0	1.0	1.0	1.0	1.0	1.0	1.0	1.0	1.0	1.0

Your task is enter meaningful probability values so that it looks something like this:

Water Level at Start of Day	Low				Medium				High			
Rainfall	None	mm	mm	mm	None	mm	mm	mm	None	mm	mm	mm
Low	1.0	0.7	0.1	0.0	0.1	0.0	0.0	0.0	0.1	0.0	0.0	0.0
Medium	0.0	0.3	0.5	0.2	0.9	0.8	0.2	0.1	0.4	0.1	0.0	0.0
High	0.0	0.0	0.4	0.8	0.0	0.2	0.8	0.9	0.5	0.9	1.0	1.0

Although a 36-cell table might be manageable, suppose that we decided to change the states of the water level nodes to be: very low, low, medium, high, very high. Then the NPT now has $5 \times 5 \times 4 = 100$ cells. If we add another parent with 5 states then this jumps to 500 entries, and if we increase the granularity of rainfall to, say, 10 states then the NPT has 1,250 cells. There is clearly a problem of combinatorial explosion.

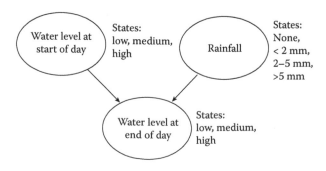

Figure 9.1   A simple BN.

Whether it is even practically possible to define the NPT for any given node depends very much on the type of the node (including how many states it has) and the types of the parent nodes. By following the principles recommended in the previous chapters you should arrive at a model in which it is practical to define the NPT for any node, but that does not mean it will always be easy to do so. The way you define the NPTs will generally differ according to the type of node and its parents.

As we show in this chapter, there are many new methods that can now be used to make the task of defining sensible NPTs far easier than before. So do not be put off by naysayers (and there are still many) who insist that BNs are unsuitable because it is too hard to define the NPTs.

9.3 Labeled Nodes and Comparative Expressions

A labeled node is one whose set of states is simply a set of labels. A number of examples of such labeled nodes are shown in Table 9.1. Irrespective of what the names of the labels are, they have no underlying semantic meaning. So, for example, in the node *Injury* there is no underlying assumption that, say, "broken bone" is better or worse than "damaged ligament" or that "broken bone" is better or worse than "other." All that is assumed by the model is that the states are different. This makes sense for the nodes *Student*, *Injury*, *Sport*, *Program language*, *Die roll*, and *Bank loss* in the table. It makes less sense for the other nodes because implicitly in those cases we would like the states of the nodes to carry some other meaning. By declaring a node like *Water level* to be labeled (which, as we will see, is possible but not recommended), there is no implicit knowledge that the state "medium" represents a higher level than the state "low." Any meaning has to be provided manually by the modeler in the way the NPT is defined. That was what we did in the way we specified the NPT in the example in Box 9.1. For example, in the NPT, if the water level at the start of the day was "medium," then the probability the water level at the end of the day was "high," given 2–5 mm of rainfall, is greater than the probability the water level is "high" if it started the day "low."

In most cases when you create a labeled node there is little alternative but to define the NPT manually cell by cell.

Table 9.1
Labeled Node Examples

Node Name	States
Student	Peter, John, Alison, Nicole
Injury	broken bone, damaged ligament, other
Sport	Tennis, Athletics, Football
Program language	C, C++, Java, Basic, Assembly, other
Die roll	1, 2, 3, 4, 5, 6, other
Bank loss	frauds only, lost transactions only, both frauds and lost transactions, no loss
Disease	true, false
Water level	low, medium, high
Rainfall	none, less than 2 mm, 2 to 5 mm, <5 mm

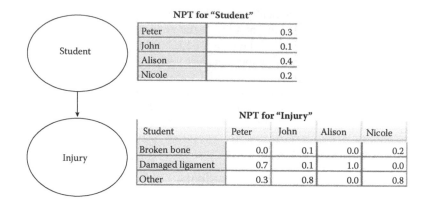

Figure 9.2   Simple BN with labeled nodes and associated NPTs.

Example 9.1

Consider the BN shown in Figure 9.2. In this model we are trying to predict the injuries incurred by a set of students who participate in college sports. The node *Student* simply represents the set of students of interest. Since it has no parents its NPT is simply the prior probability of the student's participation in sport. If all students are equally likely to participate then we could leave the NPT values as all being equal (i.e., each having a probability value 0.25). But in this example let us suppose we have prior knowledge of the relative participation as shown in the figure. So in previous activities involving these four students (assuming no overlaps), Peter was involved in 30%, John in 10%, Alison in 40%, and Nicole in 20%.

The node *Injury* is conditional on *Student* so we have to specify each of the state combinations shown. So, in this case, we judge that Peter has no chance of a broken bone, 70% chance of a damaged ligament, and 30% chance of some other injury, and so on.

Figure 9.3 demonstrates what happens when (a) we run this model in its initial state and (b) with an observation entered. So, for example, we

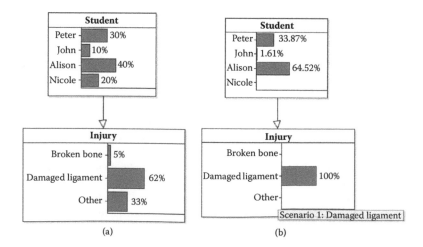

Figure 9.3   Calculating the model. (a) Probability values in initial state. (b) Probability values after entering observation.

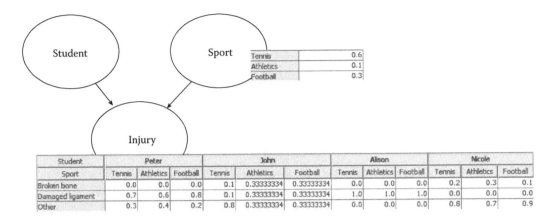

Student	Peter			John			Alison			Nicole		
Sport	Tennis	Athletics	Football	Tennis	Athletics	Football	Tennis	Athletics	Football	Tennis	Athletics	Football
Broken bone	0.0	0.0	0.0	0.1	0.33333334	0.33333334	0.0	0.0	0.0	0.2	0.3	0.1
Damaged ligament	0.7	0.6	0.8	0.1	0.33333334	0.33333334	1.0	1.0	1.0	0.0	0.0	0.0
Other	0.3	0.4	0.2	0.8	0.33333334	0.33333334	0.0	0.0	0.0	0.8	0.7	0.9

Figure 9.4   Revised BN and associated NPTs.

can see that the initial marginal probability value for damaged ligament is 62%. If we discover that the injury is definitely damaged ligament, then there is a 64% chance that Alison is the student who suffered this injury and a less than 2% chance it is John.

Example 9.2

Suppose we want to extend the model in Example 9.1 to incorporate the impact of the particular sport on the injury. Not only does the new BN shown in Figure 9.4 require a new NPT for the new node *Sport*, but we also have to revise the NPT for *Injury* to take account of the fact that it now has an extra parent node.

However, even for labeled nodes there is a way that you can drastically simplify the NPT specification, although the method is only relevant if you want to replace the probability values with deterministic values (zeros and ones). The method is to use a comparative *expression* as explained in Box 9.2. In the example presented all of the nodes are labeled. It turns out (as we shall see) that if the parents of a labeled node have a richer type, then the use of expressions becomes much more relevant.

Box 9.2   Using Comparative Expressions

In some circumstance it may be sufficient to use a simple comparative expression to define a full NPT. Let us return to the BN in Example 9.1. Let us suppose that, instead of the NPT for "Injury" defined in Figure 9.2, we have the following prior information:

- The only possible injury Alison and Nicole can suffer is a damaged ligament.
- The only possible injury Peter can suffer is a broken bone.
- The only possible injury John can suffer is "other."

In that case all the information we need for the node *Injury* is captured using the following logical expression:

```
if (student == "Alison" || student == "Nicole", "Damaged ligament",
if (student =="Peter", "Broken bone", "Other"))
```

This expression uses standard Boolean logic and is based around the following rules:

Expression	Meaning
A == B	A is equal to B. For example, the expression `student == "Alison"` means student is Alison
A || B	Either A or B is true. For example, the expression `student == "Alison" || student == "Nicole"` means the student is either Alison or Nicole.
A && B	Both A and B are true. For example, the expression `student == "Alison" && injury == "Damaged ligament"` means the student is Alison and *injury* is damaged ligament.
if (<condition>, "option1", "option2")	If <condition> is true then "option1" must be true, otherwise "option2" must be true. This is called a conditional expression. So, for example, `if (student == "Alison", "Damaged ligament", "Broken bone"))` means that if the student is Alison then "damaged ligament" is true, otherwise "broken bone" is true.

We can nest comparative expressions inside expressions to create more complex expressions. This is what we did in our example, which can be parsed as follows:

```
if (Condition 1, "Damaged ligament", Condition 2)
```

where

Condition 1 is the expression:

```
student == "Alison" || student == "Nicole", "Damaged ligament"
```

Condition 2 is the expression:

```
student =="Peter", "Broken bone", "Other"
```

In AgenaRisk if you feel it is suitable to use an expression for an NPT you simply select the option *Expression* and you can enter the comparative expression exactly like the one earlier.

Tip: In the AgenaRisk expression editor, by right clicking you can insert a parent name. Make sure the syntax is exactly correct.

AgenaRisk generates the NPT behind the scenes and uses this generated probability table in subsequent calculations. You can access this NPT directly by setting NPT editing mode to *manual* after generating the expression and the table can now be edited. In this way you can generate deterministic values in the table as zeros and ones and then replace some of them with probability values.

An alternative to comparative expressions is the *partitioned expression*, which is also available in AgenaRisk. This is explained in Box 9.3.

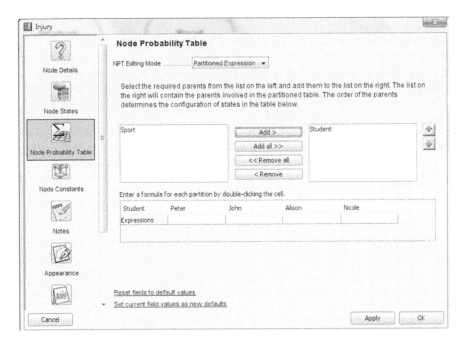

Figure 9.5   Selecting the NPT for the node "injury" as a partitioned expression conditional on "Student."

Box 9.3   Using Partitioned Expressions

If a node has more than one parent, as in Example 9.2, you can use a partitioned expression. Suppose, for example, that we have the following information:

- When Peter's sport is football his injury is a broken bone. For any other sport the injury is "other."
- When John's sport is athletics he can only suffer a damaged ligament. When football he can only suffer a broken bone.
- Alison and Nicole can only suffer a damaged ligament irrespective of what sport played.

Then what we can do is write expressions for injury that are conditioned on the *student*. This is what is meant by an NPT being defined by a partitioned expression. In AgenaRisk if you specify that an NPT for a node is to be a partitioned expression, then you can select what parent node(s) to condition on, as shown in Figure 9.5. The figure shows the blank table. We need to complete each entry with an expression. For example the entry for Peter might be:

```
if (sport == "Football", "Broken bone", "Other")
```

9.4   Boolean Nodes and Functions

In Table 9.1 we had an example of a labeled node called *Disease* whose states were true and false. In fact, such a node should be specified as a *Boolean* node. By doing so we can use, as we will see, the full power of Boolean logic to build NPTs.

In general a Boolean node is defined as a node that has exactly two states, true and false, whose meaning is that used in the chapters on probability theory (i.e., the true state is the complement of the false state). But *any* node that has exactly two states can be considered a Boolean node as explained in Box 9.4.

Box 9.4   Why Any Node with Exactly Two States Can Be Considered Boolean

Suppose the node *Disease* has exactly two mutually exclusive states: "cancer" and "bronchitis." Then if we know that the state of *Disease* is not "cancer" it follows that the state of disease must be "bronchitis." Hence, the state "bronchitis" is the complement of the state "cancer." Suppose the probability of cancer is 0.7. Then it follows by Axiom 4.2 (see Chapter 5) that the probability of bronchitis is 0.3. But then we may as well define the states of *Disease* to be "cancer" and not "cancer."

Equivalently we could simply rename the node *Disease* as *Cancer* and give it two states: *true* and *false*. In fact, it follows that any choice of node name and associated state names in Table 9.2 are equivalent.

Table 9.2
All of These Choices of Node Name and States Are Equivalent

Node Name	States
Disease	Cancer, Bronchitis
Disease	Cancer, Not Cancer
Disease	Not Bronchitis, Bronchitis
Cancer	True, False
Cancer	Yes, No
Bronchitis	False, True
Bronchitis	No, Yes

In general, suppose a node X has exactly two states: A and B. Then if X is not in state A it must be in state B, which means B is the complement of A. But then "*not A*" is equivalent to B. This in turn means that X is equivalent to a node called A whose values are *true* and *false*. In other words X is a Boolean node.

9.4.1   The Asia Model

A classic example of a BN in which every node is Boolean is the Asia model shown in Figure 9.6. The Asia model is intended to provide decision support in a chest clinic to help diagnose the likelihood of three

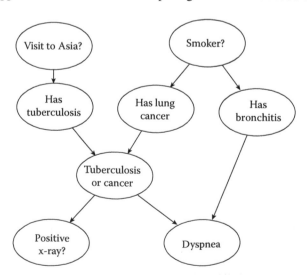

Figure 9.6   Classic Asia BN.

medical conditions—tuberculosis (TB), lung cancer, and bronchitis—in any patient coming to the clinic. It does this from two causal factors: smoking and whether the patient has visited Asia recently. Additionally, two other pieces of diagnostic evidence may be available:

- Whether the patient is suffering from shortness of breath (dyspnea)
- Whether a positive x-ray test result is available (where positive is an indicator of the disease being present).

The NPTs for the nodes are as specified in Table 9.3. Because no node in the model has more than two parents, and because Boolean

Table 9.3
Asia NPTs

Node	NPT					Justification
Smoker	True	0.5				50% of previous visitors to clinic were smokers.
	False	0.5				
Visit to Asia	True	0.01				1% of previous visitors to clinic had recently visited Asia.
	False	0.99				

Tuberculosis (TB)	Visit to Asia	True	False	Tuberculosis is 5 times more likely given Asian visit but generally very unlikely.
	True	0.05	0.01	
	False	0.95	0.99	

Lung cancer	Smoker	True	False	Figures based on previously diagnosed patients. So, for example, only 1% of nonsmokers were diagnosed with cancer.
	True	0.1	0.01	
	False	0.9	0.99	

Bronchitis	Smoker	True	False	Figures based on previously diagnosed patients. So, for example, only 60% of smokers were diagnosed with bronchitis.
	True	0.6	0.3	
	False	0.4	0.7	

Tuberculosis or cancer	Tuberculosis	True		False		This is simply the Boolean OR function. We use a conjoined node here of both medical conditions because the x-ray test cannot differentiate between them. It detects a shadow on the lung but cannot differentiate the cause.
	Cancer	True	False	True	False	
	True	1	1	1	0	
	False	0	0	0	1	

Positive x-ray	TB or Cancer	True	False	Accuracy figures based on previous patients. 5% of patients without either disease wrongly got a positive x-ray and 2% received a false-negative result.
	True	0.98	0.05	
	False	0.02	0.95	

Dyspnea	Bronchitis	True		False		Figures based on previous patients.
	Tuberculosis	True	False	True	False	
	True	0.9	0.8	0.7	0.1	
	False	0.1	0.2	0.3	0.9	

nodes have just two states, no NPT contains more than four cells. Hence, it is relatively easy to complete all the NPTs manually.

Despite its simplicity, the model is useful and is a good illustration of the power and flexibility of BNs compared with other decision support systems. This is explained in Box 9.5.

Box 9.5   Running the Asia Model

In its initial state the model is as shown in Figure 9.7. This represents our state of uncertainty at the point before a new patient arrives at the clinic. Suppose the patient complains of shortness of breath. Then we set *Dyspnea* to be "yes" as shown in Figure 9.8.

The model automatically recalculates the probabilities so you see that the chances of all three diseases increase. At this point we believe it is most likely (83%) the patient has bronchitis.

Next suppose we find the patient is a nonsmoker. If you enter the observation *Smoker* to be "no" and run the model, you will see that the belief in all diseases now falls slightly but we still believe bronchitis is the most likely (75%).

At this point, suppose the doctor conducts an x-ray and receives a positive result. So set *Positive x-ray?* to be "yes" and run the model. The belief in both lung cancer and tuberculosis now increase to about 25% each but the belief in bronchitis is still highest at 57%. Finally, suppose we find out the patient has recently visited Asia. So set Visit to Asia to be "yes" and run the model. The belief in tuberculosis jumps to 63%, while bronchitis falls to 46% and lung cancer to 13%. Hence, tuberculosis is now the most likely medical condition for that patient.

From an initial set of evidence the model arrived at a diagnosis of bronchitis, but the new evidence explained the bronchitis diagnosis away. As we saw in Chapter 7 this explaining-away property is a key feature of BNs. Essentially it means that the prior hypothesis we believed was transformed, by data, into a different posterior hypothesis. The visit to Asia and x ray results were incompatible with a high belief in bronchitis.

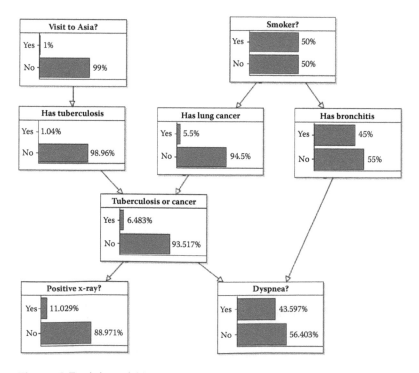

Figure 9.7   Asia model in initial state

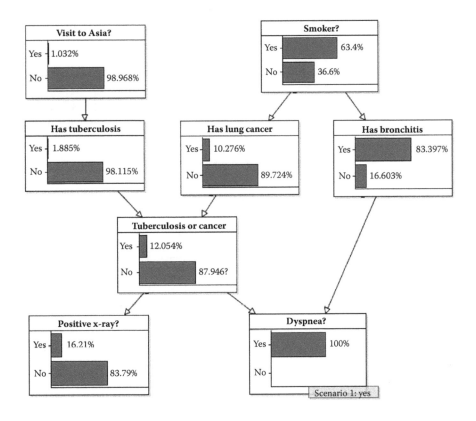

Figure 9.8 Dyspnea set to true.

Although all the NPTs in the Asia model are easily entered manually, even this model has features that suggest a more automated approach should be used, as we show in the next subsection.

9.4.2　The OR Function for Boolean Nodes

The NPT for the node *Tuberculosis or cancer* in the Asia model was defined as shown in Table 9.4. The table tells us that tuberculosis or cancer is certain to be true when either of the parent nodes is true (since the probability of true is 1 in each such case), and is certain to be false otherwise (since the probability of false is 1 in the case where this applies).

Table 9.4
NPT for Node *Tuberculosis or Cancer*

Tuberculosis	True		False	
Cancer	True	False	True	False
True	1	1	1	0
False	0	0	0	1

But there is a logical Boolean operator, called the *OR function*, that precisely captures this situation:

OR Function

In general for any two Boolean variables *A* and *B*, the expression *A* OR *B* is defined to be true if either *A* or *B* is true (meaning at least one is true) and is defined to be false otherwise.

The ‖ notation is used to ensure consistency with modern programming languages, most of which use ‖ to represent OR.

In AgenaRisk the OR function is implemented using the notation ‖ we already introduced in Section 9.3.

So to implement the expression *A* OR *B* we would write

```
if (A == "True" || B== "True", "True", "False")
```

So, for example, instead of defining the NPT for the node *Tuberculosis or cancer* manually, we could simply have used the expression

```
if (Tuberculosis == "True" || Cancer == "True",
"True", "False")
```

The real benefit of the OR operator for defining NPTs of Boolean nodes comes when a node has many parents, as the following example shows.

Example 9.3

A computer system is made up of the following hardware components: central processing unit (CPU), motherboard, hard disk, graphics card, PCI card, screen, keyboard.

A failure of any one of these components will result, as far as an end user is concerned, in a failure of the whole system. The probability of failure per session of each individual component is known. We want to build a model that enables us to answer the following kinds of questions:

- What is the probability that the system will fail in a given session?
- If the system has failed and we know that the screen, keyboard and hard disk are OK, what is the most likely cause of failure?

The solution is to build the BN model shown in Figure 9.9 where each node is a Boolean node and where in each case "true" means that component fails.

The NPT for the node System is simply defined as the OR operator—in this case the OR needs to cover each parent. Specifically in AgenaRisk you would write the expression

```
if (CPU == "True" || Motherboard== "True" || hard_disk
=="True" || graphics_card =="True" || PCI =="True" ||
Screen =="True" || Keyboard =="True", "True", "False")
```

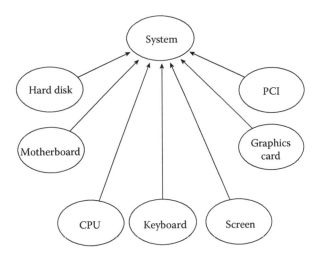

Figure 9.9 BN model for computer system failure.

For each of the parent nodes the NPT is simply defined by setting the probability of true to be the failure rate of that component. So, for example, if the failure rate of the CPU is 0.001 (i.e., 1 in a 1000) then the NPT for CPU is defined as

False	0.999
True	0.001

Figure 9.10 shows the model in its initial state (so this shows the prior NPTs for each of the components). The initial state of the model

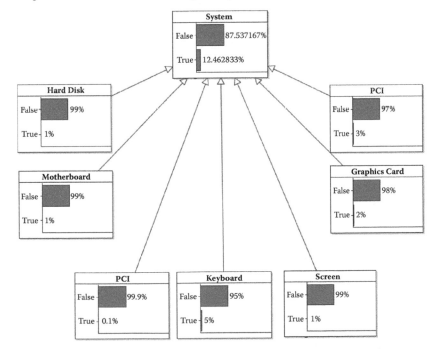

Figure 9.10 Initial priors for computer model.

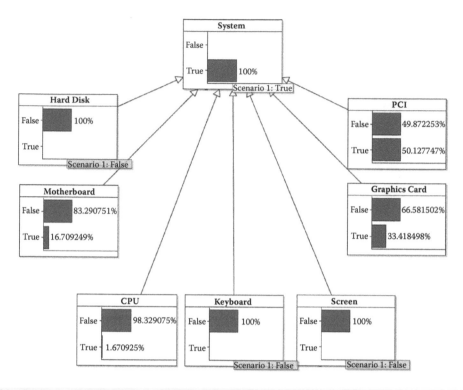

Figure 9.11   System has failed and some components are checked as OK.

If you were able to understand the discussion in Chapter 7 about conditional independence assumptions, you will know that this model assumes that all of the components fail independently. If, in practice you know that, for example, a CPU failure makes a hard disk failure more likely, then you would need to add a link from CPU to hard disk and redefine the NPT of hard disk accordingly.

The important thing is that we were able to build a useful model with minimal effort. If we had been forced to manually define the NPT for the node *System* it would involve 256 cell entries (albeit all of them either zero or one).

enables us to answer the first question of interest, namely, what is the probability that the system will fail in a given session. In this initial state the model has computed the probability of failure of the system by executing the OR function that characterizes the NPT of the node *System*. So we simply read off the calculated probability value for true—it is just over 12%.

Next suppose we know the system has failed and we know that the screen, keyboard, and hard disk are OK. To answer the question what is the most likely cause of failure, we simply enter the observations we know (as shown in Figure 9.11). The propagation revises all of the unknown probabilities and we can see, for example, that the probability the CPU has failed is less than 2%. By far the most likely cause of failure is the PCI card (just over 50%).

There are many situations in risk assessment problems that are similar to Example 9.3. Any type of system (including human systems) where correct functioning requires every one of a set of components to function properly can be modeled in the same way (examples include transport systems, management systems, and human diseases).

Example 9.4

Suppose a number of diseases (let's call them *disease1*, *disease2*, … *diseaseN*) have the same symptoms, which can be a physical manifestation or a test result (we saw examples of both in the Asia model, where lung cancer and bronchitis both have the physical symptom dyspnea and both lung cancer and TB yield a positive x-ray result). Assuming that patients can suffer from more than one of the diseases at the same time we can model this as a set of nodes: *disease1, disease2, … diseaseN*.

Assuming a symptom node can be caused by one or more disease nodes we could build a BN with only the relevant disease nodes as parents of the symptom and then define the NPT for the symptom as simply the OR operator over all relevant parent disease nodes. The symptom will then occur if one or more diseases are present.

The population rates of the diseases can be used to define the NPTs for each of the disease nodes. The resulting model will automatically provide the prior probability that a patient has the symptom. If the patient does have the symptom then we can use the model to revise our belief about the most likely disease given the symptom.

The flip side of the situation of Example 9.4 is where we are trying to diagnose one of several diseases but where the symptoms may differ between the diseases. This gives rise to a very common, and classic, type of BN called a *naïve Bayesian classifier model*. A full explanation and example is provided in Box 9.6.

Box 9.6   Naïve Bayesian Classifier Model

The three diseases cold, flu, and bronchitis (which, for simplicity, we assume are mutually exclusive and exhaustive in this context) exhibit certain similar symptoms (such as cough and headache), but to more or less degrees or uncertainty. Some symptoms (such as nausea) are much more likely in flu than in either cold or bronchitis. To model this situation we can use the structure shown in Figure 9.12.

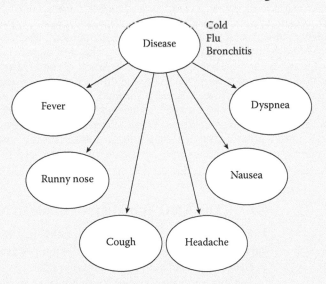

Figure 9.12   Naïve Bayes classifier model for diagnosing disease from symptoms.

Unlike the previous examples we do not have separate (Boolean) nodes for each possible disease. Instead, we simplify things by creating a single labeled node called Disease whose mutually exclusive states are the three diseases. All of the symptom nodes are Boolean, and the arc direction makes sense since the relationship is causal from disease to symptom.

The NPT for *Disease* is defined by the prior incidence rates:

Cold	0.85
Flu	0.10
Bronchitis	0.05

So what this means is that 85% of people who have one of the diseases will have a cold, 10% will have flu, and 5% bronchitis.

Examples of the NPTs for the symptom nodes are shown in Table 9.5 and Table 9.6. The model in its initial state is shown in Figure 9.13. In Figure 9.14(a) we see two symptoms (cough and dyspnea) that are especially associated with bronchitis. However, because of the low prior probability of bronchitis, and because these symptoms are also (albeit to a lesser extent) seen with people with colds and flu, the model still points to cold as the most likely disease.

Table 9.5

NPT for Runny Nose

Disease	Cold	Flu	Bronchitis
False	0.01	0.01	0.6
True	0.99	0.99	0.4

Table 9.6

NPT for Dyspnea

Disease	Cold	Flu	Bronchitis
False	0.9	0.8	0.01
True	0.1	0.2	0.99

However, things change dramatically if we know the patient does not have a runny nose (Figure 9.14(b)). Now we are almost certain that the disease is bronchitis.

It is easy to see why the naïve classifier is a popular structure for BN models. It represents a very natural combination of cause–consequence idioms and it is extremely simple to construct (no node has more than one parent and so the NPTs are easy to define manually). The Bayesian calculations can also be performed relatively easily and efficiently even without a BN tool. Indeed, this type of model has been used extensively in applications such as e-mail SPAM filters (where the parent node has states corresponding to the different SPAM classifications and the child nodes correspond to different types of SPAM characteristics) and personalization systems (where the parent node has states corresponding to the different types of preferences and the child nodes correspond to instances of each preference selected by the user).

Of course the model is limited to problems of a specific nature and with some stringent assumptions. In the example we must assume that the symptoms are conditionally independent and we must assume that no patient can have more than one disease (otherwise we would need to introduce states like cold only, cold and bronchitis, and so on, and the model then becomes messy especially when it comes to ensuring that all the NPTs are consistent).

9.4.3　The AND Function for Boolean Nodes

In Example 9.3 the computer system failed if at least one of the individual components failed, and the OR operator is exactly what was needed to enable us to model this situation in a BN.

But many modern systems (both physical and human) are designed in such a way that a system will only fail when *all* of some set of components fail. For example:

- Many safety critical systems (including transport systems, hospitals, and purely electronic systems) contain multiple power

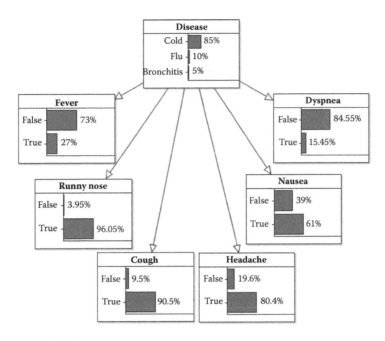

Figure 9.13 Initial state of model.

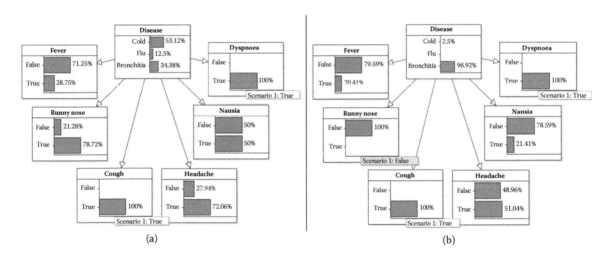

Figure 9.14 Entering observations. (a) Symptoms that are normally associated with bronchitis. (b) Patient does not have runny nose.

supplies. If the first power supply fails the system automatically switches to the second and so on. Only when every supply fails will the system fail due to lack of power.

■ A bank has several levels of control checking to ensure that fraudulent transactions are not carried out. Only if every one of these control checks fails can a fraudulent transaction take place.

- In the film *Armageddon* NASA decided to fly two separate shuttles, each with its own drilling equipment and nuclear bomb in order to destroy the asteroid. Only if both shuttle teams failed would the mission objectives fail.
- The Airbus fleet of fly-by-wire aircraft contain more than one identical computers, so that if one fails the control switches to the next.

The Boolean operator that captures this notion is the AND operator.

AND Operator

In general for any two Boolean variables *A* and *B*, the expression *A* AND *B* is defined to be true if both *A* and *B* are true and is defined to be false otherwise.

This generalizes to more than two variables. So, for example *A* AND *B* AND *C* is true if *A*, *B*, and *C* are all true and false otherwise.

So, for example, if *A* and *B* represent respectively the Boolean variables "Power supply 1 fails" and "Power supply 2 fails," then the Boolean variable "System power fails" is equivalent to "*A* AND *B*." A BN model with four power supplies is shown in Figure 9.15. Figure 9.15(a) provides the information you need to complete the NPTs for each of the parent nodes. For example, the NPT for *Power 1* has true set to 0.98 and false set to 0.02. This represents the prior belief that power supply 1 has a probability of failure of 0.02.

For the node *System* the NPT is the expression capturing the AND operator. In AgenaRisk the AND operator is implemented using the notation && we already introduced in Section 9.3 (the && notation is to ensure consistency with modern programming languages, most of which use && to represent AND). So to implement the expression *A* AND *B* we would write

```
if (A == "True" && B == "True", "True", "False")
```

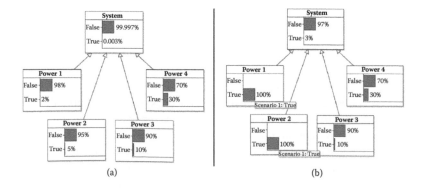

(a)   (b)

Figure 9.15 BN model of system failure given four different power supply failures. (a) Initial state of model. (b) First two power supplies fail.

Hence the expression for the NPT of the node *System* is

```
if (Power_1 == "True" && Power_2 == "True" && Power_3
== "True" && Power_4 == "True" , "True", "False")
```

It is much easier to write this expression than to complete the NPT manually, which would require 32 entries even in this simple case.

9.4.4   *M* from *N* Operator

In the AND operator the child node (such as that representing system failure) is true only when all of its parents are true, whereas in the OR operator the child node is true when at least one of its parents is true. But, in many situations we want a more general operator, one that returns true if at least a specified number of parents are true. For example, we may need to model the situation where a system will fail only if a majority of the components fail or will fail only if at least two fail.

Fortunately there is an operator called the *M from N operator* that enables us to specify any such situation (providing that M is less than N):

M from N Operator

In general for any *n* Boolean variables $A_1 A_2 \dots A_n$, the expression *M from N* $(m, n, A_1 A_2 \dots A_N)$ is defined to be true if at least *m* out of *n* of the A_i's is true.

Suppose, for example, that in Figure 9.15 the system actually fails only if at least three of the power supplies fail. Since there are four power supplies in total here the expression we need to define the required NPT for the node *System* is

```
M from N (3,4, power1, power2, power3, power4)
```

To produce this M from N operation in AgenaRisk we use the following expression:

```
mfromn(3,4, power1 == "True" , power2 == "True",
power3 == "True" , power4 == "True")
```

This works for any number of parents (so of course *N* does not have to be 4 as in this example).

Note that if *m* = 1 then the M from N operator is the same as the OR operator and if *m* = *n* then the M from N operator is the same as the AND operator.

9.4.5   NoisyOR Function for Boolean Nodes

Suppose you are trying to predict whether a person is going to have a heart attack before the age of 60 and that we identify the set of causal factors as shown in Figure 9.16.

As before the model is making a key assumption of conditional independence between the causal factors, but this assumption can be relaxed by introducing links between them. Again, what we are really interested in is how to define the NPT for the child node *Heart attack before 60*.

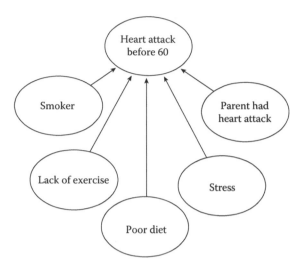

Figure 9.16   Heart attack prediction model.

In this case it does not make sense to use one of the previous Boolean expressions. This is because there will always be uncertainty about the heart attack even if we know the value of each of the parents; for example, even if all the causal factors are known to be true we cannot say with certainty that the person will suffer a heart attack before 60.

On the other hand, if we try to complete the NPT for the heart attack node manually we have 64 entries to complete, which would be extremely tedious and error prone. Moreover, imagine trying to complete any such entry, such as the entry for the child node true when *Smoker* is true, *Lack of exercise* is false, *Poor diet* is false, *Stress* is true, *Parent heart attack* is false.

If you started to do this you would quickly realize that a lot of what you are doing seems to be unnecessary. This is because the effects of the parent nodes on the child are essentially independent. We want to somehow quantify the impact of each causal factor on the heart attack node independently of considering all of the combinations of states of the other parents. It turns out that there is a special Boolean operator, called the *NoisyOR* function, that is what we need in this case (although it is important to note certain limitations described at the bottom of Box 9.7).

Box 9.7   Formal Definition of the NoisyOR Function

Let $X_1,\ldots,X_n$ be n Boolean variables. For each $i = 1,..,n$ let v_i be a number between 0 and 1 (we call v_i the **weight** associated with X_i) and let l be a number between 0 and 1, which we call the **leak factor**. Let Y be a Boolean variable with parents $X_1,\ldots,X_n$. Then we define

$$Y = NoisyOR(X_1, v_1, X_2, v_2, \ldots, X_n, v_n, l)$$

by the following conditional probability expression:

$$P(Y = true \mid X_1,\ldots,X_n) = 1 - (1-l) \times \prod_{X_i \text{ is true}} (1 - v_i)$$

For example, suppose $Y = NoisyOR(X_1, 0.4, X_2, 0.2, X_3, 0.3, 0.1)$ (so that $v_1 = 0.4$, $v_2 = 0.2$, $v_3 = 0.3$, $l = 0.1$) then:

$$P(Y = true \mid X_1 = true, X_2 = true, X_3 = true) = 1 - 0.9 \times 0.6 \times 0.8 \times 0.7 = 0.6976$$

$$P(Y = true \mid X_1 = true, X_2 = true, X_3 = false) = 1 - 0.9 \times 0.6 \times 0.8 = 0.568$$

$$P(Y = true \mid X_1 = true, X_2 = false, X_3 = false) = 1 - 0.9 \times 0.6 = 0.46$$

$$P(Y = true \mid X_1 = false, X_2 = false, X_3 = false) = 1 - 0.9 = 0.1$$

In general, if each of the X_i is false then the probability Y is true is simply equal to the leak value, since in this case:

$$P(Y = true \mid X_1, \ldots, X_n) = 1 - (1 - l) = l$$

Also, if the leak value is 0 and exactly one of the X_i is true (with the others all false) then the probability Y is true is simply equal to the weight associated with X_i, since in this case:

$$P(Y = true \mid X_1, \ldots, X_n) = 1 - (1 - 0)(1 - v_i) = v_i$$

If any X_i has a weight value of 1 then, irrespective of the leak value, Y will be true when X_i is true since in this case $1 - v_i = 0$ and so

$$P(Y = true \mid X_1, \ldots, X_n) = 1 - (1 - l) \times 0 = 1$$

If all of the weight values are equal to 1 and the leak factor is 0, then the NoisyOR function is equivalent to the standard OR function.

Just as if we were using the OR function, NoisyOR requires us to assume that we can consider each of the $X_1, \ldots, X_n$ as "causes" independently of the others in terms of their effect on Y.

It is important to note that, while NoisyOR is a useful and practical option for approximating the required relationship in many real-world situations between a set of variables that are potential causes of an effect variable, it does have limitations. In particular, one of the properties of NoisyOR is "conditional inter-causal independence" (CII). This CII property means that the "explaining away" behaviour—one of the most powerful benefits of BN inference—is not present when the effect variable is observed as *false*. Yet, for many real-world problems where the NoisyOR is a desirable solution, this behaviour would be expected in practice. The further reading includes a paper that discusses this problem and describes a simple way round it in AgenaRisk that works in most cases.

To use the NoisyOR function, you have to specify a weight value (between 0 and 1) for each of the causal factors; this value captures the probability that the consequence will be true if this particular cause (and no other cause) is true. So, for example, if we believe there is a 15% chance that being a smoker will cause a heart attack before the age of 60, providing that no other factors that can cause a heart attack are present in the smoker, then the value associated with the cause "smoker" will be 0.15. Similarly, if we believe there is a 10% chance that lack of exercise alone will cause a heart attack before the age of 60, then the value associated with the cause "lack of exercise" will be 0.1, and so on. In this

The formal, general definition of the NoisyOR function is described in Box 9.7.

example we therefore have to specify five values (one for each of the causes). But we also have to specify an additional value, called the *leak value* (between 0 and 1). This captures the amount of noise in the model and can be thought of as representing causes of heart attacks that are not in the model.

Let's examine the example model of Figure 9.16 where we have defined in AgenaRisk the NPT of the heart attack node to be

> *noisyor (Smoker,* 0.15, *Lack of exercise,* 0.1, *Poor diet,* 0.1, *Stress,* 0.05, *Parent had heart attack,* 0.2, 0.1)

In this example the leak parameter is set to 0.1, which would be the probability of a heart attack if all of the five known risk factors were absent.

To understand the way the NoisyOR function works we can experiment by entering different combinations of parent observations. Figure 9.17 shows the initial state of the model and hence also shows the prior values used for the NPTs of the causal factor nodes.

If we know that a parent had a heart attack (Figure 9.18) then, without any other evidence, the probability of *Heart attack before 60* increases from 21.9% to 35.6%.

If the only evidence we have is of stress then the impact of this evidence is much less pronounced (since the weight value associated with *Stress* was much lower than the value associated with *Parent heart attack*); the probability of *Heart attack before 60* only increases from 21.9% to 25.5%.

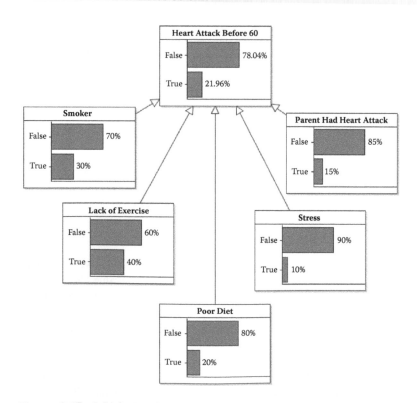

Figure 9.17   Initial state of model.

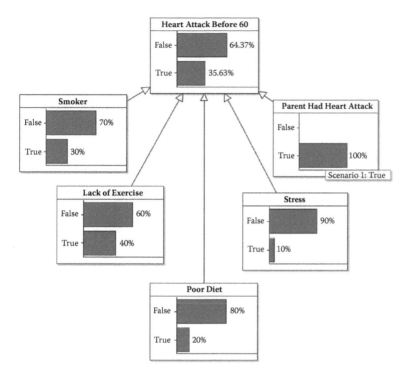

Figure 9.18 Parent had heart attack evidence entered.

The impact and meaning of the leak value can be demonstrated in a number of ways.

If we set *Smoker* to be true and all the other factors to be false (Figure 9.19) then the probability of a *Heart attack before 60* is 0.235, which is higher than the weight value 0.15 associated with the node *Smoker*. This is because the leak parameter is above zero (you can try setting it to zero and you will see that in this case the probability of *Heart attack before 60* is exactly equal to 0.15 as also proved in Box 9.7).

As we said earlier, the leak parameter can be regarded as the extent to which there are missing factors from the model that can contribute to the consequence being true. Since the leak parameter was set to 0.1 (as opposed to zero) we are saying that there are other 10% additional factors missing from the model that can cause the consequence to be true.

Indeed, if we set every causal factor to be false (Figure 9.20) then there is still a 10% chance, that is, a probability of 0.1, that a heart attack before 60 will occur. This is exactly the leak value.

So, if we are sure that the five factors here are the only ones that impact on the heart attack, then we would set the leak parameter to be zero. In this case if all of the parents are false then *Heart attack before 60* will be false with 100% probability. Also, if any one selected cause is true while all the other causes are false, then *Heart attack before 60* will be true with a probability exactly equal to the value associated with the selected cause, which is 90%. All of these results follow from the formal definiton as shown in Box 9.7.

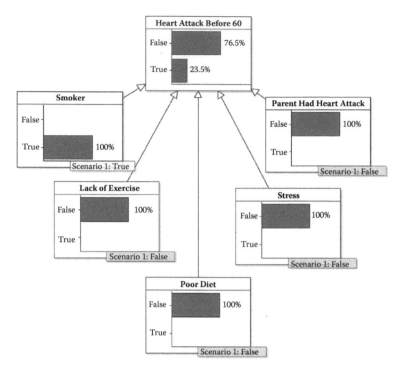

Figure 9.19 Smoker true, but all other factors are false.

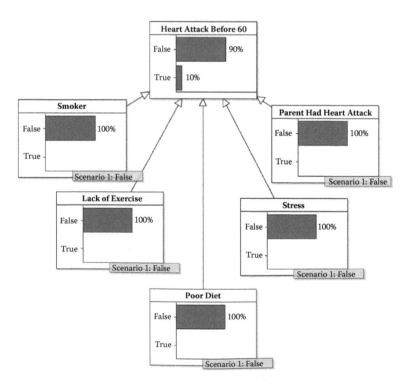

Figure 9.20 All causal factors set to false.

Smoker	False																False	
Lack of exercise	False							True									
Poor diet	False				True				False				True				False
Stress	False		True		False		True		False		True		False		True		
Parent had heart attack	False	True	False	True	False	True	False	True	False	True	False	True	False	True	False	True	False
False	0.9	0.72	0.855	0.684	0.81	0.648	0.7695	0.6156	0.81	0.648	0.7695	0.6156	0.729	0.5832	0.69255	0.55404	0.765
True	0.1	0.28	0.145	0.316	0.19	0.352	0.2305	0.3844	0.19	0.352	0.2305	0.3844	0.271	0.4168	0.30745	0.44596	0.235

Figure 9.21   Fragment of the NPT generated using the NoisyOR expression.

At the other extreme if the leak value is set to one, then no matter what the value is for any of the parent nodes the node *Heart attack before 60* will always be true. So, when the leak value is one the model tells us nothing at all since a heart attack would be inevitable regardless of the other causal risk factors.

To get a feel for the extent to which the NoisyOR function simplifies the NPT specification, Figure 9.21 shows a small fragment of the underlying NPT that is automatically generated using the NoisyOR expression in AgenaRisk. If the expression for NoisyOR was not available then this is exactly the table whose cell entries you would have to enter manually.

9.4.6   NoisyAND Function for Boolean Nodes

Just as NoisyOR provides a highly flexible "noisy" version of the OR function there is an analogous NoisyAND for the AND function (see Box 9.8 for the formal definition). Again we assume there is a set of factors (represented as Boolean variables $X_1,...,X_n$) that can be considered as "causes" independently of each other in terms of their effect on some Boolean variable Y. Consider the following example:

A soccer team generally only wins a match if all of the team "components"—the keeper, defence, midfield and attack—all play well. However, there is a "luck" factor which means that (rarely) they might win if some, or even all, components do not play well. Similarly because of "luck" there is a chance that they will not win even if all of them play well. In addition to luck some components are more critical than others. For example, although it is rare to win if even one component does not play well it is especially critical that attack plays well in order to win.

Box 9.8   Formal Definition of the NoisyAND Function

With Boolean variables $X_1,...,X_n$ weights $v_1,..., v_n$ and leak factor l defined as in Box 9.7 we define

$$Y = NoisyAND(X_1, v_1, X_2, v_2, ..., X_n, v_n, l)$$

by the following conditional probability expression:

$$P(Y = false \mid X_1,...,X_n) = 1 - (1-l) \times \prod_{X_i \text{ is false}} (1 - v_i)$$

In general, if each of the X_i is true then the probability Y is false is equal to the leak value, since in this case:

$$P(Y = false \mid X_1,...,X_n) = 1 - (1-l) = l$$

Also, if the leak value is 0 and exactly one of the X_i is false (with the others all true) then the probability Y is false is simply equal to the weight associated with X_i, since in this case:

$$P(Y = false \mid X_1,\ldots,X_n) = 1 - (1 - 0)(1 - v_i) = v_i$$

If any X_i has a weight value of 1 then, irrespective of the leak value, Y will be false when X_i is false since in this case $1 - v_i = 0$ and so

$$P(Y = false \mid X_1,\ldots,X_n) = 1 - (1 - l) \times 0 = 1$$

If all of the weight values are equal to 1 and the leak factor is 0, then the NoisyAND function is equivalent to the standard AND function.

Assuming that each of the team components are independent of each other in terms of their effect on the match result (Figure 9.22), this type of problem is easily modelled as a noisyAND function. As with noisyOR we must specify a weight value (between 0 and 1) associated with each parent representing its "criticality" and a leak value (between 0 and 1) representing the "luck" factor.

For example, using the AgenaRisk syntax, we can define the NPT for "team wins" as:

noisyand (Keeper, 0.7, Defence, 0.9, Midfield, 0.8, Attack, 1, 0.05)

If you run the model you will find that, when all of the components are true there is still a 0.05 probability (equal to the leak value) that the team will not win. To see the impact of the weights, note that when all the components are true except for "Attack plays well" which is false, then the probability of winning is 0 because the "Attack" component has weight 1 (i.e. maximum criticality). However, because "Midfield" is not as critical there is still a small probability of winning when "Midfield plays well" is false and the other components are true.

Figure 9.22 Problem suitable for noisyAND function.

Table 9.7
Risk Register Weighted Average Approach

risk	probability %	weight	normalised weights	prob times normalised weight
Risk A	10	3	0.50	5.00
Risk B	20	2	0.33	6.67
Risk C	80	1	0.17	13.33
			Weighted average	**25.00**

9.4.7 Weighted Averages

A common simple approach to quantitative risk assessment is to use a weighted average score to combine risks and produce an overall "risk score" as shown in Table 9.7. This is purely arithmetical and is easily implemented in a spreadsheet, such as Excel. Here we have identified three risks to a project: Risk A, Risk B and Risk C with respective probabilities 10%, 20% and 80% and "weights" 3, 2, and 1. This produces an overall weighted average risk score of 25%.

As we saw in Chapter 3, this is the "risk register" approach that can be viewed as the extension of the simple approach to risk-assessment in which we define risk as probability times impact. Specifically, the impacts are viewed as relative "weights."

For all of the reasons discussed in Chapter 3 we do not recommend this approach to risk assessment, but there may be many reasons why we would want to incorporate weighted averages into a BN. For example, we might wish to use a weighted average as a score to determine which new car to buy based on criteria such as price, quality, and delivery time. Although the weighted average is deterministic (and therefore can be computed in Excel) the values for the criteria could be based on a range of uncertain factors and relationships that require a BN model in which the weighted average is just a component.

Fortunately, it is possible to replicate weighted averages (using the same example probabilities and weights as Table 9.7) in a BN as shown in Figure 9.23.

Each of the risk factors is represented by a Boolean node whose "probability" is simply specified as the "True" value in the NPT—so, for example, since Risk A has probability 10% we set its NPT as "True" = 10%. The Risk Score node is also Boolean but it makes sense to replace the labels "False" and "True" with "Low" and "High," respectively. The key to ensure we can replicate the weighted average calculation is to introduce the labelled node *Weights* whose states correspond to the three risk node weights. The normalised weights are used in the NPT for this node.

The NPT for the *Risk Score* node is simply a partitioned expression conditioned on the *Weights node* with partitions:

```
if (Risk_A == "True","High","Low")
if (Risk_B == "True","High","Low")
if (Risk_C == "True","High","Low")
```

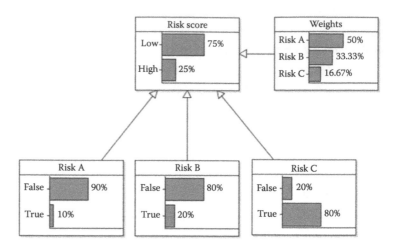

Figure 9.23 Weighted average BN.

In AgenaRisk when you enter the values 1, 2, 3, respectively, into the NPT for the *Weights* node it will automatically normalize the values to sum to one.

Reading off the probability for "High" in the *Risk Score* node gives the exact weighted average as expected (25% in this case). However, it is important to note that when evidence is entered on the *Risk Score* node all of the parent nodes change their states, including the weights as shown in Figure 9.24. At first glance this looks very surprising since, when done in a spreadsheet, the weights are immutable constants. But, from a Bayesian perspective this change makes sense since the weights must be probabilities; hence, they should change in response to new evidence and especially so in response to surprising evidence. Using continuous nodes (see Chapter 10) it is possible replicate weighted average calculations in which the weights are immutable.

It is possible to extend the scheme for weighted averages to hierarchies of nodes, but it can be tricky in this case to ensure that the weights are consistent. For example, as shown in Figure 9.25, even if the weights for three nodes A, B, and C are equal, if we decide to hierarchically

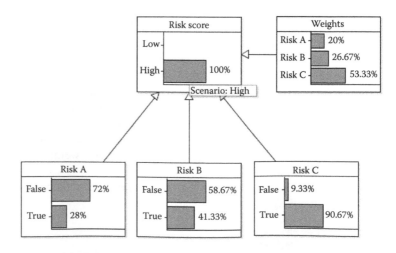

Figure 9.24 Evidence entered on risk score node.

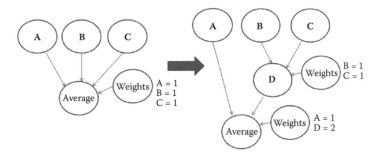

Figure 9.25 Hierarchy of weighted averages.

group nodes B and C into D then we have to readjust the weights of A and D as shown to ensure the original weighting is preserved.

9.5 Ranked Nodes

In Table 9.1 we had an example of a labeled node called *Water level* whose states were "low," "medium," "high." Intuitively, we know that there is some implicit meaning to these state values beyond the simple classification expected of a labeled node. It is not simply that the states are different. They have, additionally, order-preserving meaning. For example, if we know the current water level is "medium" and then water is added, we expect to be able to change the state to "high" but not to "low." The node Water level therefore has states representing an ordered set. Such a node is called a ***ranked node***.

Just as any node with exactly two states "true" and "false" should be defined as a Boolean node (in order to make full use of the Boolean operators available) so it is the case that any node whose states represent a ranked ordinal scale should be defined as a ranked node in order to make use of some powerful operators available to ranked nodes. In particular, as we shall see, this can make the difference between an NPT being very easy to specify and not being feasible at all to specify.

Before we explain what these operators are we need to provide some background.

9.5.1 Background

When building real-world BNs you will often arrive at fragments (which are either cause–consequence idioms or definitional idioms) such as that shown in Figure 9.26. Such fragments are characterized by the following:

- The node values are typically measurable only on a subjective ranked scale like {"very low," "low," "medium," "high," "very high"}.
- Only extremely limited statistical data (if any) is available to inform the probabilistic relationship for the child node (Y) given the parent nodes (X_1 and X_2). Yet, there is significant expert subjective judgment available.

Figure 9.26 Typical BN fragment.

Assuming each of the nodes has five states ranging from "very low" to "very high," the NPT for the child node has 125 states. This is not an impossible number to elicit exhaustively, but from extensive experience we know that all kinds of inconsistencies arise when experts attempt to do so. If the number of states increases to seven or additional parents are incorporated, then exhaustive elicitation becomes infeasible. Moreover, real-world models invariably involve dozens of fragments like these.

One potential solution would be to get samples of expert elicitation assertions like the following:

- When X_1 and X_2 are both "very high" the distribution of Y is heavily skewed toward "very high."
- When X_1 and X_2 are both "very low" the distribution of Y is heavily skewed toward "very low."
- When X_1 is "very low" and X_2 is "very high" the distribution of Y is centered below "medium."
- When X_1 is very high and X_2 is "very low" the distribution of Y is centered above "medium."

Now, and this is the crucial bit, if we can assume that each node has an underlying numerical scale (which we may as well assume is the interval [0, 1]) then such assertions suggest intuitively that there could be some kind of weighted average function to define Y. In fact, experts find it easier to understand and express relationships in such terms. Many so called self-assessment or scorecard systems are based around little more than weighted averages of attribute hierarchies. However, such systems are usually implemented in spreadsheet-based programs that have associated with them a number of problems:

- Difficulty in handling missing data
- Problems with assessing credibility of information sources
- Difficulty in using different scales

Since all of these problems are readily solved using BNs, the challenge is to capture the explicit simplicity of the weighted average approach while also preserving the intuitive properties that the resulting distributions have to satisfy. For example, simply making Y the (exact) weighted average of its parents does not work, since this does not allow us to capture any uncertainty in the distribution of Y given its parents. Also, what is especially tricky to model properly are the intuitive beliefs about the causes given certain child observations, that is, back-propagated beliefs where, for example, we have observed Y and X_1 and wish to infer the value of X_2 like

- If Y is "very high" and X_1 is "very low," then we would be almost certain that X_2 is "very high."

■ If Y is "very high" and X_1 is "average," then we would be confident that X_2 is "very high" but not as confident as in the previous case.

■ In general if Y is "very high," then the lower value that X_1 is the more confident we are that X_2 is "very high."

9.5.2 Solution: Ranked Nodes with the TNormal Distribution

By defining such nodes as *ranked nodes* it is possible to define NPTs that satisfy the criteria we described. We already indicated the key to this solution in Section 9.5.1. When a node is specified as a ranked node then, no matter what the state labels are or how many states a node has, there is an assumption that there is an underlying numerical scale that goes from 0 to 1 in equal intervals. This mapping of a ranked scale to the numerical scale is explained in Box 9.9.

Box 9.9   Mapping Ranked (Ordinal) Scale to Underlying Numerical Scale

Suppose we have a node with five states {"very low," "low," "medium," "high," "very high"}. To transform this scale to a numerical scale from 0 to 1 we have to divide the range 0 to 1 into five equal intervals. Hence, we have the mapping shown in Table 9.8. Other mappings are shown in Table 9.9 and Table 9.10.

Table 9.8
Mapping of 5-Point Ranked Scale to the Underlying Range [0–1]

Ranked Scale State	Underlying Numerical Equivalent
Very low	[0, 0.2), the range 0 to 0.2
Low	[0.2, 0.4), the range 0.2 to 0.4
Medium	[0.4, 0.6), the range 0.4 to 0.6
High	[0.6, 0.8), the range 0.6 to 0.8
Very high	[0.8, 1], the range 0.8 to 1

Table 9.9
Mapping of 3-Point Ranked Scale to the Underlying Range [0–1]

Ranked Scale State	Underlying Numerical Equivalent
Low	[0, 0.333)
Medium	[0.333, 0.666)
High	[0.6666, 1]

Table 9.10
Mapping of 4-Point Ranked Scale to the Underlying Range [0–1]

Ranked Scale State	Underlying Numerical Equivalent
Poor	[0, 0.25)
Below average	[0.25, 0.5)
Above average	[0.5, 0.75)
Good	[0.75, 1]

There is no limit to the number of states in a ranked scale; if there are n states then the mapping will be as shown in Table 9.11.

Table 9.11
Mapping of n-Point Ranked Scale to the Underlying Range [0–1]

Ranked Scale State	Underlying Numerical Equivalent
State 1 (lowest)	$[0, 1/n)$
State 2	$[1/n, 2/n)$
State 3	$[3/n, 4/n)$
...	...
State $n - 1$	$[(n - 2)/n, (n - 1)/n]$
State n	$[(n - 1)/n]$

What you, as a user, need to do when defining ranked scales is ensure that the labeling of the states is consistent with the idea of worst to best. In some cases this may mean that labels such as "very low" are the worst, but in other cases this may mean the best. So, for example, suppose that we are trying to model something like *process quality* in terms of factors such as *staff quality* and *staff turnover*. Then Table 9.12 is a consistent way to define the ranked scale because "very low" for both *process quality* and *staff quality* corresponds to the worst point of the scale, but "very high" for *staff turnover* corresponds to the worst point of the scale. In contrast, Table 9.13 is inconsistent because it would suggest incorrectly that "very high" staff turnover is best with respect to *process quality*.

Table 9.12
Consistent Ranked Scales

Process Quality	Staff Quality	Staff Turnover
Very low	Very low	Very high
Low	Low	High
Average	Average	Average
High	High	Low
Very high	Very high	Very low

Table 9.13
Inconsistent Ranked Scales

Process Quality	Staff Quality	Staff Turnover
Very low	Very low	**Very low**
Low	Low	**Low**
Average	Average	**Average**
High	High	**High**
Very high	Very high	**Very high**

Fortunately, as a user of BNs you never have to construct the mappings. All you need to know is that, irrespective of the linguistic descriptions of the states, the underlying model is working with a numerical scale.

Because it is a numerical scale we can define numerical statistical distributions on it as expressions, and there is one especially useful distribution, called the *truncated Normal* (or *TNormal*) that you can use to generate NPTs. The TNormal distribution is described in Box 9.10. In fact, this distribution is sufficiently flexible that it has been proven to be able to generate satisfactory NPTs for a very wide range of BN fragments involving a ranked node with ranked parents.

Box 9.10   The TNormal Distribution

Unlike the regular Normal distribution introduced in Chapter 2 (which must be in the range –infinity to +infinity) the TNormal has *finite* end points. For ranked nodes these endpoints are 0 and 1, respectively. Like the Normal distribution, the TNormal is characterized by two parameters: the *mean* and *variance*. This enables us to model a variety of shapes including a Uniform distribution, achieved when the variance is very large, and highly skewed distributions, achieved when the mean is not close to 0.5 and variance is close to 0. A range of such *TNormal* distributions with different means and variances is shown in Figure 9.27. The notation *TNormal* (*a*, *b*) tells us that the mean is *a* and the variance is *b*.

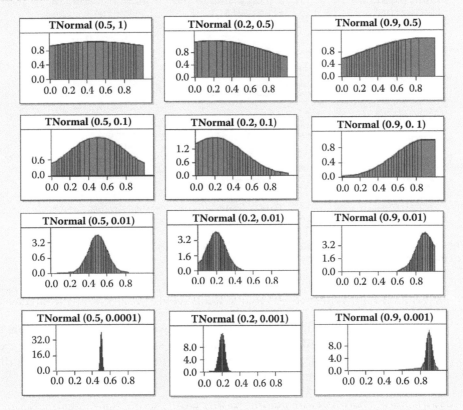

Figure 9.27   Examples of *TNormal* distributions (in range [0–1]) with different means and variances.

Note, in AgenaRisk the rank node functions are generated using TNormal distributions from samples taken from the parent nodes in order to generate "mixtures" of TNormal distributions. Obviously as the sample size changes the number of mixtures generated and the final results differ.

Figure 9.28 shows the result of defining the NPTs of ranked nodes with different scales as different types of TNormal distributions. Once a ranked node is defined by a TNormal distribution (as opposed to manually specifying the probabilities) the distribution is discretized according to the number of states in the scale. Sampling is used to determined the size of the bins.

Suppose, for example, that we have a ranked node *Testing quality* on a scale {"very low," "low," "average," "high," "very high"}. Suppose that we wish to define the NPT of this node as being slightly skewed toward low. Then, instead of specifying the individual NPT entries for the five states manually we can simply define the NPT as an appropriate

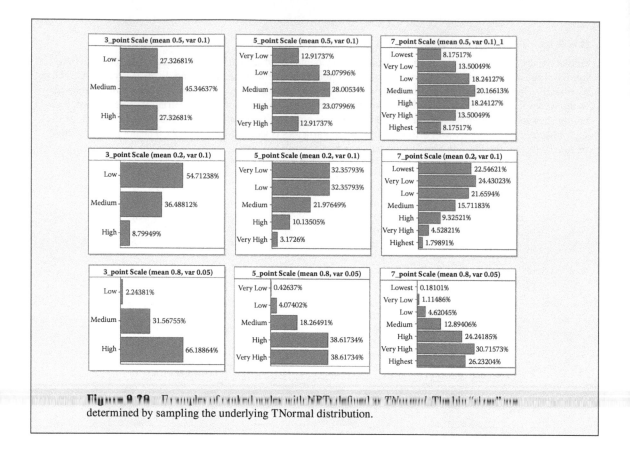

Figure 9.28 Examples of ranked nodes with NPTs defined as TNormal. The bin "sizes" are determined by sampling the underlying TNormal distribution.

TNormal expression (the mean would be below 0.5 to ensure the skew toward low). In AgenaRisk this simply means entering the expression using a dialog as shown in Figure 9.29. The resulting distribution is shown in Figure 9.30.

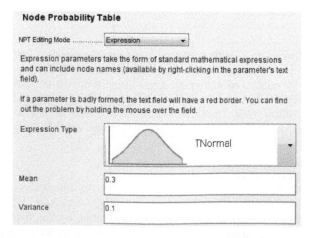

Figure 9.29 Entering a *TNormal* expression for a ranked node.

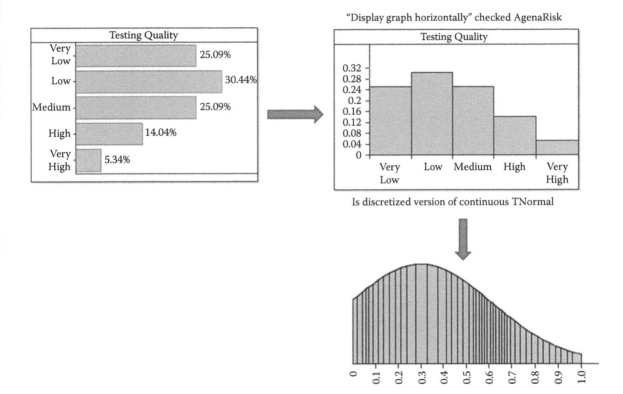

"Display graph horizontally" checked AgenaRisk

Is discretized version of continuous TNormal

Figure 9.30 Ranked node based on NPT expression *TNormal*(0.3, 0.1).

But the real power of the TNormal distribution comes when we define NPTs for nodes with parents. So, consider again the BN shown in Figure 9.26. We can define the NPT of the child node *Y* (*actual testing effectiveness*) to be TNormal with mean and variance as shown in Figure 9.31. Here the mean is a weighted average of the parents nodes X_1 and X_2 (with weights 3 and 1, respectively) and a fixed variance of 0.01.

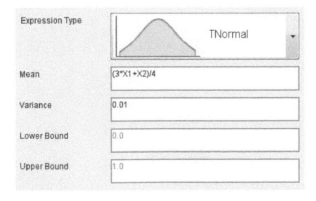

Figure 9.31 NPT for node Y defined as a *TNormal* whose mean is weighted average of parents.

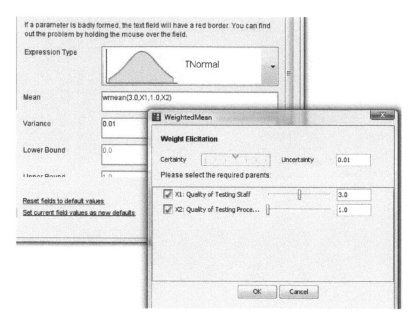

Figure 9.32   Wizard for defining the weighted mean NPT.

Using AgenaRisk this distribution can either be entered directly as an expression for the node Y or via the simple wizard as shown in Figure 9.32 (since the tool also has a built-in *WeightedMean* expression).

When we enter evidence for the parents, the prediction of child node Y has a mean value equal to the weighted average and a variance that reflects our confidence in the result. Figure 9.33 shows that the result is weighted by the importance of the parent nodes. Since X_2 is

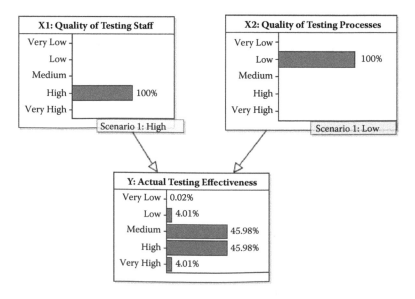

Figure 9.33   Prediction of Y given X1 high and X2 low.

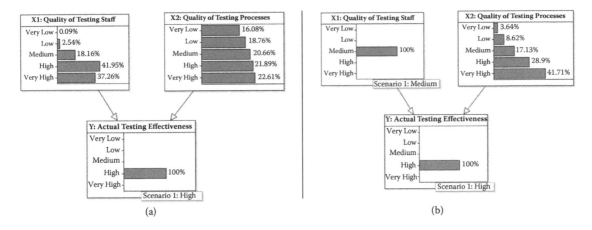

Figure 9.34 Calculating cause given evidence about effect: (a) when Y is high most likely cause if higher quality testing X1; (b) when X1 observed only as "Medium," high value of Y is explained away by high value of X2.

less important than X_1 the result on child node Y is biased toward the X_1 value.

The relative importance reflected in the weighting scheme used is also evident when we calculate causes given evidence about effects. Those nodes with higher weights will be identified as the most likely causes of the consequence. This is shown in Figure 9.34(a) where we can see that a "high" value of X_1 is identified as the most likely cause of the "high" value of Y because of its higher weight.

Note, however, that if we now observe that X_1 is in fact only "medium," the "high" value of Y is explained away by our belief in a "high" value of X_2 (shown in Figure 9.34(b)). It is this kind of expected back-propagation that could not be achieved with other functional approaches.

> The weighted mean function can be used for any number of parents (although, for reasons explained in Chapter 8, we generally do not recommend more than five) and so enables us to build very large NPTs very quickly.

9.5.3 Alternative Weighted Functions for Ranked Nodes When Weighted Mean Is Insufficient

The weighted mean is not the only natural function that could be used as the mean of the TNormal ranked node NPTs. Suppose, for example, that in Figure 9.26 we replace the node *Quality of Testing Process* with the node *Testing Effort* as shown in Figure 9.35.

In this case we elicit the following information:

- When X_1 and X_2 are both "very high" the distribution of Y is heavily skewed toward "very high."
- When X_1 and X_2 are both "very low" the distribution of Y is heavily skewed toward "very low."
- When X_1 is very low and X_2 is "very high" the distribution of Y is centered toward "very low."
- When X_1 is very high and X_2 is "very low" the distribution of Y is centered toward "low."

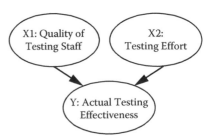

Figure 9.35 Revised BN fragment.

Intuitively, the expert is saying here that, for testing to be effective, you need not just to have good people but also to put in the effort. If either the people or the effort is insufficient, then the result will be poor. However, really good people can compensate to a small extent for lack of effort.

A simple weighted mean for Y will not produce an NPT to satisfy these elicited requirements (you can try it out by putting in different weights; you will never be able to satisfy both of the last two elicited constraints). Informally, Y's mean is something like the *minimum* of the parent values, but with a small weighting in favor of X_1. The necessary function, which we call the *weighted min function* (WMIN), is what is needed in this case. The general form of this function (together with analogous WMAX and the mixture function MIXMINMAX) is shown in Box 9.11. You need not know the details because the function is built into AgenaRisk, so it is sufficient to know what the effect of the function is with different values.

Box 9.11   The Special Functions WMIN, WMAX, and MIXMINMAX

For n variables $X_1,\ldots,X_n$ and corresponding weights $w_1,\ldots,w_n \geq 0$ the weighted min function WMIN is defined as

$$WMIN = \min_{i=1,\ldots,n}\left[\frac{w_i X_i + \sum_{j \neq i} X_j}{w_i + (n-1)}\right]$$

where $w_i \geq 0$ and n is the number of parent nodes.

The WMIN function can be viewed as a generalized version of the normal MIN function. In fact, if all of the weights, w_i, are large then WMIN is close to the normal MIN function. At the other extreme, if all the weights $w_i = 1$ then WMIN is simply the average of the X_i's. Mixing the magnitude of the weights gives a result between a MIN and a MEAN.

In AgenaRisk the syntax for the WMIN function where a ranked node has ranked node parents $X_1,X_2,\ldots,X_n$ with corresponding weights $w_1,w_2,\ldots,w_n$ is $wmin(w_1,X_1,w_2,X_2,\ldots,w_n,X_n)$. So, for example, if we have two parents X_1 and X_2 with respective weights 8 and 2, then the expression needed is $wmin(8, X_1, 2, X_2)$.

The analogous *weighted max function (WMAX)* has the following general form:

$$WMAX = \max_{i=1,\ldots,n}\left[\frac{w_i X_i + \sum_{j \neq i} X_j}{w_i + (n-1)}\right]$$

In AgenaRisk the syntax for the WMAX function is $wmax(w_1,X_1,w_2,X_2,\ldots,w_n,X_n)$.

Finally we have function *MIXMINMAX*, which is a mixture of the two functions *min* and *max* with associated weights, w_1 and w_2:

$$MIXMINMAX(X_1,X_2,\ldots,X_n) = \frac{w_1 \min(X_1,X_2,\ldots,X_n) + w_2 \max(X_1,X_2,\ldots,X_n)}{w_1 + w_2}$$

In AgenaRisk the syntax for the MIXMINMAX function is $mixminmax(w_1,w_2,X_1,X_2,\ldots,X_n)$.

In the preceding example, if we use a TNormal distribution for the NPT of Y whose mean is a WMIN function with respective weights 8 and 2 for parents nodes X_1 and X_2 and whose variance is 0.01, then we get the results shown in Figure 9.36.

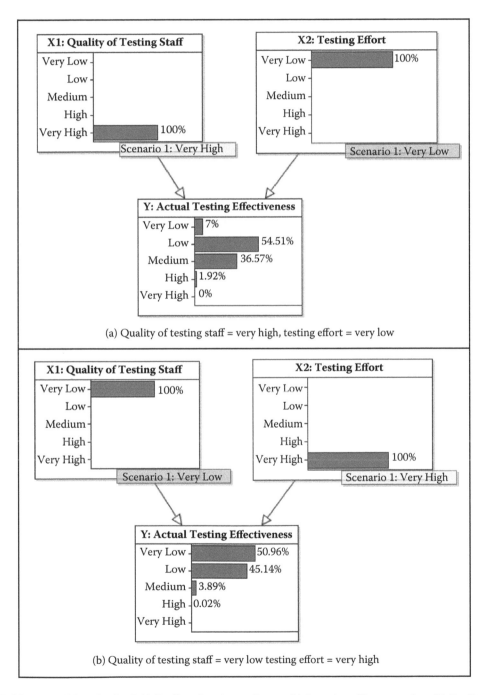

Figure 9.36 WMIN function for Y. (a) Quality of testing staff = very high; testing effort = very low. (b) Quality of testing staff = very low; testing effort = very high.

X1: Quality of Testing Staff			Very Low	
X2: Quality of Testing Processes	Very Low	Low	Medium	
Very Low	0.76095337	0.63206816	0.4790075	
Low	0.23400883	0.35233364	0.4790075	
Medium	0.005033576	0.015567329	0.0417951	
High	4.2458196E-6	3.0893458E-5	1.8985946E-4	
Very High	9.66345E-11	1.7729923E-9	2.7132453E-8	

Close to 0 but not exactly 0

Figure 9.37   Part of table generated for node with WMIN function.

Again it is important to stress that constructing the necessary NPT requires experts only to choose the weights and the value for the variance.

The results show that we can preserve the elicited constraints (really good people can compensate to a small extent for lack of effort, but really poor people cannot produce effective testing simply by putting in a lot of effort).

9.5.4   Hints and Tips When Working with Ranked Nodes and NPTs

9.5.4.1   Tip 1: Use the Weighted Functions as Far as Possible

We have found that the set of weighted functions (i.e., WMEAN, WMIN, WMAX, and MIXMINMAX) is sufficient to generate almost any ranked node NPT in practice where the ranked node's parents are all ranked.

In cases where the weighted function does not exactly capture the requirements for the node's NPT it is usually possible to get to what you want by manually tweaking the NPT that is generated by a weighted function. For example, Figure 9.37 shows a part of the table that is automatically generated for the node Y as specified in Figure 9.31.

You will note that the probability of Y being "very high" when both parents are "very low" is very close to 0 but not equal to 0. If you really want this probability to be 0 then you can simply enter 0 manually into that cell.

9.5.4.2   Tip 2: Exploit the Fact That a Ranked Node Parent Has an Underlying Numerical Scale

In many real-world models you will find that nodes that are not ranked nodes will have one or more parents that are ranked. In such situations you can exploit the underlying numerical property of the ranked node parent to define the NPT of the child node. For example, it makes sense to extend the model of Figure 9.35 by adding a Boolean node called *Release Product?* which is true when the product has been sufficiently well tested to be released and false otherwise. The extended model is shown in Figure 9.38.

Figure 9.38   Extending the testing model to help make a release decision.

We could as usual define the NPT for the new Boolean node manually (it has 10 entries). But it makes much more sense and is far simpler to exploit the fact that the node Y has an underlying numerical value

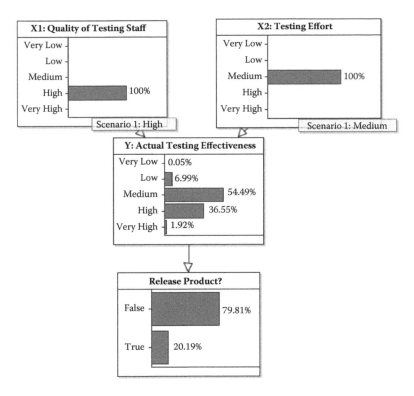

Figure 9.39 A scenario in which the recommendation would be not to release the product.

between 0 and 1. Since we have a 5-point scale we know that if Y is above 0.5 then the quality is at least "medium." If the value is 0.7 then the quality is in the middle of the "high" range. So, suppose that previous experience suggests that testing effectiveness needs to be "high" in order for the product to be released without too many problems. Then we can simply define the NPT of the node *Release product?* by the expression:

$$\text{if } (Y > 0.7, \text{ "True", "False"}).$$

The effect of running the resulting model with some observations is shown in Figure 9.39.

9.5.4.3 Tip 3: Do Not Forget the Importance of the Variance in the TNormal Distribution

The variance captures our uncertainty about the weighted function. Because the TNormal for ranked nodes is always in the range [0,1] any variance above 0.5 would be considered very high (you should try it out on a simple weighted mean example). You may need to experiment with the variance to get it just right.

In each of the previous examples the variance was a constant, but in many situations the variance will be dependent on the parents. For example, consider the BN in Figure 9.40 that is clearly based on a definitional idiom.

Figure 9.40 BN based on definitional idiom.

In this case system quality is defined in terms of the quality of two subsystems *S1* and *S2*. It seems reasonable to assume all nodes are ranked and that the NPT for *System quality* should be a TNormal whose mean is a weighted mean of the parents. Assuming that the weights of *S1* and *S2* are equal we therefore define the mean of the TNormal as *wmean(S1, S2)*.

However, it also seems reasonable to assume that the variance depends on the **difference** between the two subsystem qualities. Consider, for example these two scenarios for subsystems *S1* and *S2*:

1. Both *S1* and *S2* have "medium" quality.
2. *S1* quality is "very high," while *S2* quality is "very low."

If the variance in the TNormal expression is fixed at, say 0.1, then the System Quality in both scenarios 1 and 2 will be the same—as is shown in Figure 9.41(a) and (b). Specifically, the system quality in both cases is medium but with a lot of uncertainty.

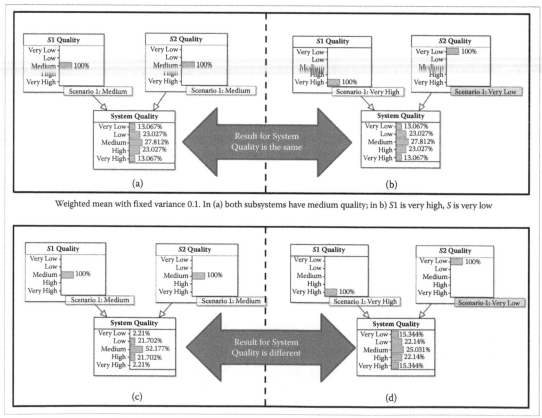

Weighted mean with fixed variance 0.1. In (a) both subsystems have medium quality; in b) S1 is very high, S is very low

Weighted mean with variance abs(S1-S2)/5. In (a) both subsystems have medium quality; in b) S1 is very high, S is very low

Figure 9.41 Using a variable variance produces more sensible result.

However, it seems logical to assume that there should be less uncertainty in scenario 1 (when both subsystems have the same, medium, quality) than in scenario 2 (when both subsystems have very different levels of quality). To achieve the required result we therefore have to ensure that the variance in the TNormal expression is a function of the difference in subsystem qualities. Setting the variance as abs($S1$-$S2$)/5 produces the required result as shown in Figure 9.41(c) and (d).

The use of a variable variance also enables us to easily implement the measurement idiom in the case where all the nodes of the idiom are ranked. This is explained in Box 9.12. The special case of indicator nodes is shown in Box 9.13.

Box 9.12   Implementing the Measurement Idiom for Ranked Nodes

In the measurement idiom, Figure 9.42(a), the assessed value is close to the actual value only when the accuracy of assessment is high. Suppose all of the nodes in the idiom are ranked nodes as shown in Figure 9.42(b).

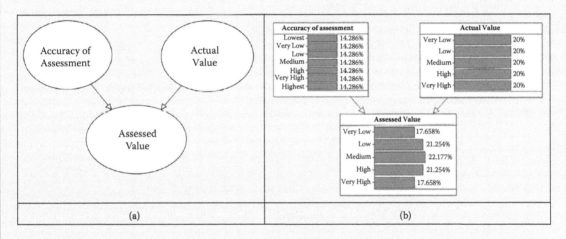

| (a) | (b) |

Figure 9.42   Measurement idiom. (a) Basic idiom, (b) idiom with ranked nodes.

A very simple and reasonable way to ensure the idiom works as required is to define the NPT for the *Assessed Value* node as

$$\text{TNormal(actual, (1-accuracy)/5)}$$

Therefore, the mean of the expression is the value "actual" (i.e. the actual value) and the variance is a function of the value of "accuracy," namely: (1-accuracy)/5. The closer the node "accuracy" is to "Highest" the closer the underlying value is to 1 and hence the closer the variance is to 1. As the node "accuracy" moves towards "Lowest" the closer the underlying value is to 0 and hence the bigger is the variance. The effect is demonstrated in Figure 9.43.

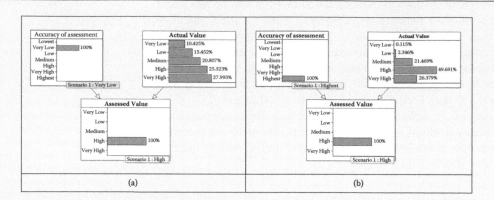

Figure 9.43 Running the measurement idiom with different "accuracy" when the assessed value is "high" (a) Low accuracy means the assessed value provides low confidence in the actual value. (b) High accuracy means the assessed value gives high confidence in the actual value.

Box 9.13   Implementing Indicators for Ranked Nodes

Where we have ranked nodes it is also especially simple to handle the case of indicators, which as we saw in Chapter 8, is a special case of the measurement idiom. We use the same example introduced there where *process quality* cannot be measured directly but can be measured indirectly by indicators *defects found*, *staff turnover*, and *customer satisfaction*. Assuming these are all ranked nodes we can use the Normal distribution to model the NPTs for the indicator nodes where the mean will simply be the parent (i.e., *Process quality* in this case). It is the variance parameter that captures the extent to which the indicator accurately measures the parent (think of it as a kind of "credibility index"—the higher the credibility index the greater the correlation between the indicator and parent node). Suppose, *Defects found* is considered an accurate indicator of *process quality*, while *Customer satisfaction* is less so, and *Staff turnover* much less so. We could capture this information by specifying the variances respectively in the NPTs as 0.01, 0.05, and 0.2.

By setting *process quality* equal to "*medium*," we can see in Figure 9.44 the correlation between the parent node and the indicators. Clearly, *Defects found* is more highly correlated than the others. Figure 9.45 shows how we can use the indicator nodes to infer the true state of the unobservable *process quality*.

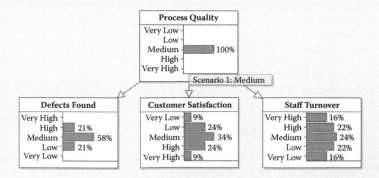

Figure 9.44   Indicator node distributions when *process quality* is "medium." (Note: For all nodes the scale goes from least favorable to most favorable, so "very high" process quality is favorable, while "very low" staff turnover is favorable.)

In Figure 9.45(a) we observe favorable values for the two weaker indicators and this leads us to believe the process quality is likely to be high (and in turn this impacts on the as yet unobserved indicator defects found). However, in Figure 9.45(b) we observe a high number of defects. Because this is a stronger indicator the effect is to more or less discount the contradictory evidence from the other indicators.

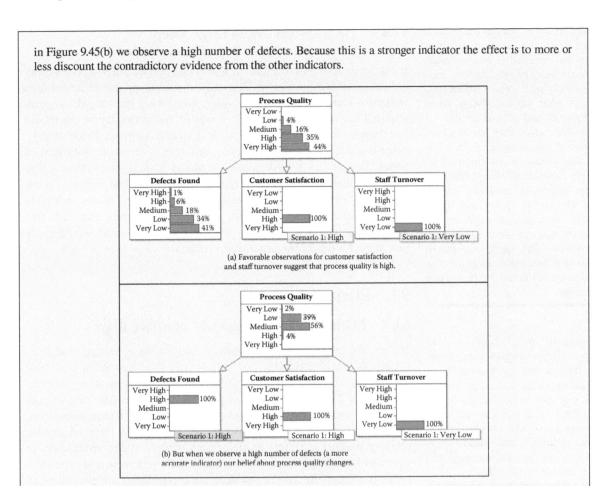

Figure 9.45   The impact of different observations of the indicators.

9.5.4.4   Tip 4: Change the Granularity of a Ranked Scale without Having to Make Any Other Changes

The ranked node on the left-hand side of Figure 9.46 is defined as a TNormal distribution with mean 0.3 and variance 0.1. It has the (default) 5-point scale. If we want to make the scale more or less granular we can simply add or remove states without having to change the NPT in any way. This is shown in the other two versions of the ranked node in Figure 9.46.

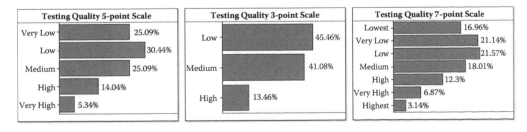

Figure 9.46   Ranked node, different scale but same *TNormal* distribution.

Ranked nodes should be used as surrogates for what might ideally be measurable as continuous quantities, and should be used sparingly with some consideration as to what the underlying model semantics and structure might be. As such, given they are not continuous quantities they lack the scalar precision we might wish from continuous states. Realistic and justifiable variances should be chosen: if these tend to be large, then, as a consequence, the model may not predict with any precision—this would of course be fine if it then reflected reality but might be of limited use in decision making.

Elicitation of probabilities and associated parameters from experts is difficult but not impossible. It requires practice and is best tackled by trial and error, with an inbuilt feedback cycle to allow each party to adjust and refine previous valuations in the light of new information or changes in understanding and perspective.

9.5.4.5 Tip 5: Do Not Create Large, Deep, Hierarchies Consisting of Rank Nodes

It can be tempting to use ranked nodes throughout a BN model simply because they are convenient. Typically, the resulting model would then contain a hierarchy of ranked nodes, mimicking the causal structure required but with the observable nodes on the periphery of the model and the nodes being predicted located far away from these. Unfortunately, there is a potential danger in using this sort of structure with ranked nodes. Because each ranked node contains a variance parameter any evidence entered at the periphery will tend to have little impact on the node being predicted since it passes through so many nodes with wide uncertainties, governed by their variance parameters: we call this unwanted property of a model "washing out." The same would apply vice versa for diagnosis rather than prediction.

9.6   Elicitation

9.6.1   Elicitation Protocols and Cognitive Biases

We are aiming to build a scientific model, so open, factual, and honest discussion of the risks, our beliefs (i.e., theories) about how they interrelate, and what the probabilities are is of the utmost importance. The elicitor (the modeler/risk analyst) and the elicitee (the subject matter expert) must be mutually respectful of each other's professionalism, skills, and objectives. Attributes of a good elicitation protocol involve elicitors making an effort to understand subject matter sufficiently to probe and challenge discussion in order to allow experts to sharpen and refine thinking. Similarly, more accurate probabilities are elicited when people are asked for reasons for them, but the BN structure supplies some or all of this, thus making this easier than when asking for probabilities alone. Without these prerequisites the elicitation exercise will be futile.

Some practical advice on how to elicit numbers from experts is provided in O'Hagan et al (2006). Box 9.14 provides some examples of what has been used, based primarily on Spetzler and von Holstein 1975 (also known as the Stanford Elicitation Prototcol).

Box 9.14   Elicitation Protocols

1. Motivating
   - Outline and understand goals of study
   - Make participants aware of cognitive biases
   - Search for motivational biases
2. Structuring
   - Use unambiguous definitions
   - Use rigorous and testable judgments
3. Conditioning
   - Identify reasons and relevant factors
   - Encourage consideration of a range of situations

4. Encoding
- Use warm-up exercises to get buy-in
- Ask for Extreme Values to counteract "central bias"
- Do not suggest "likely" values to avoid anchoring
- Sketch curves and complete NPTs

5. Validation
- Internal validation against expert
- External validation against literature
- External validation against data

You will notice that the above protocol is sequential with five steps each following after the last. Given that feedback is critical, a contrasting approach is to use feedback cycles with verification and validation steps at each stage between model artifacts produced. Such a scheme, called the BN modeling life-cycle process, is shown, diagrammatically, in Figure 9.47.

If we assume a sequential process the BN model contains six major stages from problem definition to validation of the BN. After problem definition, the modeler matches the problem description fragments provided by the expert against idioms (as described in Chapter 8). In this process the problem fragments are made concrete as idiom instantiations, which are then integrated into objects. Next the modeler elicits and refines the node probability tables for each of the nodes in each object. The objects are then integrated to form the complete BN and inferences made and tests run for validation purposes. Ideally, real test data/expert opinions not used in deriving the BN model should be used to validate the model.

At each stage a verification step takes place to determine whether the output product of the stage is consistent with the requirements of the previous stage and the original problem definition. Failure to pass a verification step results in the invocation of a feedback step that can return the process to any previous stage. For example, it might become obvious to the expert when building the NPT that the BN object may not be quite right. In such a case we may have to redefine the idiom instantiations. In practice we may frequently move between defining the NPTs and the graph structure of objects and idiom instantiations.

For verification and validation we perform a number of tests to determine whether the BN is a faithful model of the expertise and whether the expert's opinions match real data. These range from comparing empirical distributions for key nodes with the marginal distribution from the BN. Likewise we can check consistency by comparing opinions from different experts (we have successfully performed elicitation sessions with up to a dozen experts at a time) and resampling the same probabilities elicited at different points in time.

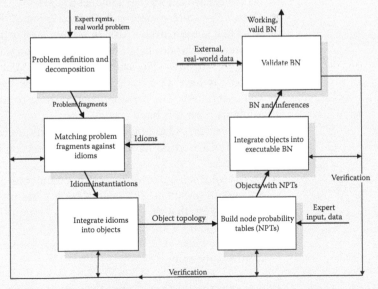

Figure 9.47 BN modeling life-cycle process.

There is plenty of advice on how not to perform elicitation from the field of cognitive psychology as pioneered by Kahneman and colleagues (1982). A summary (by no means exhaustive) of the well-known biases is listed next and we recommend that these be presented and discussed with experts as part of any pre-elicitation training:

- *Ambiguity effect*—Avoiding options for which missing information makes the probability seem unknown.
- *Attentional bias*—Neglecting relevant data when making judgments of a correlation or association.
- *Availability heuristic*—Estimating what is more likely by what is more available in memory, which is biased toward vivid, unusual, or emotionally charged examples.
- *Base rate neglect*—Failing to take account of the prior probability. This was at the heart of the common fallacious reasoning in the Harvard medical study described in Chapter 2. It is the most common reason for people to feel that the results of Bayesian inference are nonintuitive.
- *Bandwagon effect*—Believing things because many other people do (or believe) the same. Related to groupthink and herd behavior.
- *Confirmation bias*—Searching for or interpreting information in a way that confirms one's preconceptions.
- *Déformation professionnelle*—Ignoring any broader point of view and seeing the situation through the lens of one's own professional norms.
- *Expectation bias*—The tendency for people to believe, certify, and publish data that agrees with their expectations. This is subtly different to confirmation bias, because it affects the way people behave before they conduct a study.
- *Framing*—Using an approach or description of the situation or issue that is too narrow. Also *framing effect*, which is drawing different conclusions based on how data are presented.
- *Need for closure*—The need to reach a verdict in important matters; to have an answer and to escape the feeling of doubt and uncertainty. The personal context (time or social pressure) might increase this bias.
- *Outcome bias*—Judging a decision by its eventual outcome instead of based on the quality of the decision at the time it was made.
- *Overconfidence effect*—Excessive confidence in one's own answers to questions. For example, for certain types of question, answers that people rate as 99% certain turn out to be wrong 40% of the time.
- *Status quo bias*—The tendency for people to like things to stay relatively the same.

Biases operate unconsciously; they need to be confronted and removed (they might of course not be biases so at least discussing them may serve to highlight an issue very important to the modeling exercise). Simple tricks can be employed to avoid biases and built into elicitation protocols. Identification of the BN structure means much of the job is done at that stage before probability elicitation is done. Elicitation of probability values must be done rigorously and seriously. Remember every case is different, so there is a need for flexibility.

9.6.2   Validation and Scoring Rules

If elicitation from experts is difficult, then validation of their predictions and estimates can be even more challenging.

You are dealing with experts because there is much uncertainly, novelty, and often little data. Hence, validation is complex.

If you have some data and need to assess the extent to which an expert is over or underconfident in their probability assessments, then you can use a method called a *scoring rule*. Assuming we have a set of predictions from the expert and a full set of data on actual values of outcomes we can use a scoring rule to determine the extent to which the predictions deviate from the actual values.

The simplest situation is where the model is predicting an event that either happens or does not (such as Norman arriving late for work). In other words the model provides a probability, q, that the event happens (and hence $1 - q$ that it does not). In this case there is a simple and popular scoring rule called the *Brier score* that can be used. For a single prediction the Brier score is simply the square of the difference between the predicted probability, q, and the actual outcome, x, which is assumed to be 1 if the event occurs and 0 if it does not, that is, $(x - q)^2$. So

There are many different scoring rules available; the most suitable depends on the type of outcomes being predicted. The Further Reading section provides references to comprehensive scoring rules for different situations.

- If the event occurs, the Brier score is $(1 - q)^2$.
- If the event does not occur, the Brier score is q^2.

Suppose, for example, that the predicted probability for the event "Norman arrives late" is 0.1. If the event occurs then the Brier score is 0.81, while if it does not occur the Brier score is 0.01. The lower the score the more accurate the model. For a sequence of n predictions, the Brier score is simply the mean of the scores for the individual predictions.

The Brier Score (B) for a Sequence on n Predictions

$$B = \frac{\sum_{i=1}^{n}(x_i - q_i)^2}{n}$$

where q_i is the estimated probability that the outcome happens in the ith prediction and x_i is the actual observed event occurrence in the ith case (0 for nonoccurrence and 1 for occurrence), and n is the total number of predictions. The closer the Brier score is to zero the better the prediction.

Table 9.14

Predicting Rain and Brier Scores

Day	Rain x	Bob q	Bob B	Ron q	Ron B	Todd q	Todd B	Jane q	Jane B
1	1	0.8	0.04	0.5	0.25	0	1	1	0
2	0	0.2	0.04	0.5	0.25	1	1	0	0
3	1	0.7	0.09	0.5	0.25	0	1	0	1
4	0	0.1	0.01	0.5	0.25	1	1	0	0
5	1	0.9	0.01	0.5	0.25	0	1	1	0
6	1	0.8	0.04	0.5	0.25	0	1	1	0
Overall B			0.04		0.25		1		0.166

Example 9.5   Weather Forecasting and the Brier Score

Suppose we have four forecasters, Bob, Ron, Todd and Jane, who each produce forecasts over 6 days of weather. In each case their job is to forecast if it will rain the next day. By simply observing if it did rain after the forecast is made we can determine which forecaster was the most accurate, using the Brier score, as shown in Table 9.14.

Bob's Brier score over the 6 days is close to 0, so he is a very accurate forecaster. Ron has given a series of noninformative forecasts and clearly lacks conviction. His Brier score is 0.25. Todd's performance is awful (he is clearly overconfident). His predictions all have probability 1, but he is completely wrong each day, resulting in the worst possible Brier score of 1. Jane is also always totally confident in her predictions; but she is also perfectly accurate on all but one of the days. The fact that her Brier score is higher than Bob's raises some possible issues about the general suitability of the Brier score—is it really better to be nearly correct all the time or perfectly correct almost all of the time?

9.7   Summary

In this chapter we discussed one of the main challenges encountered when building a BN and attempting to complete the NPTs in the BN: that the number of probability values needed from experts can be unfeasibly large. We have described a number of functions available to help define NPTs using functions with vastly fewer parameters than values required to fill in an NPT, such as conditional functions, Boolean functions, and NoisyOR. To support this material we have introduced the Labelled, Boolean and ranked node types and made clear which types are compatible with what functions. For ranked nodes we offered a number of tips and tricks to aid completion of the NPTs as well as specification of the overall BN model. We have also introduced some ideas to make the elicitation process easier using elicitation protocols and a checklist of cognitive biases best avoided, or at least minimized, when eliciting probability numbers from experts. Where data are available we can validate or recalibrate expert predictions and to this end we, in the final section, discussed scoring rules in this context.

Further Reading

Brier, G. W. (1950).Verification of forecasts expressed in terms of probability. *Monthly Weather Review* 78, 1–3.

Constantinou, A. C., and Fenton, N. E. (2012). Solving the problem of inadequate scoring rules for assessing probabilistic football forecasting model. *J. Quant. Anal. Sport.* 8, (2012).

Fenton, N. E., Neil, M., and Caballero, J. G. (2007). Using ranked nodes to model qualitative judgements in Bayesian networks. *IEEE TKDE* 19(10), 1420–1432.
This paper provides the detailed background on ranked nodes.

Fenton, N. E., Noguchi, T., Neil, M. (2018) An extension to the *noisy-OR* function to resolve the 'explaining away' deficiency for practical Bayesian network problems
This paper explains the main limitation of NoisyOR in practice and how this limitation can be solved easily in AgenaRisk.

Gneiting, T., and Raftery, A. (2007). Strictly proper scoring rules, prediction, and estimation. *Journal of the American Statistical Association* 102(477), 359–378.
This paper provides a comprehensive overview of relevant scoring rules for different types of data.

Kahneman, D., Slovic, P., and Tversky, A. (1982). *Judgment Under Uncertainty: Heuristics and Biases*, Cambridge University Press.
This is the definitive work on cognitive biases.

O'Hagan, A., Buck, C. E., and Garthwaite, P. H. (2006). Uncertain Judgements: Eliciting Experts' Probabilities. Wiley.

Spetzler, C.-A. S., and Von Holstein, S. (1975). Probability coding in decision analysis. *Manage. Sci.* 22, 340–358.

Srinivas, S. (1993). A generalization of the noisy-OR model. In *Proceedings of Ninth Conference on Uncertainty in Artificial Intelligence*, Morgan Kaufman, pp. 208–215.

Visit www.bayesianrisk.com for exercises and worked solutions relevant to this chapter.

10

Numeric Variables and Continuous Distribution Functions

10.1 Introduction

In most real-world risk assessment applications we cannot rely on all variables of interest being of the types covered in Chapter 9, that is, either labeled, Boolean, or ranked. Inevitably, we will need to include variables that are *numeric*. Such variables could be *discrete* (such as counts of the number of defects in a product) or *continuous* (such as the water level in centimeters in a river); they generally require an infinite number of states.

In Section 10.2 we show that much of the theory, notation, and ideas encountered in earlier chapters apply equally well to numeric nodes. This section also provides a useful introduction into some of the specialist terminology used by Bayesian statisticians, who mainly use numerical variables in their inference models.

A major advantage of defining a node as a numeric node (as opposed to, say, a labeled or even ranked node) is that we are able to use a wide range of pre-defined mathematical and statistical functions instead of having to manually define node probability tables (NPTs). But there is also a problem, which until recently was truly the Achilles heel of BNs. The standard (exact) inference algorithms for Bayesian network (BN) propagation (as described in Chapter 7, and Appendices B and C) only work in the case where every node of the BN has a finite set of discrete states. Although it is always possible to map a numeric node into a pre-defined finite set of discrete states (a process called *static discretization*), the result of doing so is an inevitable loss of accuracy. We will explain why in Section 10.3. In most real-world applications this loss of accuracy makes the model effectively unusable. Moreover, as we shall see, increasing the granularity of static discretization (at the expense of model efficiency) rarely provides a workable solution.

Fortunately, breakthroughs in algorithms for BNs that incorporate both numeric and nonnumeric nodes (called *hybrid BNs*) have provided a solution to the problem using an algorithm called *dynamic discretization,* which works efficiently for a large class of continuous distributions. This is explained in Section 10.4. Users of AgenaRisk, which implements this algorithm, do not have to worry about predefining the set of states of any numeric node.

So far in this book we have defined discrete probability distributions for discrete nodes. In this chapter we will define continuous probability distributions for numerical nodes. The ideas and concepts introduced in Chapters 5, 6, and 7 still apply but there are some additional mathematical concepts that will be needed to support these numeric nodes properly.

AgenaRisk defines the use of the dynamic discretization algorithm as "simulation"—all approximate methods for solving this problem are dubbed simulation because they "simulate" or approximate the true mathematical function. Details of the dynamic discretization algorithm are given in Appendix D.

All models in this chapter are available to run in AgenaRisk, downloadable from www. agenarisk.com.

Three examples are presented in Section 10.5 to show the power and usefulness of dynamic discretization. The first example involves predicting automobile costs using a hybrid model containing mixtures of discrete and continuous nodes. The second involves a simple risk example of a bank assessing whether it can pay the interest on a loan under stressful conditions. The third example is more challenging and presents three different ways to estimate school exam performance using a classical frequentist model and two different Bayesian models. The objective here is to learn parameters from data and then use these to make a decision about which school to choose. The latter part of this example is quite challenging and introduces ideas like conjugacy, hierarchical modeling, and hyperparameters.

In Section 10.6 we describe the important problem of risk aggregation, which is easily solved using the AgenaRisk compound sum analysis tool and provides an elegant method for calculating loss distributions by taking account of both frequency and severity. Finally section 10.7 offers some tips and tricks on using simulation nodes.

10.2   Some Theory on Functions and Continuous Distributions

Before we start investigating some models that use numeric nodes it is necessary to cover some basic theory and notation needed to express the ideas involved. Many of the concepts, such as joint, marginal, and conditional distributions are common to those we met in Chapters 5 and 6 when we discussed discrete variables and Bayes' theorem.

Recall that in Chapter 5 we defined:

A *probability distribution* for an experiment is the assignment of probability values to each of the possible states (elementary events).

What happens when the number of elementary states is extremely large or even infinite? In these cases we use continuous, rather than discrete, distribution functions. Although we introduced continuous distributions in Section 5.3.1 we are going to provide a more formal treatment here. Box 10.1 formally defines the key notions of continuous variable, probability density function and cumulative density function. These definitions require an understanding of calculus (differentiation and integration).

Box 10.1   Continuous Variables, Probability Density Functions and Cumulative Density Functions

In Chapter 5, Box 5.6, we introduced the notion of probabilities defined with continuous distributions. One of the examples concerned people's height X measured in centimetres. The probability of X being equal to any exact value, say, 182 centimetres is zero, so instead we are interested in the probability that X lies within a range, such as 181.5 and 182.5. In order to calculate such probabilities for a continuous variable X there must be an associated **probability density function** (*pdf*) $f(x)$ which is an integrable function that satisfies the following properties:

$$(1)\ f(x) \geq 0 \text{ for all } x \quad (\text{so } f(x) \text{ is always non-negative})$$

$$(2)\ \int_{-\infty}^{\infty} f(x)dx = 1 \quad (\text{so the "area under the curve" is 1})$$

In general, the probability that X is between a and b is defined as the area under the curve from a to b:

$$P(a \leq X \leq b) = \int_{a}^{b} f(x)dx$$

In the height example, the function $f(x)$ was a Normal distribution (Bell curve). A Normal distribution with mean μ and standard deviation σ is the function

$$f(x) = \frac{1}{\sigma\sqrt{2\pi}} e^{-\frac{(x-\mu)^2}{2\sigma^2}}$$

which can be shown to satisfy properties (1) and (2).

A simpler example, that we considered in Chapter 5, Box 5.6, was of a number X drawn randomly between 0 and 2. The pdf associated with this continuous variable Y is easily seen to be the uniform distribution:

$$f(x) = \begin{cases} 1/2 & \text{if } 0 \leq x \leq 2 \\ 0 & \text{otherwise} \end{cases}$$

In this case property (1) clearly holds while property (2) holds because

$$\int_{-\infty}^{\infty} f(x)dx = \int_{0}^{2} \frac{1}{2} dx = \left[\frac{1}{2}x\right]_{0}^{2} = 1 - 0 = 1$$

To calculate the probability that X lies between 0.3 and 0.6:

$$P(0.3 \leq X \leq 0.6) = \int_{0.3}^{0.6} f(x)dx = \int_{0.3}^{0.6} \frac{1}{2} dx = \left[\frac{1}{2}x\right]_{0.3}^{0.6} = \frac{0.6}{2} - \frac{0.3}{2} = 0.15$$

The **cumulative density function** (cdf) F associated with a pdf $f(x)$ is defined as

$$F(x) = \int_{-\infty}^{x} f(t)dt$$

In other words, this is the probability that X is less than or equal to the value x.
Note also that this means

$$P(a \leq X \leq b) = \int_{-\infty}^{b} f(x)dx - \int_{-\infty}^{a} f(x)dx = F(b) - F(a)$$

It is also useful to know that in general the derivative of the cdf is equal to the pdf since

$$F'(x) = \frac{d}{dx}F(x) = \frac{d}{dx}\int_{-\infty}^{x} f(t)dt = f(x)$$

Whereas with discrete probability distributions we talk in terms of the probability of a particular state of a variable X, with continuous distributions (as explained in Box 10.1) we are only interested in the probability that X lies within a *range of values* $[a, b]$ rather than a single constant value. Indeed, the probability of any single constant value is zero, since in that case $a = b$, and (from Box 10.1):

$$P(a \leq X \leq b) = F(b) - F(a) = F(a) - F(a) = 0$$

For the same reason for any range $[a, b]$, the same probability is generated whether the end points a and b are included or not (given that each end point has probability zero):

$$P(a < X \leq b) = P(a \leq X < b) = P(a \leq X \leq b) = P(a < X < b)$$

Continuous random variables adhere to the axioms of probability theory as much as discrete variables do, as shown in Box 10.2.

Box 10.2   Axioms of Probability for Continuous Variables

Probability Axiom 1: The probability of any interval $[a, b]$ is a number between 0 and 1.
   Since the function $f(x)$ is a *pdf* if it satisfies properties (1) and (2) in Box 10.1, hence

$$0 \leq \int_a^b f(x)dx \leq \int_{-\infty}^{\infty} f(x)dx = 1$$

Probability Axiom 2: The probability of exhaustive intervals is 1.
   The exhaustive interval is simply $[-\infty, \infty]$ so this follows from property (2).
Probability Axiom 3: For any mutually exclusive intervals the probability of either interval is the sum of the probabilities of the individual intervals.
   This is because any two mutually exclusive intervals must be of the form $[a, b]$ and $[c, d]$ where $b < c$. But then the probability that X lies in either interval $P(a \leq X \leq b$ OR $c \leq X \leq d)$ is the area under the curve between a and b plus the area under the curve between c and d, so

$$P(a \leq X \leq b \text{ OR } c \leq X \leq d) = P(a \leq X \leq b) + P(c \leq X \leq d)$$

There are striking similarities between the notation we use in the discrete and continuous cases. In the former we use summation and in the latter we use integration and can consider the semantics of each of these operations to be equivalent. In the discrete case we would sum probabilities over piecewise uniform discrete states X_i that together compose an interval $X \in [a,b]$:

$$\sum_{i=1}^{n} P(x_i) \approx P(a < X < b) = \int_a^b f(x)dx$$

Figure 10.1 shows an example of this approximation over five subintervals. Here we have split the range of X into five locally constant

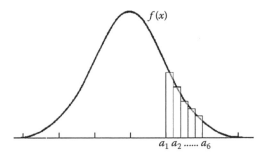

Figure 10.1   Approximating a continuous function using discrete intervals.

subintervals $[a_1,a_2],[a_2,a_3],[a_3,a_4],[a_4,a_5],[a_5,a_6]$ (where $a_1 = a$ and $a_6 = b$) and calculate the approximation:

$$P(a_1 < X < a_6) = \int_{a_1}^{a_6} f(x)dx = \sum_{i=1}^{5}\left(\int_{a_i}^{a_{i+1}} f(x)dx\right) \approx \sum_{i=1}^{5}(a_{i+1} - a_i)f\left(\frac{a_{i+1} + a_i}{2}\right)$$

Note that the uniform subintervals do not exactly coincide with the function at all points, since they are histogram boxes, and so they provide only an approximation. Hence we use the symbol $\approx$ here. The more intervals we use the better would be the approximation. We can therefore use summation as a way of approximately calculating an integral using a constant or variable number of discrete subintervals, decided in advance or during inference. We exploit this basic idea to produce algorithms that approximate these continuous functions later in this chapter and in Appendix D.

We previously defined joint and conditional probability distributions for discrete variables and now do the same for continuous variables in Box 10.3. Again the ideas carry over easily so if you understand the principles for discrete variables, you should not have much difficulty in understanding them for continuous variables.

Box 10.3   Joint, Marginal and Conditional Continuous Distributions

Suppose we have two continuous random variables X and Y. For example, X is the time (in minutes) it takes Norman to travel to work and Y is the time it takes Martin to travel to work. Then we are now interested in probabilities of events such as "X and Y both lie between 20 and 100." The associated *joint* probability density function is a function $f(x,y)$ which satisfies similar properties to those specified for a normal pdf, that is

$$1)\ f(x, y) \geq 0$$

$$2) \int_{-\infty}^{\infty} \int_{-\infty}^{\infty} f(x,y)dx\,dy = 1$$

The probability of an "event" like "$a < X < b$ and $c < Y < d$" is then defined as

$$P(a < X < b \text{ and } c < Y < d) = \int_{a}^{b}\int_{c}^{d} f(x,y)dxdy$$

The *marginal* pdfs of X and Y, respectively, are

$$f_X(x) = \int_{-\infty}^{\infty} f(x,y)dy$$

and

$$f_Y(y) = \int_{-\infty}^{\infty} f(x,y)dx$$

If it is the case that

$$f(x,y) = f_X(x)f_Y(y)$$

then the variables X and Y are defined as **independent**. This means the pdf of X given Y is equal to $f_X(x)$. However, if they are not independent then the conditional pdf of X given Y is defined as

$$g(x \mid y) = \frac{f(x,y)}{f_Y(y)}$$

All of these definitions extend to a set of multiple variables $X_1, X_2, ..., X_k$ with a joint pdf $f(x_1, x_2, ..., x_k)$. For example, the marginal pdf of X_1 is

$$f_{X_1} = \int_{-\infty}^{\infty} \int_{-\infty}^{\infty} \cdots \int_{-\infty}^{\infty} f(x_1, x_2, ..., x_k) dx_2 dx_3 ... dx_k$$

In practice we declare two types of continuous distribution functions:

In AgenaRisk conditionally deterministic functions are called "arithmetical."

- *Conditionally deterministic functions*—These are simply mathematical functions such as $Z = X + Y$. Such a function would be used if a node Z was a deterministic function of its parents.
- *Statistical distributions* (See Appendix E for a list of these.) These are special functions that govern the *conditional* relationship between one variable and others via some mathematical parameters that define the shape and scale and other properties of the function. Examples of statistical distribution functions are the Normal distribution, with parameters for mean and variance, and the Uniform distribution, with parameters start and end points, that we have encountered in earlier chapters.

The pdf for a continuous variable X will generally have parameters that may also be uncertain. Hence, these parameters are variables in their own right with their own associated pdfs. For example, in Box 10.1, we considered the example of a variable X whose pdf was a Normal distribution; this has two parameters—the mean μ and standard deviation σ. We write this as $f(x) = N(\mu, \sigma)$ or alternatively $N(\mu, \sigma^2)$ (if we consider the variance rather than the standard deviation as a parameter). When we observe evidence (data) for X, the prior distributions for these variables get updated using Bayes' theorem. Hence, we can use any BN containing continuous functions to perform either prediction or inference. Bayesian statistical inference uses some specific terminology that has a lot in common with the discrete variable case but the emphasis is more on estimation of parameters.

We can treat the integer case as the same as the continuous variable case when we have discrete probability distributions.

There are a lot of competing notations used by different textbooks for describing continuous probability functions. For brevity we will use some abbreviated forms shown in Box 10.4.

Box 10.4   Notation

The approximation symbol ≈
For continuous functions we write $\tilde{f}_X \approx f(x)$ where $\tilde{f}_X$ is an approximation to the function $f(x)$

Overloading
The pdf $f(x)$ associated with the variable X is often written as $f(X)$, f_X or as $f_X(X)$ (small x is the value and X is the variable)

The symbol ~
This symbol is used as a shorthand for representing the particular probability distribution associated with a variable X such as in

$$X \sim Binomial(n, p)$$

$$X \sim Normal(\mu, \sigma)$$

So in the first case this is shorthand for saying that

$$P(X = x \mid n, p) = \binom{n}{x} p^x (1-p)^{n-x}$$

and in the second case this is shorthand for saying that $f(X \mid \mu, \sigma)$ is the Normal distribution function.

Treating discrete and continuous cases as if they were the same

If X is a continuous variable we sometimes write the pdf of X as $P(X)$ instead of $f(X)$ whereas if X is discrete we might write $f(X)$ instead of $P(X)$

10.3   Static Discretization

The model in Figure 9.1 of Chapter 9 contained a node for rainfall, R, which represents the amount of rain (measured in millimeters) falling in a given day. We defined the set of states of a variable called rainfall to be "none," "<2 mm," "2–5 mm," ">5mm." In mathematical notation this set of states is written as the following intervals:

Box 10.5 provides a formal description of the state domain notation.

$$\Psi_R : \{[0], (0\text{–}2], (2\text{–}5], (5, \text{infinity})\}$$

Box 10.5   State Domain Notation

When we specify the discrete states of a numerical node we do so by partitioning the range into mutually exclusive subranges. Formally, we define the range of X as Ω_X, and the pdf of X is denoted by f_X. The idea of discretization is to approximate f_X by, first, partitioning Ω_X (or a subset of it) into a set of mutually exclusive intervals, Ψ_X.

For example, suppose a continuous node X has range $\Omega_X = [0, \text{infinity})$. Then two possible discretizations of this range, denoted Ψ_X^1 and Ψ_X^2 are:

$$\Psi_X^1 = \{[0-10], (10-20], (20-30], \ldots, (100, \infty)\}$$

$$\Psi_X^2 = \{[0], (0-100], (100-200], \ldots, (200, 1000]\}$$

The notation $(x, y]$ means all numbers that are greater than x but less than or equal to y.

When we approximate the range Ω_X by Ψ_X^1 or Ψ_X^2 we are making a number of design decisions about the range and the nature of the function that is being modeled. So, when choosing Ψ_X^2 we are assuming that the upper bound of 1000 is big enough to contain all of the probability mass generated by the function. Likewise, choosing to include [0] as an integer value means that this might have a special role to play in our function and so deserves a special place alongside the continuous range specified.

More detail on how AgenaRisk handles state ranges and approximates infinite ranges is given in Appendix D.

Since the objective of the model was to predict the probability of water levels rising and causing a flood, it is clear that any information about rainfall is very important. But the problem is that, because of the fixed discretization, the resulting model is insensitive to very different observations of rainfall. For example:

■ The observation 2.1 mm of rain has exactly the same effect on the model as an observation of 5 mm of rain. Both of these observations fall into the interval (2,5].
■ The observation 6 mm of rain has exactly the same effect on the model as an observation of 600 mm of rain. Both of these observations fall into the interval (5, infinity).

The obvious solution to this problem is to increase the granularity of the discretization. Including AgenaRisk there is automated support to enable you to do this relatively easily. Normally, this involves declaring the node to be of type *continuous interval* and then bringing up a dialogue (Figure 10.2(a)) that enables you to easily create a set of intervals of any specified length, as shown in Figure 10.2(b).

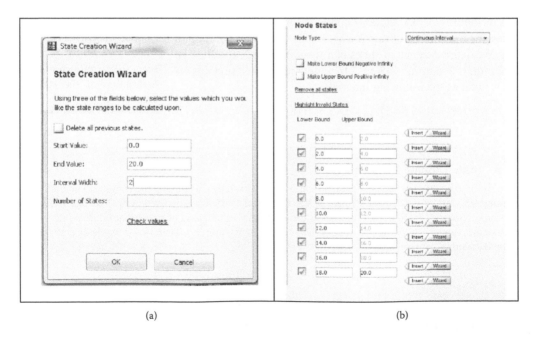

(a)   (b)

Figure 10.2 Using a state wizard tool to create a set of intervals. (a) Specifying the interval width or number of states. (b) The resulting automatically generated set of states.

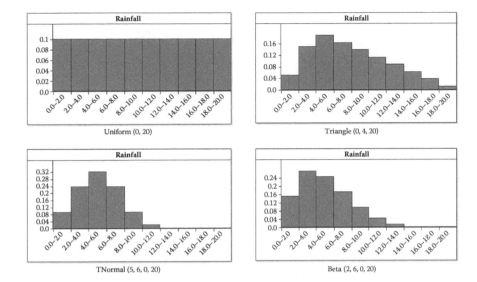

Figure 10.3 Example NPT distributions for node *Rainfall*.

It is also possible, of course, to add and edit states manually. Similar wizards are available for nodes that would be better declared as *integer* (such as a node *number of defects*).

The problem is that, although such tools are very helpful, they do not always solve the problem. No matter how many times we change and increase the granularity of the discretization, we inevitably come across new situations in which the level is inadequate to achieve an accurate estimate. But this is a point we will return to at the end of the section.

Whatever the discretization level chosen there are immediate benefits of defining a node that really does represent a numeric value as a numeric node rather than, say, a labeled or ranked node. Suppose, for example, that rainfall is defined as a numeric node with the set of states in Figure 10.2(b). Then when we come to define the NPT of rainfall we can use a full range of statistical and mathematical functions (Appendix F lists all the statistical distribution functions available in AgenaRisk). Some examples are shown in Figure 10.3.

In each case it is simply a matter of selecting the relevant distribution from a list and then stating the appropriate parameters; so, for example, the parameters of the *TNormal* $(5,6,0,20)$ distribution in Figure 10.3 are, in order: mean (5), variance (6), lower bound (0), upper bound (20).

The benefits of numeric nodes are especially profound in the case where the intervals are not all equal, as explained in Box 10.6.

Box 10.6   Plotting Densities or Absolute Probabilities

The rainfall node in Figure 10.3 was defined on a range of 10 equal intervals of length 2 from 0 to 20. Suppose that we decide that the range should be 100 rather than 20. For simplicity we add just a single extra interval 20–100 to cover this new range. By default the NPT for any node is Uniform in the sense that each of the predefined states of the node has equal probability. If you accept this default NPT, then the resulting distribution is shown in Figure 10.4(a). But there is a problem with this. From the perspective of a numerical range of 0 to 100 this

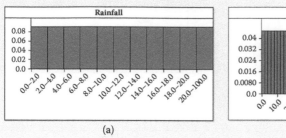
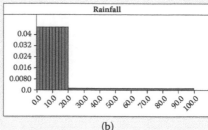

(a)   (b)

Figure 10.4   The issue of nonequal intervals. (a) Uniform distribution? (b) Distribution when viewed on continuous scale.

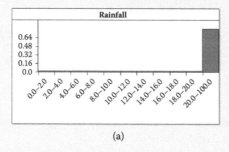
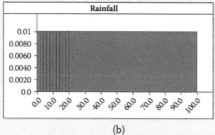

(a)   (b)

Figure 10.5   Absolute and density plots of same distribution function. (a) Genuine Uniform distribution with non-equal intervals plotting absolute probabilities, (b) Uniform distribution when plotted on continuous scale using densities.

distribution is certainly not Uniform despite the fact that the probability density of the (very wide) interval 20–100 is exactly the same as all of the other (much smaller) intervals.

To convert the absolute probability, $P(a < X < b)$, which is the cumulative probability in the interval $[a,b]$, to the probability density we simply divide the absolute probability by the width of the interval:

$$\frac{P(a < X < b)}{b - a}$$

In fact, when we plot the distribution on a continuous scale (you can choose either setting in AgenaRisk) you will see the probability density function plot shown in Figure 10.4(b). In general, we would like to define and plot our distributions in such a way that we do not have to worry about different interval lengths and to do this we choose the right distribution and plot it in AgenaRisk as "continuous," and it then plots the density instead of the absolute values.

If, for example, we believe the prior Rainfall should be Uniform across the whole range 0–100, then we simply define the NPT by the expression *Uniform*(0,100). The result of this change is shown in Figure 10.5(a) which shows the *absolute* probability distribution plotted against the (nonequal) intervals, while Figure 10.5(b) shows the *density* distribution plotted on a continuous scale.

Figure 10.6   Structure of customers model.

To get an idea for how a model with statically discretized numeric nodes works consider the model in Figure 10.6. The objective of this model is to predict the net number of new customers a company gains in a year. In other words, the model has to take account of the probability that some of the new customers are lost in the same year.

To build this model we first need to specify that the type of each of the nodes is numeric. In fact we can be more specific; the type is *integer* (the reason for making this specific will soon become clear).

Next, we need to decide on an appropriate range and discretization for the nodes. Suppose that we know from previous years that the typical number of new customers is something like 100, with the range between 40 and 160. Then we certainly need the range to include 40 to 160, but to cover the possibility that future years will be different it makes sense to ensure the range is larger, say, 0 to 300. To keep things simple we will use equal intervals of size 20, using the stage range wizard, so that the set of states is

$$\Psi_{Net\ New\ Customers} : [0, 19], [20, 39], [40, 59], ..., [260, 279], [280, 299]$$

We can use the same discretization for each of the three nodes.

To complete the model we need to define the NPTs for each node as follows:

- *New customers*—Rather than accept the default NPT in which all states have equal probability, it makes sense (as explained in Box 10.6) to define a more meaningful prior distribution. In this case, based on the previous years, the prior distribution is something like a Normal distribution with mean 100 and variance 500. To capture this, we select the TNormal distribution (since the range must be finite), with mean 100, variance 500, and range 0 to 299. Formally, this is written:

$$New\ Customers \sim TNormal(\mu = 100, \sigma^2 = 500, 0, 299)$$

- *Lost customers*—The NPT for this node is conditioned on its parent *New Customers*. In Chapter 5 we described a distribution, which is relevant for integer variables, that is very suitable in the case, namely, the Binomial distribution. Suppose that there is a 20% chance that any new customer will be lost during the year. Then it is reasonable to conclude that the distribution for *Lost Customers* is a Binomial distribution in which the parameter *Number* of *Trials* is simply the number of new customers and the parameter *Probability of Success* is equal to 0.2. Hence we define the NPT for the node *Lost Customers* simply as the expression:

$$Binomial(n = New\ Customers, p = 0.2)$$

This is shown in Figure 10.7.
- *Net new customers*—The NPT for this node is simply the arithmetic expression (i.e., a conditionally deterministic function):

$$New\ Customers - Lost\ Customers$$

When you run the model you should see the initial states shown in Figure 10.8.

Since the type is integer rather than continuous, we need to end at 299 rather than 300 to get equal intervals each containing 20 integers.

The default graph settings in AgenaRisk actually display numeric nodes on a continuous scale, so to produce the output as shown here you need to uncheck the continuous scale setting.

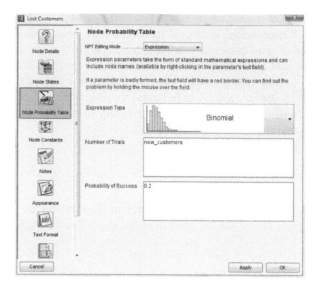

Figure 10.7   Defining Binomial distribution for node *Lost Customers*.

Now suppose in a particular year we know that the number of new customers is 85 and the number of lost customers is 35. Enter these values as observations into the relevant nodes. When you run the calculation you should see something like Figure 10.8. If you move the mouse over the graph for node *Net Lost Customers* you will see that the mean value is 60 and not the "correct" value of 50 which you might reasonably expect. The model is calculating the result of 85–35 to be a distribution whose mean is 60.

This approximation may or may not be satisfactory for your application. The reason this result (and not the "correct" result of 50) is obtained is because when you enter the value 85 for node *New Customers* all the

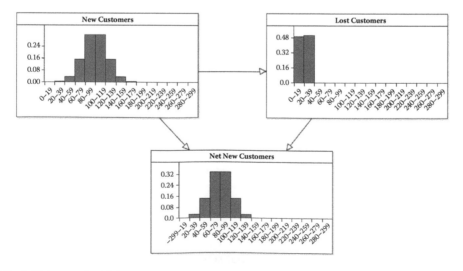

Figure 10.8   Initial state of model.

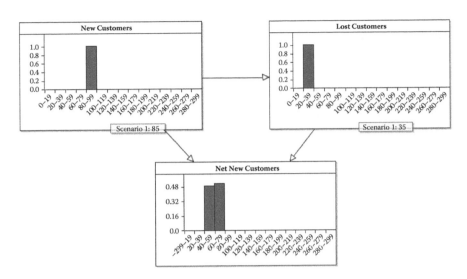

Figure 10.9   Calculation with observations entered.

model actually knows is that the value of *New Customers* lies in the interval containing 85. In other words the value is anywhere in the interval 80–99. Similarly, the observation 35 for node *Lost Customers* simply means that the value is anywhere in the interval 20–39. When the model runs, the calculation uses the intervals 80–99 and 20–39, respectively, as the evidence taking samples. For the expression

New Customers – Lost Customers

some of the results end up in the interval 40–59 and the rest end up in the interval 60–79.

If you think the calculation is inaccurate, then you will be even more disappointed to discover that things can get much worse. Change the observation for *New Customers* to be 99 and the observation of *Lost Customers* to be 21. Instead of the expected result 78 for node *Net New Customers*, when you run the calculation you will get exactly the same result as before (mean 60). This is because 99 and 21 are in the same intervals as 85 and 35, respectively.

To increase accuracy you can edit the set of states of each node to include smaller intervals. As an extreme you could make the interval size just 1 for each of the nodes. When you do this the first thing you will notice is that the model starts to run slowly. This is not surprising. Since the underlying NPT for node *Net Lost Customers* has to store values for every combination of the states this means $300 \times 300 \times 300$ different values; that is a table of 27 million values. This is simply impractical. Moreover, even if you built such a model you may discover that the range 0 to 299 was actually insufficient since in some years the values might be higher. You may even find that you cannot set a maximum number and wish to include the range 0 to infinity.

Fortunately, although static discretization is the approach you have to adopt with traditional BN methods and tools, the next section explains why this is no longer necessary.

Clearly static discretization has limitations. It can suffer from having too many states that have low probability regions and too few states for high probability regions in the results. You will always be fighting a losing battle. The more states in a model the slower its execution and the more memory it demands. A real study that highlights this problem very well is described in Box 10.7.

Box 10.7   When Static Discretization Is Never Sufficient

One of our studies involved building a BN model to predict software defects in modules developed by one organization. The model contained a key node *software size* to capture the size of each module under study as measured in KLOC (thousands of lines of code). The most typical size of a software module at the organization was between 10 and 20 KLOC, but this was by no means consistent. A part of the original BN with statically discretized nodes is shown in Figure 10.10. In the model the NPT for the node *Defects found* is defined as a Binomial distribution with parameters *p* equal to the probability of finding a defect and *n* equal to the value of *Defects inserted*. The NPT for the node *Residual defects* is simply defined by the arithmetic expression *Defects inserted* minus *Defects found*.

As with any attempt at discretization, there was a need to balance the number of states (accuracy) against computational speed. There was much discussion, agonizing, and continual refinement of the discretization. Although predictions were generally reasonable within the expected range, there were wild inaccuracies for modules whose properties were not typical. The inaccuracies were inevitably due to discretization effects. For example, the model cannot distinguish between modules whose sizes are in the range 50 to 100 KLOC, so a module of size 51 KLOC is treated identically to one of 99 KLOC, and if we observe, say, 1501 defects found, then the model cannot distinguish such an observation from 1999 defects found.

This particular problem was solved by the use of dynamic discretization.

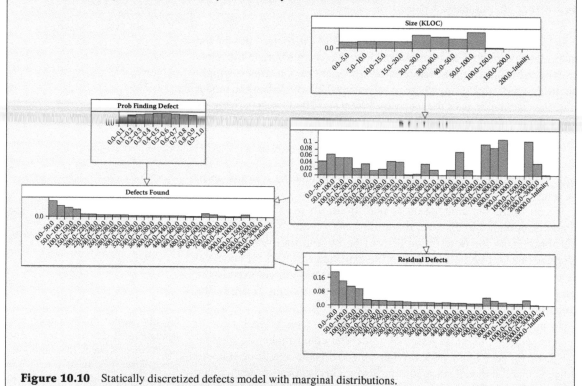

Figure 10.10   Statically discretized defects model with marginal distributions.

10.4   Dynamic Discretization

Fortunately, all of the problems of static discretization of numeric variables can be avoided now that an efficient algorithm using dynamic discretization has been implemented in AgenaRisk. A very broad overview of the way the algorithm works is provided in Box 10.8; full details can be found in Appendix D. But, as a user of BNs, you do not need to know any of the details of that algorithm (just as you do not need to know the details of the standard propagation algorithm).

Box 10.8   Overview of the Dynamic Discretization Algorithm

1. Convert the BN to an intermediate structure called a junction tree (JT) (this is a standard method used in BN algorithms and is described in Appendix C).
2. Choose an initial discretization in the JT for all continuous variables.
3. Calculate the NPT of each node given the current discretization.
4. Enter evidence and perform global propagation on the JT, using standard JT algorithms.
5. Query the BN to get posterior marginals for each node, compute the approximate relative entropy error, and check if it satisfies the convergence criteria.
6. If not, create a new discretization for the node by splitting those intervals with highest entropy error.
7. Repeat the process by recalculating the NPTs and propagating the BN, and then querying to get the marginals and then split intervals with the highest entropy error.
8. Continue to iterate until the model converges to an acceptable level of accuracy.

This dynamic discretization approach allows more accuracy in the regions that matter and incurs less storage space over static discretization.

To use dynamic discretization you simply declare that a numeric node is a *simulation* node. There is no need to consider how to discretize a numeric node. You do not even have to specify the range, since this will be determined by the distribution you define. For example if you define the node as a Uniform[0,10] distribution then the range is automatically determined as 0 to 10. By default a simulation node is set as Normal distribution with mean 0 and variance 10000000. The only additional (but optional) requirement on the user is to specify the required level of accuracy, which is done via the model simulation settings (see Figure 10.11). The most important value for this is the simulation convergence. By default this is set at 0.01, which typically provides a reasonable balance between accuracy and efficiency. If you require greater accuracy, then you decrease this number, but this will come at a cost of reduced speed of calculations and increased memory load.

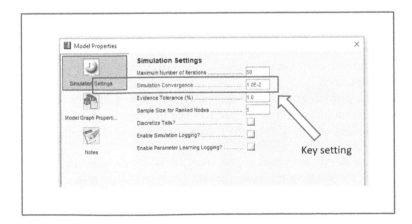

Figure 10.11   Simulation settings in AgenaRisk.

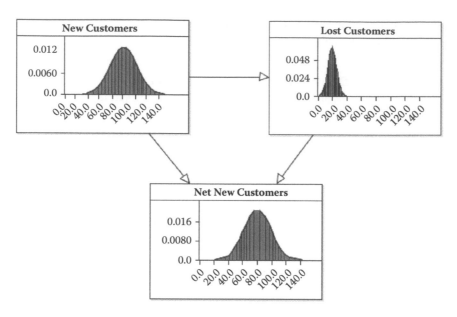

Figure 10.12   Model with simulation nodes and graphs displayed.

To demonstrate both the simplicity and impressive improvements in accuracy that come with dynamic discretization, we reconstruct the customers model from Section 10.2 replacing the (statically discretized) numeric nodes with simulation nodes.

For each of the three nodes define the NPTs exactly as in Section 10.2. When you run the model you should see something like Figure 10.12.

By default the graph settings for simulation nodes display 0 to 100 percentiles of the distribution. You should uncheck this setting and instead select an appropriate range like 0 to 200 to display.

The first thing you should do is compare this initial state of the model with the statically discretized one in Figure 10.8. Note how we no longer have the very coarse jumps.

Now enter the same observations as in Section 10.2 for nodes *New customers* and *Lost customers* (namely 85 and 35, respectively) and run the model. You should see something like Figure 10.13 (in this screenshot we put the mouse over the node *Net New Customers* to display the details of the resulting distribution).

What Figure 10.13 shows is that the distribution of *Net New Customers* is very narrowly fixed on the value 50 (there is only a tiny variation which would actually decrease as we decrease the simulation convergence setting.) Unlike in the statically discretized case this is exactly the result expected.

If you now enter the observations 99 and 21 respectively for *New customers* and *Lost customers* you will see that the distribution of *Net New Customers* is very narrowly fixed on the value 78, which again is the expected result.

Hence, not only is it simpler to build the model using simulation nodes (we do not have to worry about predefining the discretization intervals) but we have solved the problem of inaccuracy.

The problems described in the study of Box 10.7 were also easily resolved by changing the numeric nodes to simulation nodes. In this case the part of the resulting dynamically discretized model (in its initial state) is shown in Figure 10.14. The difference between this and the original in Figure 10.10 is profound.

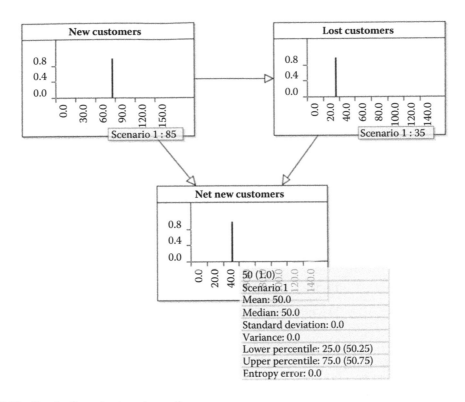

Figure 10.13 Result of entering two observations.

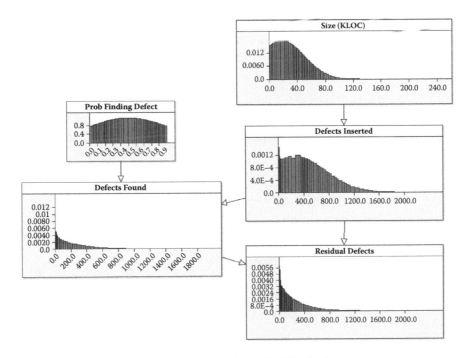

Figure 10.14 Dynamically discretized defects model with marginal distributions.

10.5   Using Dynamic Discretization

In this section we consider some example models that incorporate numeric nodes with all the different types of nodes covered in Chapter 9. The first examples show how dynamic discretization can be used to model purely predictive situations based on prior assumptions and numerical relationships between the variables in the model. We present the car costs example here to model a prediction problem, that is, to model consequence based on knowledge about causes. The next example shows how we might use the algorithm for induction, that is, as a means to learn parameters from observations, and by doing so exploiting Bayes' theorem in reasoning from consequence to cause. The final example is more challenging and presents three ways to estimate school exam performance using a classical frequentist model and two Bayesian models.

10.5.1   Prediction Using Dynamic Discretization

The objective of this model, shown in Figure 10.15, is to predict the annual running costs of a new car (automobile) from a number of assumptions. This particular example uses a number of modeling features that illustrate the flexibility and power of dynamic discretization, because it

A hybrid BN is a BN containing both discrete and continuous nodes.

- Shows how we can use mixture distributions conditioned on different discrete assumptions.
- Uses constant values that are then used in subsequent conditional calculations.
- Uses both conditionally deterministic functions and statistical distributions alongside discrete nodes as a complete hybrid BN.

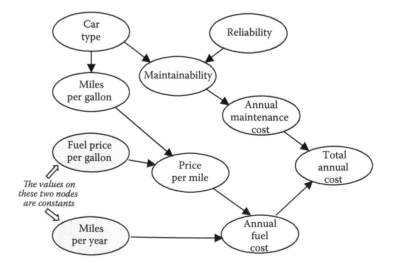

Figure 10.15   Car costs model.

Table 10.1
Car Costs Model Node Details

Node	Type	NPT
Car type	Ranked (small, medium, large)	Default (all values equal)
Reliability	Ranked (high, medium, low)	Default (all values equal)
Maintainability	Ranked (high, medium, low)	TNormal with mean *wmean(2.0,Car type,1.0,Reliability)* and variance 0.1. Hence car type is weighted as twice as important as reliability in its impact on maintainability.
Miles per gallon	Simulation (continuous interval)	Partitioned expression:
Annual maintenance cost	Simulation (continuous interval)	Partitioned expression:
Fuel price	Simulation (continuous interval)	fuel_price_constant set as a constant with value 3
Miles per year	Simulation (continuous interval)	miles_per_year_const set as a constant with value 10,000
Price per mile	Simulation (continuous interval)	fuel_price/miles_per_gallon
Annual fuel cost	Simulation (continuous interval)	price_per_mile × Miles_per_year
Total annual cost	Simulation (continuous interval)	maintenance_cost + total_fuel_cost

Miles per gallon partitioned expression:

Car type	Small	Medium	Large
Expression	TNormal (35, 50, 0, 100)	TNormal (28, 50, 0, 100)	TNormal (18, 30, 0, 100)

Annual maintenance cost partitioned expression:

Maintainability	High	Medium	Low
Expression	TNormal (100, 100, 0, 600)	TNormal (200, 150, 0, 600)	TNormal (500, 150, 0, 600)

The idea in this model is to perform prediction from prior assumptions. There is no induction or diagnostic reasoning going on. This will be illustrated in the subsequent examples.

The nodes are defined as shown in Table 10.1. The important points to note here are:

- No NPT is defined manually. They all involve expressions (arithmetical or statistical distributions) or partitioned expressions.
- The scale of *car type* goes from "small" to "large," whereas the scales of *reliability* and *maintainability* go from "high" to "low." This is to preserve the consistency discussed in Chapter 9 that we need when we combine *car type* and *reliability* to define *maintainability*.

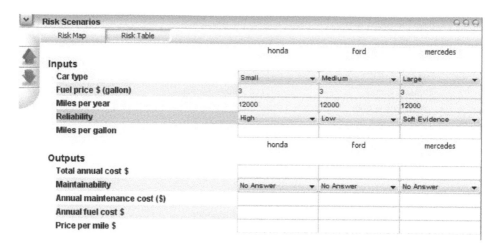

Figure 10.16   Entering evidence for multiple scenarios.

- The nodes *Fuel price* and *Miles per year* are associated with constants. This enables great efficiency savings when running the model since such nodes (unlike all other numeric nodes) never have to be discretized.

We now wish to use the model to help us make a decision about which of three cars (a set of fictitious cars from three manufacturers: Mercedes, Honda, and Ford) we might consider buying. To make a direct comparison between the three cars we create a scenario corresponding to each brand and enter the following evidence:

- Honda scenario—Car type is small, reliability is high
- Ford scenario—Car type is medium, reliability is low
- Mercedes scenario—Car type is large, reliability is soft evidence (50% medium, 50% high)

Note the constant information (about fuel price and miles per year) is automatically the same for all scenarios.

In AgenaRisk such multiple scenario evidence can be entered by switching to the table view shown in Figure 10.16.

Figure 10.17 shows the result of running the model with these three scenarios. Each of the risk graphs show the posterior predictive distribution for nodes like *Total annual costs* and *Annual fuel costs* given the prior assumptions, conditionally deterministic functions, and statistical distributions assumed.

Because of uncertainty about mileage costs and maintenance costs the predictions for total annual cost are, of course, also distributions and not point values. Figure 10.18 shows an enlarged version of the *Total annual costs* distribution, highlighting the differences between the different cars.

The Honda is not only likely to be the cheapest to run but is also the one with least potential surprises (since it also has the lowest standard deviation).

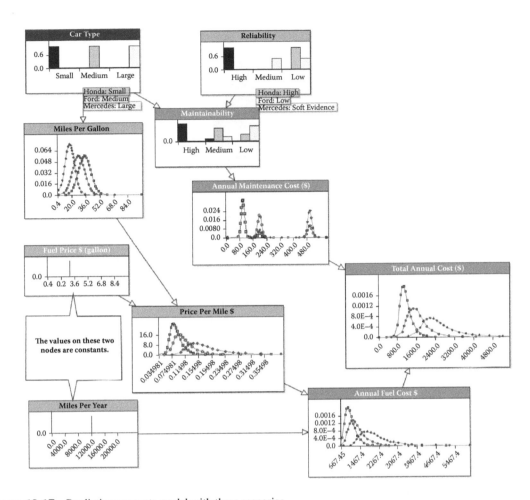

Figure 10.17 Predicting car costs model with three scenarios.

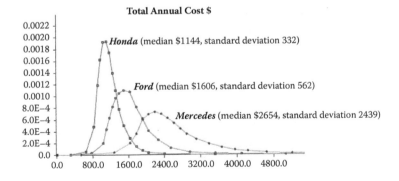

Figure 10.18 Details of the predicted total annual costs.

10.5.2 Conditioning on Discrete Evidence

Now that we have a scheme for propagating evidence through a hybrid BN let's look at a simple example involving a discrete single observation, from which we wish to derive a continuous function conditioned on this observation. This is the simplest form of propagation in a hybrid BN.

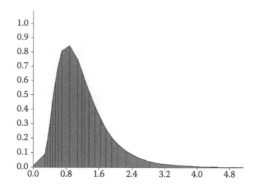

Figure 10.19   Interest rate, X, distribution (percentage).

Consider a bank that needs to predict the future costs of some loan based on the yearly interest rate charged by another bank to provide this loan. Typically, the bank would expect to be able to afford the loan should interest rates stay at a manageable level. However, should they go beyond some value, specified by the regulator as stressful (because it is unexpected or rare), there is a risk that the interest payment due may be unaffordable and the bank might default on the loan.

Here we have some simple parameters for the model:

- The capital sum, which we will treat as a constant, $100M.
- A variable percentage monthly interest rate, X, and the regulator specifies the stress interest rate threshold as any value above 4% interest ($X > 4$).
- The number of months of the loan, which we will treat as a constant, say 10. Then, assuming a single interest payment is made at the end of the 10-month loan period, the total interest payable

Y is equal to $100(1 + \frac{X}{100})^{10} - 100$.

The LogNormal distribution is described in Appendix E.

Let's assume percentage interest rates follow a LogNormal distribution with mean value 0.05, and a standard deviation of 0.5 (don't worry about the choice of distribution for now; we have simply chosen one that has an interesting shape), so

$$X \sim LogNormal(\mu = .05, \sigma = 0.5)$$

and this distribution is shown in Figure 10.19. Notice that the marginal distribution is highly skewed with a long tail of low probability, high interest rate values.

We wish to derive the function for the total interest payment but conditioned on the regulator's stress test scenario. Formally, this is shown in Box 10.9

But, while this looks intimidating, we can easily get the probabilities we need using a BN model. The way to do it is to build a model with a new Boolean variable, Z, as shown in Figure 10.20. The NPT of Z is defined as follows (in AgenaRisk we simply use a comparative expression to do this):

$$if(X > 4, \text{``True''}, \text{``False''})$$

Box 10.9   Calculating the Interest Payment

$$f(Y \mid X > 4) = \int_X \frac{f(Y, X > 4)}{f(X > 4)} dX$$

The denominator is

$$f(X > 4) = \sum_Z f(X \mid Z = True) = \sum_Z \frac{f(Z = True, X)}{P(Z = True)}$$

We know that $P(Z = True) = 1$ and that $P(Z = True \mid X) = 1$ when $X > 4$ therefore

$$\Rightarrow \sum_Z \frac{f(Z = True, X)}{P(Z = True)} = \sum_Z \frac{P(Z = True \mid X) f(X)}{1}$$

$$= \begin{cases} f(X) & \text{if } X > 4 \\ 0 & \text{otherwise} \end{cases}$$

All we now have to do is simply enter in the evidence $Z = True$ and calculate the model. This is exactly what is shown in Figure 10.20. The dynamic discretization algorithm handles all of the conditioning automatically.

You can observe from Figure 10.20 that the tail region has been isolated and the interest payment distribution, conditioned on this tail region, has been calculated. The expected value of the resulting payment is \$60M ($E(Y \mid X > 4) = 60$). Should the bank not have this capital to hand they would default on the loan interest payment and would therefore be bankrupt.

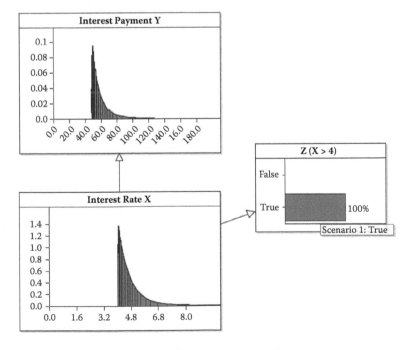

Figure 10.20 Hybrid BN model for stress test scenario.

10.5.3 Parameter Learning (Induction) Using Dynamic Discretization

10.5.3.1 Classical versus Bayesian Modeling

An education department collects exam pass rate data each year on the schools under its management. One year it decides to publish the following anonymised pass rates for a random selection of the schools:

School	Pass rate %
1	88
2	40
3	69
4	53
5	36
6	78
7	90
8	78
9	67
Average	66.4

How helpful is the overall average figure in determining a randomly selected school's success rate next year?

Recall that the objective of the induction idiom is to learn the true value of some population parameters (typically the mean or variance statistics) from observations (i.e., samples). The learned values of these parameters can then be used to make future predictions about members of the same population. This is the basis of Bayesian statistical inference (or *adaptive updating* as it is often called by computer scientists).

We can solve this problem approximately using dynamic discretization by assigning reasonable prior distributions to the parameters of interest and then, for each of the observations we have, create nodes in the BN that depend on these parameters. When we execute the model the algorithm then learns the posterior marginal distributions for the parameters conditioned on the observations we have. To illustrate the solution we use the education department example in the sidebar.

Now we know that the average school success rate prediction of 66.4% fails to take into account any variability in the data. Indeed, as a prediction it will almost always be wrong since actual values will seldom exactly match it. How then do we represent our uncertainty in this problem and how likely is it that a prospective student would actually attend a school with that success rate? Let's first look at how this would be solved using classical frequentist methods and then turn to Bayesian alternatives.

The most accessible classical frequentist approach to this problem involves using the Normal distribution and taking the mean and variance and using these as the parameter estimates. Translating the percentages into proportions (i.e. using 0.88 instead of 88% etc), this approach yields estimates for population mean and variance respectively of

$$\hat{\mu}_p = \bar{p} = \frac{\sum_{i=1}^{n} p_i}{n} = 0.664$$

and

$$\hat{\sigma}_p^2 = \frac{\sum_{i=1}^{n}(p_i - \bar{p})^2}{n-1} = 0.0391$$

This results in $N(\hat{\mu}_p = 0.664, \hat{\sigma}_p^2 = 0.0391)$ as the prediction for any randomly selected school, as shown in Figure 10.21. For example, from this distribution the probability that a randomly selected school will achieve a pass proportion of less than a half is about 0.2.

Let's now turn to the first Bayesian alternative to the classical approach. We are still going to assume that the "underlying" population distribution is a Normal distribution, but since we know the range must be from 0 to 1 we assume it is a TNormal distribution on this range. The crucial difference in the Bayesian approach is that we assume as little as possible

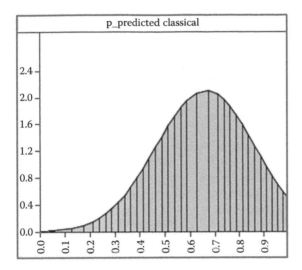

Figure 10.21 Classical school success rate prediction (marginal).

about the mean and variance of this distribution and instead learn these unknown parameters from the observed data, treating each observation as a sample from the unknown distribution. With these assumptions the structure of the model is shown in Figure 10.22.

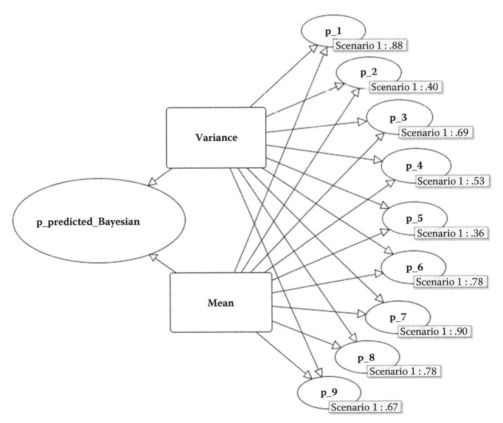

Figure 10.22 Structure of parameter learning model.

The idea of prior ignorance is very important in Bayesian statistics. An ignorant prior is noninformative about what data the model expects to predict and is therefore neutral. In some cases the ignorant prior can be easily modeled as a Uniform distribution. But in others more careful thought needs to go into choosing an ignorant prior specification for your model and sometimes producing a water-tight ignorant prior might be near impossible. However, you should not worry too much about this since we rarely wish to let the data rule our analysis alone. In any case, we are already making lots of assumptions when we choose a particular form for the model (or indeed when we select the data) and in this situation working with a range of different priors, and assessing the sensitivity of the results to these, makes more sense and can be more productive.

Table 10.2

NPTs for Parameter Learning Model

Node	NPT	Justification
mean	Uniform(0,1)	This assumes total prior ignorance. The mean is just as likely to be 0.01 as 0.99.
variance	Uniform(0,1)	This assumes a form of prior ignorance. The variance is just as likely to be 0 as 1.
$p_{predictedBayesian}$	TNormal (mean, variance, 0, 1)	This is the prediction for an unknown school.
p_1	TNormal (mean, variance, 0, 1)	This was the assumption about any sample distribution.
p_2	TNormal (mean, variance, 0, 1)	This was the assumption about any sample distribution.
etc.	...	...

All of the nodes are continuous numeric nodes on the range 0 to 1. The NPTs are defined in Table 10.2. With these assumptions the prior distributions for the model in its initial state is shown in Figure 10.23. Notice that the predicted value for any new proportion, p_i (node *p_predicted* in the figure), is almost flat with mean value of one-half, showing that we are roughly

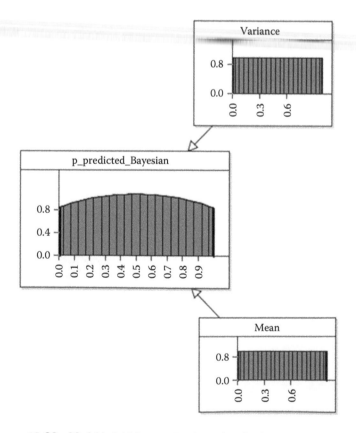

Figure 10.23   Model in initial state, showing priors for parameters.

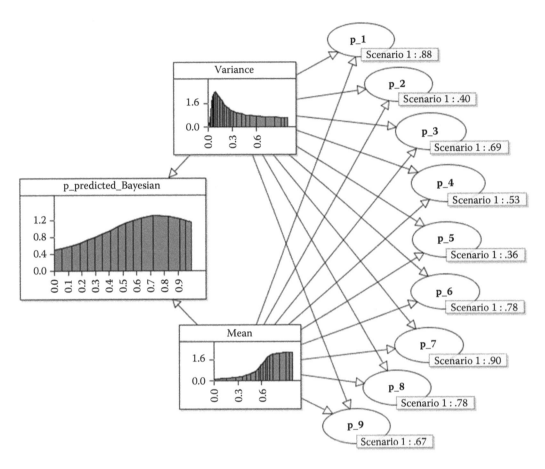

Figure 10.24 A set of similar observations. Bayesian normal prediction for *p*.

indifferent to the possibility of observing any particular value of p_i in advance of it being presented.

Figure 10.24 shows the result of running the model with the observed data for the nine sample schools. The learned posterior marginal distributions for the variables in the model provide updated estimates for the mean variance and a predicted posterior distribution for an unknown proportion, *p*. This results in a predicted value having a mean of 0.57 and variance of 0.07. By eye you can verify that the spread of this distribution is quite wide. Much wider in fact than that provided by using the classical frequent approach (which had mean of 0.664 and variance of 0.0391). The most obvious difference is in the variances. The Bayesian estimate for the variance is significantly greater than the classical one. This is primarily because a dataset of 9 samples is small - not enough to "shake off" the prior assumptions. So, whereas the classical approach predicted a 0.2 probability that a randomly selected school will achieve a pass proportion of less than a half, in the Bayesian approach the probability is 0.38.

The contrast between the Bayesian and classical estimates is stark when the posterior marginal distributions for the prediction are graphed, as shown in Figure 10.25. The Bayesian model is much more uncertain

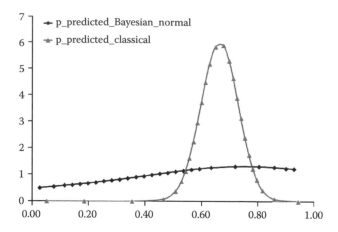

Figure 10.25 Marginal posterior predictions for classical and Bayesian approaches.

in its prediction of the success proportion than the classical one. Which approach is best, though?

One way of determining which is best is to determine whether the prediction envelope actually contains the data being modeled. This can be assessed formally using methods we will cover in Chapter 12, but informally one can simply determine whether the success proportions for each school are surprising under each model. If we look at schools 1, 2, and 7, with values 0.88, 0.4, and 0.9, we see that the p_i values are extremely improbable under the classical model since the values clearly reside in the tail regions but very likely under the Bayesian one. So, by this measure at least, the Bayesian model more properly represents the uncertainty in the underlying data and the variability between the schools. Therefore, it seems reasonable to expect that each school has its own intrinsic proportion.

It is also important to note that, as is common in classical approaches, the above analyses assume that the underlying data are normally distributed. In particular, this assumes the underlying distribution for the data is symmetric. But, in practice, the distribution for data like pass rates is not symmetric. Sometimes it may be skewed towards higher values (and other times to lower values), or it may even be like a "bathtub" with relatively large numbers of high and low scoring schools. In such situations results from both classical analyses (and also from Bayesian analyses that assume an underlying Normal distribution) will inevitably be inaccurate. Fortunately, there is a distribution called the Beta distribution (see Appendix E2 for the full definition) that is able to represent a very wide range of shapes simply by selecting appropriate values for its two parameters alpha and beta, as shown in Figure 10.26.

Instead of using the Normal assumptions in Figure 10.22 and Table 10.3, we could have assumed that the underlying distribution for the observed p_i and predicted p values was a Beta distribution with parameters *alpha* and *beta*. The structure of the BN would be the same in Figure 10.22 but with the nodes "Mean" and "Variance" replaced with "alpha" and "beta." The NPT expressions "alpha" and "beta" can be

In AgenaRisk the Beta distribution is available as one of many predefined distributions for continuous variables. Simple select the distribution and insert the parameter values. You are not restricted to a 0–1 scale; you can select any range of non-negative values.

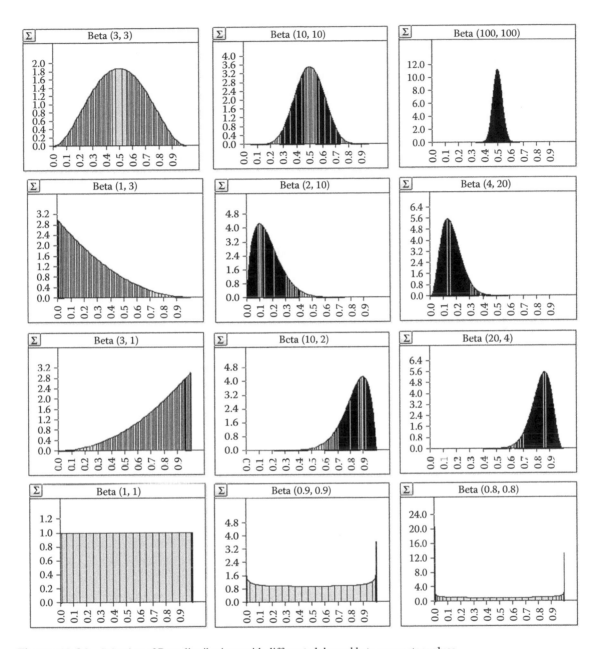

Figure 10.26   Selection of Beta distributions with different alpha and beta parameter values.

set to Uniform(0,100) and the NPT expressions for the p nodes would be Beta(alpha, beta). As in the Normal case the parameters are learned when we enter observations.

When this Beta distribution version of the model is run the distribution for the predicted p has a mean 0.66 and variance 0.031.

10.5.3.2   Bayesian Hierarchical Model Using Beta-Binomial

The simple Bayesian model is not the end of the story. Let us assume that we now get access to the additional data about the sample schools, as

Table 10.3

Exam Pass Rate per School

School i	n_i	x_i	$p_i = x_i/n_i$
1	200	176	0.88
2	100	40	0.40
3	160	110	0.69
4	190	100	0.53
5	140	50	0.36
6	460	360	0.78
7	100	90	0.90
8	90	70	0.78
9	60	40	0.67
pooled	1500	1036	0.69

Hyperparameters can be regarded as parameters of parameters.

The compatibility between the Binomial and Beta distributions is due to a concept that is termed *conjugacy* by Bayesian statisticians. Mathematically, the resulting posterior distribution has the same conjugal form as the prior used, thus ensuring analytical tractability (see Box 10.12 at the end of this section for a discussion of conjugacy). So, for example, a Beta prior distribution multiplied by a Binomial likelihood results in a Beta posterior distribution. This Beta-Binomial pairing is explained more fully in Box 10.10.

shown in Table 10.3. In particular, we now have the school size data. Note that with this data we can see the overall student pass rate is 69% - higher than the average of the schools' rate. In fact, we can use the school size information to extend the previous particular model and determine the uncertainty about exam results within each school as well as between the schools at the same time. But why bother taking account of uncertainty within schools? Well, one reason is that we will be more certain about the results for a larger school with more students than we will about a smaller school with fewer students simply because we have more data for the former. Also, being able to estimate this variability would help us compare schools directly, which is something we could not do in the previous models.

To fulfill this objective we need to build a *Bayesian hierarchical model*. This is simply a model with additional parameters, called *hyperparameters*, which model our uncertainty about the parameters themselves. In this way we can model additional, second-degree uncertainty. There is a large literature on hierarchical modeling in Bayesian statistics that we cannot hope to cover here, but we can give a flavor of it by continuing with the education example. What follows is advanced material.

In this education example using hyperparameters involves modeling the distribution of the success rates for each school as a variable rather than as an observable value. This should make sense since it is a sampled value at a given point in time after all. The natural way to model the number of successes, X_i, is as a Binomial distribution, since each is the result of a Bernoulli process (either pass or fail):

$$X_i \sim Bin(n_i, p_i)$$

We know the values (n_i, X_i) from Table 10.3, but since we now consider p_i as a variable rather than an observed value, we need some way of specifying a probability distribution on p_i that reflects our uncertainty. As it turns out the Beta distribution that we introduced above is compatible with the Binomial distribution and so makes for a natural prior distribution for p_i.

Box 10.10　Beta-Binomial Hierarchical Model Formulations

We have already encountered the Binomial distribution. The likelihood is

$$P(X \mid n, p) = \binom{n}{x} p^x (1-p)^{n-x}$$

We can model p as the Beta function where (α, β) are the number of successes and failures respectively in a finite number of trials. This is thus identical to the idea of Bernoulli successes and failures in the Binomial model. The prior model is

$$P(p \mid \alpha, \beta) = Beta(\alpha, \beta)$$

The neat thing about the Beta distribution is that the conditional posterior distribution is still a Beta distribution (proven by way of Bayes' theorem):

$$P(p \mid X, \alpha, \beta) = Beta(\alpha + x, \beta + n - x)$$

So, in practice, we can choose a Beta prior for a Binomial model and simply select parameter values that match our preconceived expectations about how much information we have, in terms of sample size, and a prior proportion that we consider reasonable.

So, how do we determine the prior distributions for α and β? One way is to choose fixed values. For example, in a coin-tossing experiment we could specify sets of values for α and β that represent ignorance, strong, or biased assumptions about the probability of a coin coming up head and then execute the model, with data, to learn the posterior parameters for p. We can do this using scenarios in AgenaRisk:

$$P(\alpha = 1, \beta = 1 \mid ignorant), P(\alpha = 10, \beta = 10 \mid strong), P(\alpha = 9, \beta = 1 \mid biased)$$

We can then add observational data for (n, X) and use the Binomial likelihood, with the prior, to calculate the posterior under a number of different conditions, combining different priors with different amounts of evidence. Figure 10.27 shows the marginal posteriors that result from fixed-value priors combined with different empirical values for (n, X).

Notice that the ignorant prior is completely flat. Likewise, also note that the strong prior is centered on a half, meaning that there is a predisposition to favor values closer to this region. In contrast, the biased prior favors values closer to one.

However, as we add data the parameter value learned changes. When there are lots of data the change is quite dramatic—adding 50 heads from 100 trials changes the biased prior to a posterior almost centered on 0.5 (but not quite). Thus in this case, the data (the likelihood) could be said to *overthrow* the prior. Also, note that the ignorant prior is steadily updated by whatever data we provide to it such that the posterior is simply equivalent to the data (likelihood); thus, the data could be said to speak for themselves. It should be clear from this that the Beta distribution provides a very flexible way of specifying and accommodating prior knowledge: witness the variety of shapes it can take for the pdf.

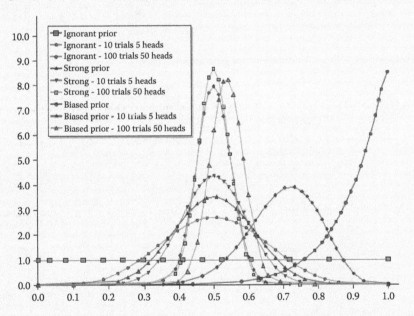

Figure 10.27 Posterior marginal distributions for Beta Binomial with fixed values and different sets of observations.

So, assuming that each p_i is a Beta(α, β) distribution and that the parameters α and β are each Uniform[0,100] we form the model shown in Figure 10.28.

With the Beta-Binomial setup we can also specify additional nodes in the model for the (α, β) parameters, as shown in Figure 10.29 (with posterior marginal distributions superimposed).

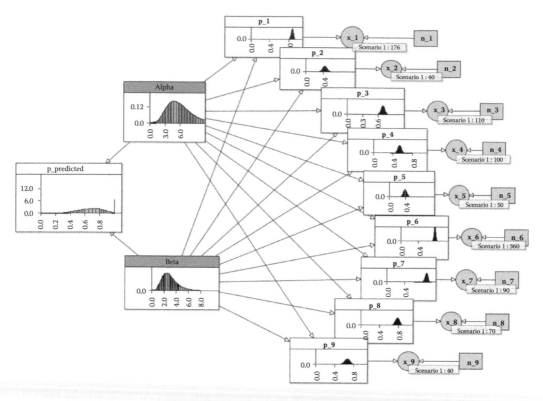

Figure 10.28   Including school size information in the parameter learning.

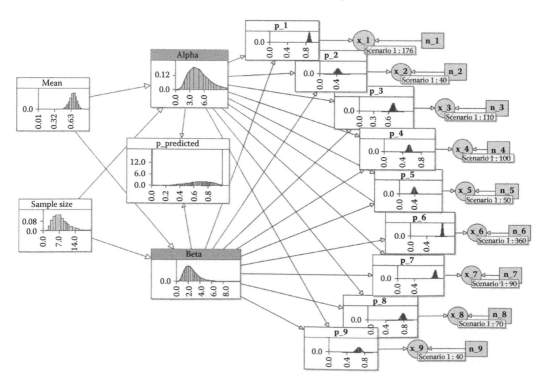

Figure 10.29   Bayesian hierarchical model superimposed with marginal distributions for exam data.

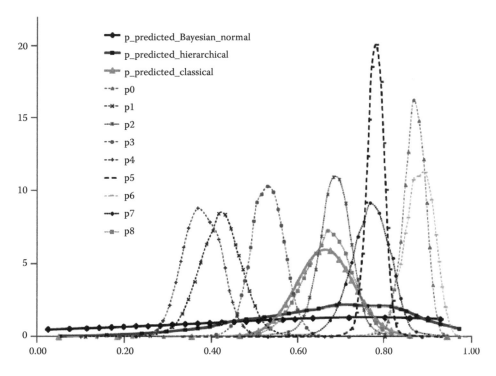

Figure 10.30 Marginal posterior estimates for each school and posterior predictions for classical and Bayesian normal and hierarchical models.

These additional nodes are hyperparameters as explained earlier. In this case (α, β) are parameters for the proportion, which in turn is a parameter for the number of successes. The details and explanation for this particular form of hyperparameter setup is given in Box 10.11.

The structure of the model forms an attractive hierarchy, with the additional benefit that we can now calculate posterior marginal distributions for the success rate of each school and predict the success rate for a new, unknown, or randomly selected school. Figure 10.30 compares

---

Box 10.11   Beta-Binomial Hyperparameter Specification

Another way of determining the prior distributions for the hyperparameters (α, β) is to choose prior distributions that reflect our uncertainty. The nice thing about the Beta distribution in this context is that we can think of the parameter α as representing the number of successes and β as representing the number of failures, so that the sample size is $(\alpha + \beta)$ and the success rate (or population mean) is $\dfrac{\alpha}{\alpha + \beta}$. The sample size and mean are realistically estimable things rather than purely mathematical abstractions and are hence more natural hyperparameters than α and β. Moreover, by making these parents of α and β we simply define α and β by the deterministic functions:

$$\alpha = mean \times sample\ size \quad since\ \alpha = \left(\frac{\alpha}{\alpha + \beta}\right) \times (\alpha + \beta)$$

$$\beta = (1 - mean) \times sample\ size \quad since\ \beta = \left(1 - \frac{\alpha}{\alpha + \beta}\right) \times (\alpha + \beta)$$

This is exactly how the nodes alpha and beta, respectively, are defined in the model of Figure 10.29. But we still have to provide sensible priors for sample size and mean. For the mean a reasonable prior distribution is Uniform[0,1] since it does not favor any particular value in this range. For the sample size it seems reasonable to use a Uniform distribution over a large range, for example Uniform[0,10000]. What we need to be wary of is that, coupled together, these should provide a prior distribution on p that is noninformative. In fact, the choice of Uniform[0,10000] for sample size does provide almost exactly a Uniform[0,1] distribution for p. We could also have chosen a prior distribution for sample size that reflects the fact that smaller samples are much more likely than larger samples. In this case a very good candidate is the Exponential distribution, with mean value equal to the notional sample size we might expect to see, because it has this shape (see Appendix E). So, assuming a notional sample size of 500, the Exponential(1/500) distribution would be reasonable. Fortunately, this choice also results in a prior distribution for p that is almost exactly uniformly flat as shown in Figure 10.31.

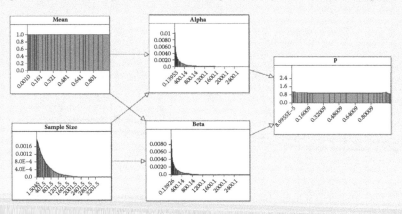

Figure 10.31   Prior marginal distributions for hierarchical model of exam results.

the classical, Bayesian normal and the estimates and predictions derived from the hierarchical model. As you can see by taking into account the variability induced by the differences in sample size, the success rates of the schools are very different and very few of them overlap much with the classical estimate at all. Likewise, note that the Bayesian hierarchical model is more informative than the Bayesian normal model but is less overconfident than the classical model, and given this looks like the best model for the data.

So, in conclusion, any student choosing a school would be wise to distrust the education department's 69% figure as an indication of the pass rate for all schools in the district since it is not a good measure of performance for any particular school. In fact choosing a school in this district, at random, would be little more than a gamble.

Box 10.12   Note on Conjugacy

Conjugacy is treated as a bit of a sacred cow in Bayesian statistics for reasons both good and bad. Commendable reasons include those mentioned earlier—ability to produce analytical results free of compromising approximation, elegance, and the benefits of provability. However, conjugacy does not come at zero cost.

Nonstatisticians cannot easily derive and build models without losing sleep about whether prior distributions are provably noninformative. This is bad news. In our opinion the closer experts can get to express their knowledge directly in models, as priors, the better we can explore the uncertainty inherent in the prediction process. Noninformative priors are just one choice from many (informative or even biased) priors we might wish to explore and these noninformative priors will, unfortunately, not have the neat mathematical properties we hope for. It therefore makes sense to recommend you use the same computational machinery for noninformative and informative priors in tandem. Hence, we eschew presenting lots of analytical formulas to demonstrate the properties of conjugacy and the derivation of conjugate relationships.

However, it makes sense to know where to use conjugate priors when you might wish to. This book is not the place to find an exhaustive list of them; but most, if not all, of these can be modeled using the dynamic discretization algorithm within the hybrid BN approach presented here. AgenaRisk does not require conjugate priors for any of the calculations and will happily produce results for any prior provided. This is neat and often helpful.

10.6 Risk Aggregation, Compound Sum Analysis and the Loss Distribution Approach

Risk aggregation is a popular method used to estimate the sum of a collection of financial assets or events, where each asset or event is modelled as a variable. There are numerous and many applications of this, such as:

- In cyber security we might estimate the number of network breaches over a year and, for each breach, have in mind the severity of loss (in terms of lost availability, lost data or lost system integrity). The network operator might wish to know how much to invest to cover possible future losses.
- In insurance we might have a portfolio of insurance policies and expect a frequency of claims to be made in each year, with an associated claim total. We would, therefore, wish to know our exposure over any year.
- In operational risk we might be able to forecast the frequency and severity of classes of events and then wish to aggregate these into a total loss distribution for all events (the so-called loss distribution approach [LDA]).

All these applications involve aggregating distributions (Section 10.6.1) and calculating a loss distribution by taking account of both frequency and severity distributions (Section 10.6.2).

10.6.1 Aggregating Distributions

Imagine an insurance company that has several different classes of risk portfolios (such as one for home insurance, one for motor vehicle insurance, one for holiday insurance, etc.). Each of these has its own different loss distribution. Suppose these distributions are $S_0, S_1, \ldots, S_n$. To compute the company's total loss distribution, T, we simply compute the sum:

$$T = S_0 + S_1 + \cdots + S_n$$

While this looks simple in principle, computing this kind of sum of probability distributions (called "convolution") is complex. Fortunately,

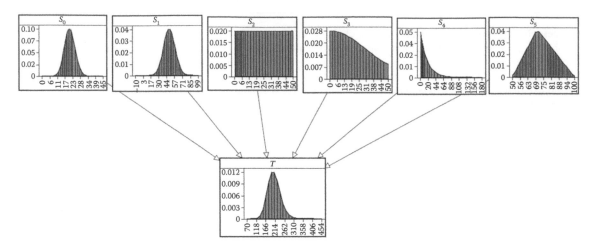

Figure 10.32   Risk aggregation for six classes of risk event or asset.

the dynamic discretisation algorithm automatically carries out the convolution operations required when performing risk aggregation. So, we can compute the sum using the sum() function in AgenaRisk. For example, given six loss distributions:

$$S_0 = Binomial(0.2,100), \; S_1 = Normal(50,100), \; S_2 = Uniform(0,50)$$

$$S_3 = TNormal(0,1000,0,50), \; S_4 = Gamma(1,20)$$
$$S_5 = Triangular(50,70,100)$$

we can simply sum these by creating a child node, T with arithmetic function equal to

$$sum(S_0,S_1,S_2,S_3,S_4,S_5).$$

The results of this summation are shown in Figure 10.32. It is interesting to note that, even when the input distributions are very varied in terms of shape and scale, the summed total tends to a normal distribution, as one might expect from the central limit theorem (see sidebar). However, it is critical to note that they converge on normal only for large summations and, in this example, there remains a significant asymmetry with a longer tail on the right-hand side.

We could build a separate BN for each S_i, perhaps representing different classes of risk events arising from different business operations and aggregate these to estimate our total risk. Furthermore, if they are dependent then common risk factors could be included to make the S_i co-dependant.

The central limit theorem states that, in most situations, as a sequence of independent random variables are added their sum tends towards a normal distribution, even if the original variables are not normally distributed. If the variables are independently identically distributed (IID), that is they have the same distribution function, and finite variance, convergence on a normal distribution for the sum is guaranteed.

10.6.2   Calculating a Loss Distribution and Using the Compound Sum Analysis Tool

Suppose, we have a portfolio of M insurance policies. In any given year each policy either makes a claim or does not. Hence, in any given year the total number of claims n is a number between 0 and M. Suppose, for example that $M = 2$. Then $n = 0$, 1 or 2.

Suppose that for each claim there is a (fixed known) severity distribution S that characterises the size of the claim (i.e. the loss/payout). Now S is usually a continuous distribution, but to illustrate the point simply, we will suppose S is discrete and that the only possible sizes of loss for a claim are 100, 200 and 500 dollars with

- $S(100) = 0.5$ (so there is a 50% chance that the payout is \$100)
- $S(200) = 0.4$ (so there is a 40% chance that the payout is \$200)
- $S(500) = 0.1$ (so there is a 10% chance that the payout is \$500)

(Note: we also assume that if there is no claim then there is no loss.)

Suppose that, based on historical data, we know that the probability distribution for n (in the case where $M = 2$) is:

- $P(n = 0) = 0.6$ (so there is a 60% chance that there will be no claims)
- $P(n = 1) = 0.3$ (so there is a 30% chance that there will be one claim)
- $P(n = 2) = 0.1$ (so there is a 10% chance that there will be two claims)

What is the distribution for total losses?
First, we consider what the loss values can be:

- When $n = 0$ there are no claims, so the loss is 0.
- When $n = 1$ there is one claim and its size can be 100, 200 or 500.
- When $n = 2$ there are two claims which could occur as any of the permutations:
    - (100, 100), that is a total of 200
    - (100, 200), that is a total of 300
    - (200, 100), that is a total of 300
    - (100, 500), that is a total of 600
    - (500, 100) is a total of 600
    - (200, 200), that is a total of 400
    - (200,500), that is a total of 700
    - (500, 200), that is a total of 700
    - (500, 500), that is a total of 1000

These possible loss sizes, and their probabilities are shown in Table 10.4.

The total loss distribution T is therefore as shown in Table 10.5.

We can construct a BN to do the above computation in AgenaRisk as shown in Figure 10.33. Note that we have been careful to represent the two separate severity claims as separate nodes since they are independent variables, but with the same distribution function.

Note how the partitioned expression involves successfully summing the severity distributions, and so relies on aggregation. Also note that we can think of the total loss distribution T as being made up of

Table 10.4
Computing Loss Probabilities

n	$P(n)$	Loss	$P(\text{Loss}\mid n)$	$P(n)*P(\text{Loss}\mid n)$
0	0.6	0	1	0.6
1	0.3	100	0.5	0.15
1	0.3	200	0.4	0.12
1	0.3	500	0.1	0.03
2	0.1	200 (100 + 100)	0.5*0.5	0.025
2	0.1	300 (100 + 200)	0.5*0.4	0.02
2	0.1	300 (200 + 100)	0.4*0.5	0.02
2	0.1	600 (100 + 500)	0.5*0.1	0.005
2	0.1	600 (500 + 100)	0.1*0.5	0.005
2	0.1	400 (200 + 200)	0.4*0.4	0.016
2	0.1	700 (200 + 500)	0.4*0.1	0.004
2	0.1	700 (500 + 200)	0.1*0.4	0.004
2	0.1	1000 (500 + 500)	0.1*0.1	0.001

distributions T_0, T_1 and T_2 which correspond to the separate loss distributions for where T_i is the distribution corresponding to the case where exactly i claims are made.

Explicitly, T_0 is the distribution:

Loss	$T_0 = P(\text{Loss})$
0	1

T_1 is the distribution:

Loss	$T_1 = P(\text{Loss})$
100	0.5
200	0.4
500	0.1

Table 10.5
Computing the Total Loss Distribution

Loss	Number of occurrences	$T = P(\text{Loss})$
0	1 (with prob 0.6)	0.6
100	1 (with prob 0.15)	0.15
200	2 (with probs 0.12, 0.025, respectively)	0.145
300	2 (with probs 0.02, 0.02, respectively)	0.04
400	1 (with prob 0.016)	0.016
500	1 (with prob 0.03)	0.03
600 (100 + 500)	2 (with probs 0.005, 0.005, respectively)	0.01
700 (200 + 500)	2 (with probs 0.004, 0.004, respectively)	0.008
1000 (500 + 500)	1 (with prob 0.001)	0.001

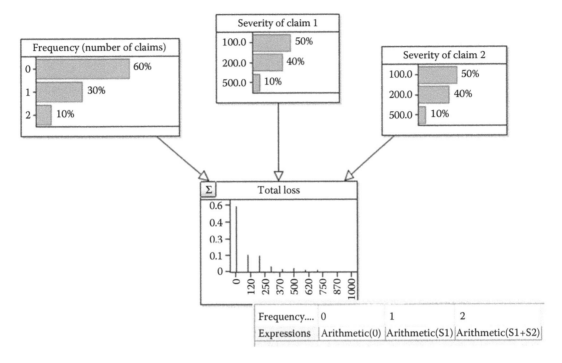

Figure 10.33　BN to calculate total loss distribution, T.

T_2 is the distribution:

Loss	$T_2 = P(Loss)$
200	0.25
300	0.4
400	0.16
600	0.1
700	0.08
1000	0.01

Note that the total loss distribution T is then

$$T = P(0)T_0 + P(1)T_1 + P(2)T_2$$

This is what is referred to as *n-fold convolution*. This process is often called a *compound sum* by non-actuaries and non-statisticians. The general form of this is shown in Box 10.13.

Box 10.13　Computing Loss Distribution Using *n*-Fold Convolution

Suppose that, in a given period a class of events (such insurance claims, security breaches, system failures, etc.) can occur n times where n is any number between 0 and M and that the events have a fixed severity distribution S. So, in any period we have double uncertainty—uncertainty about the value of n (the number of events in the period) and uncertainty about the severity S of each event.

Assuming we know the frequency distribution, $P(n)$, that is the probability there will be n events during a given period (for $n = 0,\ldots,M$), then we can compute the total loss distribution T as

$$T = P(0)T_0 + P(1)T_1 + P(2)T_2 + \cdots + P(M)T_M$$

where

$$T_0 = 0, \qquad T_n = T_{n-1} + S \quad \text{for } n = 1 \text{ to } M$$

This recursive process is called n-fold convolution, but is also often called a compound sum by non-actuaries and non-statisticians.

Although we computed the compound sum in a BN for the above example (with $n = 2$, and a discrete severity distribution) it would be laborious and time consuming to do this when n is large. Fortunately, within AgenaRisk the process is automated, and an n-fold convolution analysis calculator is available in the form of the *compound sum analysis* function.

Example 10.1

To use the compound sum analysis function to the above example, we only need to declare a single instance of the severity distribution (although note that compound sum analysis requires the severity node to be continuous). In fact, we simply have to create three (separate) nodes as shown in Figure 10.34. You then need to launch the compound sum analysis tool and declare the Frequency, Severity and Compound nodes in the dialog. After running you should see the distribution for Total loss (the Compound node) as shown in Figure 10.34 which is identical to the result we saw in Figure 10.33.

When you use AgenaRisk's compound sum analysis tool there is no need to link the frequency and severity nodes to the total node. Thus, there are no edges connecting the nodes in Figure 10.34.

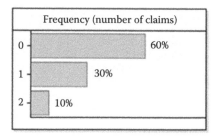

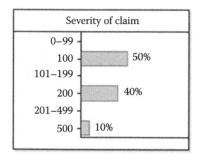

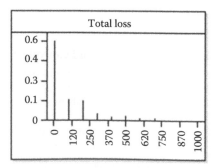

Figure 10.34 Compound sum function for example gives identical total loss distribution, T.

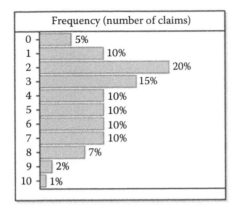

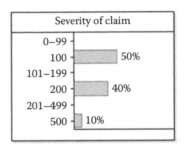

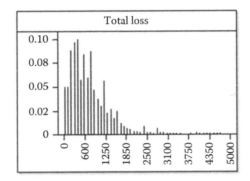

Figure 10.35 Compound sum function for example, but with extended frequency.

Whereas it would be difficult, without the compound sum analysis tool, to extend Example 10.1 to the case where frequency goes up to 10, it is trivial to do this using the compound analysis tool. The only change needed is to define the new frequency distribution as shown in Figure 10.35.

In general, we can use the compound sum analysis tool where we have any frequency function and severity distribution function to get fast results, as shown in Example 10.2.

Example 10.2

Consider where the frequency of loss, n, is a Poisson distribution with mean 100 loss events per year and a severity distribution, S, which is Triangular with parameters (5, 80, 100). Computing the compound sum yields the results shown in Figure 10.36. The marginal probability distribution for the total (aggregate) loss has a mean of 6132 and a 99 percentile interval (2852 and 9241).

Note that the frequency distribution for n must be an integer whilst the severity distribution can be continuous or integer value. Also, if you want to compute the total loss distribution for a specific frequency value n (i.e. to compute a specific n-fold convolution) simply define the frequency distribution as the constant value n (you cannot simply use the same frequency distribution and define n as an observation—this isn't allowed for the severity or frequency nodes in the compound sum analysis function).

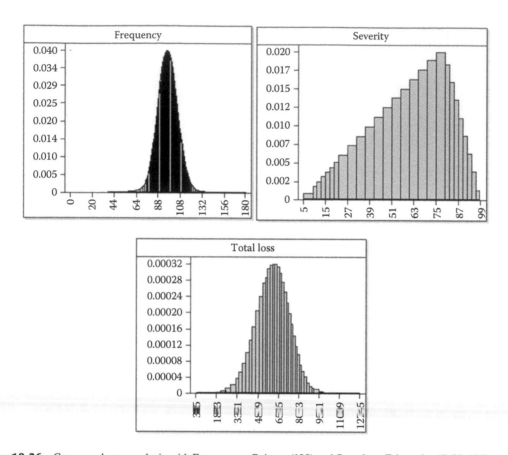

Figure 10.36 Compound sum analysis with Frequency a Poisson (100) and Severity a Triangular (5, 80, 100)

10.7 Tips and Tricks when Using Numeric Nodes

When using numeric nodes in BNs (whether you use static or dynamic discretization) there are some issues you need to be aware of that do not apply to nodes of other types. By being aware of these issues you will avoid some very common errors that are made by BN modelers and users.

10.7.1 Unintentional Negative Values in a Node's State Range

Consider the model shown in Figure 10.37 where A, B, and C are numeric nodes with range zero to infinity. The NPTs are

- $A \sim TNormal(\mu = 200, \sigma^2 = 1000000, 0, 1000000)$

- $B \sim TNormal(\mu = 0.3A, \sigma^2 = 100A, 0, 1000000)$

  Meaning that on average we will find and fix 30% of the known defects in testing but we expect that the more defects the bigger the variance.

- $C = A - B$

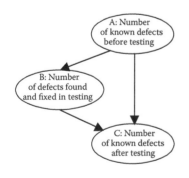

Figure 10.37 New model with simulation nodes.

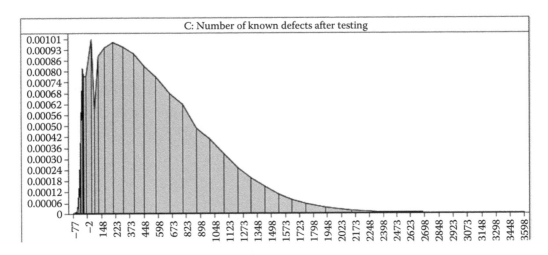

Figure 10.38 Node C after calculation.

When the model is run the marginal distribution for node C is as shown in Figure 10.38.

There seems to be something wrong here. The graph clearly shows nonzero probability for negative values, even though logically B can never be greater than A.

There is a rational explanation for this: The model knows that $C = A - B$ but it does not know that $B \leq A$. When the calculations are performed during the dynamic discretization it will actually encounter cases where $B > A$. This explains the (prior) nonzero probability of negative values. Although the model provides the correct result when observations are entered for A and B this is still unsatisfactory. For example, if we know that there are 500 defects before testing and want to use the model to estimate the number of defects after testing we will get the result shown in Figure 10.39.

Note that there is still a nonzero probability for negative values.

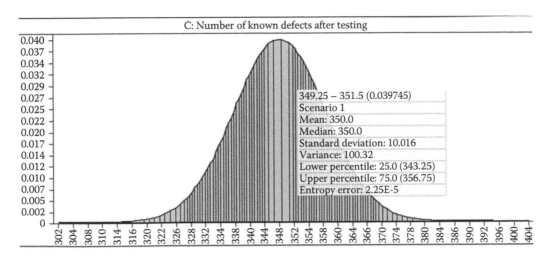

Figure 10.39 Prediction for C with $A = 500$.

The solution to this and similar problems is to force the node C to only take on nonnegative values. The simple way to do this is to use the expression $C = max(0, A - B)$ instead of the expression $C = A - B$.

10.7.2 Faster Approximation When Doing Prediction

If you are only interested in prediction then you should set the number of iterations to 1 as it gives a good approximation of the true posterior. The reason for this is explained in Box 10.14.

Box 10.14

Faster approximation is performed by examining all the distributions and functions declared in the model and determining for each what a reasonable approximation for its initial state range might be based on where the prior probability mass of the node might lie, given the parameters declared on the node and the parent node state ranges. For instance, we can easily determine an approximate initial range for a node declared as Normal(0,1) because we know the mean and standard deviation and can use this to calculate the prior ranges as mean plus or minus three/four standard deviations from the mean. From this we can then insert the range into the node and add sub intervals within this range. If this node was a parent of another node given we know the parent range, it is easy to see how we could compute the range of the child node. In this way, we propagate ranges through the BN model in the first iteration, from parents to children, computing approximate initial ranges for each node. Then, in subsequent iterations the dynamic discretisation algorithm is used to refine the result.

If you open the Car Costs model in Section 10.5.1 and set the number of iterations in model properties to be one (rather than the default value of 50) and then run it, you will get the result shown in Figure 10.40(a) for the "Total annual cost" node. For the same node, running the model under the default settings gives the results shown in Figure 10.40(b). Under default conditions, this model, which has three scenarios, is eight times slower to compute than in the "fast approximation" case (1.5 vs. 0.2 s on an Intel i7 processor, for three scenarios). You can verify that the distribution and summary statistics are very close. For instance, in the third scenario in this model ("Mercedes"), the mean and standard deviation summary statistics are very close at mean of 2286 and standard

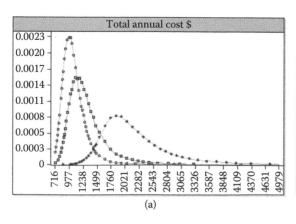

(a)

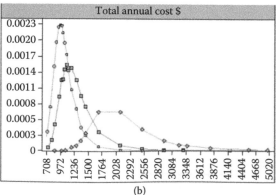

(b)

Figure 10.40 (a) Node run under default conditions, (b) node calculated using fast approximation (one iteration).

deviation of 739 in the faster approximation case, compared with a mean of 2273 and standard deviation of 717 under default conditions.

For many applications where speed has priority, and prediction is the objective, running the model with a single iteration setting should therefore suffice to give a good approximation.

10.7.3   Observations with Very Low Probability

Although a numeric node with an expression like the Normal distribution in theory has a nonzero probability value for even the smallest and most extreme intervals, you must be aware that ultimately any BN model has to store the probability distribution for any node as a finite set of intervals (this is true even of dynamically discretized nodes). In that case, you cannot expect any model to be able to store every (non-zero) probability interval without the storage being infinite. Therefore, in practice, regions of very low probability are stored as zero. So, if you attempt to enter observations in such a range you will get the *inconsistent evidence* message that we described in Chapter 7.

As an example, consider a numeric node whose NPT is a Normal distribution with mean 0 and variance 1000. Although in theory arbitrarily high or low values are possible for this distribution (such as a value within the range 100 million to 100 million plus one) such probability values are so small that if you entered an observation within such a range you might get an inconsistent evidence error message. In fact, almost all of the nonzero probability mass (99.9% of it) lies within –30 and 30. If you try to enter a value above, say, 300, you may trigger the error message.

So, you need be very careful when entering extreme observations for numeric nodes, since such observation (although theoretically possible) may, rarely, be treated as inconsistent evidence.

10.7.4   "Tail Discretisation"

In the model properties there is an option to perform Tail Discretisation, as shown in Figure 10.41. This involves examining the tails of a node's marginal probability distribution and adding more states in the region around the tail, that is around the 5th, 1st, 95th and 99th percentiles. This produces higher accuracy percentile statistics for all nodes in the model, but because of the increase in the number of states it does come with an increase in computation time.

Example 10.3

Suppose we wish to sum five standardised Normal distributions, that is where each of the five distributions is defined as Normal(0,1). Theoretically, the sum should be equivalent to Normal(0,5) since the means and variances of the Normal distribution are additive under summation. The result of summation without discretising the tails is shown in Figure 10.42(a) and when the tails are discretised the result is given by Figure 10.42(b). You can see that the tails are longer in (b) than (a). In

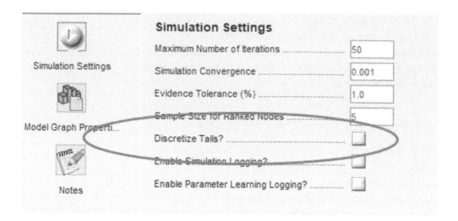

Figure 10.41   Discretise tails option in simulation settings.

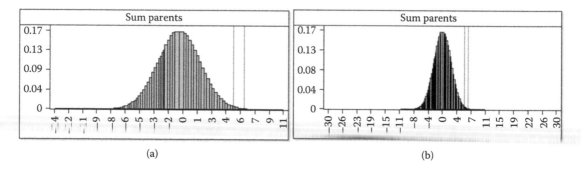

Figure 10.42   (a) Non-discretised tails and (b) discretised tails.

the non-discretised version the 99.9th percentile is 7.01 and in the discretised version it is 6.93. The analytical result is actually 6.91 so plainly the additional discretisation in the tail area has much improved the percentile estimate.

10.8   Summary

We know that discrete variables are insufficient on their own to model complex problems and that is why we need to include variables that are numeric. So far in this book we have defined discrete probability distributions for discrete nodes. In this chapter we have defined continuous probability distributions for numerical nodes. Fortunately the ideas and concepts introduced in Chapters 5, 6, and 7 still apply but we covered the new mathematical concepts needed to put us on firm ground.

We have described the drawbacks of static discretization, which was truly the Achilles heel of BNs and cover a new algorithm called dynamic discretization, which works efficiently for a large class of continuous distributions.

We demonstrated how this works using three sizeable examples. The first example involves predicting car (automobile) costs using a hybrid model containing mixtures of discrete and continuous nodes. The second example showed how we could model a financial stress testing problem using a hybrid BN. The third example was more challenging and presented three ways to estimate school exam performance using a classical frequentist model and two Bayesian models. We also described how the compound sum analysis tool in AgenaRisk easily solves the class problem of risk aggregation. Further details of this can be found in the reference Lin et al. (2014).

Further Reading

Fenton, N. E., Neil, M., and Marquez, D. (2008). Using Bayesian networks to predict software defects and reliability. Proceedings of the Institution of Mechanical Engineers, Part O, *Journal of Risk and Reliability* 222 (O4): 701–712.
> This paper describes how we used dynamic discretization to resolve the problem described in Box 10.8.

Gelman, A., Carlin, J., Stern, H., and Rubin, D. (2003). *Bayesian Data Analysis*, Chapman & Hall.

Kozlov, A. V., and Koller, D. (1997). Nonuniform dynamic discretization in hybrid networks. *Proceedings of the 13th Annual Conference on Uncertainty in AI (UAI)*, 314–325.

Lin, P., Neil, M., Fenton, N. E. (2014). Risk aggregation in the presence of discrete causally connected random variables. *Annals of Actuarial Science* 8 (2), 298–319.
> This paper compares different approaches to compound sum analysis, convolution and risk aggregation.

Neil, M., Tailor, M., and Marquez, D. (2007). Inference in hybrid Bayesian networks using dynamic discretization. *Statistics and Computing* 17(3), 219–233.
> Definitive source for the dynamic discretization algorithm.

11

Decision Analysis, Decision Trees, Value of Information Analysis, and Sensitivity Analysis

11.1 Introduction

Many risky problems involve decisions that require serious thought, reflection, and analysis. So far, we have modeled risk problems using Bayesian networks without providing any explicit means for automatically identifying those decisions you should make to optimize some notion of cost or utility. For many problems and analyses, this isn't always necessary because other methods are available to perform the financial analysis within the BN, or the decision-making involved is so obvious that it can be done tacitly. However, for more complex decision problems, mainly involving sequentially organized decisions, where your future decisions depend on your past decisions, this less formalized approach may be inadequate.

This chapter covers decision-making under conditions of uncertainty, where you can identify your possible decisions in advance and the sources of uncertain information that may become known (observable) at some point—or may remain unknown (unobservable)—and where we wish to optimize some measure of utility, either in financial terms, in the form of personal satisfaction, or perhaps in terms of some other measure, such as "riskiness."

There are three critical components in a decision analysis model:

- *Decisions*: Decisions are identified at each stage in the analysis or point in time based on what is known or may be expected to be known. Decisions are made about alternatives selected from mutually exclusive and exhaustive sets. For instance, before building a new power station, we have to decide from the alternatives "coal fired," "nuclear," or "gas burning."
- *Chance variables*: Chance variables can either be observed or not. An observable chance variable will be one where a report on its true state becomes available at some point in the analysis either as a result of a decision to identify its state or as a known input to a decision. Of course, an observable chance variable

Even mundane decisions, like whether to take an umbrella to work, implicitly involve making a selection based on a notion of "cost/utility." Taking the umbrella incurs the "cost" of the discomfort of carrying it and the utility value of "staying dry" if it rains; not taking the umbrella has the negative utility value of "getting wet" if it rains.

may be an inaccurate measure of an underlying unobserved variable that we wish to know but that is unobtainable. Note that chance variables are simply the familiar uncertain nodes in a BN, as introduced and defined in preceding chapters. For example, we may not know the performance of a new nuclear power technology but may sponsor a report that gives a reliable measure of what it is, for a fee. Depending on the decision context, the decision maker will need to identify what chance variables are indeed observable and actually observed at the point when each decision is made.

- *Utilities*: Associated with each decision we make will be actions and outcomes that have a cost and benefit. These costs and benefits will usually be expressed as economic measures but can be subjective utilities (after all, if we are fine with subjective probabilities, then we should be relaxed about subjective utilities). Ideally, as rational agents, we wish for the utility (benefit) to be maximized and the disutility (cost) to be minimized.

Here, we use a variety of methods, all available in AgenaRisk, to support decision analysis under uncertainty. First, we cover *influence diagrams*, with examples, and show how we can compute these using BNs. Typically, other BN software tools cannot handle continuous variables properly without major inconvenience and potential inaccuracy, and attempt to present the decision within the BN. We feel this is too restrictive and leads to confusion. To perform decision analysis, we instead use a method called *Hybrid Influence Diagrams* (HIDs) that can easily accommodate continuous variables. We also choose instead to present the results of HID computations as *Decision Trees* (DTs), which we recommend as more powerful ways of communicating the results. Unlike previous algorithms, our method is based on dynamic discretization and provides a fully automated solution for HIDs that contain continuous chance nodes with virtually any probability distribution, including non-Gaussian types, or any conditionally deterministic function of these distributions.

A critical potential step in decision analysis is determining how much value you are willing to pay for a source of information. If you could buy perfect information about an uncertain variable, how much should you pay to know its value given its role in the decision analysis? To address this, we use a computation method called *Value Of Information* (VOI) analysis to focus the attention on those variables that matter in terms of seeking further or better information about them than others.

A critical step, even in the absence of decisions and utilities, is determining how sensitive a target variable is to changes in those other variables that affect it. This is called *sensitivity analysis* and can be carried out on a wide variety of styles of BNs, including decision analysis models.

Influence and hybrid influence diagrams and decision trees are covered in Sections 11.2 to 11.4. Section 11.5 deals with value of information analysis. Section 11.6 covers sensitivity analysis.

11.2   Hybrid Influence Diagrams

A hybrid influence diagram is a graphical probabilistic model that offers a general and compact representation of decision-making problems under uncertainty.

As discussed in Section 11.1, a decision analysis model involves decisions, chance variables (that can be either observed or unobservable), and utilities. In an HID, we associate nodes with each instance of these. The convention is that rectangles represent decisions, ellipses represent chance variables, and diamonds represent utilities. Each decision node represents a decision-making stage, each chance node represents a random variable, and each utility node has an associated table or continuous probability distribution that defines the utility values based on the states of its parents. There is also always an "ultimate" utility node (with at least one parent) that we seek to optimize. Generally, the chance, decision, and utility nodes of an ID can be discrete or continuous variables.

In an HID, incoming arcs to chance or utility nodes represent causal, deterministic, or associational relations between the node and its parents. Incoming arcs to decision nodes (shown by a dashed line) are "informational" arcs, representing the possibility that the state of any parent node might be known before the decision is made. Informational (dotted-line) arcs also specify a partial sequential order of decisions and observations. An HID cannot be computed without this (at least implied) sequential order.

Example 11.1 describes a simple HID problem based on Winkler's classic "umbrella" example, with the HID model shown in Figure 11.1.

Example 11.1   Consider the Following Problem

You have to decide whether to take an umbrella when you leave home in the morning. There is a main decision with two action states, "take umbrella" or "do not take umbrella." There is one uncertain (Boolean) variable: "Rain" (it either rains or it doesn't). If you know it will rain, you will take the umbrella; if you know it won't rain, you will not. However, you do not know this with certainty, but you can listen to the weather report to find out, knowing that this forecast is imperfect. Let's assume there is some discomfort in carrying an umbrella when it doesn't rain, some benefit in doing so when it does (you will not get wet), and a cost of listening to the boring weather forecast.

We model this problem with an HID in Figure 11.1. The decision nodes are "D1 Test," representing the decision to listen to the weather forecast or not, and "D2 Umbrella," representing the decision to take the umbrella or not. The unobservable chance node is "Rain," which cannot be observed at the point of making any of the decisions. The observable chance node is "Test Result," representing the weather forecast. There are two utility nodes, "u test" and "u." The first represents the utility cost of the "D1 Test?" decision, and the second represents the ultimate utility combining the payoffs from the second decision, "D2 Umbrella" and "Rain."

In Figure 11.1, notice that there is an information arc, denoted by a dotted line, between Decisions D1 and D2 representing the fact that D1

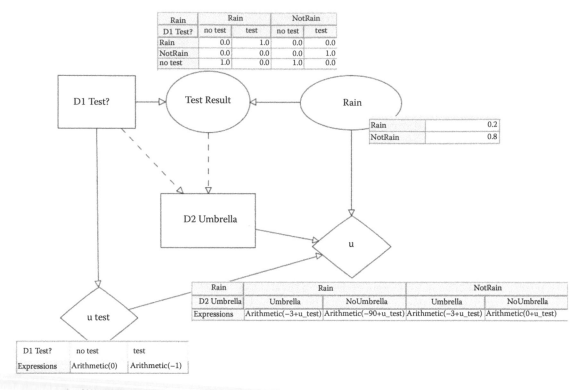

Rain	Rain		NotRain	
D1 Test?	no test	test	no test	test
Rain	0.0	1.0	0.0	0.0
NotRain	0.0	0.0	0.0	1.0
no test	1.0	0.0	1.0	0.0

Rain		0.2
NotRain		0.8

Rain	Rain		NotRain	
D2 Umbrella	Umbrella	NoUmbrella	Umbrella	NoUmbrella
Expressions	Arithmetic(−3+u_test)	Arithmetic(−90+u_test)	Arithmetic(−3+u_test)	Arithmetic(0+u_test)

D1 Test?	no test	test
Expressions	Arithmetic(0)	Arithmetic(−1)

Figure 11.1 HID for Example 11.1 umbrella decision.

is made before D2 and the result is known at D2. Likewise, the "Test Result" is also known at D2, and this is also represented as an information arc. For the decision node D1, we have assigned a state "NA" (meaning "not applicable") to represent the fact that the test is not carried out because the weather forecast is not listened to.

We choose not to show the NPTs for the decision nodes because these are simply uniform valued, representing our open mindedness about the alternatives and the fact that the decisions are unconstrained (more on this later when we talk about asymmetry). The NPT for "Rain" shows the probability for rain, while that for "Test Result" shows that the forecast is perfect (it doesn't have to be, of course—there could be uncertainty here). Also, notice that the states for "Test Result" include the state "no test"—this is very important and will be discussed later.

The utility nodes in this example are declared using arithmetic functions representing the payoffs for each combination of parent node. These could instead be distributions. For the utility node "u test," the utility distribution reflects the cost of listening to the forecast, and for the ultimate utility node "u," it models the discomfort from being caught in the rain without the umbrella (−90), taking an umbrella regardless of whether it rains (−3), and not taking an umbrella on a dry day (0), plus the parent node "u test" value.

Typically, the aim in decision analysis is to maximize utility (or minimize disutility) given the decisions and evidence in the BN. The *maximum expected utility principle* says that we should choose the alternative (decision states) that maximizes the expected utility. Box 11.1 provides the formal explanation together with the algorithm that is implemented in AgenaRisk.

Box 11.1 Algorithm to Evaluate Influence Diagrams

The maximum expected utility for a set of decisions, $\bar{D} = \bar{\delta}$, is computed by finding the optimal policy, given observable nodes, $\bar{O}$; unobservable chance nodes, $\bar{N}$; and utility nodes, $\bar{U}$.

The optimal policy set, $\bar{D} = \bar{\delta}$, is the set of decision states on each decision node that maximizes the utility function, $f\left(U|\bar{D}\right)$, such that:

$$E_{\delta}\left[f\left(U|\bar{D}\right)\right] = argmax_{\bar{D}}\left[\sum_{N,O,U} f(U|\bar{D},\bar{N},\bar{O},\bar{U})f(\bar{D},\bar{N},\bar{O},\bar{U})\right]$$

Given that this is a factorization of a joint probability distribution, we can use BN calculation machinery, such as propagation, to factorize and calculate the maximization operations in an efficient way. The algorithm for this first requires us to convert the ID to a BN as follows:

1. Record the sequential order of the decisions $\bar{D} = D_1,...,D_k$ and observable chance nodes $\bar{O} = O_1,...,O_p$ according to the informational arcs in the HID.
2. If only a partial sequential order is defined by the informational arcs, transform this into a total sequential order by randomly ordering the nodes in the same level.
3. Convert the decision nodes $\bar{D} = D_1,...,D_k$ to the corresponding BN nodes $\bar{\Delta} = \Delta_1,...,\Delta_k$ representing decisions. In this conversion, a decision state d_{ij} of a decision node D_i is converted to a state δ_{ij} of the corresponding BN node Δ_i. If there is no asymmetry, all states $\delta_{i1}, ... , \delta_{im}$ of Δ_i have equal probabilities in its NPT.
4. Convert the utility nodes $\bar{U} = U_1,...,U_l$ to corresponding BN nodes $\bar{\gamma} = \gamma_1,...,\gamma_l$. The nodes $\gamma_1,...,\gamma_l$ can have child nodes that aggregate the utilities by a deterministic equation. Other types of child nodes, however, are forbidden for utility nodes.

From the transformed BN, we can determine the order of all calculations we need to perform on the BN in order to traverse all possible (non-zero probability) paths through the BN:

1. Define the sequential order of decision and observable chance nodes.
2. Build a decision tree by visiting each node in sequential order:
   a. For each node state in the BN, add an appropriate node type (decision or chance) to the decision tree, or
   b. Add a decision or chance node to the last arc added to the decision tree.
   c. If the current state is from an observable chance node, calculate the posterior probability from the BN. If this results in an inconsistency, then remove the arc from the decision tree (this path is not possible).
   d. If the state is from the last node in the sequential order, then add a utility node in the decision tree and calculate the posterior distribution of the utility node using the BN.
3. Evaluate the decision tree using the standard "averaging out and folding back" algorithm, a.k.a. the rollback algorithm. This starts from the utility nodes and rolls back toward the root node by computing the weighted average of chance nodes and maximum of decision nodes.

In AgenaRisk, the full utility distribution is calculated, not just the expectation. There is an option to maximize or minimize utility, where both seek the min or max value of the expectation of the utility distribution. But you can also specify your own function containing other summary statistic values such as variance or a percentile value. This can be useful if you are attempting to minimize "risk." Knowing the complete distribution of outcomes enables us to analyze and compare the optimal decisions for each of those functions. This can be more useful than the probability distribution of the expected utility (i.e., the risk profile) alone.

11.3 Decision Trees

Decision trees (DTs) have traditionally been used to choose an optimal decision from a finite set of choices, which are sometimes called policies. Typically, the value being optimized is some utility function expressed for each possible outcome of the decision. A DT represents the structure of a decision problem by modeling all possible combinations of decisions and observations, usually in the particular sequence in which one would expect observations and decisions to be made. DTs are composed of three types of nodes: chance nodes, decision nodes, and utility nodes. Each outgoing arc from a chance node represents an outcome and is labeled with the name and probability of this outcome. Each outgoing arc from a decision node is labeled with a decision alternative.

Now that we have introduced HIDs and represented our decision analysis problem in HID form, we can translate HIDs into a DT in order to do the computation and identify the optimal decisions.

Computing this model using AgenaRisk, as a BN, requires the use of the hybrid influence diagram analysis function. The decision nodes, observable chance nodes, ultimate utility node, and evaluation policy are identified in the analyzer and then calculated. The analyzer determines the decision/information sequence between the nodes in the model. Providing there are no ambiguities in the model (AgenaRisk provides warnings if there are), it computes the optimal decision policy and presents the results as a decision tree, as shown in Figure 11.2 (for the Example 11.1 problem).

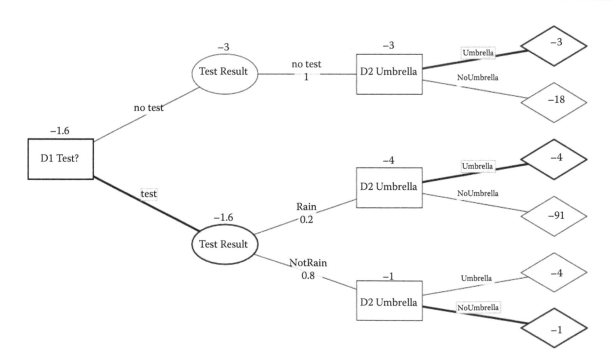

Figure 11.2 DT for Umbrella (Example 11.1).

In Figure 11.2, the rectangles on the DT represent decisions, observable chance nodes are represented by ellipses, and diamonds are the expected utility values. The edges represent either decision alternatives (states) or states for observable chance nodes. These edges are annotated with the states and probabilities where appropriate. Labels above each decision and observable chance node show the expected utility on that path through the DT. The optimum decisions are shown with bold arcs, and the expected utility is also shown in bold. The optimum decisions are to perform the test (D1 Test? = test; that is, listen to the weather forecast) and take the umbrella (D2 umbrella = umbrella). These maximize the expected utility, which remains negative at −1.6 (there are no positive utilities in this example—just like the British weather).

Notice that the decision tree is asymmetric in that if there is no test, then the test result can only be "no test." Similarly, if a test is done and the forecast heard, the "no test" result cannot happen and "Rain" and "NotRain" are the only valid states to consider. This asymmetry is automatically detected by the HID analyzer when it computes the result.

So, what is the optimal decision for Example 11.1? From the DT, if we listen to the weather forecast and receive a dry day forecast, then we should not bother to take the umbrella since the only cost here is the inconvenience of listening to the boring weather forecast for an expected utility of −1.

Let's look at a slightly larger example:

Example 11.2

A wildcatter is searching for oil, and has to decide whether to "Drill" (D) a particular site. He is uncertain about the "quantity of Oil available" (O). The wildcatter can make a "seismic Test" (T), which can reveal more information about the presence of oil, but the "Result of this test" (R) is not totally accurate. Note that the chance node O generally cannot be observed, whereas the chance node R may be observed if the wildcatter decides to undertake the "seismic Test."

Figure 11.3 shows the HID for Example 11.2.

The decision tree generated for Example 11.2 is shown in Figure 11.4. Here, the optimal decision is to conduct the seismic test, T, which has an expected utility of 22.5, compared to the alternative of not conducting the seismic test, which has an expected utility of 20. If the seismic test result detects no oil, the optimum decision is not to drill, since drilling results in an expected loss (−40.49), compared to a lower loss (−10) of not drilling, reflecting the result of the seismic testing costs. If the seismic test shows that the well is open, the best option is to drill (with utility 22.86); otherwise, if closed, the optimal option is 77.5, generating a high profit. If the wildcatter decides not to conduct a seismic test, the best option is to drill all the same, since the return would be an expected utility of 20.

Asymmetric Influence Diagrams

A decision problem is asymmetric if the number of possible decision scenarios is less than the cardinality of the Cartesian product of the state spaces of all decision and chance variables. Often, such asymmetry arises because the states in one decision or observable chance node are constrained by the states in others; that is, they are not mutually exclusive. There are two kinds of asymmetry.

1. *Functional asymmetry* is present if the availability of some decision alternatives depends on preceding observations or decisions.
2. *Structural asymmetry* is present if the possibility of an observation or decision depends on the preceding observations or decisions. In other words, a chance or decision node is defined based on the preceding nodes in this case.

Therefore, an entire decision or chance node becomes impossible when structural asymmetry is present, whereas only a decision state becomes unavailable when functional asymmetry is present. For example, there is a structural asymmetry in the oil wildcatter example (Figure 11.4) because if the seismic test (T) is not done, it is impossible to observe test results (R).

Functional asymmetry can be encoded directly in a decision node's NPT. Unlike with other algorithms, which assume a uniform distribution, the HID analyzer can handle this correctly, thus giving the decision analyst added flexibility.

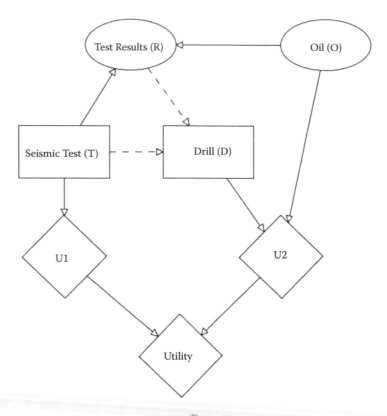

Figure 11.3 HID for Oil Wildcatter (Example 11.2).

11.4   Advanced Hybrid Influence Diagrams

Traditionally, influence diagrams and decision trees have only been applied to decision problems where the chance and utility nodes are discrete. Likewise, they have also only been applied when the expectation of the utility is being optimized. This is because the previous available algorithms were not able to conveniently and accurately handle nodes with continuous distribution functions. The HID analyzer in AgenaRisk can, however, handle continuous chance node types as well as utility nodes.

Let's consider an example where we have a nonlinear utility function modeling a trade-off between expected payoff and the variance in the payoff, such as when a decision maker has an aversion to risk (in the form of uncertain returns). In this case, the decision maker would prefer a lower expected utility if it were less volatile to one that had a higher expected utility but was more volatile. This might be motivated by a fear of an extreme negative utility outcome, which might cause financial ruin.

Example 11.3

Let's revisit the oil wildcatter example in Example 11.2. In Example 11.2, we made our decisions based on those options that maximized expected utility. To minimize the risk of ruin, we would need to devise a function

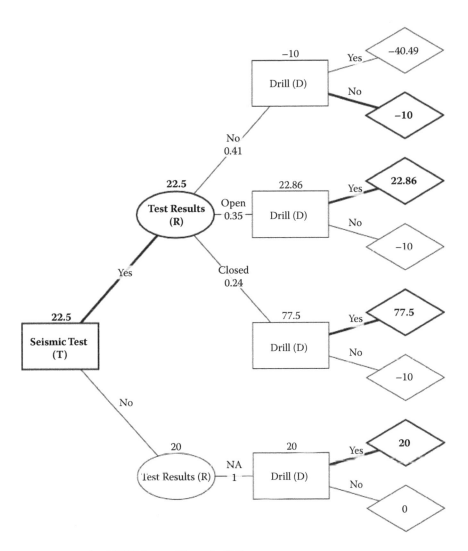

Figure 11.4   Decision tree for Oil Wildcatter (Example 11.2).

that modeled the trade-off between expected utility and the variance of the expected utility. So, a given increase in variance of the expected utility would be offset by an increase in the mean (average) utility. What might this function be? Well, if we examine the Wildcatter BN and inspect the marginal probability distribution for the ultimate utility node, we find that the expected utility, μ, is 5 and the variance, σ^2, is 5.525. Given the ratio between these values is 1000:1, we could "normalize" the trade-off between them by dividing the variance by 1000. In this way, a change of 1000 in variance is equivalent to a change of 1 in the expectation. Next, we might want to represent the risk averseness of the decision maker and apply a risk aversion factor—the higher the value, the more averse to high variance they might be. Let's denote this as λ.

The function to be maximized is therefore:

$$\mu - \lambda \frac{\sigma^2}{1000}$$

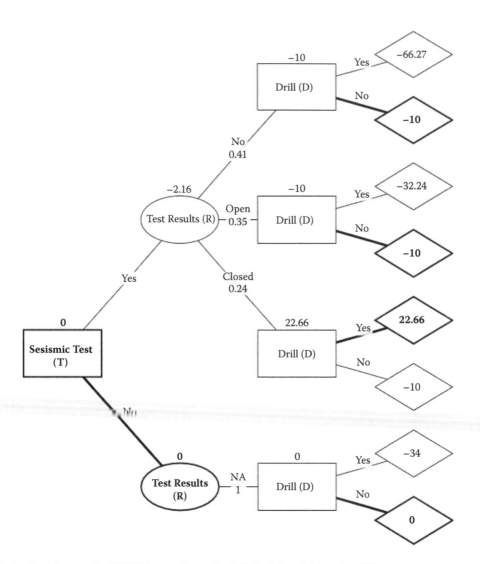

Figure 11.5 Decision tree for Oil Wildcatter (Example 11.3), including "risk trade-off."

Let's set $\lambda = 5$ for our decision maker and use this in the HID analyzer to generate the DT, which is shown in Figure 11.5. The optimal decision here is not to conduct a seismic test and not to drill for oil. Therefore, in this scenario, the decision maker is very risk averse, choosing the option with zero utility.

Note that in Example 11.3, if we set $\lambda = 1$, we get a decision tree result with decision options approximately equal to those shown in Figure 11.4, with only minor differences in optimal values in the DT nodes.

As we have said, we can also consider decision-making situations that involve continuous chance and decision nodes. Let's consider an example:

Example 11.4

In this example, we expand the wildcatter example and introduce the "Oil Volume," V. This has a mixture distribution conditioned on the type of the site. This is modeled by conditioning the probability distribution of

V on the Oil site, O. If O is "dry," the volume has a point value of zero, but if the site is "wet" or "soaking," then V follows a normal distribution with means of 4 and 13.5 and variances of 1 and 4, respectively. The price of drilling, represented by utility node U3, follows a normal distribution with mean 70 and variance 100. The oil price follows a lognormal distribution where the mean and variance of the underlying normal distribution are 2.75 and 0.5. The probability distributions of nodes "Seismic Test," R; "Oil," O; and "Drill," D, are exactly the same as the discrete model shown in Figure 11.3.

We can extend the wildcatter model by making the "Test Result," R, node a continuous node, which is a function of whether the "Seismic Test" was carried out, resulting in a zero value, whatever the value for O, if the test was not done and a mixture of beta functions if it was. Each beta function models the accuracy of the test with respect to the O node.

To finish, we have amended the utility functions for U2 such that $U2 = V \times P$ when D = Yes, and zero otherwise.

Figure 11.6 shows the graphical structure and parameters of the BN representation of this revised oil wildcatter model. Continuous nodes are denoted by the sigma icons.

The computation of the optimal decision policy is done using the same methods as before with the HID analyzer, and a fragment of the resulting DT is shown in Figure 11.7.

Notice one major change in how the "Drill" (D) node is defined compared to the discrete case in Example 11.3. Because it now has a continuous node, it requires a uniform distribution, but you cannot declare this manually in the node. So, for this purpose, a special function is required to be inserted as the expression for decision nodes that are continuous. The special function needed is DecisionUniform() and should be applied in all cases where a decision node has continuously valued parent nodes.

Seismic Test (T)	Yes			No		
Oil (O)	Dry	Wet	Soaking	Dry	Wet	Soaking
Expressions	Beta(1,9,0,1)	Beta(5,5,0,1)	Beta(9,1,0,1)	Arithmetic(0)	Arithmetic(0)	Arithmetic(0)

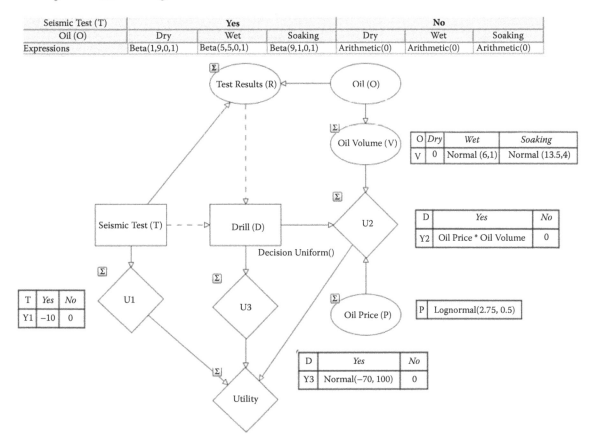

Figure 11.6 Wildcatter HID with continuous nodes and statistical distributions.

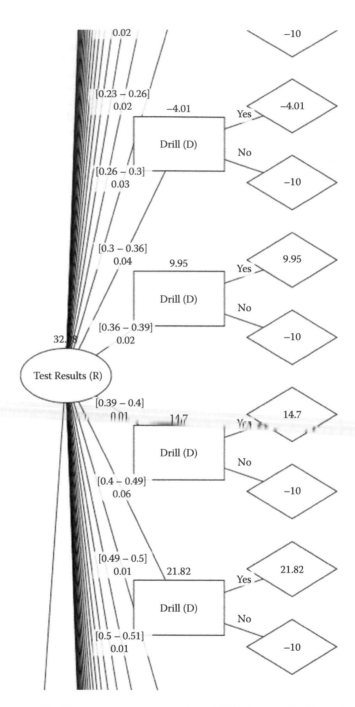

Figure 11.7 Fragment of DT generated from Oil Wildcatter BN with continuous nodes with simplify option off.

The DT fragment in Figure 11.7 is difficult to read simply because of the high number of regions generated for the test result node (R), but nevertheless, it does provide a very good fine-tuned view of where the different decision regions lie. To produce a simplified version, the HID analyzer can be executed again using the "simplify" option. This merges

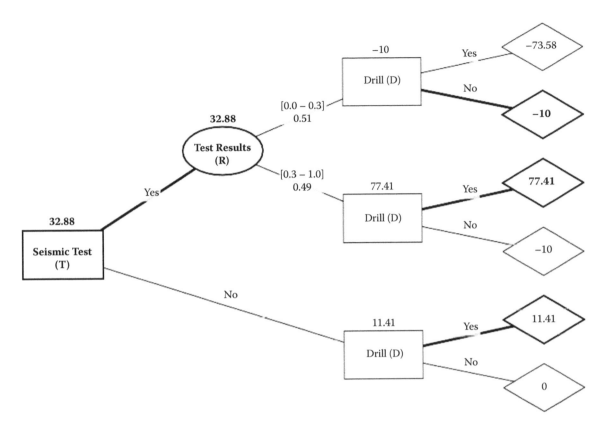

Figure 11.8   DT generated from Oil Wildcatter BN with continuous nodes with simplify option on.

continuously valued chance node DT branches if they share a descendant optimal decision. This transforms the DT, resulting in the more digestible Figure 11.8.

The optimal policy shown in Figure 11.8 remains the same as it was in Example 11.2, given the expected utility is similar overall, reflecting the small differences in assumptions.

Our final example in this section illustrates more fully a problem involving asymmetry arising from the structure of an HID imposed by the sequence of decisions, and also involves continuous variables for the utility nodes.

Example 11.5

Your team is playing in a big match, but you do not have a ticket. You could go (it costs money and time to get there) and try to bribe the gateman (a cheap but potentially dangerous option) or try to buy from a tout (expensive option if successful). While all your choices have a cost, there is a major benefit of getting in to the match if your team wins. However, this becomes negative if your team loses, as there is nothing worse than being there when they lose a big game.

In this example, the observable chance nodes are: "Outcome" (this is the result of any decision to attempt to bribe the gateman, with states "success," "arrested," "turned away," NA), "Get ticket," and "Get into match."

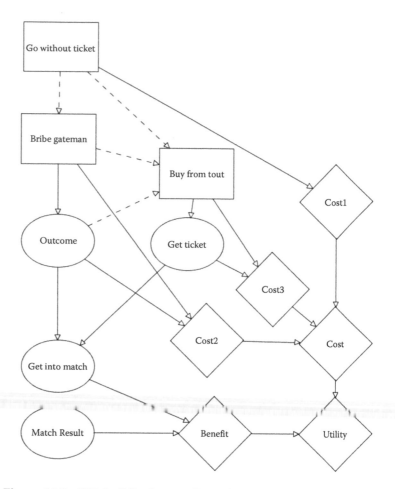

Figure 11.9   HID for Bribe Gateman (Example 11.5).

The HID for Example 11.5 is shown in Figure 11.9. The key difference between this HID and the preceding examples is that the decision nodes are functionally asymmetric. If you do not go without a ticket, then you do not need to bribe the gateman at all. Similarly, if you do successfully bribe the gateman, then you won't need to buy a ticket from a tout. Each of these conditions makes the probability tables for the decision nodes mutually exclusive under certain conditions, and this is a challenge that other influence diagram algorithms cannot handle. Instead, they assume all decision states are reachable from all other decision states in all decision nodes in sequential order. These asymmetric constraints need to be declared directly in the NPTs of the decision nodes.

The asymmetry can easily be seen in Figure 11.10 when the decisions that precede "Buy from tout" have these states: "Go without ticket" = Yes, "Bribe gateman" = Yes, and "Outcome" = Arrested. Under this scenario, you can see the only possible state that "Buy from tout" node can now have is "Buy from tout" = No.

When we run this example in the HID analyzer, we get the DT shown in Figure 11.11. Notice that this DT clearly reveals the asymmetry encoded in the decision sequence and does not compute branches in the DT that

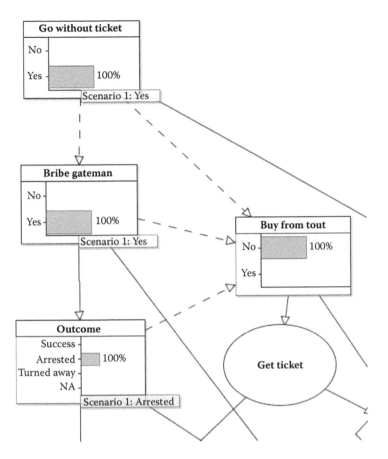

Figure 11.10 Functional asymmetry in Example 11.5.

are impossible. Here, the optimal decision is to go without a ticket, avoiding bribing the gateman, and attempt to buy a ticket from a tout.

11.5 Value of Information Analysis

Often, when making a decision, decision makers face the vexing question of whether to pay for better information about key uncertain variables that they suspect might heavily influence the viability of the outcome. For instance, before launching a new product, a marketing manager might want to conduct a survey of potential customers to determine likely customer demand, or an investor might be motivated to estimate economic growth before recommending investment portfolio options to clients. A medical researcher might wish to make decisions about medical interventions and quantify the effects of different clinical medical strategies over populations of patients. The question, therefore, for the decision maker is: How much should they invest in obtaining answers to these questions before making their decision?

One way of answering these kinds of questions involves using *value of information (VOI) analysis*. VOI analysis aims to identify the

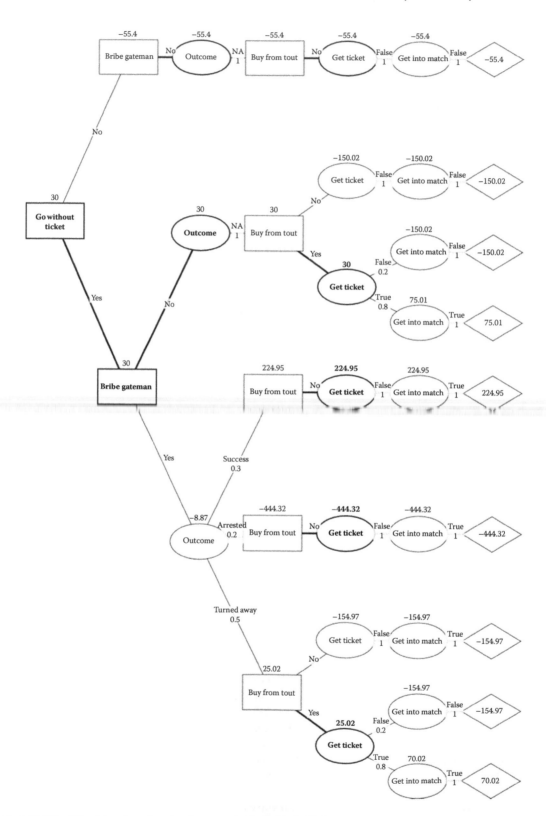

Figure 11.11 DT arising from functional asymmetry in Example 11.5.

Box 11.2 Formal Definitions for Value of Information Analysis

a. *Expected maximum value (EMV)*: The maximum expected value over a decision, D, when there is no perfect information of any uncertain variable, O, from a set, $\{\bar{N}, \bar{O}\}$, of observable and unobservable chance nodes, accompanied by utility nodes, $\bar{U}$. This has the same definition as we had in an HID, except the name is usually slightly different and we typically only have one decision node, D (assuming discrete nodes, but it easily generalizes to continuous).

$$EMV_D = argmax_D \left[\sum_{\bar{N}, \bar{O}, \bar{U}} P\left(U \mid D, \bar{N}, \bar{O}, \bar{U}\right) P\left(D, \bar{N}, \bar{O}, \bar{U}\right) \right]$$

b. *Expected value given (partial) perfect information (EV|PI)*: The expected utility when you perfectly know (i.e., observe) the state of a single chance node, O:

$$EV \mid PI = \sum_{O} P(O = o) EMV_{D\mid O=o}$$

c. *Expected value of partial perfect information (EV(P)PI)*: This is quantified as the highest price the decision maker will be willing to pay for being able to know (i.e., observe) the information for the uncertain node under assessment. It is formally defined as the difference between expected utility given perfect information and expected utility:

$$EV(P)PI = EV \mid PI - EMV$$

maximum value the decision maker should be willing to pay for perfect information about one or more chance nodes in the BN model.

To motivate the formal definitions involved in VOI (which are provided in Box 11.2), we consider the following problem.

Example 11.6

An investor wishes to calculate the expected monetary value for "Profit," P, given "Economic growth," E (which has states "Negative," "Even," or "Positive"), and "Investment decision," D (which can be "Bonds," "Stocks," or "Gold"). Table 11.1 (the payoff table) shows that the financial returns from bonds are independent of fluctuations in economic growth (over some fixed period of time). On the other hand, fluctuations in economic growth are expected to affect the returns from investing in stocks and gold. The decision maker is interested in determining whether to seek information from an expert forecaster about economic growth (E) before making the investment decision.

The BN for Example 11.6 is shown in Figure 11.12, where "Profit" is a utility node, "Investment" is a decision node, and "Economy growth"

Table 11.1
Profit for Different Values for Economic Growth, E, and Investment, I

E	Negative			Even			Positive		
D	Bonds	Stocks	Gold	Bonds	Stocks	Gold	Bonds	Stocks	Gold
P	£30	–£1000	–£300	£30	£50	£75	£30	£400	£150

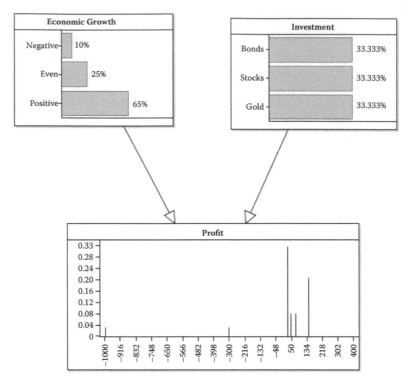

Figure 11.12 BN with superimposed marginals for Example 11.1 simple investment example.

is a chance node. The NPT for "Profit" is simply Table 11.1; that is, it is encoded as a partitioned table. Notice that the NPT values for the "Investment" node are uniform, which is a requirement for VOI analysis, and the prior for "Economic growth" favors positive values.

The first thing we want to compute is the expected maximum value (EMV) for the decision node D ("Investment"), which we write as EMV_D. We compute this by instantiating each and every state of the "Investment" node, calculating the BN, and recording the expected value for the utility node profit, P. When we do this, we get:

$$E(P|D = Bonds) = 30$$
$$E(P|D = Stocks) = 172.5$$
$$E(P|D = Gold) = 86.25$$

The maximum occurs when the decision maker chooses stocks, and so the value $EMV_D = E(P|D = Stocks) = 172.5$.

However, we want to identify the decisions that maximize the expected utility when the "Economic growth" rate (E) is known; that is, when it is no longer uncertain. To calculate this, we need to determine EMV_D for each of the possible values of E (we write this as $EMV_{D|E=e}$) and see which decision maximizes profit, P (the maximum utility combination results are highlighted in bold):

$$E(P|E = Negative, D = Bonds) = \mathbf{30}$$

$$E(P|E = Negative, D = Stocks) = -1000$$

$$E(P|E = Negative, D = Gold) = -300$$

$$E(P|E = Even, D = Bonds) = 30$$

$$E(P|E = Even, D = Stocks) = 50$$

$$E(P|E = Even, D = Gold) = \mathbf{75}$$

$$E(P|E = Positive, D = Bonds) = 30$$

$$E(P|E = Positive, D = Stocks) = \mathbf{400}$$

$$E(P|E = Positive, D = Gold) = 150$$

So, if we know the "Economic" growth rate is "Negative," we should choose "Bonds." If it the growth rate is "Even," we should choose "Gold," and if "Positive," we should choose "Stocks." Any other choice would be irrational under this decision criterion. Now, we know the probabilities of P(E) are {0.10, 0.25, 0.65}, so we can calculate what is called the *expected value given perfect information*, which we write as *EV|PI*, by multiplying these probabilities by the corresponding $EMV_{D|E-e}$ values:

$$EV \mid PI = \sum_{E} P(E = e)EMV_{P|E=e} = 0.1(30) + 0.25(75) + 0.65(400)$$

$$= 281.75$$

So, if we perfectly know the state for E, the expected value is 281.75. How much, then, should we be willing to pay for this information? The answer is the *expected value of partial perfect information*, which we write as *EV(P)PI*, and this is simply the difference of the expected value given we know nothing about E, and the expected value given we purchased the information needed to confirm the value $E = e$ before we made the decision. In our example, this is:

$$EV(P)PI = EV \mid PI - EMV = 281.75 - 172.5 = 109.25$$

So, our decision maker should be prepared to pay up to 109.25 for perfect information about E.

Fortunately, we do not need to perform this kind of laborious analysis by hand and can instead use AgenaRisk's VOI analyzer. This takes as input the single decision under analysis and a list of observable chance nodes whose VOI we wish to measure, along with the required ultimate utility node from which the maximum expected utilities will be derived.

For our Example 11.6, the output from the VOI analyzer is shown in Table 11.2 in its fully expanded form, showing all combinations of relevant node states.

In VOI analysis, we define (partial) perfect information as perfect information without error or inaccuracy about a given individual observable variable in the model. Of course, in many situations, this is unrealistic and may even seem fanciful given there may be a true value that can never really be observed without great expense and inconvenience. In this situation, the measurement idiom can be used to represent the relationship between an estimate and the underlying latent unobservable variable. In these circumstances, the VOI analysis would thus be performed on the estimate directly to find the value of the measurement process. This is legitimate since the underlying unknown variable would remain unobserved.

Expected value of perfect information (EVPI) is the potential gain associated with having perfect information on *all* variables in a model. EVPI is often not a useful tool to analyze a model, as the analyst is often interested in the value of information of individual variables or groups of variables rather than all variables. A more useful tool for analysis is expected value of partial perfect information (EVPPI). EVPPI focuses on the *joint* value of information of individual or groups of variables. This isn't covered in this chapter but is worth mentioning for completeness. In this chapter, we have described a means to evaluate the marginal value of information for each chance node, O, individually and independently of the others.

Table 11.2
VOI Analysis for Example 11.6

EMV			172.5
Economic Growth (Growth)		EV|PI	281.75
		EV(P)PI	109.25

		Investment		
		Bonds	Stocks	Gold
	Negative	30	−1000	−300
Economic Growth	Even	30	50	75
	Positive	30	400	150

EV\|PI = 0.1 * 30 + 0.25 * 75 + 0.65 * 400 = 281 .75

EV(P)PI = 281 .75 − 172.5 = 109.25

Visually, a BN model subject to a VOI analysis looks just like an HID and the same diagrammatic conventions apply. However, there are some important differences in the kind of analysis undertaken, as highlighted in Box 11.3.

Box 11.3   Differences between Value of Information and Hybrid Influence Diagram Analysis

VOI analysis:

- Focuses on a set of chance nodes and a single decision
- Analysis is "one-at-a-time" assessment of the VOI of each chance variable identified (i.e., it is not the "joint" value of information)
- Does not generate a DT

HID analysis:

- Uses many decisions that might influence and be influenced by many chance variables
- Identifies optimum decision based on maximizing or minimizing expected utility and other alternative objective functions.
- Has many measurement decisions about measured observable variables before actions taken
- Generates a decision tree

11.6   Sensitivity Analysis

Sensitivity analysis can be used for a number of purposes:

- As an extremely useful way to check the validity of an expert built model, whereby it is possible to see diagrammatically which nodes have the greatest impact on any selected (target) node.

■ As a means of determining how sensitive the results of a decision analysis are to changes in related observable variables. Thus, it can be used as an adjunct and supporting step after decision analysis or value of information analysis.

In the discrete case, assessing sensitivity involves identifying a target node, T, the node we are interested in assessing the sensitivity of, and the set of source nodes, $\bar{X}$, we want to assess in terms of their joint or single effects on T. Box 11.4 describes sensitivity analysis formally, but AgenaRisk automatically provides these computations.

There are many types of sensitivity analysis, some of which are extremely complex to describe and implement. AgenaRisk has a Sensitivity Analysis tool that provides a range of automated analyses and graphs. Here, we consider only the simplest type of sensitivity analysis with discrete nodes and tornado graphs.

Box 11.4   Formal Definition of Sensitivity Analysis

Formally, sensitivity is given by $S(\bar{X},T)$ where:

$$S(\bar{X} = \bar{x}, T = t) = \frac{p(T = t \mid e, \bar{X} = \bar{x})}{p(T = t \mid e)}$$

where $p(T = t \mid e)$ is the current probability value for T, given evidence, and $p(T = t \mid e, \bar{X} = \bar{x})$ is the new value taken by T when values for the set of source variables, $\bar{X}$, are entered into the BN model.

Note that sensitivity values are likely to be highly responsive to changes in the evidence, e, and any analysis will be contingent on the evidence currently entered on the model. Note also that currently instantiated nodes, denoted by e, will be excluded from the set $\bar{X}$.

Assessing the joint sensitivity of the target to perturbations in the source nodes is exponential in time and space. Therefore, here we will restrict ourselves to assessing pairwise sensitivity, but the approach can easily be generalized:

$$S(X = x, T = t) = \frac{p(T = t \mid e, X = x)}{p(T = t \mid e)}$$

Pearl points out two ways of assessing sensitivity: *inwards analysis*, involving setting values on all source variables, $\bar{X}$, and assessing the change in T, and *broadcasting*, which involves changing only the target node, T, and assessing the changes in the source set, $\bar{X}$. Broadcasting evaluates changes to source nodes in parallel, and this has obvious time-saving advantages. The results are equivalent since:

$$\frac{p(T = t \mid e, X = x)}{p(T = t \mid e)} = \frac{p(X = x \mid T = t, e)}{p(X = x \mid e)}$$

In the continuous case, sensitivity analysis is achieved by running dynamic discretization beforehand and conducting all sensitivity analysis on the resulting statically discretized model. Given continuous variables, we can calculate sensitivity of statistics and percentiles from the full distribution and report the changes in those values. To derive these measures, we broadcast the change in values for T and store the full distributions: $p(T = t \mid e, X = x)$, $p(T = t \mid e)$, $p(X = x \mid e)$, as is required by the discrete case. The statistics and percentiles will then be derived at the end of the process, giving results for:

$$S(X = x, E(T)) = \frac{E(T \mid X = x, e)}{E(T \mid e)}$$

$$S(X = x, V(T)) = \frac{V(T \mid X = x, e))}{V(T \mid e)}$$

$$S(X = x, s.d.(T)) = \frac{s.d.(T \mid X = x, e))}{s.d.(T \mid e)}$$

$$S(X = x, p_{\alpha}(T)) = \frac{p_{\alpha}(T \mid X = x, e)}{p_{\alpha}(T \mid e)}$$

$$S(X, p_{1-\alpha}(T)) = \frac{p_{1-\alpha}(T \mid X = x, e)}{p_{1-\alpha}(T \mid e)}$$

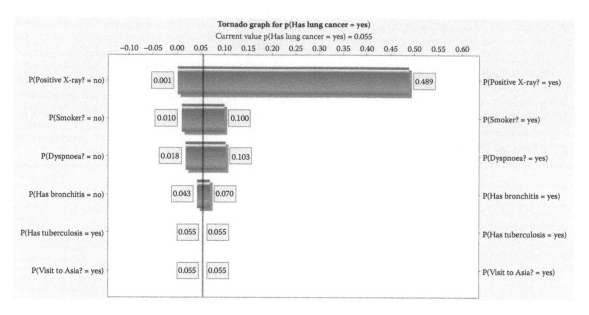

Figure 11.13   Tornado graph showing which nodes most impact lung cancer being true.

Consider the Asia model example in Chapter 9. It would clearly be interesting to know, based on the overall BN definition, which nodes have the greatest impact on the node "Lung Cancer." In theory, we could find this out manually by running through various scenarios of the model setting different combinations of true and false to all the other nodes and observing the different resulting values for "Lung Cancer" being yes. Fortunately, AgenaRisk does this automatically by allowing us to select a target node and any number of other nodes (i.e., sensitivity nodes).

So, setting "Lung Cancer" as the target node, we can automatically obtain the tornado graph in Figure 11.13.

From a purely visual perspective, you can think of the length of the bars corresponding to each sensitivity node in the tornado graph as being a measure of the impact of that node on the target node. Thus, the node "Positive x-ray" has by far the most impact on lung cancer.

The formal interpretation is that the probability of lung cancer given the result of positive x-ray goes from 0.001 (when positive x-ray is no) to 0.49 (when positive x-ray is yes). This range (from 0.001 to 0.49) is exactly the bar that is plotted for the tornado graph. The vertical bar on the graph is the marginal probability for lung cancer being yes (0.055).

11.7   Summary

Hybrid influence diagrams are an important extension of BNs because, unlike standard BNs, using the algorithm described in this chapter (which is implemented in AgenaRisk), it is possible to automatically compute decision choices that optimize utility. HIDs require us to specify decision nodes, utility nodes, and observable chance nodes. The same conventions are required for value of information, which enables us to determine those variables that matter most in terms of seeking further or better information in terms of impact on a decision. The key differences between HIDs and VOI were summarized in Box 11.3.

Even in the absence of decisions and utilities, it can be necessary to determine how sensitive a target variable is to changes in other variables that affect it. We explained how such sensitivity analysis can be carried out in any BN model.

Further Reading

Winkler, R. L. (1972). *An Introduction to Bayesian Inference and Decision*. New York and Chicago, Holt, Rinehart and Winston, Inc. (Series in Quantitative Methods for Decision Making).

Yet, B., Constantinou, A., Fenton, N., Neil, M. (2018). Expected value of partial perfect information in hybrid models using dynamic discretization, *IEEE Access*, 6, 7802–7817. https://doi.org/10.1109/ACCESS.2018.2799527.

Yet B., Neil M., Fenton N. E., Dementiev E., Constantinou A. (2018), An improved method for solving hybrid influence diagrams, *International J Approx Reasoning*, 95, 93–112. https://doi.org/10.1016/j.ijar.2018.01.006.

12

Hypothesis Testing and Confidence Intervals

12.1 Introduction

Hypothesis testing and confidence intervals play a prominent role in classical statistical texts. We have already indicated (in Chapters 2 and 6) some concerns about the way these concepts are often interpreted, but there is no doubt that they are extremely important for building models for risk assessment. Although the term *hypothesis testing* is not prominent in books and articles on Bayesian reasoning, this is largely because the evaluation of alternative hypotheses, in the form of causes, explanations, and predictions, is such an integral part of the Bayesian approach, that it tends to be implicit and therefore goes unmentioned. Likewise *confidence intervals* may go unmentioned because, with the Bayesian approach, we often have access to the full probability distribution.

It is important to present the Bayesian approach to hypothesis testing. In Section 12.2 we describe the fundamental issues with hypothesis testing (including p-values) using a simple coin tossing example that demonstrates the simplicity and clarity of the Bayesian approach compared with the classical approach. In Section 12.3 we focus on the important problem of testing for hypothetical differences. In Section 12.4 we present the notion of Bayes' factors and model comparison. In Section 12.5 we deal with confidence intervals. In not explicitly tackling the topic most Bayesian texts may give the reader the impression that Bayesians don't do hypothesis testing or confidence intervals. But we do and in spades.

We already introduced the idea of hypothesis testing informally in Chapter 5 where we used Bayes to explain why the classical statistical technique of *p*-values could lead to conclusions that were demonstrably wrong. A primary objective of this chapter is to show how to do it properly using a Bayesian network (BN) approach.

12.2 The Fundamentals of Hypothesis Testing

Hypothesis testing is the process of establishing which out of a number of competing hypotheses is most plausible (i.e., has the greatest probability of being true) given the data or judgments available. Examples

---

All models in this chapter are available to run in AgenaRisk, downloadable from www.agenarisk.com.

The default Bayesian position is that there may be many different hypotheses, $\{H_1, H_2,...,H_n\}$. For example, competent and lucky might not be the only relevant hypotheses for the soccer coach. He might also be bad or unlucky. However, in many cases there will be just two competing hypotheses which, as we saw in Chapter 5, are normally referred to as the null and alternative hypothesis, written H_0 and H_1, respectively.

of the kind of hypotheses we might be interested in when assessing risk include

- Is a soccer team coach competent or merely lucky?
- Is drug "Precision" better than drug "Oomph" in causing weight loss?
- Is investment in U.S. government bonds riskier than investment in gold?
- Is one scientific weather model better able to predict floods than another?
- Are calorific intake and lack of exercise a joint cause of obesity or is it calorific intake alone?

In this section we introduce many of the key principles and issues about hypothesis testing using the following very simple example:

It is suspected that a coin being used in a gambling game is 'biased towards Heads'. You have the opportunity to test the hypothesis by observing the coin being tossed 100 times and counting the number of heads you observe. Suppose you observe, say, 61 Heads. What do you conclude?

The main objective of this section is to compare and contrast the classical and Bayesian approaches.

12.2.1 Using p-Values and the Classical Approach

As we discussed in Chapter 6, in classical hypothesis testing, we have to determine whether the observed data (the 61 heads out of 100 in this case) provides us with sufficient evidence to reject the so-called "null hypothesis" H_0 that the coin is unbiased.

Being a good experimenter you will have set a threshold—called a p-value—in advance. A typical p-value is 0.05, or 5% when written as a percentage. The p-value is the probability of obtaining a result at least as extreme as the one that was actually observed (in this case 61 heads out of 100), assuming that the null hypothesis is true.

There is a simple "correct" way to calculate the p-value but, for reasons we will explain later, the p-value is often calculated differently in the "classical statistical method" (and that way of calculating it is often wrongly referred to as the definition of p-values). The correct way to calculate the probability of obtaining a given number of heads on 100 tosses assuming the null hypothesis is true is as follows:

First note that the null hypothesis (the coin is fair) is simply the assumption that the probability of tossing a head is equal to 0.5. The probability of tossing any given number of heads in 100 tosses—assuming the probability of tossing a head is 0.5—is simply computed from the binomial distribution. For example, the probability of tossing exactly 61 heads is (to 3 decimal places):

$$\binom{100}{61} 0.5^{61} \times 0.5^{39} = 0.007$$

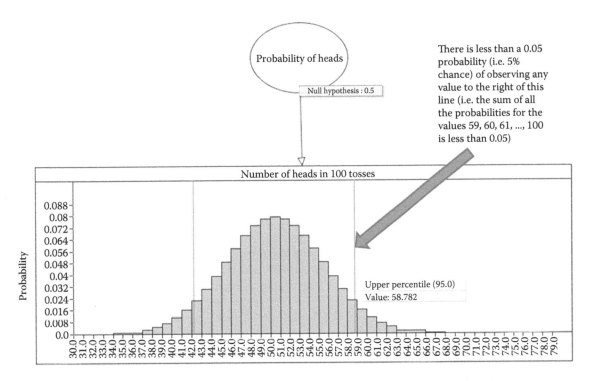

Figure 12.1 Under the null hypothesis there is less than a 5% chance of observing more than 58 heads.

But what we really want to know is the probability of tossing *at least 61 heads* and for this we need the full binomial distribution shown in Figure 12.1. The probability of observing at least 61 heads is the sum of the probability values 61, 62, 63, etc. This is actually 1.8%.

It turns out that 59 heads is the necessary "tipping point" for the p-value of 5%. That is because the probability of observing at least 58 heads is just over 5% while the probability of observing at least 59 heads is 4.5%. Since 61 heads is more than 58 we can reject the null hypothesis at the 5% level. We cannot reject the null hypothesis at the 1% level (because, as stated above, the probability of observing at least 61 heads is actually 1.8%). The tipping point needed for the 1% p-value is 62 heads.

So that is the "correct way" to compute the p-value. But that is not the classical frequentist way. To understand why, we return to the problem discussed in Chapters 2 and 6:

Knowing the p-value tells us nothing directly about the probability the null hypothesis is true or false

Although we can, by definition, reject the null hypothesis at the 5% p-value all we can actually say is that:

There is less than a 5% chance we would have observed what we did if the null hypothesis were true. Or, equivalently, there is more than a 95% chance we would not have observed what we did if the null hypothesis were true.

Thus, we cannot (contrary to what many people assume) conclude any of the "typical experimental conclusions" shown in Sidebar 12.1.

To create the BN model in Figure 12.1 ensure both nodes are created as simulation nodes. Define the NPT of the node "probability of heads" (p) as uniform(0,1) and define the node "number of heads in 100 tosses" as an integer interval node with NPT binomial(100, p). Enter the observation $p = 0.5$ and run the model.

Sidebar 12.1

Typical experimental conclusions

"There is less than a 5% chance that the null hypothesis is true"

"There is greater than a 95% chance that the null hypothesis is false (i.e. there is a greater than 95% chance the coin is biased in this case)"

"We are more than 95% confident than the null hypothesis is false (i.e. we are more than 95% confident that the coin is biased in this case)"

Even though the meaning of the "unbiased coin" hypothesis is reasonably unambiguous, **the meaning of the alternative—"biased coin" hypothesis is not**. As an extreme example, suppose that the reason for your suspicion about the coin being "biased" in the first place is because it is known that there are some double-headed coins in circulation. All coins, other than these are assumed to be fair. Then the null hypothesis ("coin is unbiased") does not change. But if you observe 61 heads in 100 tosses of a coin then it is NOT the case that "There is less than a 5% chance that the coin is unbiased (null hypothesis is true)." On the contrary, **it is 100% certain that the coin is unbiased.** This is because, although 61 heads is unlikely for an unbiased coin, **it is actually impossible for a biased coin**. In fact, the same is true even if you observe 99 heads in 100. Only if you observe 100 heads would you be justified in rejecting the null hypothesis.

All these three statements are about the probability of the null hypothesis H_0 being true having observed the data D. But all we actually know is the probability of observing the data given the null hypothesis being true. In other words, the statements wrongly assume $P(H_0 \mid D)$ is the same as $P(D \mid H_0)$ which, as explained in Chapter 6, is known as the fallacy of the transposed conditional (or the "prosecutor's fallacy").

In fact, as we will show later *the typical experimental conclusions are all demonstrably false* except under quite specific (and not necessarily realistic) assumptions about the null and alternative hypotheses (see sidebar for an obvious example).

If we are testing whether a null hypothesis is true or false, then we want to be able to make a direct statement about the probability the null hypothesis is true or false. To do this using classical frequentist statistics a completely different method for calculating a "p-value" is used. In fact, although they are also called p-values they are actually simply *percentiles* of a particular distribution (percentiles are defined formerly in Box 12.1).

To avoid confusion, we will refer to them as p'-values. Most importantly, it turns out that with this approach (and its underlying assumptions) the p'-value really does tell us something directly about the probability of the null hypothesis. In our example let:

pH = the "true" probability of tossing a head with the coin

Then a p'-value of, say 5%, really does tell us that there is a 95% chance that pH is above 0.5, so there is a 5% chance the null hypothesis is true.

Box 12.1   Percentiles

For a numeric variable, X, the α-percentile is the specific value X_α for which $P(X \leq X_\alpha) = \dfrac{\alpha}{100}$.

For example, the 50-percentile (written 50% percentile) is the value for which $P(X \leq X_\alpha) = 0.5$; this is the median. A percentile is 100th of the probability mass.

AgenaRisk produces percentile values automatically.

So how is this p'-value defined? Classical statistical hypothesis testing is based on the idea of treating the observed data as a sample from which we estimate the "true" statistics/probabilities. In this case we want to determine pH, the "true" probability of tossing a head with that coin. If we could toss the coin an infinite number of times, the proportion of heads would converge to pH. The 100 tosses of a coin are assumed to be a sample from the set of all possible tosses of the coin. However, because 100 is not a very big number there will be "uncertainty" that the proportion of heads observed really is equal to pH. The classical approach provides a way of handling that uncertainty. Specifically, in the classical approach:

- No explicit "prior assumptions" are made about the "true" probability *pH*. Instead, the observed data alone is used to estimate *pH*.
- The sample mean and variance are typically used to estimate the "true value" and "uncertainty" about the true value, respectively. The sample mean is assumed to be an unbiased estimator of the true value. So, because we observe 61 heads out of 100 the sample mean probability of heads is 0.61 and this is assumed to be the best estimate of *pH*. The uncertainty about *pH* is approximated as a Normal distribution whose mean is equal to *pH* and whose variance is the sample variance (which is 0.2379 in this case—see sidebar) divided by the number of samples (so in this case we get 0.002379). The bigger the sample the lower the variance (converging to 0).

Figure 12.2 shows this distribution.

Using this distribution we can, for any number *x* between 0 and 1, calculate the probability that *pH* is at most *x*. For example, the probability that *pH* is at most 0.53 is 5%. But this is the same as saying that the 5% percentile is 0.53.

Now here is the crux: the *p'*-value in the classical approach is **defined** to be the *p*-percentile. So the *p'*-value is, by definition, the value for *x* which the probability $(pH \leq x) = p\%$.

So the *p'*-values are simply the percentile values from the resulting Normal distribution.

Calculating the Sample Variance

For each individual coin toss the "observed" probability of a head is 1 if the toss is a head and 0 if it is a tail. The sample mean is therefore simply the total number of heads observed in the 100 throws divided by 100. The sample variance is

$$\frac{1}{100} \sum_{i=1}^{100} (observation_i$$

$$- sample\ mean)^2$$

Since there are 61 heads and 39 tails, the observed outcome is 1 for a head and 0 for a tail therefore the formula we need is

$$\frac{61(1-0.61)^2 + 39(0-0.61)^2}{100}$$

$$= \frac{23.79}{100} = 0.2379$$

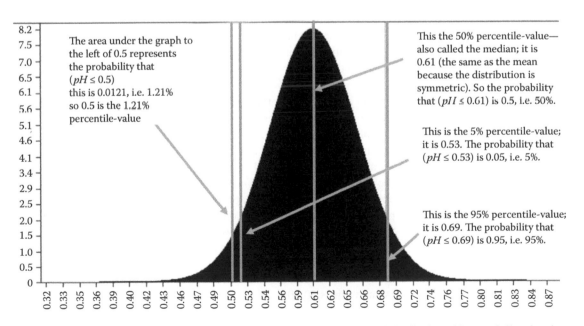

Figure 12.2 Frequentist estimate of *pH* based on 61 heads in 100 tosses: Normal distribution with mean 0.61 and variance 0.002379. We can also use the z-score approach to calculate the percentile values.

Since 0.5 is less than 0.53 (the 5% value) we can reject the null hypothesis ($pH \leq 0.5$) at the 5% p'-value. In fact, the distribution shows there is only a 1.21% chance that ($pH \leq 0.5$), so the p'-value associated with 0.5 is 1.21%. So, given our observation of 61 heads in 100 tosses, we can actually reject the null hypothesis at the 1.21% level (but not at the 1% level).

Crucially, unlike the case for p-values the p'-value really is the same as the probability that the null hypothesis is true.[*]

What is clear from the above is that the classical approach in this case produces similar, but certainly not identical results when it comes to considering whether to reject the null hypothesis. In the "correct" approach, the p-value for observing 61 heads in 100 given the null hypothesis is 1.8%. In the classical approach it is 1.21%. Nevertheless, in both cases the null hypothesis is rejected at the 5% level but not at the 1% level and in both cases the tipping point for rejecting the null hypothesis at the 1% level is 62 heads.

12.2.2 The Bayesian Approach Avoids p-Values Completely

What we really want to do is determine the probability that the null hypothesis is true given the data observed. This is precisely what the Bayesian approach provides. Moreover it enables us to:

- Achieve a completely rational and unifying approach to hypothesis testing.
- Avoid all ambiguity about the meaning of the null and alternative hypothesis, including what assumptions are being made about them.
- Expose potential flaws in the classical approach.
- Identify precisely what assumptions in the classical case are needed for it to "make sense."

The generic BN for the coin tossing hypothesis problem (which will enable us to capture every possible type of assumption and also extends to arbitrary number of coin tosses) is shown in Figure 12.3.

The Boolean nodes H and "$pH > 0.5$?"—which would not normally be made explicit in a BN of this kind—are included since these are the nodes that clarify the assumptions being made in the different approaches. In a Bayesian approach we are allowed to condition the prior for the unknown pH on our background knowledge. Moreover, as we shall see, *this is the (only) way of capturing exactly what we mean by a biased/non-biased coin*. In the Bayesian approach we are also, of course, allowed to incorporate any prior knowledge about whether or not the coin is biased or not (although in most of what follows we will assume this is 50:50).

---

[*] Although the null hypothesis is $pH = 0.5$ we include values below 0.5 since the alternative hypothesis is $pH > 0.5$.

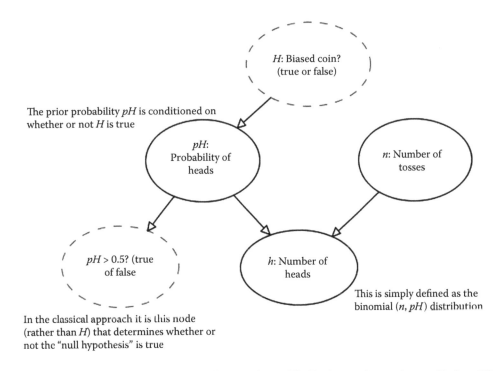

The prior probability *pH* is conditioned on whether or not *H* is true

In the classical approach it is this node (rather than *H*) that determines whether or not the "null hypothesis" is true

This is simply defined as the binomial (*n*, *pH*) distribution

Figure 12.3 Basic network structure for coin tossing experiment. The Boolean nodes are shown with dotted lines.

All of the key differences and assumptions are captured in the way we define the prior probability of node *pH given the hypothesis*. We consider some different possible assumptions in the following cases.

Case 12.1 A coin is considered unbiased if
$pH = 0.5$ and biased if $pH > 0.5$, that is:

H	Meaning
False	$pH = 0.5$
True	$pH > 0.5$

In this case it is easy and non-controversial to define the prior *pH* when *H* is False—it is simply the point value 0.5. But trying to define the prior for *pH* when *H* is True already exposes problems with the classical approach. A reasonable interpretation would be to define *pH* as being "any number greater than 0.5 and less than or equal to 1 with any number in this range being equally as likely as any other." Formally, this is the Uniform distribution:

$U(0.5, 1)$ excluding the endpoint 0.5.

A sensible way of approximating this is to use a number just above 0.5 as the starting point, for example

$U(0.500000001, 1)$

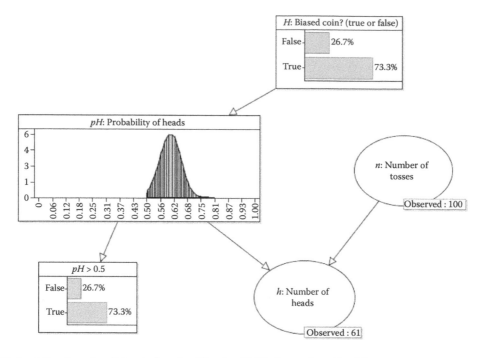

Figure 12.4   After observing 61 heads there is still only a 73.3% chance that *pH* > 0.5.

It turns out that with this explicit assumption we get quite different results to what we saw above. Indeed, Figure 12.4 shows that the probability that *pH* > 0 in this case is only 73.3% compared with the approximation 98.79% in the classical statistical case.

This is also summarised in Table 12.1.

Table 12.1

Summary of Differences and Results in Case 12.1 When 61 Heads in 100 Tosses are Observed

Approach	Prior Assumptions	Resulting Probability That *pH* > 0.5	Probability of Observing at Least 61 Heads if Coin is Unbiased	Probability Coin is Biased
Simple (binomial) test	*pH* = 0.5 for unbiased coin	Not given	1.8%	Not given
Classical statistical hypothesis test	*pH* is a Normal distribution whose mean is the sample mean, and whose variance is sample variance divided by *n*	98.79%	1.21%	98.7%
Bayesian hypothesis test	*pH* = 0.5 for unbiased coin, *pH* = *U*[0.5000001, 1] for biased coin	73.3%	1.8%	73.3%

In fact, there is a rational explanation why the probability is lower in the Bayesian case. The assumption that the probability of a head on a biased coin is uniformly distributed between 0.5 tells us, for example that:

> Before seeing any observation at all, if the coin is biased then the EXPECTED probability value of *pH* is exactly half way between 0.5 and 0.1, that is 0.75.

Since we are assuming a 50:50 chance the coin is biased, the prior marginal distribution for number of heads in 100 tosses is a multimodal distribution as shown in Figure 12.5. Its mean is 62.5 (half-way between the mean of the expected values for a biased and non-biased coin, respectively).

Thus, when we observe 61 heads this is actually not very strong evidence to support the biased coin hypothesis. It still tips in favour of the biased coin because it make values of *pH* more than 0.70 (all of which are only true under the biased coin hypothesis) very unlikely.

The difference is really made clear **if we observe 58 heads, instead of 61**, as shown in Table 12.2.

In this case observing 58 heads **actually supports the non-biased coin hypothesis over the biased coin hypothesis**. That is because, as unlikely as it is to observe 58 heads under the unbiased coin hypothesis (when the expected value is 50), it is even more unlikely to observe 58 heads

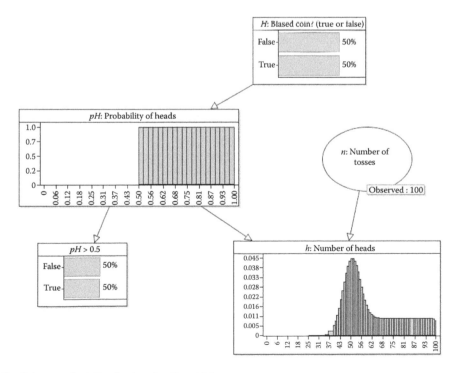

Figure 12.5 Prior marginal distributions for Case 12.1.

Table 12.2

Summary of Differences and Results in Case 12.1 when 58 Heads (Rather Than 61) in 100 Tosses are Observed

Approach	Resulting Probability That $pH > 0.5$		Probability of Observing at Least 58 Heads if Coin is Unbiased		Probability Coin is Biased	
	61 heads	58 heads	61 heads	58 heads	61 heads	58 heads
Simple (binomial) test	Not given	Not given	1.8%	4.4%	Not given	Not given
Classical statistical hypothesis test	98.79%	96.6%	1.21%	3.4%	98.79%	96.6%
Bayesian hypothesis test	73.3%	45.6%	1.8%	4.4%	73.3%	45.6%

under the biased coin hypothesis (when the expected value is 75). Note that the classical test gets the conclusion completely wrong in this case.

Case 12.2 A coin is considered unbiased if $pH =$ is "close to 0.5" and biased if $pH > 0.5$.

This case is more realistic. It is an empirical fact that it is very rare for any coin to have a pH exactly equal to 0.5. So a value very close to it would not be considered "biased." There are different ways we could represent this, but we will choose pH to be a Normal distribution with mean 0.5 and variance 0.00001 (this has 5% percentile 0.495 and 95% percentile 0.505).

Since we acknowledge in this case that a coin for which pH is very slightly above 0.5 may not really be biased it makes sense to slightly change the assumption for a biased coin. So, whereas in Case 12.1, we assumed a biased coin had $pH = U(0.500000001, 1)$ here we will use $pH = U(0.5001, 1)$.

Figure 12.6 shows the result for this case when 61 heads are observed in 100 tosses.

There is something important to note here: The probability that H is "True" is no longer equal to the probability "$pH > 0.5$" is "True." That is because there is now an acknowledged possibility that a non-biased coin could have pH very slightly more than 0.5.

Again the results are very different from the conclusions that would be drawn from the p-value and classical tests.

Case 12.3 The closest Bayesian equivalent to the classical statistical approach.

In the classical approach no explicit prior assumptions are made about pH since the core idea is that we

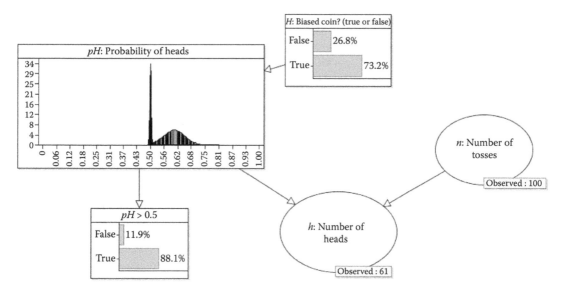

Figure 12.6   Resulting probabilities in Case 12.3a where 61 heads are observed.

approximate it from the observed data. In the Bayesian approach, the closest equivalent to "no explicit prior" is to assume a uniform prior for the distribution we are trying to learn from the data. So, in this case we do not need any explicit Boolean hypothesis node, and instead use a $U(0,1)$ prior for pH; this is equivalent to saying that any value between 0 and 1 is equally likely before we have observed any data.

The result after observing 61 heads in 100 tosses with this "uniform prior" assumption is shown in Figure 12.7.

Despite its uniform prior, the posterior distribution for pH is now quite close to the Normal (0.61, 0.00238) distribution calculated in the classical approach. In fact,

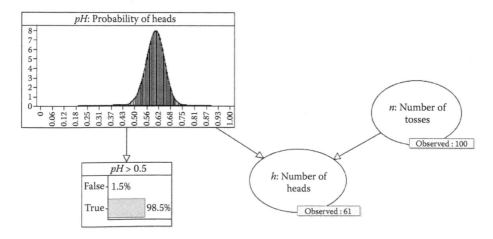

Figure 12.7   Case 12.3 after observing 61 heads.

All the models in this section are in the AgenaRisk examples, plus extra ones with different prior assumptions. The models can be easily adapted to incorporate prior knowledge about whether or not the coin really is biased (e.g. suppose it is known that 1 in a 1000 coins is known to be biased in the sense of having a 0.7 probability of heads) and a range of probability distributions for *pH*.

it is approximately Normal(0.608, 0.00232). Hence, the probability that *pH* > 0.5 is also quite close to the result obtained in the classical approach: 98.59% compared with 98.79% in the classical approach. The reason why the mean learnt in the Bayesian approach is slightly less than the sample mean is because with the uniform prior there was a "starting" mean of 0.5 that had to be "shifted." The 61 heads in 100 throws is not sufficient evidence to completely overturn the prior. However, if we observe 610 heads in 1000 throws then the posterior probability for *pH* is almost exactly Normal(0.61, 0.0000238) which is the distribution that would result from such data in the classical approach.

Therefore, while it turns out that we can replicate the classical hypothesis testing in the Bayesian approach, in doing so we have shown the greater meaningfulness, power and simplicity of the latter; moreover, the Bayesian approach is able to easily incorporate realistic assumptions about what a biased and non-biased coin actually means which, if ignored as in the classical approach, produces erroneous or misleading conclusions. In the rest of this chapter we shall see further convincing examples.

12.3 Testing for Hypothetical Differences

There are many situations where we are interested in the magnitude of difference between attributes and might wish to test a hypothesis that one is better than the other in some sense. In Section 12.3.1, we focus on the general approach to this problem, again comparing and contrasting the Bayesian and classical statistical approaches. It turns out that, in practice, for many important problems people focus on the difference between means (i.e. averages) of attributes when they should be focusing on the difference between the distributions of the attributes. We address this important point in Section 12.3.2.

12.3.1 General Approach to Testing Differences between Attributes

In general, for a pair of numeric variables, X and Y, we will be interested in situations where the hypothesis can be expressed as

$$H_0 : X \geq Y$$

$$H_1 : X < Y$$

Example 12.1   Quality Assurance

A production manager needs to select a material for use in the production of a new product. He has a choice of material A or material B, and has conducted tests on samples of each material to determine whether it is faulty.

The hypothesis concerns the probability of faultiness for each material, p_A and p_B, respectively:

$$H_0 : p_A \geq p_B$$

$$H_1 : p_A < p_B$$

Let's assume that the testing yielded 10 faults in 200 samples of material A and 15 faults in 200 samples of material B.

Before producing a solution here we first need to make an assumption about the prior probability of the materials being faulty. It seems reasonable to be indifferent here and select an ignorant prior where all probabilities of faultiness are equally likely. So we choose the Uniform(0,1) distribution as the prior distribution for both p_A and p_B.

As we explained in Chapter 6, the Binomial (n,p) distribution (with n equal to the number of samples and p equal to the probability of faultiness) is a reasonable choice for modeling the likelihood of observing a number of faults in a sample of size n. We can use this for both material A and material B.

We can represent all of this information in a Bayesian network:

- Clearly we need nodes p_A and p_B to represent the probability of faultiness for each of the materials (the node probability tables [NPTs] for these nodes will be the U(0,1) distribution).
- We clearly also need nodes to represent the number of faults in the respective samples (the NPTs for these nodes will be the Binomial distributions).

All that remains is to specify the nodes associated with our hypothesis. This is easily done by recasting the hypotheses as a difference, since $p_A \geq p_B$ is equivalent to $p_A - p_B \geq 0$:

$$H_0 : p_A - p_B \geq 0$$

$$H_1 : p_A - p_B < 0$$

We therefore need to add an additional node to represent the function $(p_A - p_B)$ with parent nodes p_A and p_B, and a Boolean child node of $(p_A - p_B)$ to represent the hypothesis itself.

The resulting BN is shown in Figure 12.8 with the prior distribution for the hypothesis node displayed. As you can see the hypothesis is 50:50 for $H_0 : H_1$.

We can now enter the testing data into the BN as evidence (shown in Figure 12.9) and can see that the distributions for p_A, p_B and $(p_A - p_B)$ have all been updated and that the percentage probability that material A is better than material B is 84%.

Note that when comparing differences we need to be careful that we are doing so fairly and that the variables involved represent as much information as possible about them. For example, if we did no testing at all of material B and simply used the prior distribution for B our conclusion would be that material A is better than B. This is because most of the posterior probability mass for p_B would be to the right of the mass for p_A and so $(p_A - p_B)$ would be less than zero, thus supporting material A even though we know nothing about material B.

Clearly it always makes sense first to assess whether the data we have is sufficient to support our model, and if not we should not proceed to test for difference. We will discuss this later in the chapter.

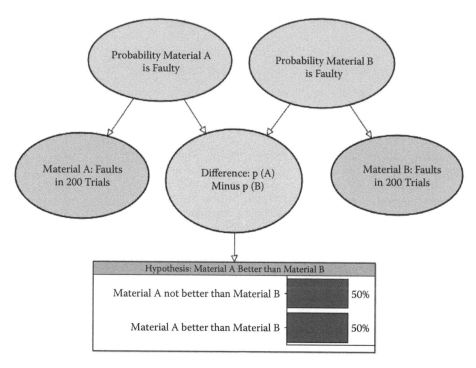

Figure 12.8   BN for quality assurance example.

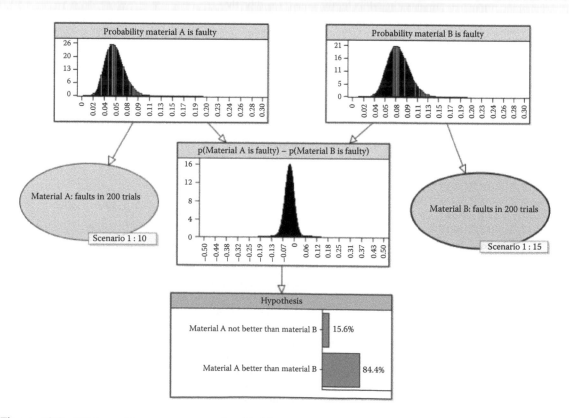

Figure 12.9   BN for quality assurance example with data.

Example 12.2   Drug Testing

Here we dig deeper into the example (taken from Ziliak and McCloskey, 2008) first introduced in Chapter 2, Box 2.1. We are interested in comparing two weight-loss drugs to determine which should be sold. Experimental data on the drugs Oomph and Precision have shown that, over a one-month period for a number of test subjects:

- Weight loss for Precision is approximately Normally distributed with mean 5 pounds and variance 1. So we define the NPT for the node Precision as a Normal(5,1) distribution.
- Oomph showed a much higher average weight loss of 20 pounds but with much more variation and is estimated as a Normal (20, 100) distribution.

We might be interested in two things here: one is predictability of the effect of a drug on weight loss and the other is the size or impact of the drug on weight loss. Let's adopt a similar formulation to before:

$$H_0 : \text{Precision} > \text{Oomph}$$

$$H_1 : \text{Oomph} \geq \text{Precision}$$

Figure 12.10 shows the AgenaRisk model used to test this hypothesis. Rather than create a node for the difference, Precision–Oomph, directly in AgenaRisk we have simply declared a single expression for the hypothesis node as

```
if(Precision > Oomph, "Precision", "Oomph")
```

rather than use two separate nodes as we did in the quality assurance example.

Notice that the hypothesis is approximately 93% in favor of Oomph over Precision, since only 7% of the time is Precision likely to result in more weight loss than Oomph. Interestingly we have included here two additional nodes to cover the risk that any of the drugs actually cause negative weight loss, that is, weight gain for the good reason that some may be fearful of putting on weight and might choose the drug that was less effective but also less risky. In this case someone

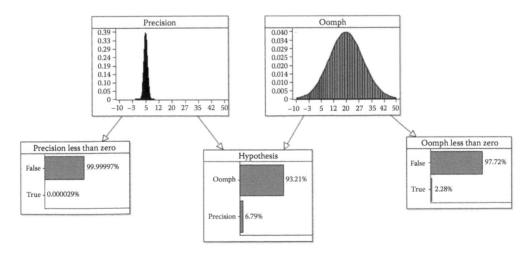

Figure 12.10   BN for Oomph versus Precision hypothesis.

might choose Precision since Oomph has a 2.28% chance of weight gain, where the chance of weight gain with Precision is negligible.

In the classical statistical approach to hypothesis testing in the drug testing example we would be interested in assessing the *statistical significance* of each drug with respect to its capability to reduce weight, that is, whether any weight reduction is likely to have occurred by chance. In doing this we compare the results for each drug against the null hypothesis: zero weight loss. So, for both Precision and Oomph we test the following hypotheses, where μ is the population mean weight loss in each case:

$$H_0 : \mu_0 = 0$$

$$H_1 : \mu_1 > 0$$

Box 12.2 describes the classical statistical approach to testing this hypothesis in each case (i.e. the approach introduced in Section 12.2.1).

Box 12.2   Classical Statistical Hypothesis Testing for Weight Loss Drugs Example

As we saw in Section 12.2.1, classical statistics instead of testing the hypotheses with the variables directly we are reduced to testing unobservable parameters, in this case the mean, for a conjectured population of test subjects. In statistics we are interested in making inferences from, well obviously, statistics so we then need to consider the sample distribution of the statistics themselves, not of the population distributions.

So working only with the sample distribution (and assuming that the mean and variance of this distribution are satisfactory estimates of the true unknown population mean and variance), statistical hypothesis testing comes down to calculating the probability of rejecting a true null hypothesis. This probability is calculated from the sample distribution using what is called a Z-score. If the Z-score is below some predefined *significance level* (this is the *p*-value as discussed in Section 12.2.1), say, 1%, then the null hypothesis is rejected. Statisticians would say that "the null hypothesis is rejected at the 1% significance level."

Assuming that the sample distribution is a Normal distribution the Z-score is calculated as

$$Z = \frac{\text{sample mean } - \text{ null hypothesis mean}}{\text{sample standard error}}$$

where the *sample standard error* is simply the square root of the sample variance.

So, applying these assumptions to the sample distribution for Precision (which is a Normal distribution with mean 5 and variance 1) we get the Z-score:

$$Z_P = \frac{5-0}{1} = 5$$

For Oomph the sample distribution is a Normal distribution with mean 20 and variance 100, so the Z-score is

$$Z_O = \frac{20-0}{10} = 2$$

At a significance level of 1%, the table of standardized Z-scores yields a critical value of 2.326, which is greater than the Z-score for Oomph but less than the Z-score for Precision. Hence we get the counterintuitive conclusion discussed in Chapter 2 whereby the drug Oomph is "rejected" as an effective weight loss treatment, but Precision is accepted.

Classical hypothesis testing involves a lot more than the brief overview provided here and gets a lot more complicated when the assumptions deviate from normality or when the population variances involved differ. In fact, there are more appropriate tests we could apply here or more sensible values for the null hypothesis, but this simple significance test is invariably applied as an entrance condition before subsequent analysis is performed with the result that if a variable is not statistically significant it is discarded.

So, it turns out that, using the classical statistical approach to hypothesis, we must:

Choose Precision in preference to Oomph because Precision is statistically significant with respect to weight loss (the null hypothesis can be rejected) whereas Oomph is not statistically significant with respect to weight loss (the null hypothesis cannot be rejected).

Although the process looks nice and scientific there are some serious flaws here. The result is the opposite of the Bayesian result we saw earlier. What then is going on? The classical case actually represents a test of sampling reliability rather than impact. The signal-to-noise ratio of Precision is five times that of Oomph, thus the hypothesis favors Precision.

In fact, Figure 12.10 already provided a visual explanation for the statistical hypothesis testing result. Look again at the node labeled *Oomph less than zero*. There is a 2.28% probability that this is true. You can think of this as the probability of rejecting a true null hypothesis. This explains why, at the 1% level, the null hypothesis for Oomph cannot be rejected.

The statistical test is determining whether an effect exists rather than what is the actual size of the effect. But to someone who wants to lose weight Oomph is much more significant in terms of size of impact than Precision and beats it hands down.

12.3.2 Considering Difference between Distributions Rather Than Difference between Means

Consider the following problem scenario:

A new drug is trialled which is believed to increase survival time of patients with a particular disease. It is a comprehensive randomised control trial lasting 36 months in which 1000 patients with the disease take the drug and 1000 with the disease do not take the drug. The trial results show that the null hypothesis of "no increase in survival time" can be rejected with high confidence; specifically, there is greater than 99% chance that the mean survival time of people taking the drug is higher than those who do not. You are diagnosed with the disease. Should you take the drug and, if so, what are your chances of surviving longer if you do?

As in the previous problems we are comparing two attributes—the survival time with the drug and the survival time without the drug. Unfortunately, the 99% probability that the mean of the former is greater than the mean of the latter tells us nothing about the probability that any given person will survive longer if they take the drug. So, it does not help us to answer the question. The problem is that if we have reasonable size samples—as in this case—there will actually be very little uncertainty about the mean. All the "interesting" uncertainty is about the *variance*. As an extreme example suppose you could measure the height of every adult male in the United Kingdom. Then, despite wide variance in the results, the mean height will be an exact figure, say 176 cm. Even if you were only able to take a sample of, say 100, the mean of the sample would be a very accurate estimate of the true mean of 176 (i.e. with very little uncertainty). So, in the drug example, if the mean survival time of the 1000 patients taking the drug is 28.5 weeks, then this will be very close to the true, but unknown, survival time with little uncertainty. Yet the mean survival time will inevitably "hide" the fact that many of the patients survive the full 36 weeks while many die within the first 3 weeks.

Suppose that, based on the trial data, we get the following estimates for the mean and variance of the survival times:

	Mean survival time	Variance of survival time
With drug	Normal(28.5, 0.15)	Normal(150, 100)
Without drug	Normal(27.0, 0.1)	Normal(350,100)

Then we can construct the necessary BN model as shown in Figure 12.11.

The nodes with the estimated means and variances of the survival times are defined using the Normal distributions in the above table. Each of the "true" survival time nodes is defined simply as a Normal distribution with mean equal to the (parent) estimated mean and variance equal to the (parent) estimated variance. The Boolean node "Mean survival time no greater with drug" has its NPT defined as

if(mean_with > mean_without, 'False', 'True')

The Boolean node "survive no longer with drug" has its NPT defined as

if(with > without, 'False', 'True')

Note that the null hypothesis "Mean survival time no greater with drug" is easily rejected at the 1% level, and so the drug would certainly be accepted and recommended for patients with the disease. However, the situation for the null hypothesis "survive no longer with drug" is very different. There is actually a 47.45% chance that your survival time will be *less* if you take the drug than if you do not.

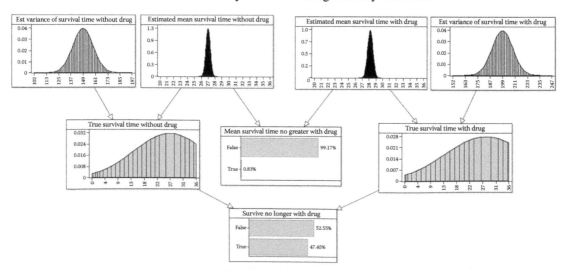

Figure 12.11   Taking account of full distributions of survival time rather than just means.

Therefore, the BN model with the full distributions is able to answer the question about your chances of surviving longer with the drug than without. It does not, however, answer the question about whether you should take the drug. For this we need to extend the BN model to an influence diagram as shown, for example, in Box 12.3.

Box 12.3   Answering the Question: "Should I Take the Drug" Using an Influence Diagram

In Figure 12.12, we have extended the drug survival time model into an influence diagram by adding a Boolean decision node "Take drug" (yes/no) and the utility nodes. For the definition of the utility nodes, we assume that each month of survival time has a monetary value of 50 units and that if the drug is taken then there are variable costs and side effects which, on average "cost" 100 units. Specifically, the NPTs are defined as follows:

	Drug Not Taken	Drug Taken
Value of extra life	$50 \times$ survival_time_without	$50 \times$ survival_time_with
Costs and side effects	0	Normal(100, 10)

The NPT for the node called "utility" is simply defined as: value—costs_side_effects

When we run the HID analyser we get the output shown in Figure 12.13, suggesting that the optimal decision is **not** to take the drug. Clearly, the costs and side effects are not sufficiently compensated by the slight increase in expected survival time.

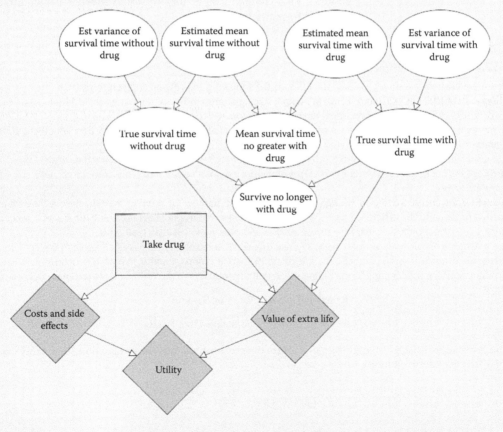

Figure 12.12   Influence diagram extension to drug survival time model.

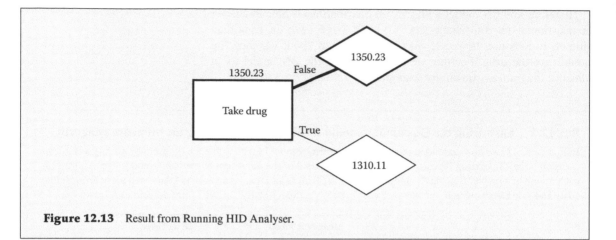

Figure 12.13 Result from Running HID Analyser.

The example highlighted how misleading it can be to rely on the difference between means in order to decide how different two populations are. As the sample size increases—and the means become more certain – we get to the situation whereby we are likely to find a "statistically significant" difference between two populations that are essentially the same. This phenomenon is also known as Meehl's conjecture and is described formally in Box 12.4.

Box 12.4 Meehl's Conjecture

The eminent clinical psychologist Paul Meehl made the following interesting conjecture in 1978:

"In non-experimental settings with large sample sizes, the probability of rejecting the null hypothesis of nil group differences in favor of a directional alternative is about 0.50."

A number of experiments have been conducted, before and after the conjecture was first documented, that lend support to his conjecture, using very large samples of social science data.

Let us consider Meehl's conjecture for the following example hypothesis given by Gigerenzer (2002):

"Protestants are better than Catholics in terms of memory span, reaction times, and have larger shoe sizes and testosterone levels."

Meehl's conjecture says that if we take a sufficiently large number of samples we will have a 50% chance of discovering a fictitious difference *where there is in fact none* and might conclude that Protestants have better memory spans, reaction times, and larger shoe sizes or testosterone levels than Catholics.

Let's simply consider one variable here, say reaction time, and assume we have been able to collect sample data from an (almost) infinite number of people drawn from two groups A and B. We wish to determine whether the group A reaction time is significantly higher than that of group B. We therefore need to test these hypotheses:

$$H_0: \text{Reaction Time A} = \text{Reaction Time B}$$

$$H_1: \text{Reaction Time A} - \text{Reaction Time B} > 0$$

The conventional statistical machinery to perform this test involves the Z-score (assuming Normality) and a test of differences between means:

$$c = (\bar{x}_A - \bar{x}_B) + Z_{1-\alpha}\sqrt{\frac{\sigma_A^2}{n_A} + \frac{\sigma_B^2}{n_B}}$$

where α the probability of rejecting H_0 when it is in fact true (i.e., the significance level). Figure 12.14 below shows the critical region involved.

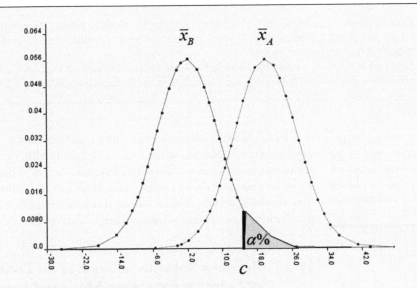

Figure 12.14 Distribution curves for two groups A and B with critical region.

In Meehl's conjecture n_A and n_B are both assumed to be very large; we can consider them to be (asymptotically) infinite, written as $n_A \to \infty$, $n_B \to \infty$. Under these conditions we get:

$$Z_{1-\alpha}\sqrt{\frac{\sigma_A^2}{n_A} + \frac{\sigma_B^2}{n_B}} \to 0$$

since dividing the variances by a very large number will be near zero.

And so the test reduces to a straightforward difference in means: $c = (\bar{x}_A - \bar{x}_B)$ Now, there is really no difference between the reaction time of Catholics and Protestants—the groups are actually identical. So $\bar{x}_A = \bar{x}_B$ and the resulting critical region is shown in Figure 12.15. Clearly the critical region is 50% and the chance of rejecting the null hypothesis is 50%.

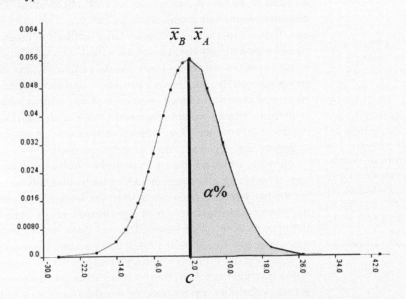

Figure 12.15 Distribution curves for two identical groups A and B with critical region.

So we have shown Meehl's conjecture to be true as the sample size tends to infinity. In practice, increasing the sample size makes the hypothesis much more sensitive and makes us more likely to reject the null hypothesis; much more than researchers expect.

In the Bayesian case it is easy to demonstrate that the two hypotheses, $H_0 : p(X < Y), H_1 : p(X > Y)$ are equiprobable, i.e. $p(X < Y) = p(X > Y) = 0.5$ by using AgenaRisk to test the difference between the distributions.

Anyone engaged in data mining should be very worried by the implications of Meehl's conjecture. It is unfortunate that, despite its importance, Meehl's conjecture has been under-reported.

Clearly, Meehl's conjecture has significant implications for data mining. Trawling a database in the hope of detecting hitherto undiscovered relationships between variables runs a serious danger of detecting differences between populations where there is, in fact, no natural difference. You should keep this in mind when you read the next media headline proclaiming a newly discovered gender, social, or medical inequality.

12.3.3   Bayesian Network Solution to the Problem of Learning Population Mean and Variance from Sample Mean and Variance

In the examples above where we have a large sample of data for some attribute of interest we have assumed classical statistical methods for estimating the mean and variance of that attribute in the population. These methods typically assume the population data are normally distributed, so an estimate of the mean and variance is sufficient to get an estimate for the full population distribution. However, in Chapter 10, we described a much more powerful method of parameter learning (and hence full distribution learning) from data providing examples using both Normal and Beta distribution assumptions. This involved creating a model in which we add a node for each observation. This works fine, although when there are a very large number of observations it is necessary to use import and/or programming methods to create the necessary model automatically (and the model will also be very large). However, in many situations we do not know the data explicitly, but instead are only provided with the sample data summary statistics (typically the sample mean and variance). Indeed , in most typical "comparison of two means" hypothesis testing we are presented ONLY with the respective sample means and variance. The sidebar provides an example of an actual undergraduate 2nd year statistics exam question.

Mr Baker thinks his bread sales will increase if he changes the packaging. On 10 days with the old packaging he sells an average 188 loaves per day with variance 156. On 10 days with the new packaging he sells on average 202 loaves per day with variance 115. Assuming data are random samples from normal populations with the same variance, investigate Mr Baker's belief by performing an appropriate hypothesis test at the 5% level of significance.

Now the way students are expected to "solve" this problem involves lots of weird assumptions (including the completely unnecessary need for the two population variances to be the same), the use of the t-statistic and the meaningless notion of significance level. There is no way of incorporating any prior knowledge, and there is the fundamental ambiguity (discussed in Section 12.3.2) about whether the conclusion should be based on the difference between the population means or the population distributions. And, on top of all that, you do not end up with an actual probability for whether or not the hypothesis is true. Even the brightest students find this sort of problem awkward and the conclusions unsatisfactory and even bewildering. Fortunately, it turns out that (under the normal population assumptions) there is a completely Bayesian solution to this problem which is far simpler and more powerful than the classical statistical "solutions" that are used.

The solution rests on the following theorem:

$$\text{Sample variance} = \text{Chisquared}(n-1) \times \text{variance}/(n-1)$$

where n is the sample size.

Knowing this, the BN in Figure 12.16 enables us to learn the population distribution. Here the NPTs are:

- sample mean = Normal (popMean, popVariance/n)
- chisquared = Chisquared($n-1$)
- sample variance = chisquared($n-1$) × popVariance/($n-1$)

Using the exam question example data for the new packaging we get the results shown in Figure 12.17 (so, as expected we "learn" that the population mean is about 188, but more importantly we learn the full distribution—its variance is about 375).

To compare two population means and distributions (such as with and without the new packaging in the exam question), we simply use a generic BN that contains two copies of the BN in Figure 12.16 and two additional Boolean nodes that determine, respectively, whether the first population mean is greater than the second and (more importantly) whether an arbitrary member of the first population is greater than the second (i.e. comparing the full population distributions). This solution is shown in Figure 12.18 which is calculated with the data in the exam question.

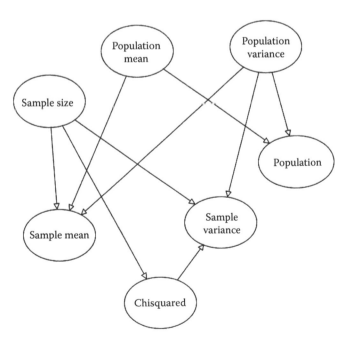

Figure 12.16   BN solution to Problem 1.

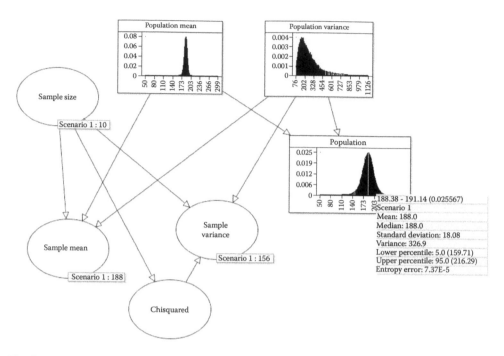

Figure 12.17   Population mean and variance (and hence population distribution) learnt from sample mean and variance.

We can see the probability that the mean with packaging is less than the mean without packaging to be 2.02% so we can "reject the null hypothesis at the 5% level." However, we can also see from the BN that the probability that on any given day more loaves will be sold with packaging than without is only 74.46%.

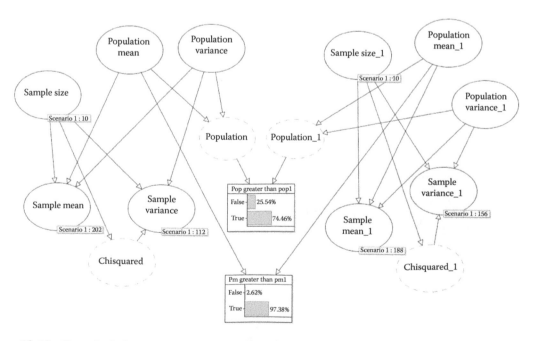

Figure 12.18   General solution to comparison of two means hypothesis test.

12.3.4 Summary of the Issues Raised Comparing Classical Statistical Hypothesis Testing and the Bayesian Approach

Sadly, throughout the sciences there has been a tendency to deify statistical hypothesis testing and follow its prescriptions in a purely ritualized sense where seeming objectivity is prized over value. The consequences of this have been catastrophic in many cases. This does not mean it is wrong to ask whether our hypothesis can be explained by chance results alone; we should ask the question, but significance testing is often the wrong tool.

> The method we recommend is Bayesian model comparison, which we cover later.

In summary the Bayesian approach discussed here differs from what you would see in a classical statistics test in the following main respects:

- Hypotheses of differences are often expressed as hypotheses about statistical parameters, such as means and variances rather than as hypotheses about differences between the variables themselves. The latter sits more easily in the Bayesian framework, is easier to understand, and involves the variables themselves rather than some unobservable summary statistic. In the Bayesian approach parameters are unknown (random) and the data is known (fixed). In the classical approach the data is random and the parameters are fixed.

- In classical statistics the onus is placed on the researcher to collect sufficient evidence to warrant rejecting the null hypothesis, say, with probability less than 1% or 5%—the so-called significance level. In the Bayesian approach we are interested in probabilities of the truth of hypotheses without any arbitrary cutoff between truth or falsehood at 1%, 5% or some other number.

- Much of hypothesis testing is concerned with separating noise from signal and determining whether the results of an experiment are statistically significant; as a consequence less reliable variables tend to get rejected. In contrast, scientific enquiry often hinges on whether the impact of one variable is greater than another, even if the variable with greater impact is less reliable. In practice reliability and impact are two separate questions deserving careful formulation of research hypotheses. Meehl's conjecture highlights the great dangers of relying on classical statistical significance tests to determine differences and relationships between variables.

12.4 Bayes Factors and Model Comparison

In this section, we describe the formalised approach to Bayesian analysis and model comparison. We first define Bayes factors, which then brings us neatly onto how we choose between competing models (hypotheses) on the basis of how well they predict events or data. We next show how we can accommodate expert opinions into the hypothesis testing process in a way that complements data. For completeness we show how the "Bayesian model choice" approach can be used as a coherent alternative to the "distribution fitting and testing" that comes as a standard part of the classical statistical armory. Finally, we talk about complex causal hypotheses and how we might test different causal explanations of data in hypothesis testing terms.

12.4.1 Bayes Factors

Sidebar 12.2

In other texts, the Bayes factor is defined as

$$\frac{P(D \mid H_1)}{P(D \mid H_2)}$$

This is simply the likelihood ratio defined in Chapter 6. The relationship between the two alternative versions—and why we feel our definition is preferable—is explained in Box 12.5

Typically we will assess whether the data/evidence we have on hand supports one hypothesis over another and by how much. In Bayesian terms, for two competing hypotheses, this is simply expressed using the Bayes factor; this is the ratio of the conditional probability of one hypothesis, H_1, being true given the data, D, and the conditional probability of the competing hypothesis, H_2, being true given the same data.

Bayes Factor

$$\frac{P(H_1 \mid D)}{P(H_2 \mid D)}$$

The ratio provides the following very simple and straightforward approach to hypothesis testing.

- If the ratio equals one, then $P(H_2 \mid D) = P(H_1 \mid D)$, and so we are indifferent to the hypotheses (neither is more likely than the other given the data available).
- If the ratio is greater than one, then $P(H_1 \mid D) > P(H_2 \mid D)$, and so the data supports H_1 over H_2.
- If the ratio is less than one, then $P(H_1 \mid D) < P(H_2 \mid D)$, and so the data supports H_2 over H_1.

So compared to classical methods of hypothesis testing Bayesian hypotheses are expressed in plain terms with undemanding interpretations. This stands in stark contrast to the classical approach with p-values described above.

Box 12.5 explains the relationship between the Bayes factor and the *likelihood ratio* introduced in Chapter 6. Note that the two are identical if the prior probabilities for the different hypotheses are equal.

Box 12.5   Bayes Factor and the Likelihood Ratio

In Chapter 6 we considered the special case where the two hypotheses were simply H and not H when we defined the likelihood ratio as

$$\frac{P(D \mid H)}{P(D \mid \text{not } H)}$$

But, in general, for any two hypotheses, H_1 and H_2, we define the likelihood ratio as

$$\frac{P(D \mid H_1)}{P(D \mid H_2)}$$

Note that the alternative definition of the Bayes factor described in Sidebar 12.2 is therefore simply the likelihood ratio. Using the odds ratio form of Bayes (see Chapter 6 for details) we get

$$\text{Bayes factor} = \frac{P(H_1 \mid D)}{P(H_2 \mid D)} = \frac{P(D \mid H_1)P(H_1)}{P(D \mid H_2)P(H_2)}$$

$$= (\text{likelihood ratio}) \times \frac{P(H_1)}{P(H_2)}$$

Note that when $P(H_1)$ and $P(H_2)$ are equal the Bayes factor is equal to the likelihood ratio. One of the benefits of using our definition of the Bayes factor (rather than the likelihood ratio) is that when the two competing hypotheses H_1 and H_2 are not exclusive and mutually exhaustive (i.e. when H_2 is NOT the same as "not H_1") then—as explained in Chapter 6—the Bayes factor does not provide information about the posterior probability of the hypotheses. Indeed, it may be possible that the likelihood ratio provides strong support for H_1 over H_2 even though the posterior probability of H_1 given the evidence D decreases.

We can also express the Bayes factor in logarithmic form, as log likelihood, as a scoring rule.

Here we have a prior belief in each hypothesis that is independent of the data and the likelihood of observing the data under each hypothesis. All other things being equal a hypothesis will be penalized if it is not supported by the data, that is, its predictions or outcomes are poorly matched by the data.

Example 12.3   Soccer Coaches

We want to determine whether a soccer coach is a skilled manager (hypothesis H_1) or merely lucky (hypothesis H_2). If we have no reason to believe one hypothesis over the other, then we can set the prior probability as

$$P(H_1) = P(H_2)$$

Let us assume that a skilled coach has a 0.6 probability of winning a game compared to a 0.4 probability for a lucky coach (both of these probabilities are higher than the standard expected win probability of 1/3). Now, suppose that after five games we have observed three wins (this is our data, D). Then, by the Binomial theorem

$$P(D \mid H_1) = 0.34 \quad P(D \mid H_2) = 0.23$$

The Bayes factor is therefore calculated as follows (since we are assuming that $P(H_1) = P(H_2)$):

$$\frac{P(H_1 \mid D)}{P(H_2 \mid D)} = \frac{P(D \mid H_1)P(H_1)}{P(D \mid H_2)P(H_2)} = \frac{P(D \mid H_1)}{P(D \mid H_2)} = \frac{0.34}{0.23} = 1.48$$

So the odds are roughly 3:2 that the coach is competent rather than just lucky. However, suppose we started with a stronger belief that the coach is lucky, say,

$$P(H_1) = 0.3 \text{ and } P(H_2) = 0.7$$

Then with the same observed data as above the Bayes factor is

$$\frac{P(H_1 \mid D)}{P(H_2 \mid D)} = \frac{P(D \mid H_1)P(H_1)}{P(D \mid H_2)P(H_2)} = \frac{0.34 \times 0.3}{0.23 \times 0.7} = 0.63$$

So in this case the odds still favor the coach being lucky rather than competent. However, suppose that after 10 matches we have observed 7 wins. In this case

$$P(D \mid H_1) = 0.215 \quad P(D \mid H_2) = 0.042$$

So the Bayes factor is

$$\frac{P(H_1 \mid D)}{P(H_2 \mid D)} = \frac{P(D \mid H_1)P(H_1)}{P(D \mid H_2)P(H_2)} = \frac{0.215 \times 0.3}{0.042 \times 0.7} = 21.94$$

This means that, despite the prior belief in favor of the coach being lucky, once we have observed the data, the odds in favor of the coach being competent rather than lucky are about 20:1.

12.4.2 Model Comparison: Choosing the Best Predictive Model

Bayesian model comparison involves the empirical evaluation of a set of competing hypothetical models in terms of their ability to predict actual data. The aim is to select the best model or produce a "new" model that is a weighted average of the possible models considered. In each case we can think of the models as hypotheses and consider model comparison as a form of hypothesis testing. We next explain how we can use the Bayesian framework to hypothesis testing to answer questions like

- Which of two soccer prediction models is superior?
- Is one scientific weather model better able to predict floods than another?

Box 12.6   Comparing Models as Hypothesis Testing

Suppose that we have two models: $Model_1$ and $Model_2$. We can think of these as two competing hypotheses. To determine which model is best with respect to some observed data, D, we simply calculate the Bayes factor:

$$\frac{P(Model_1 \mid D)}{P(Model_2 \mid D)}$$

In the case where we assume that the priors for each hypothesis (model) are equally likely then we know from Box 12.5 that the Bayes factor is equal to the likelihood ratio:

$$\frac{P(D \mid Model_1)}{P(D \mid Model_2)}$$

So, the better of the two models is the one with the highest likelihood for the observed data, D.

In general, for two models (M_1 and M_2) we may have a number of observed data ($D_1, ..., D_n$) that we can assume are independent. In that case we can compute

$$\frac{P(D \mid M_1)}{P(D \mid M_2)} = \frac{P(D_1,...,D_n \mid M_1)}{P(D_1,...,D_n \mid M_2)} = \frac{\prod_{i=1}^{n} P(D_i \mid M_1)}{\prod_{i=1}^{n} P(D_i \mid M_2)}$$

Example 12.4   Comparing Sports Prediction Models

This is an example of the general case explained in Box 12.6. The Reds and Blues compete regularly against each other, with the Reds winning 90% of the matches and the Blues 10%. Suppose two statistical models, M_1 and M_2, are developed to predict the outcome of individual matches between the Reds and Blues to help gamblers place appropriate bets.

- Model M_1 is a perfect frequentist model (developed from data only) in the sense that, for every game, its prediction is always:

  Probability of Reds winning: 0.9
  Probability of Blues winning: 0.1

  For simplicity we write this prediction as (0.9, 0.1).

- Model M_2 on the other hand is a clever model that combines the data with expert judgment. This expert judgment incorporates an understanding of the factors that lead to the rare Blues victories (for example, the experts may know that Reds tend to play badly when certain key players are not available).

So, suppose we have data on 10 matches. Table 12.3 shows the predictions from models M_1 and M_2 and the actual results. If you only look at the model performance over the entire sequence of matches frequentists would argue that both models are perfect in the sense that they both

Table 12.3
Model Predictions and Actual Outcomes over 10 Matches

Match	Model M_1 Prediction	Model M_2 Prediction	Actual Result
1	(0.9, 0.1)	(0.99, 0.01)	Reds
2	(0.9, 0.1)	(0.99, 0.01)	Reds
3	(0.9, 0.1)	(0.99, 0.01)	Reds
4	(0.9, 0.1)	(0.99, 0.01)	Reds
5	**(0.9, 0.1)**	**(0.09, 0.91)**	**Blues**
6	(0.9, 0.1)	(0.99, 0.01)	Reds
7	(0.9, 0.1)	(0.99, 0.01)	Reds
8	(0.9, 0.1)	(0.99, 0.01)	Reds
9	(0.9, 0.1)	(0.99, 0.01)	Reds
10	(0.9, 0.1)	(0.99, 0.01)	Reds

predict the Reds will win 90% of the matches and that is exactly the proportion that they win. However, the gamblers know that model M_2 is superior because it wins them far more money. We are not interested in the long run but only at how well the models do on any given game. The Bayesian model comparison confirms what the gamblers know.

Since we can assume no prior preference between the models we know from Box 12.6 that the Bayes factor is

$$\frac{P(D \mid M_1)}{P(D \mid M_2)} = \frac{\prod\limits_{i=1}^{10} P(D_i) \mid M_1}{\prod\limits_{i=1}^{10} P(D_i) \mid M_2}$$

where D_i is the result of the ith match. With the exception of match 5 we know that

$$P(D_i \mid M_1) = 0.9 \quad \text{and} \quad P(D_i \mid M_2) = 0.99$$

However, for match 5 we have

$$P(D_5 \mid M_1) = 0.1 \quad \text{and} \quad P(D_5 \mid M_2) = 0.91$$

Hence the Bayes factor is equal to

$$\frac{0.9^9 \times 0.1}{0.99^9 \times 0.91} = \frac{0.0387}{0.831}$$

which means that model M_2 is superior to M_1 by a factor of more than 20.

It is also worth noting that if we had a third model, M_3, which won similar to M_2 but predicted (0.9, 0.1) when Reds won and (0.1, 0.9) when Blues won, then this model would also be superior to model M_1 by a factor of 9. Since it predicts exactly the same as model M_1 except in the case when Blues win, the Bayes factor criteria for model evaluation is demonstrably sensible. Yet, from a frequentist aspect, model M_3 would be scorned on the basis that it is not properly calibrated because, over the sequence of matches, it predicts that Reds will win only 82% of the time rather than the observed 90%.

Example 12.4 provides a very powerful case for why, for model comparison, the Bayesian approach is superior to the frequentist approach.

Box 12.7   Producing a New Model from Several Competing Models

Suppose we have a set of competing models $(M_1, \dots , M_n)$ that have different features and benefits. Instead of attempting to determine a single "winning" model we might instead choose to produce a new model that averages all of the competing models M_i and produces a single posterior prediction. This is done as follows:

1. Derive the distribution for the hypotheses from our data $D_1, D_2, ..., D_m$, yielding $P(M_i \mid D_1, D_2, ..., D_m)$..
2. Marginalize out the data to get a new prior weighting, $P(M_i)$, where

$$P(M_i) = \sum_{D_1, D_2, ..., D_m} P(M_i \mid D_1, D_2, ..., D_m)$$

3. Use this new prior as the basis for weighting the predictions, O_j, from each model:

$$P(O_j) = \sum_{M_i} P(O_j \mid M_i) P(M_i)$$

It is important to note that well-performing models, with respect to past data, will have higher weights, $P(M_i)$.

Example 12.5   Extreme Weather Model Comparison

This is an example of the general case explained in Box 12.7. A particular challenge faced by insurance companies is how to forecast the risk of extreme weather events such as hurricanes, typhoons, volcanic eruptions, and floods. To this end they use a myriad of different scientific models, sourced from many experts to predict these events and set insurance premiums.

There is rarely one single, trusted, scientific model for these phenomena. Therefore, insurers face the challenge of deciding which particular scientific model to trust in making these predictions. One way to decide is to compare the models or use some weighted average prediction.

Let's look at the example where we have two competing models for the total losses arising from flood risk (in millions of dollars):

1. Model A is a newly developed highly granular model to estimate floods for every city block using data on height above sea level, flood defenses, and so on.
2. Model B may have a lot more provenance, having been used for some time, but uses a less detailed regional form purely derived from historical flood data alone.

Typically, for these types of models there may be different parameters for different regions of the country, so we can think of regions as "submodels" of the super model and will want to compare models across regions. The submodels and summary statistics for Regions X, Y, and Z by Models A and B are shown in Table 12.4.

The summary statistics, and in particular the 99th percentiles, show that both distributions are long tailed in the sense that small loss events are much more frequent than extremely high loss events. Model B is especially pessimistic about very large losses.

We have three years worth of loss data (in millions of dollars) for each region and want to use this to compare the models and then make a weighted average prediction for each region. The data are

Region X = {5, 10, 100}
Region Y = {25, 40, 55}
Region Z = {75, 38, 94}

Once we enter this into the AgenaRisk model, shown in Figure 12.20, we can see that Model B has a posterior probability of approximately 95% compared to Model A with 5%. So, if we had to choose a single model we would go for Model B as the best choice. Its prior distribution better accommodates the likelihood of the data than Model A.

Extreme weather events are an example of events known as *high impact/low probability* events. Although they might be rare, their financial and human impact may be considerable; indeed a single extreme weather event can dent an insurance company's profits or even exceed their ability to pay, thus making them insolvent.

The 99% figure, for example, the value 255 for Model A Region X, is simply the *99th percentile* as defined in Box 12.1 above

So in Year 1, Region X suffered $5 million losses, in Year 2 $10 million, etc.

Figure 12.19 illustrates the difference in prediction made by each model. Notice that Model A's predictions of losses are hunched toward the lower end of the loss scale while Model B's distribution has a longer tail.

Table 12.4
Models for Regions X, Y, and Z

	Region		
	X	**Y**	**Z**
Model A	*LogNormal(2, 1.5)*	*LogNormal(3, 1.5)*	*LogNormal(3, 2)*
	Median 7	Median - 20	Median - 20
	99% - 255	99% - 731	99% - 2116
Model B	*Exponential(1/60)*	*Exponential(1/80)*	*Exponential(1/100)*
	Median - 41	Median - 55	Median - 69
	99% - 279	99% - 371	99% - 471

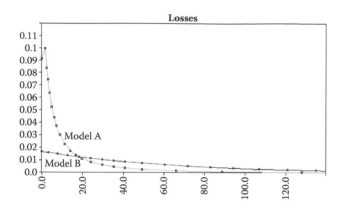

Figure 12.19   Model A and Model B predictions for Region X.

Figure 12.20 also shows the posterior loss prediction for the next year for each of the Regions X, Y, and Z. A close inspection confirms that these look like a reasonable weighted average of the Models A and B.

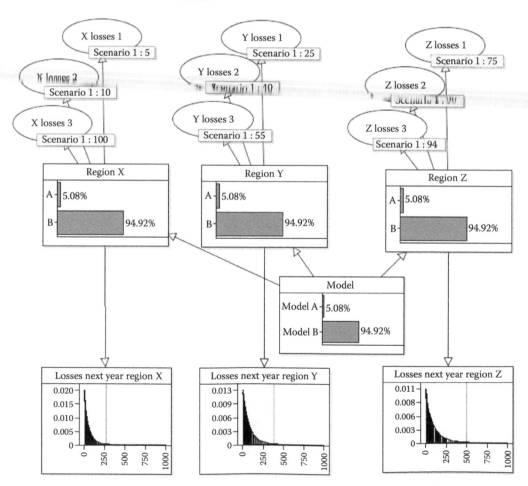

Figure 12.20   Extreme weather model.

It is important to note that the Bayesian approach to model comparison is equivalent to that produced by a technique called *maximum likelihood estimation*. A basic introduction and overview of maximum likelihood estimation is presented in Box 12.8.

Box 12.8   Maximum Likelihood Estimation

Imagine an urn contains a large number of blue and white balls. If you know that 30% of the balls are white, then we have seen in Chapter 5 that there is a standard method—the Binomial distribution—for calculating the probability of selecting X white balls from n. More generally the distribution is

$$P(X = x \mid n, p) = \frac{n!}{x!(n-x)!} p^x (1-p)^{n-x}$$

where p is the proportion of white balls (which you can think of as the probability of a randomly selected ball being white).

What we have done here is to calculate the probability of seeing certain data given a known value for p. But in many situations, the problem is that we do not know the value of p. What we want to do instead is to infer the value of p from the data we have observed. Suppose, for example, we have selected 2 white balls out of a sample of size 5. The question we ask is: given the data (i.e., the observation of 2 white balls from 5) what is the most likely value of p?

Now, if we have observed 2 white balls in a sample of size 5, then we can use the Binomial distribution formula to ask the question: What is the most likely value of p that would give rise to finding 2 white balls in a sample of 5? In other words, out of all possible values of p, which one makes the following function of p the greatest?

$$10p^2(1-p)^3$$

This function of p is called the *likelihood of p* (often written as $L(p)$) and in this example the maximum value occurs when p is 0.4, which happens to be the same as the sample proportion. This is not a coincidence. In general, if we observe x white balls in a sample of size n the likelihood function of the unknown parameter p is

$$L(p) = \frac{n!}{x!(n-x)!} p^x (1-p)^{n-x}$$

Our task is to find the value of p for which the function $L(p)$ is greatest. For those of you who remember your school calculus, this is actually quite easy. As long as the function has some nice properties (which we will not go into here) the maximum is the value for which the differential is equal to zero. In other words we just differentiate the function $L(p)$ with respect to p, set $L(p)$ to zero and solve the resulting equation for p.

Noting that the expression:

$$\frac{n!}{x!(n-x)!}$$

in $L(p)$ is simply a constant, which we can write as K, the differential of $L(p)$ (written $\frac{dL}{dp}$) is

$$\frac{dL}{dp} = \frac{d}{dp} K p^x (1-p)^{n-x}$$

$$= K p^x (x-n)(1-p)^{n-x-1} + K x p^{x-1}(1-p)^{n-x}$$

$$= K p^{x-1}(1-p)^{n-x-1}[p(x-n) + x(1-p)]$$

So when $\frac{dl}{dp}$ is set to 0 we get

$$0 = p(x - n) + x(1 - p)$$
$$= x - pn$$

and $p = x/n$ which is the sample mean.

12.4.3 Accommodating Expert Judgments about Hypotheses

In most examples thus far in this chapter we have assumed that the prior probabilities for competing hypotheses and models were equal and have proceeded to compare models and test hypotheses on the basis of empirical data alone. While the prior assumption that competing hypotheses are equally likely is convenient (it means, for example, that the Bayes factor is equal to the likelihood ratio) it is overly restrictive and certainly does not exploit the full potential of the Bayesian approach.

If there is prior expert knowledge or data to inform us about the hypotheses or models under comparison it certainly makes sense to incorporate that information into the relevant prior probabilities. We could choose a more informative prior for a variable of interest, perhaps favoring it over another variable. So, for example, in our weight loss drug example we might favor one drug over the other by simply stating a prior distribution that gives a higher probability of losing more weight to one drug over another. This is especially useful when testing for differences between variables. We could also set prior probabilities on the hypotheses variable itself by giving higher weights to some hypotheses over others, an approach that is well suited when making model comparisons.

Let's look at each of these situations in turn with specific examples.

This assumes the same quality assurance problem as in Example 12.1.

Example 12.6   Quality Assurance with Expert Judgment

Before testing starts we get information from a trusted expert, with knowledge of the composition of the materials. This expert feels able to express a prior belief in the probability distribution of each material. Whereas before we assumed an ignorant prior whereby all probabilities of faultiness are equally likely (the Uniform distribution) here the expert feels that material B is much more likely to be better than A. This expertise might be expressed by choosing these priors:

$$P(p_A) \sim Beta(2,8)$$
$$P(p_B) \sim Beta(1,9)$$

This gives a prior mean failure probability of 0.2 for A and 0.1 for B.

As before assume that the testing yields 10 faults in 200 samples of material A and 15 faults in 200 samples of material B. Once we enter this data into the model with the expert judgment priors we achieve 78%

probability in $H_0 : p_A > p_B$, whereas before it was approximately 84%. If the target was 90% confidence clearly more testing would be needed to accumulate more evidence that might support material A.

Example 12.7 Extreme Weather Model Comparison with Expert Judgments

When dealing with rare events you may have little or even no relevant data—after all that is why the events are rare. But without such data how do you decide between hypotheses? One way is to model expert judgments about the attributes and features of a model or hypothesis in comparison with others on the basis of empirical or theoretical grounds. After all, it is common throughout science, engineering, and medicine for experts to make judgments on these matters, so what could be more natural than expressing these directly in the Bayesian model?

One way to do this is to weight the factors being considered and model these as "indicator" variables for the hypothesis node (see Chapter 8 for a discussion of indicator nodes), where each indicator allocates higher likelihood to one hypothesis or another based on the strength of the feature or attribute associated with the hypothesis. This is explained formally in Box 12.9.

We could go further and produce a tree of indicator nodes reflecting the decomposition of features and attributes into subfeatures and sub-attributes, ultimately arriving at a kind of probabilistic checklist to assess each hypothesis or model.

Box 12.9 Weighting the Factors

Formally, each feature, F_j, acts as a Bayes factor function where the ratio determines the degree to which the presence of the features favors one model (hypothesis) over another:

$$\frac{P(M_1 | F_1, F_2, ..., F_m)}{P(M_2 | F_1, F_2, ..., F_m)} \propto \frac{\prod_{j=1}^{m} P(F_j | M_i, W_j) P(W_j)}{\prod_{j=1}^{m} P(F_j | M_{i+1} W_j) P(W_j)}$$

where $P(W_j)$ is the weight allocated to reflect the importance of the factor.

The symbol $\propto$ means "is proportional to."

So, suppose we have three features relevant to our opinion about what makes a good extreme weather model:

- Hazard feature
- Geographical feature
- General feature

We could simply score each model as follows:

$$\frac{P(\text{Model A} | \text{Hazard,Geographical,General})}{P(\text{Model B} | \text{Hazard,Geographical,General})}$$

$$\propto \frac{P(\text{Hazard} | \text{Model A})}{P(\text{Hazard} | \text{Model B})} \frac{P(\text{Geographical} | \text{Model A})}{P(\text{Geographical} | \text{Model B})}$$

$$\times \frac{P(\text{General} | \text{Model A})}{P(\text{General} | \text{Model B})}$$

We can choose any probability values here so long as the desired ratios for the Bayes factors are respected. For example, we might want to allocate weights to the hazard, geographical, and general features in the ratio: 3:2:1. Then for each of these features we "score" the models on the extent to which the desired feature is satisfied by the model.

Let's assume that we favor Model A to Model B in ratio 3:1, 3:2, and 1:2 for each of the hazard, geographical, and general features respectively. We model this in AgenaRisk as shown in Figure 12.21.

Note that we do not need to assess the extent to which the feature is not satisfied. If both models are equally poor you might wish to invent a third model representing "random model" or "neither" and include this in the hypothesis set.

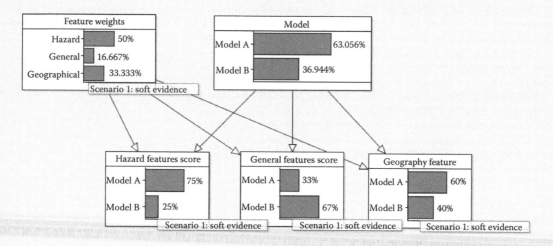

Figure 12.21　Extreme weather model comparison using expert judgment.

Associated with each factor node in the BN is the NPT with crisp 1 and 0 values for where the feature node corresponds to the state value for the same feature weight and indifferent values elsewhere (in this case 0.5 for two models; for four models the values would be 0.25). Thus, if one feature is favored entirely then only that element of that part of the single NPT associated with that feature would be relevant, since all others would have zero weight or have indifferent values. In this case the NPT for the Hazard feature score node in Figure 12.21 is shown in Table 12.5.

In Figure 12.21 we have entered the score for each feature as "soft evidence" into the risk table feature in AgenaRisk in such a way as to exactly reflect the ratios required. When executed the BN shows results that favor Model A with probability 63%.

Table 12.5
NPT for the Hazard Feature Score Node

Feature Weights	Hazard		General		Geographical	
Model	Model A	Model B	Model A	Model B	Model A	Model B
Model A	1.0	0.0	0.5	0.5	0.5	0.5
Model B	0.0	1.0	0.5	0.5	0.5	0.5

When this expert sourced information is combined with the empirical data from the three regions X, Y, and Z, the posterior probability distribution still favors Model B by 92% rather than 95%, so in this case the empirical data is of such strength that it overrides the results of the expert analysis.

12.4.4 Distribution Fitting as Hypothesis Testing

Statistical packages are popular for performing what is commonly called *distribution fitting*. This is the process of determining the parameters for the typical statistical distributions (e.g., Normal, Gamma) and selecting the distribution that best fits the data. The process proceeds in two stages:

1. Parameter selection
2. Goodness-of-fit test

> Goodness-of-fit tests are simply another type of hypothesis test with accompanying test statistic; there are many varieties in classical statistics including the chi-square test.

In Bayesian terms we can use parameter learning to determine the parameters and then produce a model to carry out a Bayes factor test. Parameter learning is covered in Chapter 9. For the Bayes factor test we can consider each model (i.e., distribution with its accompanying parameters) and perform model comparison as discussed earlier, but with the original data used to fit the parameters. Let's look at an example.

Example 12.8 Distribution Fitting and Goodness of Fit

Let us assume that there are three competing models—Normal (H_1), Exponential (H_2), and Binomial (H_3)—that are proposed for predicting system failure time. Suppose also that we have no prior preference for any one model, that is, $P(H_1) = P(H_2) = P(H_3)$. Then we know that to determine the model that best fits a number of actual observations $O_1,...,O_m$ we simply have to compute the likelihoods for each model:

$$P(O_1,...,O_m \mid H_i) = \prod_{j=1}^{m} P(O_j \mid H_i)$$

Let's assume we have the following observations for the time (in days) until failure of a system: 27, 32, 15, 25, 20, 7, 30, 8, 33. So, on one occasion the system failed after 27 days, on another occasion the system failed after 32 days, and so on.

If we "fit" that data to the three different models corresponding to the three hypotheses we would get something like the following distributions (note that the sample mean is about 22):

H_1: Normal(22, 100)

H_2: Exp(22)

H_3: Binomial(220, 0.1)

So, for example, the probability of observing O if H_1 is "true" is assumed to be the distribution Normal(22,100) etc.

In AgenaRisk we can build the model required to determine which is the best fit for the data by constructing a labeled node for the hypothesis

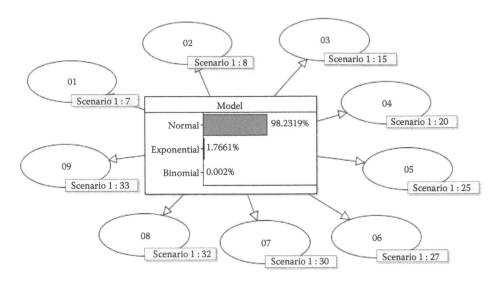

Figure 12.22   Goodness-of-fit model.

set and observation nodes for each of the data points conditioned on the above distributions. The results are shown in Figure 12.22. You can see that the hypothesis that has highest probability, by some margin, is the Normal distribution with 98.4% probability.

Note that the approach to assessing goodness of fit takes no account of the number of parameters in the model. In fact it is pretty much irrelevant. The next section digs a little deeper and addresses this issue.

12.4.5   Bayesian Model Comparison and Complex Causal Hypotheses

We can think of competing causal explanations for some phenomena as yet another example of hypothesis testing and the automated discovery of causal relations, typified by machine learning, as an exercise in hypothesis testing. Here we look at Bayesian model comparison of complex causal hypotheses and strongly recommend this approach as an alternative to the standard statistical significance testing approach. This is because of its generality and rigor and also because it forces the analyst to consider and compare different models that might explain the data, rather than simply perform ritualized testing against fictitious naive hypotheses.

The symbol $\propto$ means "proportional to."

As before our goal is to assign probabilities or scores to hypotheses, H_i, based on data available, D:

$$P(H_i \mid D) \propto P(D \mid H_i)P(H_i)$$

Now each hypothesis (model), H_i, has a number of parameters associated with it, represented as a vector, $\mathbf{w}_i$, and these parameters represent the causal variables entertained by the hypothesis. Box 12.10 describes the steps involved.

Box 12.10 Steps Involved

Step 1: Model fitting. We fit the model parameters using the available data by Bayes theorem:

$$P(\mathbf{w}_i \mid D) = \frac{P(D \mid \mathbf{w}_i, H_i) P(\mathbf{w}_i \mid H_i)}{P(D \mid H_i)}$$

$\propto P(D \mid \mathbf{w}_i) P(\mathbf{w}_i)$ if we assume the hypothesis, H_i, is true.

If we consider the data, D, as a vector of observations we can decompose the model fitting step as:

$$P(\mathbf{w}_i \mid D_1, D_2 ..., D_m) = \frac{P(\mathbf{w}_i \mid H_i) \prod_{j=1}^{m} P(D_j \mid \mathbf{w}_i, H_i)}{P(D_1, D_2 ..., D_m \mid H_i)}$$

Step 2: Model comparison. We compute how well the hypothesis explains the data, in competition with other hypotheses. To do this we first compute how likely the data is under a given hypothesis and parameter set to get a Bayesian score:

$$P(D \mid H_i) = \sum_{\mathbf{w}_i} P(D \mid \mathbf{w}_i, H_i) P(\mathbf{w}_i \mid H_i)$$

which is the marginal likelihood of the data.

Now we use Bayes theorem, $P(H_i \mid D) \propto P(D \mid H_i)$, as before where the prior hypotheses are equally probable.

Again, if we consider the data as a vector of observations then

$$P(D_1, D_2 ..., D_m \mid H_i) = \sum_{\mathbf{w}_i} \left[P(\mathbf{w}_i \mid H_i) \prod_{j=1}^{m} P(D_j \mid \mathbf{w}_i, H_i) \right]$$

This looks complicated, but if you look at the example in AgenaRisk you will see that this is simply the marginal distribution for each data node.

The Bayesian score includes the prior belief about the parameters and, as such, is generally more pessimistic than that achieved by maximum likelihood estimation. Noisier or more variable prior distributions for the parameters will lead to less accurate predictions for the data and smaller scores. Clearly those hypotheses containing parameters that better fit the data will have scores that fit the data less well than others. This ability to discriminate between models based on their parameters means that Bayesian methods automatically contain a widely accepted principle of good scientific practice, called Occam's razor.

Some claim that parsimony is therefore built into the Bayesian approach: the more parameters one has in a hypothesis, and so the more complexity in a model, the more data is needed to support it (again, all other things being equal). To put it another way, a hypothesis containing more unknowns, or more ignorant priors, may be less well identified than one with fewer unknowns given the same data. So, when learning parameters the variance of each parameter learned, and thus the total noise, will be higher the more parameters we include. Therefore, a model with many parameters or more vague priors runs the risk of being less

Occam's Razor

Occum's razor is the assertion that we should prefer the simplest scientific explanation to other more complex ones, all other things being equal.

A classic example of overcomplexity is where statisticians use an nth degree polynomial function to fit all data points in a data set. This approach is marvelously successful until you encounter the next data point. This danger is called "over fitting the model" where a parametric form or set of explanatory variables is chosen that exactly, or near exactly, explains the data to hand and gives a false impression that the model successfully predicts the phenomena when it actually does not. The remedy to this is to use the approach described here which would naturally penalize overfitted models.

During model comparison we can use the same data or perform cross-validation using a training set of data to fit the parameters and a testing set to test the hypotheses.

Interestingly, this looks like another example of Simpson's paradox, as covered in Chapters 2 and 6.

accurate than a model with fewer parameters. This does not mean to say that more complex models will never succeed over simpler models, rather it means that one cannot necessarily get away with adding more and more parameters until every data point is explained.

So what does the Occam's razor have to do with assessing complex causal hypotheses? Well, when assessing different causal hypotheses (including the null hypothesis that there is no identifiable cause) we can think of each cause as a parameter; when we add more and more causes we run the risk of discovering complex relations where simple ones might do. The Occam's razor therefore penalizes more complex hypotheses and sets the bar that bit higher.

Example 12.9   Race Equality in U.S. Prison Sentencing

The influence of race on the imposition of the death penalty for murder in the United States has been widely debated. Of particular prominence in the debate are three variables:

- Victim's race (V)
- Defendant's race (D)
- Whether the defendant was sentenced to death (S)

Table 12.6 classifies 326 cases in which the defendant was convicted of murder.

There has long been a concern that the race of the defendant affects the chances of being given the death penalty. But studies have actually shown that the significant distinction is the race of the victim. From Table 12.6 it looks like the death penalty was applied much more often when the victim was white than when the victim was black, so there may be evidence here of racial bias in the U.S. justice system. However, we also need to examine whether the race of the defendant is relevant or indeed whether either variable is relevant.

Table 12.6
Classification of Murder Cases

Total	Race of Defendant	Death penalty = Yes	Death penalty = No	
	White	19	141	11.88%
	Black	17	149	10.24%
Victim White	Race of Defendant	Death penalty = Yes	Death penalty = No	
	White	19	132	11.58%
	Black	11	52	17.46%
Victim Black	Race of Defendant	Death penalty = Yes	Death penalty = No	
	White	0	9	0%
	Black	6	97	5.83%

Source: M. Radelet, 1981, Racial characteristics and imposition of the death penalty, *American Sociological Review* 46, 918–927.

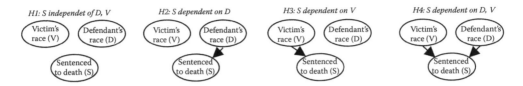

Figure 12.23   BN models for the four causal hypotheses.

Our first task is to identify hypotheses for each of the possible combinations of the three causal factors D (defendant's race), V (victim's race), and S (death penalty):

H_1 : S independent of V and D $P(S,V,D) = P(S)P(V)P(D)$

H_2 : S dependent on D $P(S,V,D) = P(S|D)P(V)P(D)$

H_3 : S dependent on V $P(S,V,D) = P(S|V)P(V)P(D)$

H_4 : S dependent on V and D $P(S,V,D) = P(S|V,D)P(V)P(D)$

The corresponding hypothesized causal BN models are shown in Figure 12.23.

Most books on Bayesian analysis don't provide a blow-by-blow account of the calculation steps in problems like this, so for completeness and as an aid to understanding we have carefully detailed each step in turn in Box 12.11. All of the detailed results can be obtained simply by running the relevant components of the AgenaRisk model.

Box 12.11   Detailed Calculation Steps Involved in Analyzing Each Model

Step 1. Fit the model for each hypothesis.

In this example we can assume that the Binomial distribution is an appropriate choice, where q is the probability of the death penalty and is the only parameter we need to fit. We can fit the parameters from the data $\{n, r\}$ for the number of cases, n, and the number of death penalties applied, r

Let's take each hypothesis in turn.

$\underline{H_1 : S \text{ independent of } V \text{ and } D}$

We need to consider the case where V and D are independent of S: Unlike when data fitting we need to predict all of the cells in the original data table, i.e. every combination of states being considered, in order to assess whether the model is surprised by the data:

$$P(r=19, n=151 \,|\, H_1) = \sum_q [P(q \,|\, r=19, n=151, H_1)P(r=19, n=151 \,|\, q, H_1)] = 0.072$$

$$P(r=11, n=63 \,|\, H_1) = \sum_q [P(q \,|\, r=11, n=63, H_1)P(r=11, n=63 \,|\, q, H_1)] = 0.049$$

$$P(r=0, n=9 \,|\, H_1) = \sum_q [P(q \,|\, r=0, n=9, H_1)P(r=0, n=9 \,|\, q, H_1)] = 0.345$$

$$P(r=6, n=103 \,|\, H_1) = \sum_q [P(q \,|\, r=6, n=103, H_1)P(r=6, n=103 \,|\, q, H_1)] = 0.036$$

The Joint probability of the data is therefore:

$$P(r=19, n=151 \,|\, H_1)P(r=11, n=63 \,|\, H_1)P(r=0, n=9 \,|\, H_1)P(r=6, n=103 \,|\, H_1)$$

and this equals 4.32 E-05.

H_2 : S dependent on V

This is similar to the previous case except we need to condition each calculation on the particular combination of parent variables we are considering:

$$P(r=19, n=151 \mid H_2) = \sum_{q_{D-White}} \left[P(q_{D-White} \mid r=19, n=151, H_2) P(r=19, n=151 \mid q_{D-White}, H_2) \right] = 0.069$$

$$P(r=0, n=9 \mid H_2) = \sum_{q_{D-White}} \left[P(q_{D-White} \mid r=0, n=9, H_2) P(r=0, n=9 \mid q_{D-White}, H_2) \right] = 0.315$$

$$P(r=11, n=63 \mid H_2) = \sum_{q_{D-Black}} \left[P(q_{D-Black} \mid r=11, n=63, H_2) P(r=11, n=63 \mid q_{D-Black}, H_2) \right] = 0.042$$

$$P(r=6, n=103 \mid H_2) = \sum_{q_{D-Black}} \left[P(q_{D-Black} \mid r=6, n=103, H_2) P(r=6, n=103 \mid q_{D-Black}, H_2) \right] = 0.052$$

The joint probability of the data is therefore:

$$P(r=19, n=151 \mid H_2) P(r=0, n=9 \mid H_2) P(r=11, n=63 \mid H_2) P(r=6, n=103 \mid H_2)$$

and this equals 4.75 E-5.

H_3 : S dependent on V

Again we condition appropriately:

$$P(r=19, n=151 \mid H_3) = \sum_{q_{D-White}} \left[P(q_{D-White} \mid r=19, n=151, H_3) P(r=19, n=151 \mid q_{D-White}, H_3) \right] = 0.067$$

$$P(r=11, n=63 \mid H_3) = \sum_{q_{D-Black}} \left[P(q_{D-Black} \mid r=11, n=63, H_3) P(r=11, n=63 \mid q_{D-Black}, H_3) \right] = 0.094$$

$$P(r=0, n=9 \mid H_3) = \sum_{q_{D-White}} \left[P(q_{D-White} \mid r=0, n=9, H_3) P(r=0, n=9 \mid q_{D-White}, H_3) \right] = 0.576$$

$$P(r=6, n=103 \mid H_3) = \sum_{q_{D-Black}} \left[P(q_{D-Black} \mid r=6, n=103, H_3) P(r=6, n=103 \mid q_{D-Black}, H_3) \right] = 0.119$$

The joint probability of the data is therefore:

$$P(r=19, n=151 \mid H_3) P(r=0, n=9 \mid H_3) P(r=11, n=63 \mid H_3) P(r=6, n=103 \mid H_3)$$

and this equals 4.32 E-04.

H_4 : S dependent on V and D

Again we need to consider all combinations:

$$P(r=19, n=151 \mid H_4)$$

$$= \sum_{q_{V-White/D-White}} \left[P(q_{V-White/D-White} \mid r=19, n=151, H_4) P(r=19, n=151 \mid q_{V-White/D-White}, H_4) \right]$$

$$= 0.068$$

$$P(r=11, n=63 \mid H_4)$$

$$= \sum_{q_{V-Black/D-White}} \left[P(q_{V-Black/D-White} \mid r=11, n=63, H_4) P(r=11, n=63 \mid q_{V-Black/D-White}, H_4) \right]$$

$$= 0.092$$

$$P(r=0,n=9\,|\,H_4)$$

$$=\sum_{q_{V-White/D-Black}}\left[P(q_{V-White/D-Black}\,|\,r=0,n=9,H_4)P(r=0,n=9\,|\,q_{V-White/D-Black},H_4)\right]$$

$$=0.525$$

$$P(r=6,n=103\,|\,H_4)$$

$$=\sum_{q_{V-Black/D-Black}}\left[P(q_{q_{V-Black/D-Black}}\,|\,r=6,n=103,H_4)P(r=6,n=103\,|\,q_{V-Black/D-Black},H_4)\right]$$

$$=0.116$$

The joint probability of the data is 3.81 E-4.

Step 2. Calculate the Bayesian scores for all hypotheses:

Let's take each hypothesis in turn.

H_1 : S independent of V and D

We need to consider the case where V and D are independent of S. Unlike when data fitting we need to predict all of the cells in the original data table, i.e. every combination of states being considered, in order to assess whether the model is surprised by the data:

$$P(r=19,n=151\,|\,H_1)=\sum_q\left[P(q\,|\,r=19,n=151,H_1)P(r=19,n=151\,|\,q,H_1)\right]=0.071$$

$$P(r=11,n=63\,|\,H_1)=\sum_q\left[P(q\,|\,r=11,n=63,H_1)P(r=11,n=63\,|\,q,H_1)\right]=0.049$$

$$P(r=0,n=9\,|\,H_1)=\sum_q\left[P(q\,|\,r=0,n=9,H_1)P(r=0,n=9\,|\,q,H_1)\right]=0.034$$

$$P(r=6,n=103\,|\,H_1)=\sum_q\left[P(q\,|\,r=6,n=103,H_1)P(r=6,n=103\,|\,q,H_1)\right]=0.036$$

The joint probability of the data is therefore:

$$P(r=19,n=151\,|\,H_1)P(r=11,n=63\,|\,H_1)P(r=0,n=9\,|\,H_1)P(r=6,n=103\,|\,H_1)$$

and this equals 4.26E-06.

From the example AgenaRisk mode you will see that this is simply the marginal probability of the data nodes (a trick for getting these is to switch off dynamic discretization in these nodes and insert the state for the data value you need to calculate the probability of).

H_2 : S dependent on D

This is similar to the previous case except we need to condition each calculation on the particular combination of parent variables we are considering:

$$P(r=19,n=151\,|\,H_2)=\sum_{q_{D-White}}\left[P(q_{D-White}\,|\,r=19,n=151,H_2)P(r=19,n=151\,|\,q_{D-White},H_2)\right]=0.068$$

$$P(r=0,n=9\,|\,H_2)=\sum_{q_{D-White}}\left[P(q_{D-White}\,|\,r=0,n=9,H_2)P(r=0,n=9\,|\,q_{D-White},H_2)\right]=0.315$$

$$P(r=11,n=63\,|\,H_2)=\sum_{q_{D-Black}}\left[P(q_{D-Black}\,|\,r=11,n=63,H_2)P(r=11,n=63\,|\,q_{D-Black},H_2)\right]=0.042$$

$$P(r=6,n=103\,|\,H_2)=\sum_{q_{D-Black}}\left[P(q_{D-Black}\,|\,r=6,n=103,H_2)P(r=6,n=103\,|\,q_{D-Black},H_2)\right]=0.052$$

The joint probability of the data is therefore:

$$P(r=19,n=151\,|\,H_2)P(r=0,n=9\,|\,H_2)P(r=11,n=63\,|\,H_2)P(r=6,n=103\,|\,H_2)$$

and this equals 4.67 E-5.

H_3 : S dependent on V

Again we condition appropriately:

$$P(r=19,n=151\,|\,H_3)=\sum_{q_{D-White}}\left[P(q_{D-White}\,|\,r=19,n=151,H_3)P(r=19,n=151\,|\,q_{D-White},H_3)\right]=0.067$$

$$P(r=11,n=63\,|\,H_3)=\sum_{q_{D-Black}}\left[P(q_{D-Black}\,|\,r=11,n=63,H_3)P(r=11,n=63\,|\,q_{D-Black},H_3)\right]=0.042$$

$$P(r=0,n=9\,|\,H_3)=\sum_{q_{D-White}}\left[P(q_{D-White}\,|\,r=0,n=9,H_3)P(r=0,n=9\,|\,q_{D-White},H_3)\right]=0.575$$

$$P(r=6,n=103\,|\,H_3)=\sum_{q_{D-Black}}\left[P(q_{D-Black}\,|\,r=6,n=103,H_3)P(r=6,n=103\,|\,q_{D-Black},H_3)\right]=0.119$$

The joint probability of the data is therefore:

$$P(r=19,n=151\,|\,H_3)P(r=0,n=9\,|\,H_3)P(r=11,n=63\,|\,H_3)P(r=6,n=103\,|\,H_3)$$

and this equals 1.93 E-4.

H_4 : S dependent on V and D

Again we need to consider all combinations:

$$P(r=19,n=151\,|\,H_4)$$

$$=\sum_{q_{V-White/D-White}}\left[P(q_{V-White/D-White}\,|\,r=19,n=151,H_4)P(r=19,n=151\,|\,q_{V-White/D-White},H_4)\right]$$

$$=0.068$$

$$P(r=11,n=63\,|\,H_4)$$

$$=\sum_{q_{V-Black/D-White}}\left[P(q_{V-Black/D-White}\,|\,r=11,n=63,H_4)P(r=11,n=63\,|\,q_{V-Black/D-White},H_4)\right]$$

$$=0.092$$

$$P(r=0,n=9\,|\,H_4)$$

$$=\sum_{q_{V-White/D-Black}}\left[P(q_{V-White/D-Black}\,|\,r=0,n=9,H_4)P(r=0,n=9\,|\,q_{V-White/D-Black},H_4)\right]$$

$$=0.525$$

$$P(r=6,n=103\,|\,H_4)$$

$$=\sum_{q_{V-Black/D-Black}}\left[P(q_{q_{V-Black/D-Black}}\,|\,r=6,n=103,H_4)P(r=6,n=103\,|\,q_{V-Black/D-Black},H_4)\right]$$

$$=0.115$$

The joint probability of the data is 3.78 E-4.

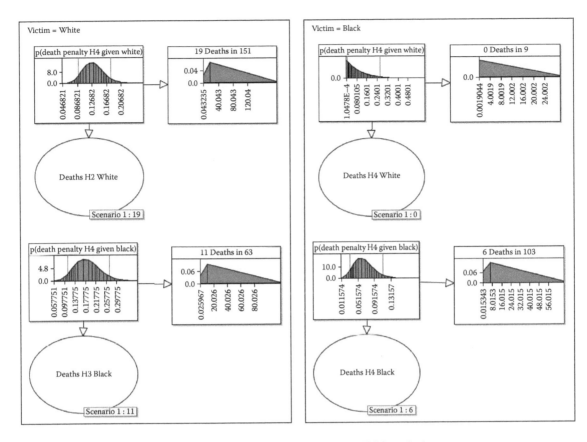

Figure 12.24 Data fitting and Bayesian score calculations for Example 12.9 hypothesis scores.

Figure 12.24 shows the results of the calculations in the AgenaRisk model.

The results for all of the hypotheses, after normalization, (by adding up the joint probabilities and dividing by the total to ensure they sum to one), are

$$P(D \mid H_1 : S \text{ independent of } V \text{ and } D) = 0.05$$

$$P(D \mid H_2 : S \text{ dependent on } D) = 0.05$$

$$P(D \mid H_3 : S \text{ dependent on } V) = 0.48$$

$$P(D \mid H_4 : S \text{ dependent on } V \text{ and } D) = 0.42$$

So given that H_3 has the greatest score we might conclude that only the victim's race is a significant causal factor in determining whether the death penalty was applied. Of course, there may be other factors that are unaccounted for, so we cannot conclude that race is the only factor. Neither can we conclude that the race factor is not a surrogate for any other unobserved factors.

We recommend you run the model and examine the sensitivity of the death penalty results given the different causal factors in order to get a feel for the results.

The U.S. Supreme Court has ruled that statistical evidence of bias in the sentencing is not sufficient, and that it is necessary to prove actual discrimination in the handling of the case. This may be sensible given the court has access to the relevant evidence but, in any case, the statistical evidence of a causal connection with race looks the stronger hypothesis of the four.

12.5   Confidence Intervals

There is a good reason why statistics textbooks do not properly explain what frequentist confidence intervals actually are. When you look at what they really are, as we do here, you will realize that they are somewhat irrational and may not correspond to what you thought they were.

Although confidence intervals are a standard technique that is taught and used in classical frequentist statistics for all kinds of data analysis and decision making, few people using them understand what they actually mean. This is because most standard texts provide a method for computing confidence intervals without actually explaining what they are.

We address this problem in Section 12.5.1 and then describe in Section 12.5.2 the natural Bayesian alternative to confidence intervals.

12.5.1   The Fallacy of Frequentist Confidence Intervals

You will all have seen or heard statements like the following from pollsters during an election campaign: "Support for candidate Joe Bloggs now stands at 43%. The margin of error is ±3%." But what exactly does the statement mean? Most people assume it means that the real level of support for Joe Bloggs must lie somewhere between 40% and 46% with 43% being most probable. But this is wrong, because there is always an unstated level of confidence about the margin of error. Typically, the level is assumed to be 95% or 99%. If pushed, a statistician would therefore expand the earlier statement to something like

- Statement A—Support for candidate Joe Bloggs now stands at 43%. The margin of error is ±3%, with confidence at the 95% level.

This combination of the margin of error and the level of confidence is what statisticians mean by a confidence interval. Unfortunately, even this more complete statement about the confidence interval is highly misleading. That is because most people incorrectly interpret the statement as being about probability, that is, they mistakenly assume it means something like

- Statement B—There is a 95% probability that support for candidate Joe Bloggs lies between 40% and 46%.

It turns out that confidence intervals, as in Statement A, are really rather complex to define and understand properly. If you look at standard statistical textbooks on the subject you will see what we mean. So we will now attempt a proper explanation that is as non-technical as possible. If you do not like the explanation (or cannot follow it) you do not need to worry too much because we will show afterward that there is a Bayesian alternative that is far simpler and more intuitive.

Statement B is a statement about the probability of an unknown population proportion, p (which in this case is the proportion of voters who support Bloggs). Most problems of statistical inference boil down to trying to find out such unknowns given observed data. However, this is where there is a fundamental difference between the frequentist approach and the Bayesian approach. Whereas a statement about the probability of an unknown value is natural for Bayesians, it is simply not allowed (because it has no meaning) in the frequentist approach. Instead, the frequentists use the confidence interval approach of Statement A, which is not a statement of probability in the sense of Statement B.

Being a standard tool of frequentist statisticians the confidence interval actually involves the idea of a repeated experiment, like selecting balls repeatedly from an urn. Suppose, for example, that an urn contains 100,000 balls each of which is either blue or white. We want to find out the percentage of white balls in the urn from a sample of size 100. The previous polling example is essentially the same—the equivalent of the urn is the set of all voters and the equivalent of a white ball is a voter who votes for Joe Bloggs. As we have seen, the frequentist approach to this problem is to imagine that we could repeat the sampling many times (that is, to determine what happens in the long run), each time counting the percentage of white balls and adding ±3 to create an interval. So imagine a long sequence of sample intervals (expressed as percentages):

[39–45], [41–46], [43–48], [44–49], [42–47], [41–46], [43–48],
[39–45], [38–44], [44–49], ...

The 95% confidence interval actually means that in the long run 95% of these intervals contain the population proportion, p.

Now while that is the technically correct definition of a confidence interval it does not shed any light on how statisticians actually calculate confidence intervals. After all, the whole point about taking a sample is

1. You don't know what p is. You want the sample to help you find out.
2. You can't afford to do long runs of samples. The whole point of a sample is that you only take as many as is economically feasible.

And this is where things get weird. It turns out that in order to turn your sample proportion into a confidence interval about the (unknown) population proportion, p, statisticians have to make certain kinds of assumptions about both the nature of the population and the value of the unknown p. This is weird because frequentists feel uncomfortable about the Bayesian approach precisely because of having to make similar kinds of prior assumptions.

Of course, both the size of the sample and the value of p will be heavily influenced by the confidence interval. As two extreme cases consider the following:

■ Case 1—The sample size is 100,000, that is, we sample every single ball. In this case the sample proportion, q, must be exactly equal to the population proportion, p. Hence, we can conclude that the population proportion is exactly $p = q$ (meaning p plus or minus zero) with confidence at the 100% level (since every time we repeat the procedure we will get this result).
■ Case 2—The sample size is one. In this case the sample proportion, q, must be either zero (when the sampled ball is blue) or one (when the sampled ball is white). So a sequence of sampled intervals (based on the plus or minus 3, ignoring numbers above 100 and below 0) might look like this when expressed using percentages:

[0–6), [0–6), (94–100], [0–6), (94–100], (94–100],
(94–100], [0–6), (94–100], ...

Unless p is either between 0 and 6 or between 94 and 100, and remember we do not know what it is, there are no confidence intervals that actually contain the population proportion, p.

■ Case 3—This is the same as case 2, but now we use plus or minus 50 for our intervals. Then we might sample these intervals:

[0–50), [0–50), [50–100], [0–50), [50–100], [50–100], [50–100], [0–50), [50–100], ...

Suppose you observe that 80% of these intervals are of the type [0–50). Then we would be able to conclude the following: "The proportion of white balls is 25% (since this is the midpoint of the interval [0–50)). The absolute margin of error is ±50%, with confidence at the 80% level."

This statement is not especially useful, since it suggests that unless we already have a good idea about what p is, we will need a large size sample to conclude anything helpful.

Each of these cases highlights a real dilemma regarding the use of confidence intervals. Although the formal definition assumes some notion of a long run of similar samples, such long runs are not intended to take place (that would otherwise defeat the whole objective of doing sampling). So what the frequentists have to do is come up with some analytical methods for predicting what might happen in the long run. For case 1 this is easy. Analytically we know that every sample of size 100,000 is guaranteed to produce a sample proportion, q, that is exactly equal to the population proportion, p. But what can we say if the sample is, say, 1000? Suppose that, in that sample, we find that the proportion is 30%. How do we turn this into a confidence interval?

This is where statisticians have to make certain kinds of assumptions about the nature of the population and the value of the unknown, p.

As explained in Box 12.8 (maximum likelihood estimation) with certain assumptions about the population and the sampling strategy, we can use maximum likelihood estimation to show that the most likely value of the population proportion, p, is actually equal to the sample proportion, q. For a sample of size n we can calculate the probability that the sample proportion, q, will be within some interval of p (for example, $p \pm 10\%$). Suppose that our sample size is 10 and that in our sample we found three white balls. Then, assuming the population proportion is $p = 0.3$, we can ask: For a sample of size 10 what is the probability that the sample proportion, q, is between 0.2 and 0.4? Or to put it another way, what is the probability of X, the number of positive samples, lying in the interval $X = [0.2n, 0.4n]$? This process turns out to be governed by the Binomial distribution that we introduced in Chapter 5.

Figure 12.25 shows the Binomial distribution for positive results, X, for four values of n: 10, 50, 100, and 1000. In Figure 12.25(a), when $n = 10$, it turns out that the probability the sample proportion will lie between 2 and 4 (inclusive), that is, equivalent to a margin of error of 20% to 40%, is approximately 0.70. So in the long run we expect that 70% of all samples of size 10 will have a sample proportion of between 0.2 and 0.4. Hence, if we observe 3 white balls from a sample of size 10,

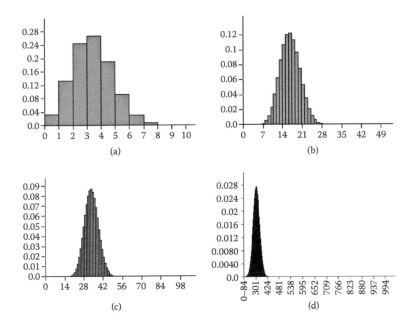

Figure 12.25 Binomial distribution for different values of n when $p = 0.3$. (a) $P(X \mid p = 0.3, n = 10)$. (b) $P(X \mid p = 0.3, n = 50)$. (c) $P(X \mid p = 0.3, n = 100)$. (d) $P(X \mid p = 0.3, n = 1000)$.

we can conclude the following: The proportion of white balls in the urn is 30% ± 10%, with confidence at the 70% level.

As we increase the sample size, n, (see Figure 12.25(b),(c),(d)) note how the variation in the number of successes, X, decreases. The effect of this is that if we found 300 white balls when the sample size $n = 1000$ then we can conclude the following: The proportion of white balls in the urn is between 27% and 33%, with confidence at the 95% level.

So, as the sample size n increases we can make stronger statements in the sense that the interval around the sample mean decreases while the confidence level increases. However, these statements are overly complex because they mix margin of sampling error and confidence in a way that defeats intuition.

12.5.2 The Bayesian Alternative to Confidence Intervals

So, as we have just discussed, whereas most people assume that a confidence interval is a statement about the probability that the unknown population proportion lies within certain bounds, it turns out not to be the case at all, but something much more complex.

Although frequentists cannot speak about the probability of unknown parameters, this notion is natural in the Bayesian approach. In particular a statement like Statement B ("There is a 95% probability that support for candidate Joe Bloggs lies between 40% and 46%") is exactly the kind of statement the Bayesian approach yields; such statements are much more meaningful and easier to understand than the previously discussed frequentist approach that was complicated and unnatural.

To explain the Bayesian approach we will consider the same example covered in the previous section (namely, the proportion of white balls in a bag containing only white and blue balls).

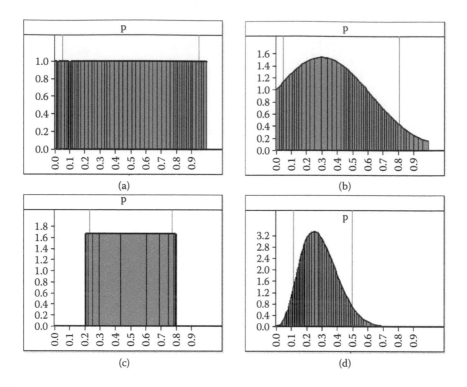

Figure 12.26 Different prior distributions for *p*. (a) Prior: Uniform across the range 0,1. (b) Prior: Truncated Normal with mean 0.3, variance 0.1. (c) Prior: Uniform in range 0.2 to 0.8. (d) Prior: Beta with alpha parameter 4, beta parameter 10.

In the Bayesian approach we start with some prior assumptions about the population proportion, which we express as a probability distribution. Examples of very different assumptions and their distributions are shown in Figure 12.26. For example, Figure 12.26(a) assumes that any value across the whole range is equally likely (this is the *Uniform*[0,100] distribution). Figure 12.26(b) assumes a truncated Normal distribution with a mean of 30 with a variance of 50. Figure 12.26(c) assumes a more constrained Uniform distribution, while Figure 12.26(d) assumes a nonsymmetric distribution (a Beta distribution). It does not matter so much here what these distributions are (they are covered elsewhere in the book) but we can readily see from their shape and where they are on the scale that they represent radically different states of knowledge about the proportion. Some are flat where any population proportion might be expected, while others "peak" around a particular value indicating a belief that this value is more likely.

In the Bayesian approach we simply revise the probability of the unknown population proportion from our observed data, using Bayes' theorem. This is very different from the frequentist approach because we are assuming the data is fixed and known, but the parameter of interest, in this case *p*, is unknown. The number observed is defined as a Binomial distribution where *n* is the sample size and *p* is the proportion.

We can implement the necessary BN model in AgenaRisk as shown in Figure 12.27. In Figure 12.27 we see the calculated posterior distribution

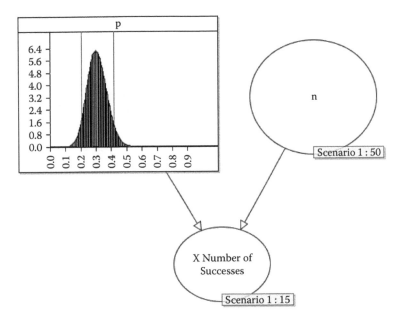

Figure 12.27 Unknown proportion, p, calculated using Bayes.

of the (unknown) population proportion, p, in the case where the prior for the population proportion is *Uniform*[0, 1] (i.e., where we know nothing) and where we have we entered data, $n = 50$ and $X = 15$.

The lines on the graph for p in Figure 12.27 show the 2.5% and 97.5% percentiles (see Box 12.12), so that 95% of the distribution lies between these lines. In this case the left-hand percentile is about 0.2 and the right hand about 0.4. This means that there is a 95% probability the value for p lies within 0.3 ± 0.1.

Box 12.12 Bayesian Confidence Intervals as Percentiles

In Box 12.1 earlier in the chapter we defined percentiles.

We can express a Bayesian confidence interval using percentiles $\alpha\%$ and $100{-}\alpha\%$. For example, the 95% confidence interval is equivalent to the distribution between the 2.5% percentile and the 97.5% percentile. Hence, in Figure 12.27 the 95% confidence interval is between $X_{2.5\%} = 0.2$ and $X_{97.5\%} = 0.4$.

AgenaRisk produces percentile values automatically.

The posterior distribution, for p, resulting from alternative priors and observations using (n, X) is shown in Figure 12.28, along with the corresponding 95% confidence interval in each case.

Some key things to note about these results are

- For the Uniform distribution assumption, shown in Figure 12.26(a),(b) the 95% region in each case is approximately the same as the confidence interval results using the same data as shown in Figure 12.25 for the frequentist case.

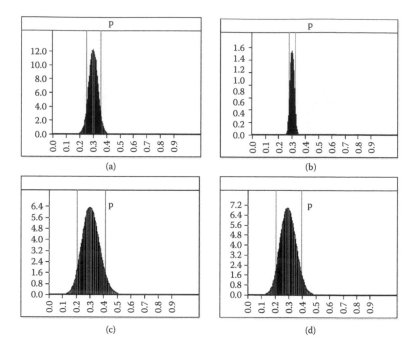

Figure 12.28 Predicted population proportion under different prior assumptions and observations. (a) $P(p \mid n = 200, X = 60)$ (with *Uniform*[0,1] prior). (b) $P(p \mid n = 1000, X = 300)$ (with *Uniform*[0,1] prior). (c) $P(p \mid n = 50, X = 15)$ (with *Normal*[30,50] prior). (d) $P(p \mid n = 50, X = 15)$ (with *Beta*(4,10] prior).

- As the amount of data (information) increases the 95% confidence region gets increasingly close to the mean average, so our confidence increases.
- There is greater power and expressiveness in the Bayesian approach where we can see the full distribution for the population parameter and can take account of very different prior assumptions.

12.6 Summary

The aim of this chapter was to present the Bayesian approach to hypothesis testing and compare it with classical statistical methods. We hope to have convinced you that hypothesis testing is elegantly handled within the Bayesian approach and that using Bayesian networks enables us to both do the computations easily and to incorporate features not possible in the classical approach. We showed how the Bayesian network approach provides more powerful and useful analyses when testing hypotheses about differences in populations. We showed how the Bayesian approach can handle distribution fitting and testing as well as offering model comparison as a better alternative to significance testing. We explained the problems with the classical approach to confidence intervals, showing the contrived way they are generated and interpreted.

Table 12.7

Summary of Key Differences in Approaches to Inference

Frequentist Approach to Inference	Subjective Approach to Inference
• Data are drawn from a distribution of known form but with an unknown parameter.	• Probability distributions are assigned to the unknown parameters.
• Often this distribution arises from explicit randomization.	• Inferences are conditional on the prior distribution and the observed data.
• Inference regards the data as random and the parameter as fixed.	• Inference regards the data as known (fixed) and the parameter as unknown (variable).

Hopefully you now appreciate that the Bayesian approach is simply better, as well as simpler. You should also be aware of the limitations of classical hypothesis testing in general and, in particular, Meehl's conjecture and its implications for data mining and scientific research in the main.

Table 12.7 summarizes the key differences between the frequentist and Bayesian (subjective) approaches to inference that we have covered in the chapter.

Further Reading

Gigerenzer, G. (2002). *Reckoning with Risk: Learning to Live with Uncertainty*, Penguin Books.

Meehl, P.E. (1978). Theoretical risks and tabular asterisks: Sir Karl, Sir Ronald, and the slow progress of soft psychology. *Journal of Consulting and Clinical Psychology*, 46, pp. 806–834.

Ziliak, S. T., and McCloskey, D. N. (2008). *The Cult of Statistical Significance*, University of Michigan Press.

13

Modeling Operational Risk

13.1   Introduction

Operational risk is the risk arising from an organization's business or operating functions. It covers people, processes, and systems that together or individually cause risk, whether these are external or internal to the organization.

Because operational risk is relevant to so many different sectors, different terminologies for the same underlying risk concepts have evolved in different key domains as shown in Table 13.1. In this chapter we will be discussing operational risk in all these domains and will use the terminology relevant to the domain.

The analysis of operational risk recognizes that risk events are not solely to be blamed on human fallibility but are supported by organizational features that fail to defend against all-too-human mistakes, slips, and (in the case of fraud or terrorism) malicious acts. Human error might be seen as an onset of a catastrophic risk event, but without latent weaknesses within the organization the event would not reach catastrophic proportions. From this we conclude that operational risk prediction is inextricably entwined with good management practices and that measurement of operational risk can only meaningfully be done if the effectiveness of organizational-specific risk management and control processes is regularly assessed and included in the modeling.

In this chapter we take a *systems* perspective to modeling risk and broadly consider a system to be a series of interconnected, interacting parts that together deliver some desired outcome, but which can also lead to unintended consequences. Thus, we do not consider systems to be solely hardware based but also consider people, process, culture, and environment to be crucial to understanding, explaining, and modeling risk. We can consider these systems to be *soft* to contrast with hard engineering-based or mechanistic systems where the focus is often strictly on the machine. This perspective is very helpful to frame the situations we wish to model, in terms of interfaces between processes, components, and organizations so that we can identify the boundary of our system and how it might affect other systems. Doing this also helps counter the potential bias of framing the problem so narrowly that we miss important aspects and underestimate (or overestimate) risk.

The term *operational risk* is primarily used to describe the financial or reputational risks faced by financial institutions, such as banks and insurance companies where the consequences are typically credit risk, market risk, and counterparty default risk. But operational risk problems are faced by all organizations. Situations involving high stakes, such as in the safety domain (including oil and gas, transport, and the nuclear sectors), are especially important candidates for operational risk analysis because some risk events may be catastrophic, involving harm, loss of life, or danger to the environment.

All models in this chapter are available to run in AgenaRisk, downloadable from www. agenarisk.com.

Table 13.1

Terminology in Different Key Domains of Operational Risk (Last Three Columns) Corresponding to the Generic Concepts Introduced in Chapter 3

Generic Risk Terminology as Introduced in Chapter 3	Financial Domain	Safety Domain	System Reliability Domain	Cyber Security Domain
Trigger	Risk	Fault	Fault	Vulnerability or attack
Risk event	Loss event	Hazard	Failure	Compromise
Control	Control	Control or defense	Control	Control or defense
Catastrophic risk event	Catastrophic loss	Accident	Crash	Breach
Mitigant	Mitigant	Mitigant	Mitigant	Mitigant
Consequence	Consequence	Consequence	Consequence	Consequence

There is a common, but misinformed, view in the banking sector that operational risk modeling solely involves the investigation of statistical phenomena. This view is synonymous with risk models that are purely data driven. In these situations management of operational risk within the banks' business processes are too often detached from the models developed to quantify risk.

Section 13.2 introduces the Swiss cheese accident model and uses it to explain how rare catastrophic risk events arise and how we might represent, explain, and quantify them. Section 13.3 introduces the idea of bow-tie modeling (in the safety domain) in the presence of identifiable hazards that threaten the system and make it vulnerable. We cover fault tree analysis (FTA) and event tree analysis (ETA) here and show how Bayesian networks (BNs) can be used to model these approaches to risk quantification and in doing so explain why BNs are so much more powerful and flexible. In Section 13.6 we cover soft systems, causal models, and risk arguments as a way of modeling operational risk in complex situations that demand a less granular approach to modeling and quantification. KUUUB factors are described in Section 13.7 as a way of embedding subjective judgments about the uncertainty behind our risk estimates to support the generation of operational models with long tail characteristics. Section 13.8 talks about operational risk in finance and discusses a BN modeling approach that captures the essence of the Swiss cheese model. This is used to present a rogue trading example. In Section 13.9 we talk about scenarios, stress testing, and quantifying value at risk (VaR). Finally, in Section 13.10 we apply the methods discussed in other areas of operational risk to cyber security risk modelling.

When combined, we postulate that these techniques can go some way to help model the kind of black swan events introduced in Chapter 2.

13.2 The Swiss Cheese Model for Rare Catastrophic Events

One of the great challenges in risk assessment is to estimate appropriately small, and hopefully reliable, probabilities for rare catastrophic events. As discussed in Chapter 2 the rarity of the event makes it unlikely that we can accumulate and apply sufficient data to help us determine the frequency of the event. We could simply ask an expert directly for his or her subjective probability estimate of the frequency of the event. But, as explained in Chapter 3, most experts cannot answer that question directly without offering a list of caveats and premises upon which their estimate is conditioned. The longer the list the less faith you might have in the reliability of the estimate and the less confident the expert will be in defending it. However, as we saw, this shows that the estimate (the number) is based on some causal theory by which the phenomena are explained and the event generated. This insight provides an opportunity for us to exploit this theory, articulated as a causal model, to arrive at numbers that we are perhaps more confident about (or can at least test against reality).

So how can we get a small, rare probability number based on a causal model? Well, the first thing to realize is that the causal model represents a number of connected processes that prevent, control, or mitigate the rare catastrophic event whose probability you wish to estimate.

Example 13.1

Suppose a driver wishes to quantify the probability that she will be involved in a head-on automobile collision in good weather conditions. She determines that a head-on collision must arise from the combination of a skid on a patch of oil, poor driver control, and the presence of an oncoming vehicle, whose driver might not able to take evasive action. Each of these events individually may have a relatively high probability but in combination their joint probability of occurring together will be small because of the product law of probability for independent events. Let's assume we can estimate these events reasonably confidently as

$$P(oil\ patch) = 0.01$$

$$P(skid \mid oil\ patch) = 0.10$$

$$P(poor\ control \mid skid) = 0.1$$

$$P(presence\ of\ vehicle) = 0.5$$

$$P(no\ evasive\ action) = 0.5$$

The probability of the head on collision is then simply the product of these probabilities:

$$P(collision)=0.01\times0.1\times0.5\times0.5=0.00025$$

Clearly getting to this number in one mental step would be difficult and error prone. By going through this exercise we might identify which constituent evaluations are more difficult for the expert. For example, we might be less certain of the oil patch probability. But we now have a causal narrative to help guide data collection from events that contribute to the accident but also are observable by other means. These take the form of "near misses," events that did not ultimately lead to the accident. So, for example, we might have observed oil patches on the road, routinely, and this gives rise to our estimate. Likewise, we know from driving in winter conditions that a skid is relatively hard to control, but that roughly 90% of the time you succeed. Also, the near misses happen more often than the accident, so paying attention to them helps in the estimation process (and also in other ways to help manage and control the risk).

This approach to modeling accident causation allows us to estimate rare events in a coherent and rational way. It was termed the *Swiss*

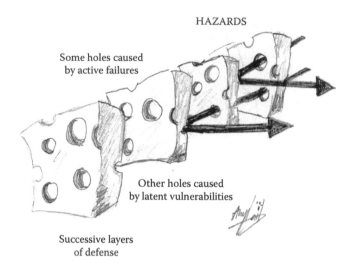

Figure 13.1   Swiss cheese model, accident successfully prevented.

cheese model by James Reason (1997). It is represented in Figure 13.1. The idea is that a complex system is defended by a number of layers, a so-called defense in depth, where the role of each layer is to prevent, control, inhibit, or mitigate a hazard. However, each defensive layer may contain holes that represent failures in the defense, perhaps caused by malign actions or latent vulnerabilities in the system. Providing that the holes in the cheese do not line up the hazard trajectories will be blocked by a successful defense.

However, should the holes line up as shown in Figure 13.2, a hazard trajectory might pass through all of the holes. In that case there is the potential for an accident. To determine whether an accident actually occurs we often need to consider factors that interact with the system under study. For example, an automobile on a railway line is a hazard, but it takes a train to arrive for an accident to occur.

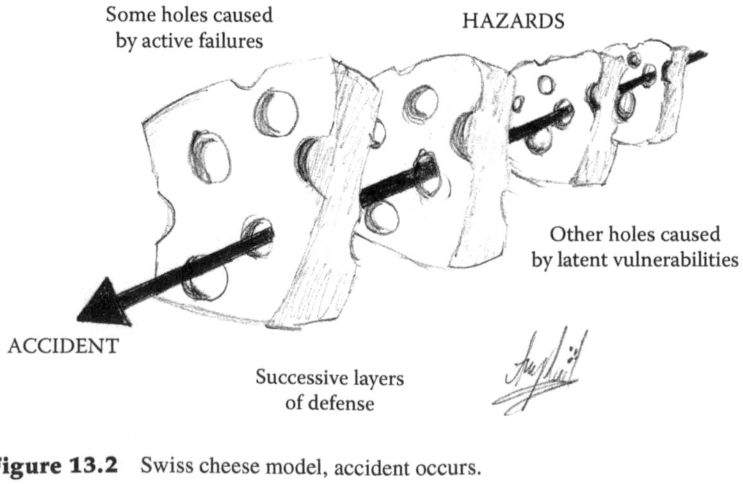

The Swiss cheese model has been used for decades in the safety domain especially in industries such as nuclear, aerospace, and petrochemicals.

Figure 13.2   Swiss cheese model, accident occurs.

Like all models the Swiss cheese model is a simplification: we do not always know where the holes are; they might move around; the layers of the cheese might not be independent as suggested. A good model of the system must be flexible enough to recognize and accommodate these factors when estimating the overall risks.

When coupled with Bayesian ideas, the Swiss cheese model provides a very flexible means of modeling risk in complex situations and systems.

13.3 Bow Ties and Hazards

Since hazards, in conjunction with other events, can cause accidents it is crucial in a safety-related system to be able to identify, analyze, control, and manage hazards. Clearly, the first step in any analysis is to identify the hazards; this can be a real challenge for novel or highly complex systems where we lack the experience or resources to pin down every reasonably possible way in which a system might fail. Doing this exhaustively for even modestly sized systems can be infeasible, so what we tend to do is analyze and define "equivalence classes" of hazards (and by implication faults and consequences), where each class contains hazards that, although possibly unique, are similar enough in terms of the properties they possess to consider them as equivalent. If we could not classify hazards in this way then all events would be considered unique and we could not learn and generalize anything.

The two most important ways of modeling safety and risk problems, involving hazards are

- Causal models that lead to the hazard, called *fault trees*.
- Consequence models contingent on the hazard having occurred, called *event trees*.

Other Related Techniques

HAZOP (hazard and operability)—This attempts to identify the hazards by systematically searching for possible malfunction by exhaustive use of keywords that might determine where things could go wrong.

FMECA (failure modes effects criticality analysis)—This looks at the modes of component failure in the system and assigns to each a simple frequency-severity ranking. However, it does not consider combined failures.

We saw informal examples of both of these models earlier in the book (event trees in Chapter 8 and fault trees in Chapter 9), while the causal approach to risk described in Chapter 3 informally combined both.

Other related, but less widely used, approaches are summarized in the sidebar.

Visually these two types of model come together in the "bow tie," as shown in Figure 13.3. The left part of the bow tie is a fault tree showing the causal mechanism by which the hazard might occur. The right part of the bow tie shows the consequence mechanism, modeled as an event tree. The "knot" in the bow is at the center and represents the hazard. Risk controls and inhibitors are involved on the left-hand side of the bow tie. Accident mitigation and recovery or repair actions would be represented on the right-hand side.

Recall that we might have many hazards in the Swiss cheese model (different holes in the cheese), caused by faults in multiple fault trees. This, combined with the fact that the consequences may be different for different hazards, suggests that for a reasonably large system we need to manage many fault trees and event trees for different hazards. Likewise, these will interact and change over time.

Both fault and event trees can be represented and manipulated as BNs, and there are many advantages of doing so. The perspective taken

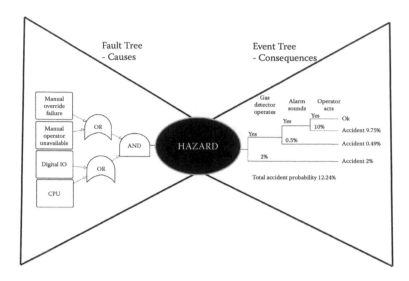

Figure 13.3   Bow tie model relating fault trees, the hazard, and event trees.

by the bow tie model assumes that causes and consequences are separated by hazards. This is artificial, but convenient fiction to make the statistical/probabilistic modeling easier for previous generations of tools. Components represented in the fault tree might also interact with events in the event tree: this kind of relationship can easily be accommodated in a BN. We discuss how we go about doing this in the next two sections.

13.4   Fault Tree Analysis (FTA)

Fault tree analysis (FTA) involves quantifying risk from knowledge of how risk events (faults or failures) in systems propagate to cause accidents or hazards. The idea is that the functional specification for a system can be decomposed into intermediate components/functions that should they fail, whether individually or together, would lead to the accident/hazard occurring. Fault trees (FTs) are therefore composed from events with relationships that connect the events together reflecting what we know about how faults are likely to interact and propagate. The tree reflects the design or expected use of the system.

Box 13.1 describes how an FTA for a system can be represented as a BN. It also describes the classic FT diagrammatic notation, where special shapes are adopted to visually indicate the type of Boolean logic under consideration.

Primary events in a fault tree are represented at the bottom of the tree. The top event is the hazard event, usually defined as systems failure.

The scenarios are those combinations of primary events that, together, would cause the top event. We use Boolean logic to describe the interactions for each scenarios, including AND, OR, NOT, and mFROMn, and simply express the logic gates as comparative statements involving Boolean logical tests. Because BN tools like AgenaRisk implement expressions for the Boolean operators it is simple to perform FTA analysis using a BN.

FTA is widely used in safety critical industries, such as nuclear power, chemical engineering, aerospace, space, and off-shore energy production.

The special shapes for nodes in a standard FT are called logic *gates*, reflecting their similarity with electronic circuits. A series circuit failing is equivalent to an *OR* Boolean; a parallel circuit failing is equivalent to an *AND* condition.

Box 13.1 Fault Tree Analysis (FTA) Represented as BN

Figure 13.4 shows an example fault tree for a system implemented as a BN. Figure 13.5 shows a classic fault tree, for comparative and reference purposes only; we make no further use of it from here (you can build these symbols by editing the node shape using AgenaRisk, should you wish).

In this example it is assumed a train is controlled by a computer. However, in the event of a computer system failure (meaning that either the computer or its power supply fails) there is a manual override that allows the (human) driver to take control. A top-level failure of this system occurs when both the computer system and the manual alternative fail. The computer fails if either its CPU unit or its digital I/O device fails. The power failure has some fault tolerance built-in in the form of three independent identical supply units. A power failure occurs only if at least two of these three supply units fail.

The comparative expressions for each of the (nonprimary event) nodes use the relevant Boolean operators, specifically (remember AND is && in AgenaRisk and OR is ||):

System failure:

```
if(Comp_System_Failure == "True" && Manual_backup_failure == "True", "True",
"False")
```

Computer system failure:

```
if(Power_Failure == "True" || Computer_Failure__OR == "True", "True", "False")
```

Computer failure:

```
if(CPU == "True" || Digital_I_O == "True", "True", "False")
```

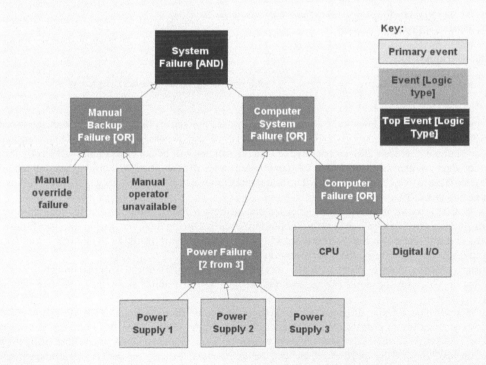

Figure 13.4 Fault tree represented as a BN.

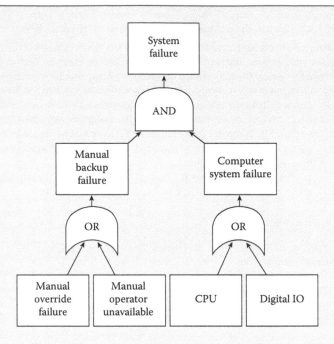

Figure 13.5 Fault tree using special logic gate symbols.

Power failure:

```
mfromn(2, 3, Power_Supply_1 == "True",Power_Supply_2 ==
"True",Power_Supply_3=="True")
```

Manual backup failure:

```
if(Manual_override_failure == "True" || Manual_operator_unavailable == "True",
"True", "False")
```

The prior failure probabilities for the primary event nodes are shown in the initial state of the model in Figure 13.6.

So, for example, there is a 20% probability the manual operator will be not be available sufficiently quickly to respond when the manual override kicks in (and there is a 5% probability it will fail to operate); the power supplies each have a 10% chance of failing. The initial model shows that the probability of a system failure during that period is 1.253%.

Let's look at how we might use the model for diagnostic reasoning. Suppose that the manual operator successfully intervenes after a computer system failure occurs. During an investigation into this failure it is found, for certain, that power supply 3 was working OK. What is the most likely cause then of the failure? We can find this out by entering the observations as shown in Figure 13.7.

By simply entering in the evidence we have, calculating, and then querying the marginal probability values it follows that the CPU was the most likely cause of failure, with 57% probability since this is the state with maximum probability.

Note that when performing diagnosis we are computing a maximum *a posteriori* (MAP) estimate, which is the posterior maximum probability state of each variable in the model given the evidence. For marginal probabilities of single variables this is easily done by eye, but care should be taken when considering the joint explanation involving two or more variables because the maximal values over the set of variables might be mutually exclusive.

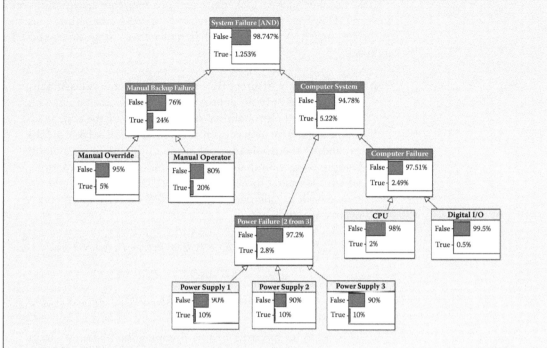

Figure 13.6 Initial state of fault tree model.

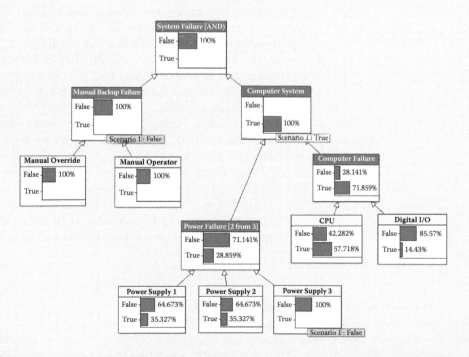

Figure 13.7 Investigation after computer system failure.

BNs subsume classic fault trees and are therefore inevitably more powerful. However, there are a number of specific compelling reasons for performing FTA in terms of BNs rather than using the classic FT approach:

- Calculations in discrete BNs are exact, whereas classic fault trees provide only an approximate method, called *cut sets*, to calculate the probability of occurrence of the top, intermediate, and primary events. This involves algebraically expanding the Boolean event space of the model, inevitably leading to a combinatorial explosion of event interactions, which then need to be truncated or "cut." For example, we might wish to calculate $P(A\ OR\ B\ OR\ C)$ from the primary events, which is

$$P(A\ OR\ B\ OR\ C) = P(A) + P(B) + P(C) - P(A\ AND\ B)$$
$$- P(B\ AND\ C) - P(A\ AND\ C)$$
$$+ P(A\ AND\ B\ AND\ C)$$

This could be approximated by dropping the "smallest" term: $P(A\ AND\ B\ AND\ C)$, leading to inaccuracy.

- Unlike classic FTs a BN can be computed in diagnostic as well as predictive mode. Therefore, given evidence of failure at the top or intermediate events, we can diagnose which of the primary, or other, events is the most likely cause of this failure. This is useful in fault finding and accident investigation.
- Classic FTA assumes a Boolean specification for the states of all variables but in a BN we need not. We can easily extend {Open, Closed} to {Open, Closed, Ajar} in a BN, resulting in a richer, more realistic model.
- Classic FTA assumes that the primary events are independent. This is seldom the case, especially in the presence of common causes of failure and also where components suffer from shared design faults. Box 13.2 shows and extends the FTA example of Box 13.1 with the addition of some logic to deal with a common cause and a shared design fault. We will show (in Chapter 14) that using BNs enables us to extend classic FTA into dynamic fault tree analysis by taking account of the time when assessing risk.

Box 13.2   Common Causes, Diversity, and Resilience

Figure 13.8 extends the BN implemented fault tree for the example in Box 13.1 with some additional nodes. In this case we assume that there is some common cause that might influence the manual override and the manual operator's action at the same time. This is represented by a new node called *Common cause*, which if true will make the manual override and operator fail together. However, should the common cause not occur

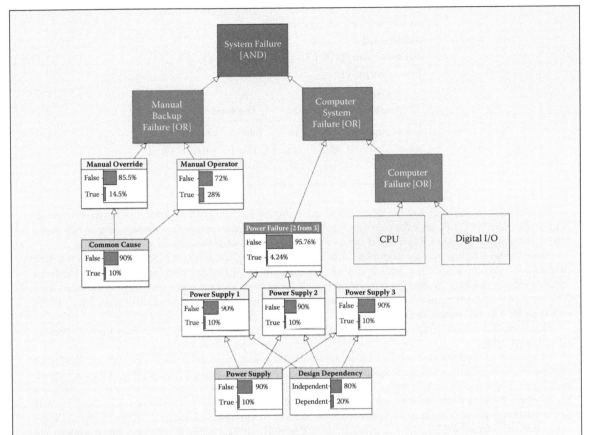

Figure 13.8 FT as BN with common cause and design dependency nodes.

the prior probabilities of failure for both of these events would be the same as in Box 13.1. This logic is shown by the new NPTs in Table 13.2 and Table 13.3.

We assumed before that the power failure has some fault tolerance built-in in the form of three independent identical supply units. However, what happens if they share a design fault (or some other dependency in how they operate) that means that 20% of the time when one fails they all fail? We can model this situation using two additional nodes. The first is a generic power supply node and the other a node called *Design dependency* with two states: dependent and independent. If the power supplies are dependent (i.e., identical) then should one fail all will fail together at the same time. If they are independent each can fail independently of each other. Under dependence we can assume we have nothing more than one generic power supply and under independence we have three. The NPT logic for each of the power supplies is now defined by Table 13.4.

Table 13.2

P(Manual override | Common cause)

Common Cause	False	True
False	0.95	0.0
True	0.05	1.0

Table 13.3

P(Manual operator | Common cause)

Common Cause	False	True
False	0.8	0.0
True	0.2	1.0

Table 13.4
P(Power supply X | Design dependency, Power Supply)

Design Dependency	Independent		Dependent	
Power Supply	False	True	False	True
False	0.9	0.9	1.0	0.0
True	0.1	0.1	0.0	1.0

If the designs are independent the reliability of power supply 1, 2, or 3 is independent of the generic power supply since the NPT column values are invariant. Thus, we can treat *Design dependency* as a switch in the model and assign a probability to it based on the extent to which the power supplies are diverse or not.

Given the prior beliefs that the common cause has a 10% probability of occurrence and that there is a 20% probability of design dependency there is now an increase in the marginal probabilities of failure for the intermediate events compared to those shown in Figure 13.6. The probability of power supply failure increases from 2.8% to 4.24%, and the manual override and manual operator failure probabilities increase from 5% to 14.5% and from 20% to 28%, respectively.

Therefore, if common causes and design dependencies are neglected in the FTA it is very likely to provide optimistic results.

This notion of diversity is crucial to achieving dependable complex systems and has been recognized since ancient times, especially when such "systems" were designed to protect against enemies. For instance, a medieval knight might design a castle with a high wall, spearmen to man the wall, a moat, an inner bailey, slit holes for bowmen, burning pitch, stone foundations to prevent undermining, and a portcullis. Each of these protects against different threats (hazards) and each might be defeated by appropriate countermeasures (failures). Some might fail at the same time because of a common cause (the wall and the moat both fail due to undermining) and some might fail because they share the same design feature (your bowmen and spearmen are prone to hunger under siege). No knight worth his keep would rely on one single means of defense alone. In fact this analogy carries directly across to areas like computer security and, to some extent, a modern analyst faces similar intellectual challenges as the noble knight. So, when faced with a variety of different hazards, we can use BNs to establish whether one system design is more resilient than others when faced by multiple threats.

13.5 Event Tree Analysis (ETA)

The analysis of accidents must consider both the state of the system and of its environment when the hazard event occurs. The analysis is made more difficult when the environment of a system is complex or variable.

Event trees are used in quantified risk analysis to analyze possible accidents occurring as consequences of hazard events in a system. As discussed, event trees are often used together with FTA, which analyzes the causes of the hazard event that initiates the accident sequence.

The most serious accident may be quite improbable, so an accurate assessment of the risk requires the probabilities of possible accident scenarios to be determined.

An example of an event tree is shown in Figure 13.9. Notice that the initiating event, the hazard on the left-hand side, is assumed to have happened and that all, or some, of the subsequent events are conditioned on this. In this example the hazard is a gas release. This may or may not be automatically detected. If it is then an alarm should sound and an operator would act to shut down the gas valve. If this fails we have an accident. Conditional probabilities are assigned to the branches

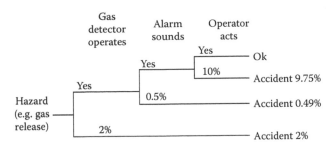

Total accident probability 12.24%

Figure 13.9   Event tree for a gas release accident.

in the event tree, such that we can calculate the marginal probability
of the different accidents as simply the sum of the joint probability of
the conditional events. So, in Figure 13.9 the probability of the accident
would be the sum of all of the different conditions that might lead to it,
and in this case it would be 0.0975 + 0.0049 + 0.02 = 0.1224.

Event trees model an accident as a strict sequence of conditional
events: this is an intuitive approach, but it fails to explicitly represent
the state of the system and its environment and how these influence the
causal sequence. Also, some events are not conditional on the hazard but
are dictated by other factors and this means the semantics of the event
tree break down. It is not possible to modularize the logic with an event
tree. However, we can address these limitations of event trees by using
BNs, since BNs can easily model the conditional event sequence as a
graph and the conditional probabilities as NPTs.

Box 13.3 compares an event tree with a BN for a train derailment
accident.

Box 13.3   Event Tree Analysis (ETA) Example Represented as a Bayesian Network

Let's adapt an existing event tree that models a train derailment in different circumstances. A train derailment
study investigated seven types of accidents that might occur as a consequence of a train derailing. These are
described in Table 13.5. The event tree using this derailment event set is shown in Figure 13.10.

Note also that conditional independence can be uncovered in the model when the conditional event probabilities are identical given any value of the preceding event. This is not always obvious but is a weakness in event
trees. For example,

$P(collision = no \mid falls = yes) = P(collision = no \mid falls = no) = 0.90$.

We can exploit this type of information when structuring the BN. Other events are clearly conditionally dependent, such as $P(hits\ structure \mid falls)$.

Using a BN these factors can be shown explicitly in a model of the possible accidents, which is still based on
the event-sequence intuition but can be used when the state of the system, or its environment at the time of the
accident, varies. The equivalent BN model for the event tree is shown in Figure 13.11.

Here the train hazard states, of which there are three, are obviously represented by mutually exclusive states
of the *Train state* node. We know that should a train fall on the adjacent side of the track this leads to a different
accident event sequence than from where it falls in the cess. This is clear from the BN but not from the event
tree. Likewise, we can use Boolean logic to combine events that are simple conjoins of previous events, such as
"carriages fall AND hit secondary structure." These are purely logical consequences and can be treated more
coherently as such in the BN (as is done in FTA).

Table 13.5
Derailment Event Set

	Event	Description
1	Derailment containment controls the train.	An extra raise containment rail, if fitted, limits the movement sideways.
2	The train maintains clearance.	The train remains within the lateral limits and does not overlap adjacent lines or obtrude beyond the edge of the track area.
3	Derails to cess or adjacent line (cess is the space between an outermost rail and the rail corridor boundary).	The train can derail to either side of the track, derailing to the cess, or outside, may lead to a collision with a structure beside the line, while derailing to the adjacent side brings a risk of colliding with another train.
4	One or more carriages fall on their side.	The carriages may remain upright or fall over.
5	Train hits a line side structure.	The train hits a structure beside the line.
6	The train structure collapses.	Collision with a line side structure causes the train structure to collapse.
7	Secondary collision with a passenger train.	A following or oncoming train collides with the derailed train.

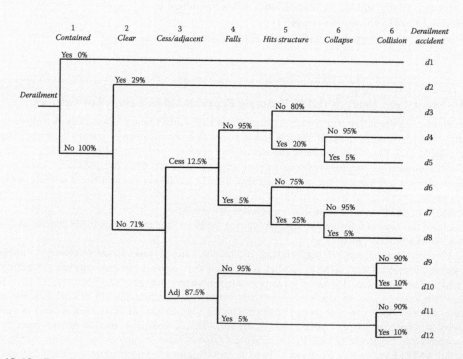

Figure 13.10 Event tree for derailment example.

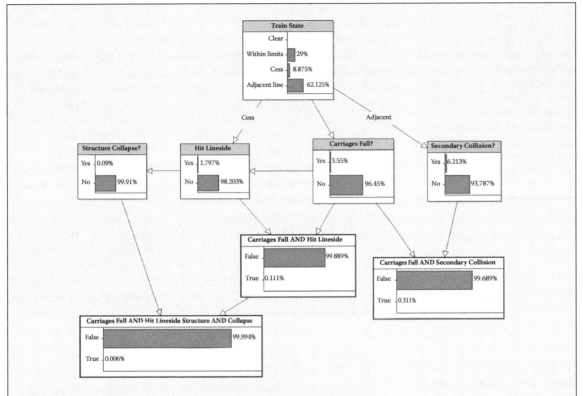

Figure 13.11 BN model representing event tree for derailment study.

However, a tricky issue does arise in the BN and it relates to the asymmetry problem previously encountered in Chapter 8. This is rooted in the fact that some variables only exist conditionally on previous events occurring and if those events fail to occur the probability of the variable states are irrelevant. In this example the events "carriages fall," "hit line-side structure," and "secondary collision" only occur if the train is in a certain state. For example, the train can only hit a line-side structure if it derailed to the cess. We therefore need to include "not applicable" states in the BN model, though in this example these are already represented by the "false" states, and repeat NPT entries to deal with these.

13.6 Soft Systems, Causal Models, and Risk Arguments

A complementary approach to a strictly causal analysis of a system or product is one that analyzes the risk or safety based on soft factors relating to how the system is designed, manufactured, or used. This approach implicitly recognizes key properties of systems that can make analysis of them difficult or intractable (as per the sidebar).

Taken at face value this list implies that it is hopeless to rely on causal explanations for risky phenomena and that attention to root causes is futile. We believe that this view is too extreme. Obviously, it would be difficult to exhaustively perform a fault tree or event tree analysis at a manageable level of granularity; but clearly it is not futile since all

What Makes Systems Analysis Difficult?

- The principles by which the component parts function are partially unknown or changeable.
- The parts are interdependent and the interdependencies might change over time.
- Processes are elaborate and human oriented.
- Teamwork involving people and machines is needed for success, rather than simply clockwork functioning.
- The presence of feedback.

For many systems a granular causal analysis, even as a fault tree, might be a daunting undertaking, especially when starting from scratch. In these circumstances it makes sense to model the high level interactions without worrying about the causal mechanisms, but with an eye to replacing these with causal descriptions over time, but only where it is deemed worthwhile.

systems of any reasonable complexity have these properties as a matter of degree. The pertinent question is whether causal analysis is worth the effort given the payback we might derive from it.

So, for a given complex system we might use a BN to articulate a higher level, less granular analysis. In doing so a *soft systems* approach to risk and safety modeling is necessary. We think of our models as arguments concerning the effects of the above on how a system might operate, rather than articulations of the hard systems themselves, in isolation, or indeed a granular description of the causal interaction of every part in the system.

The variables we might include in our soft systems analysis would typically involve processes, people, roles, incentives, protocols, procedures, and prevailing culture, as well as "hard" factors like systems, machines, and tools, and the interactions between all of these. It might still be difficult to articulate a causal understanding, and easily collectable metrics, but we need to model the situation all the same. To help we can agree to broad categories of failure, covering many separate and definable hazards and modes of failure, but where we only wish to articulate the general disposition or distribution of the hazards rather than model the details of each one.

Example 13.2   Assessing the Effect of Black Box Components

The BN in Figure 13.12, like all the BNs in this book, is available in the AgenaRisk model library. You can see all the NPTs and run the model.

Consider the following problem: You are in charge of a critical system, such as a transport system or a nuclear installation. The system is made up of many components that you buy as black boxes from different suppliers. When you need a new type of component you invite a dozen suppliers to tender. If you are lucky you might be able to get some independent test results or even operational test data on the components supplied. Your task is to accept or reject a component. One of your key acceptance criteria is the *safety* of the component. This might, for example, be measured in terms of the predicted number of safety-related failures that the component can cause in a 10-year lifespan when integrated into your system. How do you make your decision and justify it?

This is a classic risk assessment problem in which you have to come up with a quantified figure by somehow combining evidence of very different types. The evidence might range from subjective judgments about the quality of the supplier and component complexity, through to more objective data like the number of defects discovered in independent testing. In some situations you might have extensive historical data about previous similar components, whereas in other cases you will have none. Your trust in the accuracy of any test data will depend on your trust in the providence of the testers. Having little or no test data at all will not absolve your responsibility from making a decision and having to justify it. A decision based only on gut feel will generally be unacceptable and, in any case, disastrous in the event of subsequent safety incidents with all the legal ramifications that follow.

The BN in Figure 13.12 is a simplified version of a model that we have used to solve exactly this kind of problem. The square nodes represent variables that we might expect to know for a given component. The "System safety" node will never be directly observed, but the node "Number of faults found in test" might be. Similarly, we might find it

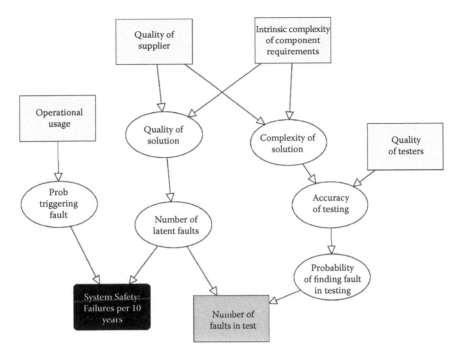

Figure 13.12 BN for risk argument for black box components.

difficult to model the quality of the solution or the complexity but we might infer it from what we know about the supplier and our intrinsic knowledge of the complexity of the component's requirements. Thus, where we lack information we supplement what we have by refining our model and regressing down the causal chain adding more variables, but with perhaps less precision. The guiding principle is that is better to be in the right ballpark than precisely wrong.

Example 13.3   Telecommunications Project Design Risk

A manufacturer and supplier of telecommunications equipment wished to assess the risk of new designs as early as possible in the systems development and use life cycle. A key element in their thinking was that often the design verification and validation work (V&V) was inadequate, defects were slipping through and being discovered by customers (lack of containment to the design phase), and designs were inherently risky in terms of architecture novelty and scope change in functional specification during the project. Likewise, mitigation measures to reduce the effects of poor design were ineffective because the right people were not involved early enough in the process. In sum, they were looking to use soft, second-order factors to analyze the risk of the design process, as well as the riskiness of the risk analysis being done within the process. Figure 13.13 shows the resulting BN for this part of the risk assessment process and Figure 13.14 shows the overall model architecture for other sources of risk including:

- Component risk—Failures originating from components contained on the unit. Indicative of poor component sourcing processes.

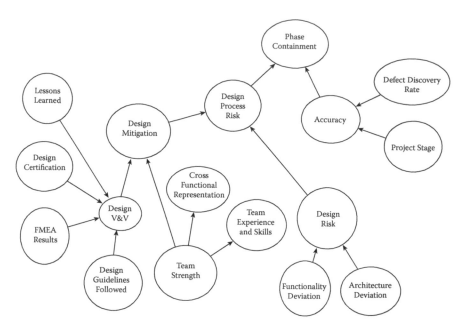

Figure 13.13   BN for telecommunications project design risk.

- Manufacturing process risks—Failures resulting from poor manufacturing processes.
- Damage from transport and use—Units damaged during installation or use. This indicates problems in instruction, training, and support.
- Failure misdiagnosis risk—Units returned wrongly as being faulty but where no faults subsequently were found (NFF, no fault found).
- Field reliability assessment—Actual field return rates used to update the reliability estimate of the system, or to make predictions thereof.

The model in Figure 13.14 was used to anticipate problems during a project before they occurred, detect wayward projects at inception, and act as a means to determine the risk of a portfolio of projects across the company, identifying extreme high-risk projects in this portfolio that might ruin the company.

Typically these models are developed from an analysis of the importance and connectivity of factors in the situation at hand rather than with explicit use of the idiom framework, presented in Chapter 8, to help structure the model and make sense of the semantics of the relationships and nodes. In terms of node types such models tend to be labeled or ranked types with use of rank node functions or KUUUB factors (see next section) to model the uncertainties.

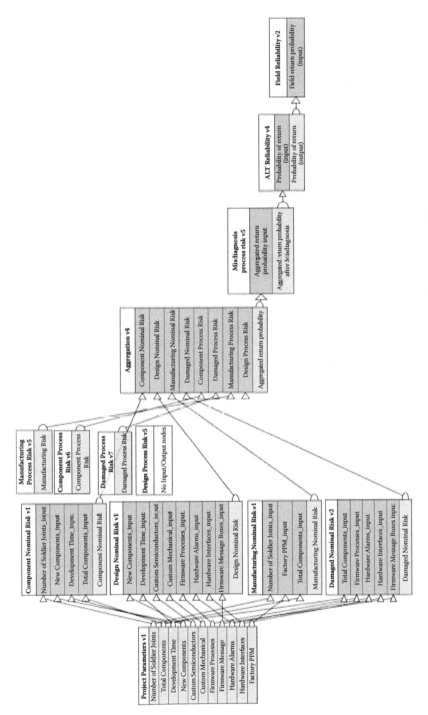

Figure 13.14 BN architecture, composed of multiple interacting risk objects, for telecommunications company risk assessment.

13.7 KUUUB Factors

The term KUUUB analysis was coined by Chapman and Ward (2000) and the acronym stands for the analysis of

The key idea is being able to make a delta change assessment (which might be based on historical data of the situation) to reflect factors not already incorporated into the estimate we have.

- *K*nown *U*nknowns
- *U*nknown *U*nknowns
- *B*ias

We can use KUUUB analysis to adjust a quantitative estimate to take account of key risks that might affect, positively or negatively, the estimate at hand.

Typically, we will consider a range of scenarios:

- Status quo—The estimate remains unchanged because the situation is unchanged
- Degradation—The estimate is directionally biased or less/more variable than it should be because the situation has degraded.
- Improvement—The estimate is biased but in the opposite direction from degradation.

We can then produce a new estimate using a simple conditionally deterministic expression, adjusting the original estimate, E, by the delta change, Δ, to give a KUUUB adjusted estimate, K:

$$K = \Delta E$$

The function Δ is usually a statistical function for the different scenarios of degradation or improvement in the situation, some of which may be more or less likely. Many will be assessed using expert judgment alone, simply because they may reflect rare conditions. The following example shows how we might use the KUUUB factor in practice as a statistical mixture model, where the low probability "extremes" are mixed with the high probability "normal" events.

Example 13.4 Adjusting a Financial Loss Estimate

An organization has suffered a historical operating financial loss over the preceding years and has an estimate of those losses, E, in millions of dollars:

$$E \sim TNormal(50,200,0,200)$$

(a truncated Normal distribution over the range 0 to 200, with mean 50 and variance 200).

However, they are required to project next year's profit or loss based on different scenarios of likely external market and internal operating conditions. These are reflected in three key risk indicator (KRI) measures, which are described on a seven-point scale from major improvement to major degradation with "no change" in between. As shown in Figure 13.15, these are used to determine a *Change* variable, defined as a ranked node, with the same scale (the NPT for this node is simply a

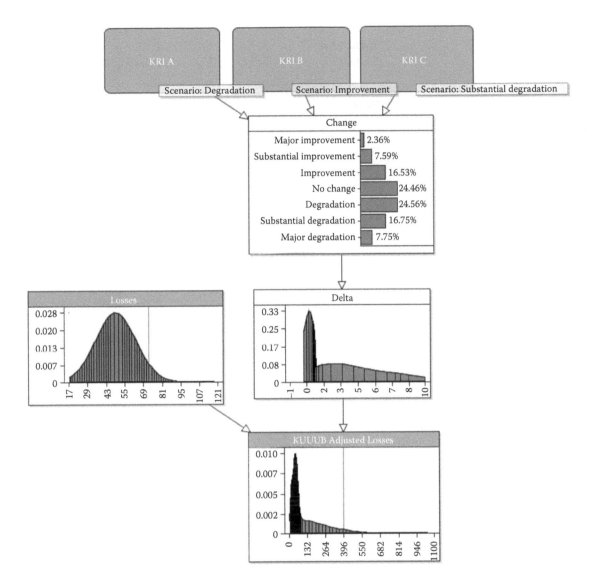

Figure 13.15 KUUUB model example, adjusting a financial loss estimate.

weighted mean as described in Chapter 9). The challenge is to map this ranked node into an associated continuous scale for the delta node needed for the KUUUB analysis. Specifically, the delta node is a child of the *Change* node. The NPT for delta is defined as a *TNormal*(mean, variance, lower bound, upper bound) distribution where the parameters are conditioned on the states of the *Change* node as shown in Table 13.6. Obviously when the *Change* state is "no change," the distribution for the delta collapses to a single constant value of one.

In the BN model for this example in Figure 13.15 we have entered observations for each of the three KRI's, which results in a slight overall degradation. You can see that the resulting KUUUB adjusted profit or loss distribution, *K*, has a long tail with a 95th percentile value of $416 m, much larger than the $200 m upper limit assumed in the loss estimate, *E*, thus showing how the extremes have been grafted onto the original estimate. The median loss, however, has only increased to $64 m.

Table 13.6
Delta Parameter Values Conditioned on Change Factor

Change factor	Mean	Variance	Lower	Upper
Major improvement	0.1	0.2	0.1	1
Substantial improvement	0.5	0.1	0.1	1
Improvement	0.7	0.1	0.1	1
No change	1	0	1	1
Degradation	2	4	1	10
Substantial degradation	5	4	1	10
Major degradation	8	2	1	10

13.8 Operational Risk in Finance

The Basel Committee on Banking Supervision, in reaction to a number of well-publicised financial disasters (e.g. Barings bank 1995, Daiwa bank 1995 and Allied Irish Bank 2002), has mandated a system of regulation addressing the issue of operational risk and its assessment and management. Key to the regulatory process is the need for financial institutions to model their operational risk, in terms of a variety of loss event types (including unauthorised trading) in order to effectively manage risk and establish an appropriate regulatory capital charge. Similar regulations govern the European Insurance Industry in the form of Solvency II

A key regulatory requirement for financial organisations is that they manage and model their operational risks. Such risks are defined in terms of a variety of loss event types which may be classified as attritional losses (the everyday costs of doing business) and catastrophic, albeit rare, events such as the losses resulting from rogue and unauthorised financial trading.

Based on what has been practiced for more than half a century, in safety critical industries, modeling for the analysis of operational risk always should strive to support decisions by providing answers to the following questions:

- Is the risk high or low; is it acceptable?
- Which causal factors are most critical?
- What are the differences in risk considering alternative solutions?
- What risk-reducing effect can be achieved considering implementation of alternative risk-reducing measures?

It is considered fundamental in operational risk management to be able to create quantitative risk models that constitute a sound basis for fruitful discussions on organization-specific risk.

Thus, detailed causal modeling at the business process level is required for the model to absorb organization specific input, highlight the criticality of causal factors, identify the potential lack of controls/barriers, and quantify the overall risk level in terms of expected and unexpected losses. Such a model obviously creates more value than a risk model based solely on actuarial techniques.

13.8.1 Modeling the Operational Loss Generation Process

Inspired by the Swiss cheese model, let's first look at the process that might generate an operational loss. This is the "loss event model," which models the potential loss events and how these dynamically evolve over time as they are influenced by controls embedded within the business process. This is a dynamic time-based evolution of an event given the controls exerted at discrete points in time. The performance of each control is modeled as a function of a set of operational failure modes. These failure modes are in turn influenced by a set of causal factors, which in isolation or combined initiate the operational failure. Dependence between operational failure modes for different controls is modeled through dependency factors.

Thus, an operational failure mode may be modeled as a function of both causal factors and dependency factors. Operational failures that have no dependency relationship with other failure modes are simply modeled as a function of causal factors. The business process is then represented by a sequence of discrete time-dependent events such that we have a BN as shown in Figure 13.16.

Each of the state transitions $P(E_t \mid E_{t-1}, C_t)$ is governed by an NPT that models the transition probability of the loss event from an undetected (unsafe) to a detected state, dependent on the control state. Should the control variable be operating correctly at time t, the loss event at E_{t-1} would transit to a correct operating state at E_t using the logical conditional transition probabilities shown in Table 13.7.

It is easy to see how we could generalize this to cope with more loss event states, including those showing a severity scale.

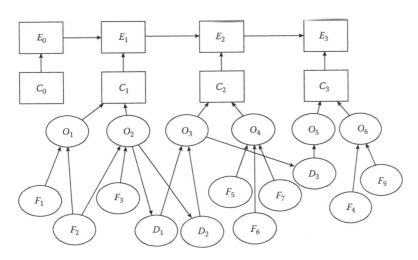

In Figure 13.16, E nodes are loss events, C nodes are controls, O nodes are operational failure modes, F nodes are causal factors, and D nodes are dependency factors.

Figure 13.16 BN for loss event model.

Table 13.7
Logical Conditional Transition Probabilities

E_{t-1}	fail		OK	
C_t	fail	OK	fail	OK
E_t = fail	1	0	0	0
E_t = OK	0	1	1	1

Monte Carlo Simulation Models

Monte Carlo simulation is a commonly used statistical approach in which the distribution you are trying to predict is generated by randomly sampling inputs whose probability distributions are known (or assumed). For example, if $Z = X + Y$ where $X \sim N(20, 100)$ $y \sim N(30, 50)$ then we randomly generate pairs (x, y) from the distributions $N(20, 100)$ and $N(30, 50)$ respectively and plot the resulting values $x + y$.

So, if the initiating loss event has not occurred, or has been detected by a previous control at a previous time, the question of whether this or later controls operate correctly become irrelevant. Thus, the chance of detection and transition to a correct (safe) operating state is determined by the number of implemented controls as well as their individual performance (reliability).

For simplicity, we have chosen to represent all control, operational failure, causal factor, and dependency factor variables using Boolean nodes, for which we can model $P(C_t \mid \mathbf{O}_{C_t})$ and $P(O_j \mid F_{O_j}, D_{O_j})$ using Boolean logic to represent the type and form of failure, that is, FTA logic. This modeling choice allows use of Boolean operators, such as AND, OR, XOR, and NOT, within a discrete BN model. The NPT for node C_1 with parents O_1 and O_2 is then generated according to Table 13.7. Alternatively, we can manually declare an NPT to represent the form of dependencies that best matches the process by assigning probabilities directly.

The presented methodology is by no means restricted to the use of Boolean variables. One of the benefits of using BNs is that the modeler may declare customized variables and state spaces to fit the problem domain. Choice of variable types and state space is ultimately left to the modeler developing a model best suited for the process being analyzed.

In contrast to Monte Carlo simulation models (see sidebar), which are commonly used, this model provides the following advantages:

- All computations are exact so it is more suitable for calculating ultra-low probability events.
- Can deal with multiple, interacting, common causes.
- In addition to strong mathematical foundations the graph representation of the BN model maps neatly onto the underlying process and clearly shows how losses propagate.
- We can generate a set of loss event probabilities for different severities of events at different stages of propagation through the process and use this to calculate our severity and loss models.

Note that some clarification of what we mean by "causes" and how we identify them is needed here since there is much understandable dispute and confusion of how these terms might be productively applied to operational risk and other problems. We recognize that there are an infinite number of actual and possible interactions between unique events and that it is, therefore, impossible to represent these "atomic level" interactions in any model. Instead, we seek to model broad classes of events that have characteristics, behavior, and outcomes that are similar enough to be considered homogenous and which can reasonably be expected to interact in a similar way (see sidebar for an example). Given this, any distinction between classes of causal events is a practical decision as much informed by process understanding, documentation, data collection standards, and budget as any refined philosophical position.

In a practical situation we might class operational information technology (IT) failure as a single causal class covering security lapses, systems failures, and performance degradation simply because the consequential effect on a control that depends on the IT infrastructure would be similar in all cases.

Example 13.5   Rogue Trading

One of the major risks identified within the trading process is that of unauthorized trading. Unauthorized trading is essentially trading outside limits specified by the bank and the result is, simply put, an overexposure to market risk.

The process of trading can be described by the following steps: (1) trade request, (2) conduct trade, (3) registration of trade, (4) reconciliation check, (5) settlement and netting, and (6) entering of trade on trading books. Any bank involved in trading also conducts continuous monitoring of results and positions, that is, monitoring of market risk exposure.

Controls are implemented to prevent the traders (or other employees) from conducting trades that violate established regulations, limits, and contracts. Specifically:

1. There are controls in the front office intended to monitor and restrict the trading activity so that it is kept within the limits set by the bank.
2. Once the trade has been processed and entered in the trading system by front-office personnel, a number of consecutive checks are performed by the back office.
3. The middle office performs continuous positions and results monitoring, including independent price checks and calculation of market risk.
4. There is the control of periodical or random audits of the trading operation.

Any trade conducted is assumed to be in one of a finite set of mutually exclusive states corresponding to potential loss events, E_i. Each of the implemented controls performs a check that may change or confirm (i.e., the state remains as before) the state of the trade given its true state. We define the following loss event variables:

E_0: State of trade when entered in the trading system by the trader, C_0

E_1: Loss event discovered during reconciliation check, C_1

E_2: Loss event discovered during market risk check, C_2

E_3: Loss event discovered by audit, C_3, or which has escaped the control system

The respective states assigned to each variable are:

E_0 = {Authorized, Accidental, Illegal Fame, Illegal fraud}

E_1 = {Discovered, OK, Accidental, Illegal Fame, Illegal fraud}

E_2 = {Discovered, OK, Accidental, Illegal Fame, Illegal fraud}

E_3 = {Discovered, OK, Accidental, Illegal Fame, Illegal fraud}

The meanings of each of the states are as follows:

- "Authorized/OK" trade—An authorized trade permitted and approved by management beforehand.
- "Accidental"—An unauthorized trade that is accidental due to mistakes by the trader, that is, a trade that was not intended to be unauthorized and, of course, not intended to incur losses for either the bank or its clients.
- "Illegal Fame" trade—An unauthorized trade with the intent to further the trader's career, status within the bank, or size of bonuses. For this category of trades the trader intends to make money for the bank and subsequently also provide benefits for himself. The important aspect is that the trade is not intended to directly benefit the trader but indirectly through providing success for the bank.
- "Illegal Fraud" trade—An unauthorized trade with the sole intent of benefiting the perpetrator. If the bank or anyone else suffers losses as a result of the trade it is irrelevant as long as the perpetrator makes money off the trade.
- "Discovered"—A trade of any unauthorized category is revealed by a control and is thus a discovered unauthorized trade.

At the start of the process the loss event state is unknown to us. As the trade progresses through the controls process and checks are performed, the state of the trade may change to "discovered" if an "unauthorized" trade is uncovered, otherwise it will remain in the "OK" state as the trade is assumed authorized. There is also a possibility that an "unauthorized" trade will not be discovered and continue to be in an unknown unauthorized state and escape the control system and our model of the process. Such loss events typically have higher severity and their impact felt much later when subsequent frauds are discovered and forensic analysis throws up unauthorized past trades. We may also use escaped unauthorized trades to assess the severity of loss where severity is assumed dependent on the occurrence of previous unauthorized trades (this is more of a research topic, beyond the scope of the example). The described process can be modeled as a BN and the corresponding loss event BN model is presented in Figure 13.17.

In Figure 13.17 each control is influenced by a series of operational failure nodes where each operational failure influences the performance of one or more controls. Here the operational failures are particular to each control with the exception of the operational failure "Active Disruption" (AD), which is shared by all of the controls, except the front office. However, since the controls are normally separated from each other, different actions, and alliances are needed by the trader to disable a complete set of controls. Thus, we have included an active disruption node, O_{ADt}, for each control.

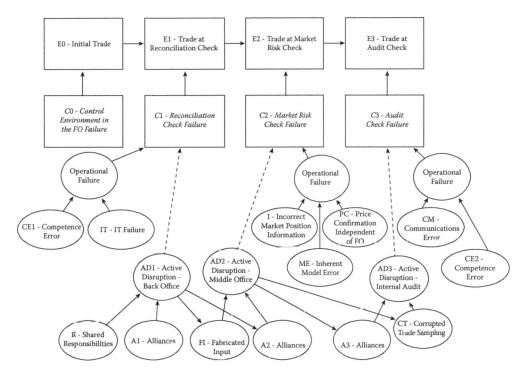

Figure 13.17 BN loss event model for rogue trading process.

The probability of active disruption in one control is dependent on whether there has been active disruption in the preceding control, assuming a trader who has successfully disrupted one control is likely to attempt to disrupt the next. Hence, given active disruption in the back office, there is an increased probability of active disruption in the remaining controls. Such dependency is modeled using dependency factors.

In the case of active disruption the dependency factors influencing controls failure include:

- Failed segregation of duties (between front and back office)
- Fabricated input data (in the positions monitoring)
- Corrupted trade sampling (in the audit process)
- Alliances between staff, within and without, the institution

The fact that this operational failure affects all back-office, middle-office, and audit controls means a rogue trader could negate or reduce the effectiveness of all controls across the board. Also, modeling the active disruption of controls as suggested enables the analyst to account for the complexity in the deliberate acts to negate or reduce the efficiency of controls. In contrast with active disruption, each of the other controls are modeled using causal factors, such as competence failure, IT failure, and position/pricing mistakes.

We have assigned probability values to each of the causal factors as well as the front-office control for two scenarios, as shown in Table 13.8. The first scenario represents status quo on a trading floor, whereas the second represents the case where the control environment in the front office has failed and there is active disruption in the back office.

Dependency factors are an important feature and a clear advantage of the BN approach is being able to account for common causes of failure that undermine the whole process.

Table 13.8
Failure Probabilities for Causal Factors in Two Scenarios

Variable (true = "failed")	Probabilities for Scenario 1	Probabilities for Scenario 2
$P(C_0 = true)$	0.01	1.0
$P(F_{CE1} = true)$	0.05	0.005
$P(F_{IT} = true)$	0.01	0.001
$P(F_R = true)$	0.05	1.0
$P(F_{A1} = true)$	0.01	1.0
$P(F_1 = true)$	0.01	0.001
$P(F_{PC} = true)$	0.05	0.005
$P(F_{ME} = true)$	0.01	0.001
$P(F_{CM} = true)$	0.01	0.001
$P(F_{CE2} = true)$	0.01	0.001

For the dependency factors resulting in active disruption in the middle office and the internal audit, which are also dependent on previous active disruption, $D_{O_{ADt}} \mid O_{ADt}$, the assigned probabilities are:

$$P(D_{A2} = true \mid O_{AD2} = false) = 0.001$$
$$P(D_{A2} = true \mid O_{AD2} = true) = 0.7$$
$$P(D_{FI} = true \mid O_{AD2} = false) = 0.001$$
$$P(D_{FI} = true \mid O_{AD2} = true) = 0.8$$
$$P(D_{A3} = true \mid O_{AD3} = false) = 0.001$$
$$P(D_{A3} = true \mid O_{AD3} = true) = 0.6$$
$$P(D_{CT} = true \mid O_{AD3} = false) = 0.001$$
$$P(D_{CT} = true \mid O_{AD3} = true) = 0.6$$

NPT normalisation

AgenaRisk will automatically normalise the column of values that you put in an NPT to ensure they sum to 1. So, if you specify a table such as:

———
99000
1000
100
1
———

then this will automatically be normalised to the first column shown in Table 13.9. Similarly, the second column of Table 13.9 is generated from:

———
99000
1000
2000
100
———

Note that the probability of occurrence of the dependency factors is assumed to increase dramatically when there is active disruption in the preceding control.

For the initial event state $E_0 \mid C_0$, the NPT is shown in Table 13.9 (see the sidebar for an explanation of why the values appear as they do). Notice that, when front-office controls are working the odds for "illegal fame"

Table 13.9
NPT Probabilities for $P(E_0 \mid C_0)$ (note the table values are not entered manually as specified here but are normalised as explained in the sidebar)

	C_0 = false	C_0 = true
E_0 = Authorised	0.98900111	0.96964
E_0 = Accidental	0.00998991	0.00979
E_0 = Illegal Fame	0.00099899	0.01959
E_0 = Illegal Fraud	0.00000999	0.00098

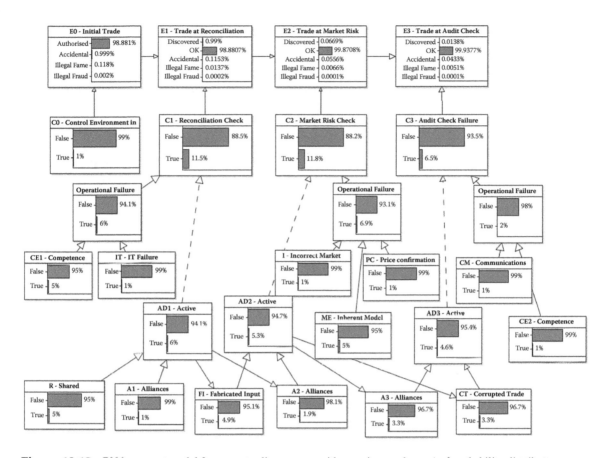

Figure 13.18 BN loss event model for rogue trading process with superimposed marginal probability distributions.

and "fraud" events are in the region of 1:1000 for "illegal fame" and 1:000000 for "illegal fraud". This rises to 1:100 and 1:10000, respectively, when controls have failed.

When we execute the BN, inference is carried out on all variables and marginal distributions produced as shown in Figure 13.18 for the first scenario.

From Figure 13.18 we obtain the loss event probabilities for scenario 1, and these are listed in Table 13.10. These loss probabilities give the

Table 13.10
Loss Event Probabilities Computed for Two Scenarios

Loss event	Probabilities for Scenario 1	Probabilities for Scenario 2
Discovered at Reconciliation check	0.0099002	0.0
Discovered at Market Risk check	0.0006688	0.0039578
Discovered at Audit check	0.0001384	0.0043812
Undiscovered Accidental	0.0004328	0.0071043
Undiscovered Fame	0.0000513	0.0142086
Undiscovered Fraud	0.0000009	0.0007104

probability that a single trade will belong to that particular loss class. From these we can calculate that the "escape" probability for a loss, that is, the probability the trade is unauthorized and evades all of the controls, is 0.00049 in scenario 1 and 0.0220233 in scenario 2. The probability of a trade being unauthorized and discovered in scenario 1 is 0.0107074 and 0.008339 in scenario 2.

13.8.2 Scenarios and Stress Testing

A major challenge in operational risk is stress testing the situation to determine the effects of credible worst-case scenarios that might affect the viability of the operation or indeed the organization. In practice these could be stress tests of the assumptions underlying the business and how variable these might be during periods of stress, whether this stress originates internally (say, from IT failure or controls vulnerabilities) or externally from changes in the market (such as financial market instability).

How do we identify stress tests and determine their consequences? First, we have to identify (hopefully) rare but possible situations where one or more hazardous events occur jointly. This is the Swiss cheese model again, involving the identification of events (holes in the cheese) that, should they coincide or align, will lead to failure. Identifying these requires imagination and judgment, and hence subjective probabilities and valuations, for both the prior probabilities of these events and the likely financial or other consequences.

Armed with a stress-testing scenario we are then interested in the distribution of operational losses arising from the scenario *in extremis*. This involves calculating some high percentile value to represent the worst-case outcome and typically regulations characterize this as the *Value at Risk* (VaR), which will typically be the 99, 99.5, or 99.9 percentile of this loss distribution (i.e., 1 in 100, 1 in 200 or 1 in 1000). An example operational loss distribution is shown in Figure 13.19, which shows two classes of loss: expected losses, involving the "cost of doing business" losses in normal time; and the unexpected losses that might occur in stressful times. One might expect the median loss value to be representative of normal operating conditions and the VaR value, in

Notice that the loss distribution in Figure 13.19 is unimodal with a single peak and a long tail. Most research on operational risk and stress testing focusing on the fatness of this tail since the VaR summary statistic will heavily depend on this. However, there is no reason to assume that the loss distribution actually follows this shape in practice, except for naturally occurring risk phenomena, such as hurricanes and floods. For societal risks one might reasonably expect "lumpy" multimodal distributions reflecting discrete, regime-changing behavior where the internal or external stresses switch on or off rather than reflect a continuous change or drift.

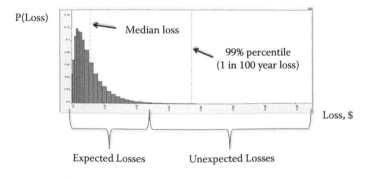

Figure 13.19   Loss distribution with 99% VaR and median losses.

this case 99%, to be the stress loss, such as the loss one would expect every 100 years.

To model the operational risk distribution in a BN we simply need to identify a series of interrelated events, their prior probabilities, and the consequences of their occurrence, usually in financial terms. Typically we will assign one financial loss distribution to each discrete event in our stress scenario set and then sum these loss distributions to arrive at a total loss distribution. Calculating this sum involves the application of Loss Distribution Analysis (LDA), risk aggregation, and compound summation as described in Chapter 10 Section 10.6.

Example 13.6 Stress Testing

We consider a scenario comprising of two events: market crash (M), and a rogue trading R event (R). Each of these is judged to have a percentage probability of occurring of 10% and 3% respectively. Each of these events is represented by a Boolean node in the BN with "true" probabilities (i.e. probability of occurring) judged to be 10% and 3% respectively. We also associate with each a child (simulation) node representing the loss distributions node L_M and L_R respectively. For L_M this is defined as:

$$P(L_M \mid M = False) \sim N(10,100)$$

$$P(L_M \mid M = True) \sim N(500,10000)$$

So, under normal trading conditions, we might expect a loss distribution average of $10 m with a variance of $100 m. Under stressful conditions the mean losses rise to $500 m but the variance increases by two orders of magnitude, reflecting market unpredictability, to $10,000 m.

The loss distribution for L_R is defined as:

$$P(L_R \mid R = False) \sim 0$$

$$P(L_R \mid R = True) \sim N(250,10000)$$

So under normal conditions there would be zero losses since there would be no rogue trading events. However, should a rogue trading event occur we might believe that mean losses rise to $250 m with a variance of $10,000 m.

For this stress test the BN model is shown in Figure 13.20. The 99.5% VaR for the market crash event is $667 m and for the rogue trading event it is $347 m. The combined 99.5% VaR is $687 m.

The big assumption in this stress test is that the consequential losses are independent, that is, there is no interaction between market crash events and rogue trading events. This may be optimistic given that in periods of market turmoil rogue traders may be subject to larger swings in position and the resulting greater losses. We next investigate the effects of this causal interaction on the model.

In the joint case the rogue trader and market crash events occur at the same time. To consider this we need a new variable called R AND M for this joint event, which will have the single events as parents and be true

Note that we cannot simply add the VaR values since the sum of VaRs is not guaranteed to be the same as the VaR of the sum of the distributions.

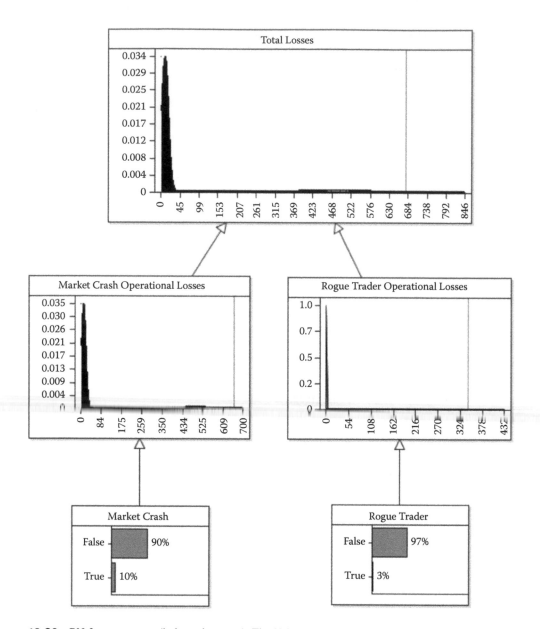

Figure 13.20 BN for stress tests (independent case). The VaR values are shown by the vertical line on each marginal distribution.

only if both parents are true. The loss distribution for this joint event is defined as:

$$P(L_{R\ \text{AND}\ M} \mid R\ \text{AND}\ M = False) \sim 0$$

$$P(L_{R\ \text{AND}\ M} \mid R\ \text{AND}\ M = True) \sim N(750,10000)$$

So the joint losses for rogue trading during a market crash are higher with a mean of $750 m. The resulting BN model is shown in Figure 13.21.

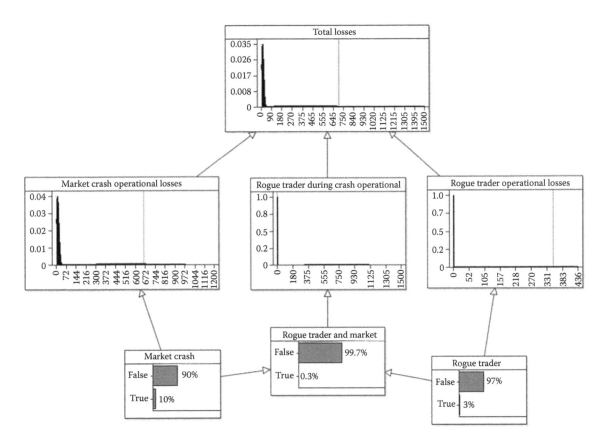

Figure 13.21   BN for stress tests (dependent case). The VaR values are shown by the vertical line on each marginal distribution.

The 99.5% VaR for the market crash event remains at $667 m and for the rogue trading event it is $347 m. The combined event, R AND M, 99.5% VaR is $0 (at first glance this looks very surprising but the probability of the joint event not occurring is 99.7%, which means there is an 0.997 chance of zero losses and 0.997 > 0.995).

The VaR for the total operational losses in the dependent case is $704 m. This is greater than in the independent case, which was $687 m, reflecting the additional risk of a jointly occurring rare event (the probability of rogue trading and market share occurring together was 0.3%).

As the probabilities of the stress test events increase the VaR will increase nonlinearly in the dependent case, as is shown in Table 13.11.

Table 13.11
Stress Test Results for Varying
Priors for Market Crash

$P(M)$	$P(R$ AND $M)$	99.5% VaR ($M)
0.1	0.003	704
0.2	0.006	1327
0.3	0.009	1475

13.9 Cyber Security Risk Modelling

Cyber security analysis involves the modelling of vulnerabilities in an organisation's information infrastructure that might be maliciously attacked, compromised and exploited by external or internal agents. An information infrastructure generally contains physical assets (networks, servers, mobile phones, etc.), people, processes and procedures that might contain security "holes," bugs, unpatched/un-updated systems and other features that might present themselves as vulnerabilities.

For instance, an email might be sent to a government security establishment disguised as an official email with a hyper link purporting to be a bona fide document, whereas it is, in fact, linked to downloadable malware which, when executed, attempts to infiltrate the security establishment's network. If successful this malware might then exploit other vulnerabilities, bypass controls and degrade the system, leak confidential data and download additional malware. Other attacks might focus on the behaviour of an individual's mobile phone, or home PC and use other paths and mechanisms to target that person in a specific personalised way to get access to the information infrastructure of interest.

Clearly, cyber security modelling exhibits many of the features common to those risks we have analysed in other areas of operational risk. Specifically, in cyber security analysis we would focus on the following objectives when using BNs for modelling purposes:

- Improve technical and economic decision support, by:
  - Allowing simulation of future attacks and defences
  - Supporting diagnosis of unknowns for knowns
  - Providing means for cost–benefit analysis
  - Aligning notations with system of systems descriptions
  - Defining set of attackers, vulnerabilities, control and their properties and relationships
  - Recognising trust boundaries and scope
- Support improved modelling, by:
  - Providing reusable template models
  - Enabling specialization from taxonomies
  - Providing means for refinement of abstract or concrete entities
  - Handling of recursion (BNs link to other BNs in a "chain")

The properties of BNs are highly applicable in cyber security modelling given that there is often an absence of information/data, but there may be access to relevant subject matter opinion. Likewise, data on known cyber attacks might be shared or reported in the public domain and these data can be used to perform "what if" and scenario modelling. However, cyber attacks and their consequences are readily amenable to causal interpretation and analysis because the semantics of the language used to describe cyber risk reveals the following attributes and entities that can be used in a BN to model cyber risks events:

- Entity—A logical or physical part of the system required for normal service (e.g. organisation, process, actor, system, sub-system, function)
- Attack—An attempt to misuse an entity by exploiting an entity's vulnerability
- Capability—The wherewithal to carry out an attack on a specific entity's vulnerability
- Vulnerability—A weakness in an entity that, when exploited by an attacker, leads to a compromise
- Control—A function designed to deter, prevent or protect against attack or mitigate a consequence of an attack (negates vulnerability)
- Compromise—An entity whose actions/use is controlled by an attacker that acts as a pre-condition enabling attacks on other entities or leads to a loss
- Breach—Actual loss of availability, confidentiality or integrity (i.e. compromise at systems boundary)

A popular approach for cyber risk analysis are "kill chains." Kill chains attempt to model the propagation of events, consequent to attacks, through a cyber security infrastructure as a chain of events. Clearly, it doesn't take much imagination to replace "chain" with graph and we replace kill chains with BNs in such a way that we can now model more complex dependencies and also quantify the propagation probabilities and conditional dependencies between events in the BN.

Figure 13.22 shows an example kill chain transformed into a kill graph (a BN). There is a superimposed set of levels commonly used in cyber security modelling to provide labels that organise the various steps in the attack, from reconnaissance, delivery of an attack and compromise, C & C (Command and Control) entities compromised "Action on objectives" (exploitation of data, systems, etc.) and subsequent movement to other parts of the information infrastructure for consequent exploitation.

The probabilities in the kill chain can represent either deterministic events (e.g. a successful infiltration will always result in an attempt to move laterally) or a probabilistic event subject to the effects of controls, mitigations and lack of attacker capabilities. In this way we could add additional layers to the kill graph to represent the effectiveness of controls, attacker capabilities and our expectations about the presence or absence of vulnerabilities in our system.

Finally, to bridge the gap between the kill chain analysis and a financial analysis of the risks presented, we can apply the kind of loss aggregation and loss distribution approach (LDA) described in Chapter 10, Section 10.6 to compute the financial loss that might occur for sets of cyber security loss event, in terms of frequency and severity of loss. This would then provide the basis for decision making about cyber security investment, trade-offs and insurance.

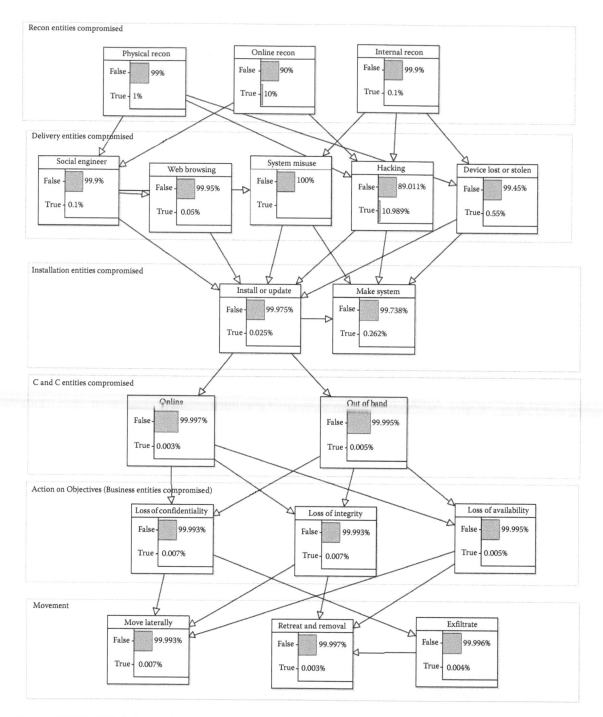

Figure 13.22 Kill chain translated into kill graph (BN).

13.10 Summary

In this chapter we set out to model operational risk for any organization and focused on how we can use BNs for safety and financial risks. In doing so we hope we have convinced you that the quantification of operational risk must take a systems perspective if we are to model how risks are produced and evolve. The role of risk prevention and detection in the form of controls and mitigation strategies should be clear as should the vulnerabilities that might undermine those controls. Clearly, systems vulnerabilities might be systematic or triggered by common causes that affect many controls, thus endangering the whole system. We have argued that BNs provide useful, accurate, and reliable means for modeling such complex dependencies in a realistic way and can replace a plethora of existing techniques, such as FTA, ETA, and VaR analysis that have developed separately across different modeling traditions. Together these methods go some way, when coupled with insight and imagination, to helping identify and quantify black swan events.

Further Reading

Bobbio, A., Portinale, L., Minichino, M., and Ciancamerla, E. (2001). Improving the analysis of dependable systems by mapping fault trees into Bayesian networks. *Reliability Engineering and System Safety* 71(3), 249–260.

Chapman, C., and Ward, S. (2000). Estimation and evaluation of uncertainty: A minimalist first pass approach. *International Journal of Project Management* 18, 369–383.

Fenton, N. E., and Neil, M. (2006). Expert elicitation for reliable system design. *Statistical Science* 21(4), 451–453.

Khodakerami, V., Fenton, N. E., and Neil, M. (2007). Project scheduling: Improved approach to incorporating uncertainty using Bayesian networks. *Project Management Journal* 38(2), 39–49.

Langseth, H., Nielsen, T. D., Rumí, R., and Salmerón, A. (2009). Inference in hybrid Bayesian networks. *Reliability Engineering and System Safety* 94, 1499–1509.

Marsh, D. W. R., and Bearfield, G. (2008). Generalizing event trees using Bayesian networks. *Proceedings of the Institution of Mechanical Engineers, Part O: Journal of Risk and Reliability* 222(2), 105–114.

Neil, M., Fenton, N. E., Forey, S., and Harris, R. (2001). Using Bayesian belief networks to predict the reliability of military vehicles. *IEE Computing and Control Engineering* 12(1), 2001, 11–20.

Neil, M., Fenton, N., and Tailor, M. (2005). Using Bayesian networks to model expected and unexpected operational losses. *Risk Analysis Journal* 25(4), 963–972.

Neil, M., and Hager, D. (2009). Modeling operational risk in financial institutions using hybrid dynamic Bayesian networks. *Journal of Operational Risk* 4(1).

Neil, M., Malcolm, B., and Shaw, R. (2003). *Modelling an air traffic control environment using Bayesian Belief networks*. Presented at the 21st International System Safety Conference, August 4–8, Ottawa, Ontario, Canada.

Reason, J. (1997). *Managing the Risks of Organisational Accidents*. Ashgate.

14

Systems Reliability Modeling

14.1 Introduction

Traditionally, assessing the reliability of a complex system has been based on considering physical failures of hardware components. Although this remains an important aspect of reliability assessment, it is becoming increasingly important to also consider the impact of design faults. Design faults are deviations from expected or intended requirements, and they have the potential to trigger failures in operation and can dominate overall reliability.

In this chapter we outline ways of assessing the reliability of complex systems in terms of both their component failure behavior and also in terms of the extent to which they contain design defects. We show how Bayesian networks (BNs) can be used to model systems' reliability, from failure data collected during test or operation, to predict future systems' reliability or to diagnose faults in such systems. Moreover, we also show how we can take account of the structure of the system in the analysis, especially the fault tolerance, redundancy, and other reliability enhancing methods used in the design.

We distinguish two broadly different scenarios when assessing a system's reliability:

1. *Discrete (probability of failure on demand)*—In this scenario we are interested in reliability associated with a finite set of uses of the system. For example, if the system is a military aircraft carrying out a number of missions during a conflict, then we would be interested in the probability of failure for a given mission. This scenario is discussed in Section 14.2.
2. *Continuous (time to failure)*—In this scenario we are interested in reliability associated with a system operating in continuous time. For example, for a military aircraft we would be interested in the number of miles it could fly before a failure occurs. This scenario is discussed in Section 14.3.

Section 14.4 gives an example of system fault monitoring (in real time or near real time) using a dynamic Bayesian network (DBN) to monitor, diagnose failures in, or control a working system.

The boundaries between hardware and software are becoming increasingly blurred in the complex mechatronic (mechanical, electronic, electrical, and software) and embedded systems used today.

The literature on reliability modeling is too large to do it justice here, so we focus on examples of using BNs for reliability modeling for a limited number of commonly encountered scenarios, including dynamic fault trees (DFTs), a more recent innovation.

All models in this chapter are available to run in AgenaRisk, downloadable from www.agenarisk.com.

In Section 14.5 we discuss the role of defect modeling, with an emphasis on defects in software, but given that software is the embodiment of design, the ideas covered here are just as relevant to all design artifacts regardless of their physical embodiment.

14.2   Probability of Failure on Demand for Discrete Use Systems

A basic model for probability of failure on demand (pfd) for discrete use systems is shown in Figure 14.1.

We wish to estimate the pfd from some test data. This will be in the form of an observed number of failures, f, in an observed number, n, of trials (demands on the system).

Although the pfd is the unknown distribution we are trying to learn, whatever it is we can assume the number of failures is a Binomial distribution, since each discrete demand can be considered a Bernoulli trial, with either success or failure as a result. Hence, it makes sense to define the NPT for *number of failures* as the distribution Binomial(n,*pfd*).

As discussed in Chapter 10 this popular approach to estimating the possibility of failure on demand (pfd) is called the *Beta-Binomial formulation* (because we use a Beta prior for the probability density function [pdf] and a Binomial for the resulting number of failures).

Of course we still need to specify an appropriate prior distribution for the pfd, using expert judgment or by comparison with previous similar systems. A very flexible approach is to use a Beta distribution, $Beta(\alpha, \beta, 0, 1)$, where we choose values for (α, β) that reflect our experience and disposition about the prior, as described in Chapter 10. If you have previously seen a number of demands then you should set α to be the number of failures and β to be the number of nonfailures. If we are ignorant about the prior, then we simply set both α and β to be equal to one, which is equivalent to a Uniform distribution.

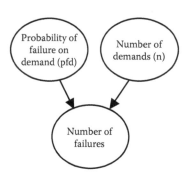

Figure 14.1 Basic BN model for probability of failure on demand (pfd).

If we had used maximum likelihood estimation (as explained in Chapter 12) as our inference method our result for the failed and successful tests would have been zero and one, respectively, for the pfd. Clearly, these results fail to reflect the uncertainty remaining after so few tests.

Example 14.1   Determining Reliability from Observing Failures in *n* Demands

We can use the model in Figure 14.1 to answer questions like:

1. If we observe f failures in n demands what is the pfd?
2. How reliable is the system given that we have observed zero failures in n demands? (This is often the question of interest for safety-critical or ultrareliable applications such as for nuclear or aviation applications.)

Early in development one test is done on the system and it succeeded. From this what can we conclude about its reliability? When we enter these observations into the BN and calculate we get the result shown in Figure 14.2(a) where the mean value for the pfd is one-third. So we cannot assume very much about the pfd from this single success. Similarly, for one failed test, we get an equally uninformative result, as shown in Figure 14.2(b) where the mean value for the pfd is two-thirds.

In contrast, turning to our second question, had we observed zero failures in one thousand tests, as depicted in Figure 14.2(c), we would be much more confident that the reliability is high (the mean pfd here is 0.001).

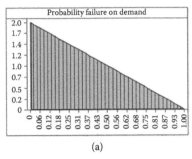 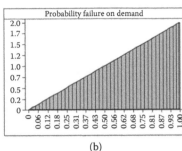 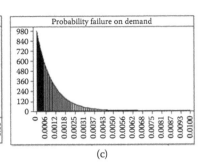

(a)   (b)   (c)

Figure 14.2   "Posterior *pfd*". (a) After successful test. (b) After one failed test. (c) After 1000 successful tests.

Example 14.2   Deciding When to Stop Testing

In addition to estimating the pfd for a complex system from test data, we also want to determine whether the system meets a reliability target and how much testing we need to do to meet some level of confidence in the system's reliability. Suppose, for example, that our target reliability requirement is a pfd no greater than 0.01, i.e. once per hundred demands. The BN model for this is shown in Figure 14.3. It is exactly the same as that in Figure 14.1 but with an additional Boolean node *pfd requirement* whose NPT is specified as

$$\text{if}(pfd < 0.01, \text{``True''}, \text{``False''})$$

that is, it is "True" if it fails less than the required pfd (0.01 in this case, but of course, this value can be changed to any other pfd requirement). We can use this model to answer the following type of question:

How much testing (i.e., how many demands) would be needed for us to be, say, 90% confident that the pfd is less than 0.01, assuming zero failures observed? This is the sort of question that might be asked by the quality assurance or test manager, and is known as a "stopping rule" as it can be used to determine when to stop testing.

Note that in this case we have both a pdf requirement target t (0.01 in this case) and a confidence level α (0.9 or 90% in this case).

We can use the sensitivity analysis feature in AgenaRisk (see Sidebar) to calculate the result as shown in Figure 14.4, which shows that 229 successful tests are needed to meet the confidence target of 90%.

Example 14.3   What Is the Probability That a System Will Not Fail at All in the Next *n* Demands?

While Example 14.2 demonstrated how to calculate the number of failure free demands needed to meet a particular reliability requirement with a given confidence level, for many safety-critical systems, such as nuclear reactors and civilian aircraft, the expectation is that the system will not fail at all. While we know that, in practice, such perfection is impossible the designers of such systems nevertheless would like to know the probability of observing 0 failures in the next *n* demands. The BN model of

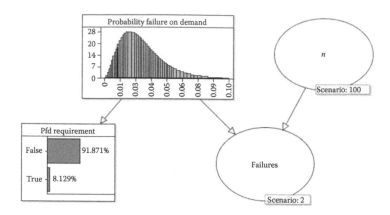

Figure 14.3 Reliability model with pdf requirement, showing results after observing two failures in 100 demands (in this example required pfd < 0.01).

Figure 14.3 can easily be adapted to answer this question as shown in Figure 14.5. In fact, this BN can be used to answer the general question:

> If a system has failed f times in n_1 demands what is the probability it will not fail in the next n_2 demands?

Of special interest is the classic reliability assessment problem for high-integrity systems considered by Littlewood & Strigini (1993) in which $f = 0$ and $n_2 = n_1$. For example, suppose we have observed 10^6 failure-free demands. What is the probability that the next 10^6 demands will be failure free?

Littlewood & Strigini (1993) proved the result (which was surprising to many) that the answer is 0.5. In fact, by using the model and entering $f_1 = 0$, $n_2 = n_1 = 10^6$ (or any other large number), you will see that indeed the result is 0.5. You can also inspect the resulting statistics for the nodes pfd and f and note that their means are respectively 10^{-6} and 1. Moreover, you can also use the model to confirm another result proved by Littlewood & Strigini (1993):

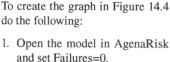

To create the graph in Figure 14.4 do the following:

1. Open the model in AgenaRisk and set Failures=0.
2. Select the sensitivity analysis tool.
3. Select "pfd requirement" as the Target node and "n" as Sensitivity node.
4. Select both "Table" and "Tornado Graph" in Sensitivity Report options.
5. After selecting "Run" you will get html output showing a large table with the False and True values for "pfd requirement" corresponding to increasing values of n. You can simply copy and paste this table into Excel and then create an x-y line plot where the x-axis is the column with the n values and the y-axis is the column with the pfd requirement True values.

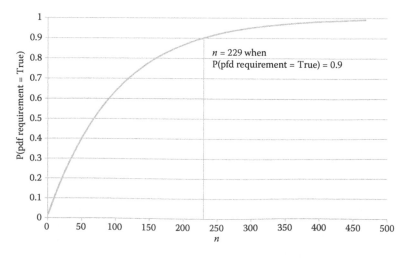

Figure 14.4 Sensitivity of $P(pfd\ requirement = True \mid f = 0, n)$.

In order to get to 90% confidence that there will be no failures in the next n demands, you will need to have observed $10n$ failure-free demands previously (i.e. an order of magnitude greater number).

In other words, for there to be a 90% chance that a system will be failure free in 10^n demands we will need to have seen a failure-free period of 10^{n+1} demands.

To confirm this result, for any n enter the values $10n$ and n, respectively, into nodes $n1$ and $n2$.

These results are considered to be problematic for systems with very high reliability requirements. For example, any new aircraft has a 10^{-9} pfd requirement in order to be "certified" for commercial use. That means less than 1 failure is expected per 10^9 flights. What the above work demonstrates is that we would need to first observe 10^9 failure-free test flights (an impossible task) in order to get to the required mean pfd. And if we wanted to be 90% confident of no failure in the first 10^9 commercial flights we would need to have observed 10^{10} failure-free test flights. For aircraft such as the Airbus fly-by-wire aircraft series (including the original A320 series and today's A380s) the critical software systems on which these depend therefore must be demonstrated to have even high reliability.

The only practical way to demonstrate that such very high reliability requirements have been met is to incorporate factors other than large numbers of failure-free tests. As we will see later in the chapter, this can include information about the quality of the processes and people involved in the design. Alternatively, it has been proposed that a system (or critical parts of it) could be "proved correct" mathematically in the sense that no failures are possible. Inevitably, such proofs of correctness may themselves be error prone so there will never be total certainty associated with them. However, we could extend the model in Figure 14.5 in cases where a proof of correctness has been claimed. Such a model is shown in Figure 14.6.

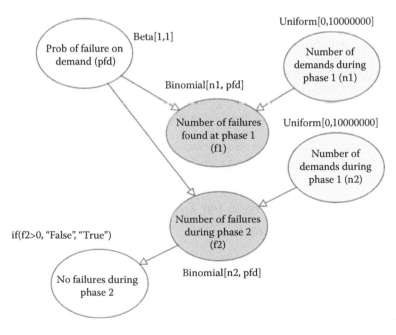

Figure 14.5 BN reliability model for two-phase testing.

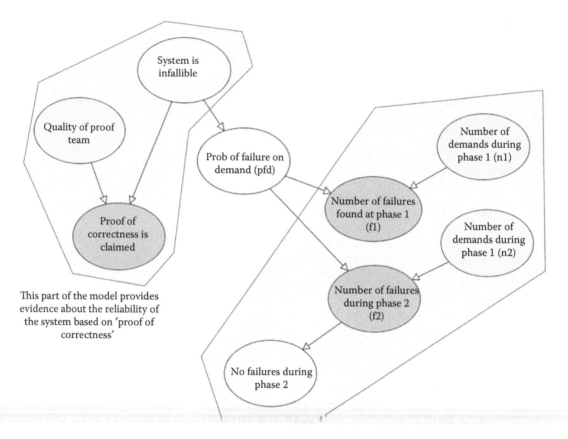

Figure 14.6 Adding different evidence to support reliability assessment (claims of a "proof of correctness").

The model assumes that very few systems are infallible and that only systems that are infallible can be proved correct, but that proofs of correctness may be (and often are) incorrectly claimed even by high-quality teams. Nevertheless, if you run this model in AgenaRisk observing 0 failures in, say, 1000 demands and then additionally observe a proof of correctness claimed there is a significant decrease in the pfd (even without knowing anything about the quality of the proof team). This is because the proof of correctness claim provides an additional "explanation" for observing 0 failures in 1000 demands—namely that the system is infallible. Obviously, if any failure is observed then the system is not infallible and hence any proof of correctness claim is bogus—the claim will have no impact on the pfd.

14.3 Time to Failure for Continuous Use Systems

In the continuous case there are various reliability measures that are used, such as:

- Probability of failure within a specified amount of time
- Mean time to failure (MTTF)
- Mean time between failures (MTBF)

We will focus on learning the time to failure (TTF) distribution. The MTTF is the summary statistic of the TTF distribution. Box 14.1 provides the mathematical formulation of the TTF distribution.

Box 14.1   Time-to-Failure Distribution Notation

For a system, S, we want to determine the reliability, $P_t(S = fail)$, for a given operating time target, t, in terms of a time to failure (TTF) distribution, τ_S:

$$P(S_t = fail) = P(\tau_S \leq t) = \int_0^t f_{\tau_S}(u)\,du$$

The TTF is often expressed as the inverse of the failure rate of the system, which itself is either constant, and characterized by the Exponential probability density function, or varies with time, and represented by some other distribution such as the Gamma or Weibull probability density function. Why these functions? Well, the Exponential distribution models situations where the systems do not wear out or degrade with time but nevertheless have a constant failure rate. However, in practice, for many physical systems, wear out occurs over time and the failure probability increases with time and is therefore not constant. In fact wear-out (or at least degradation in reliability) can occur even in software systems that degrade over time because of the presence of newly introduced design faults.

Definitions for the various distributions discussed here can be found in Appendix E.

Once you have made a choice of the appropriate distribution (or selected a number of possible choices and decided to perform hypothesis testing, as described in Chapter 12, to choose the best one) you need to collect and use failure data either from testing or operational field use. This might come from identical systems to the ones you need to estimate and in this case learning the TTF, or failure rate, simply involves parameter learning as discussed in Chapters 10 and 12. Failure data for continuous systems would be the TTF recorded in seconds, hours, and so forth.

Things get more complex when we consider repairable systems, but we will not cover this here.

However, for novel systems you will often need to look more widely to ask what similar systems might be able to tell you about the reliability of the system at hand. This approach suggests the use of meta-analysis and hierarchical modeling to mix data from specific systems and between families of similar systems, as described in Chapter 10. Here we need to estimate the failure rate parameters for each family or class of system and for the whole superclass of systems.

In addition to this challenge we often encounter data quality problems, one of which is *censoring*, involving TTF data:

- *Right censoring*—The system under study has survived longer than the time available to observe it. Therefore our data is not a specific TTF value but an observation that $P(\tau_S > t) = 1$, that is, the actual failure time is missing.

- *Type II censoring*—The system has failed at some point in time before we inspect or observe it, thus all we know is that $P(\tau_S < t) = 1$.
- *Mixed censoring*—We only know the interval (t_1, t_2) during which the system failed. This is simply an observation, $P(t_1 < \tau_S < t_2) = 1$.

We now turn to an example that combines all of these challenges:

- Estimate TTF for a single system class
- Handle censored data for one class of system
- Estimate TTF for the superclass of all systems
- Predict TTF and reliability for future, new, system

For example, the system of interest might be a 2-liter automobile engine and the historical data might be the TTF of similar engines, but ones with slight design variations.

Example 14.4   Learning from Similar Systems with Imperfect Data

Here we show how a hierarchical Bayesian data analysis can be carried out to compute the unknown failure rate of a new system, A, using historical data gathered from tests conducted on different sets of units of five types of systems, $B_1, \ldots, B_5$, with similar failure behavior to system A. Thus, the untested system A is considered to be exchangeable with the tested systems.

In order to assess the failure distribution of the similar systems, we assume that a series of reliability tests have been conducted under the same operational settings for the first four systems, $B_1, \ldots, B_4$, but that the failure data from component B_5 is right censored. In the case of $B_1, \ldots, B_4$ the data resulting from these tests consist of the observed TTFs, $\{t_{ij}\}, i = 1, \ldots, 4$ and $j = 1, \ldots, n_i$, after a fixed period of testing time, of n_i items of type i. In the case of B_5 the data consists of time to failure intervals $\{t_{5j} > 2000\}$ and $j = 1, \ldots, n_5$, that is, the tests were suspended after 2000 time units. For nonrepairable systems the order of the data is immaterial.

Our aim is to use the sequence of observed failure times and intervals to assess the failure distributions of each one of the similar systems, from which we wish to estimate the posterior predictive failure distribution for A. Thus, if n_i independent tests were conducted on components of type i for a defined period of time, T, the data result in n_i independent TTF, with underlying Exponential population distribution:

$$\{t_{ij}\}_{i=1}^{n_i} \sim \exp(\lambda_i), \quad i = 1, \ldots, 5$$

The unknown failure rates, λ_i, of the B_i similar systems are assumed exchangeable in their joint distribution, reflecting the lack of information—other than data—about the failure distribution of the systems. The parameters λ_i are thus considered a sample from the conjugate Gamma prior distribution, governed by unknown hyperparameters (α, β):

$$\{\lambda_i\}_{i=1}^{5} \sim Gamma(\alpha, \beta)$$

To complete the specification of the hierarchical model, we need to assign a prior probability distribution to the hyperparameters (α, β). Since no joint conjugate prior is available when α and β are both assumed unknown, their prior distributions are specified independently. In the absence of any additional information about the hyperparameters, we can assign them vague prior distributions, for example, by defining vague priors such as

$$P(\alpha) \sim Exp(1.0)$$

$$P(\beta) \sim Gamma(0.01, 0.01)$$

However, because reliability data can be sparse, additional information in the form of expert judgment plays an important role in the definition of statistical reliability models. Here we choose illustrative distributions for shape and scale parameters, but in practice these might be elicited from experts:

$$\alpha \sim Triangular(0, 1, 10)$$

$$\log_{10}\beta \sim Triangular(-6, -3, -1)$$

Figure 14.7 shows the BN graph capturing the above assumptions, with the marginal posterior distributions superimposed for each of the hyperparameters, (α, β); the failure rates of $B_1...B_5$, and A; and the posterior predictive distribution of the TTF of A, TTF_A, shown in larger size. The relevant summary statistics for TTF_A are $MTTF_A = 552$ and $s.d._A = 64631$ (however, note the sidebar comment about the use of median rather than mean time to failure). The high standard deviation clearly shows the result of pooling the diverse data from similar components. Had we simply pooled the data as if it were from a common population with one single unknown parameter, the estimate of the standard deviation would have been overoptimistic.

The reliability, for a target of 100 hours of service, is only 37.8%, that is, there is only a 37.8% chance of the system surviving for this long. Note that we also have, from Figure 14.7, the failure rate distributions for each class of system $B_1,...,B_4$ and have easily accommodated the censoring of failure data on system class B_5.

14.4 System Failure Diagnosis and Dynamic Bayesian Networks

Automatically determining the true state of a system from telemetry and other sensor readings is useful for complex safety critical and high stakes operations. However, it can be difficult and challenging to do this when sensors themselves might be prone to failure in such a way that they provide erroneous readings.

Example 14.5

We monitor the position of an aircraft using radar, but there is a concern that the radar may have failed at some point in time. Hence, our estimate of the aircraft's true position should be based on

- Readings produced by the radar
- Some estimate of when we think it might have failed
- Some knowledge of the aircraft's trajectory before the radar had failed

The prior distributions used for the hyperparameters might be based on past experience, and in this particular case:

- Asking experts to use a Triangular distribution is relatively easy compared to using other more complex distributions.
- The parameters can be interpreted in terms of TTF estimates. So $Triangular(0, 3, 10)$ has a modal failure count around 3 and decreasing probability of experiencing up to 10 failures, and this might be the range of values observed in past practice.

Mean or Median Time to Failure?

Although the MTTF (mean time to failure) is the most commonly used "average" measure in reliability modelling it has been argued that it is highly misleading when the true underlying distribution has a very long tail. In such circumstances, the *median* time to failure is more meaningful. In Figure 14.7 the node "TTF A" (which represents the time-to-failure distribution for A) has a very long tail. The AgenaRisk statistics show that the 99th percentile is 2281, which explains the high mean of 558. However, the median is only 151. This means that 50% of the time A will fail within 151 hours despite the more comforting high mean time to failure of 558 hours.

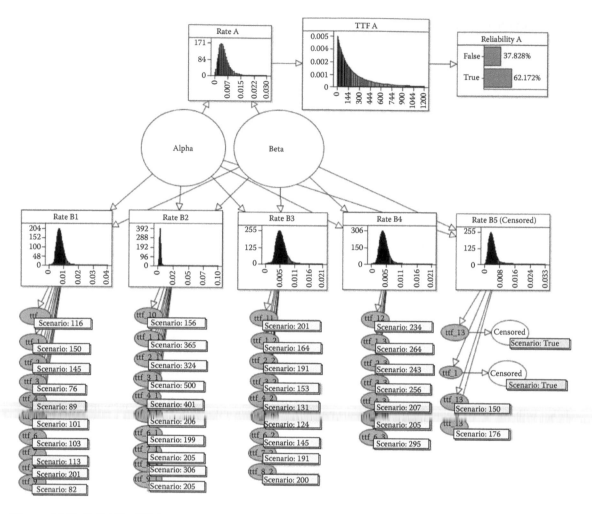

Figure 14.7 Reliability BN for the *A* engine showing marginal posterior distributions for rate λ_i, λ_A, hyperparameters, (α, β), TTF_A, and reliability for *A* (assuming target of 100 hours service).

By state we mean properties like position, velocity, and so forth.

We can use a particular form of BN called a dynamic Bayesian network (DBN). A DBN is a BN that represents the evolution of some state-space model through time, as a stochastic process, where the objects change their state dynamically.

At various points in time our sensors provide us with readings (measures) that we can use to update our estimates of the properties of interest. Examples of state-space models might be time series, Markov or Kalman filter models, and such models can either contain just continuous variables, discrete variables only, or a mix of both.

A state-space model consists of a prior state, a state transition function and one or more observation functions:

$P(X_{t=0})$: Prior state at $t = 0$

The purpose of this section is to provide a flavor of the modeling style involved in the use of DBNs and provide some insight into how they can be used in a reliability application. We recommend you look at the Further Reading section to dig deeper.

$P(X_t \mid X_{t-1})$: State transition from times $(t{-}1)$ to t

$P(Y_t \mid X_t)$: Observation, Y_t, made of true state, X_t

In formulating the model in this way we are making an assumption that it is *first-order Markov*. This means that the conditional probability distribution of future states, given the present state, depends only on the current state, that is, it is conditionally independent of past states. So, $P(X_t \mid X_{t-1})$ represents the fact that X_t depends only on X_{t-1} and not on X_{t-2}. Of course, in practice this might not be true and it is important to consider how cause–effect is represented over time.

Our state transitions might be conditionally deterministic. For example, an object traveling through one-dimensional space might be represented by these "difference equations" connecting object position, P_t, with velocity, V_t:

$$P_t = P_{t-1} + V_{t-1}$$

$$V_t = V_{t-1}$$

with some initial conditions:

$$V_0 = N(0, \theta_1)$$

$$P_0 = N(0, \theta_2)$$

and an observation model, producing observations from a sensor, O_t, that has some noise, σ^2, typically determined by the state of the sensor, such as

$$p(O_t \mid P_t) = Normal(P_t, \sigma^2)$$

In DBN parlance the states we do not observe are called *hidden* or *latent* states, so in this case these would be all the V_t and P_t for $t > 0$. *Inferring only the most recent hidden state is known as filtering; inferring past states is known as smoothing.*

Figure 14.8 shows an example state space DBN with hidden nodes governed by the above difference equations and observation nodes, with evidence, $O_1 = 10$, $O_2 = 20$, $O_3 = 17$, $O_4 = 50$ collected over four time periods. As you can see the inference is carried out and the estimates for (V_t, P_t) are produced for all time periods based on past, future, and current observations. So observation O_t updates our estimates of the previous time step's hidden variables, (V_{t-1}, P_{t-1}), as the smoothing process, at the same time as it updates (V_t, P_t), as the filtering process.

The process here is recursive—each time "slice" of the underlying process takes the same structural form as the previous or next slice. This can be exploited when designing and executing models but is advanced material.

Example 14.6   Diagnosing Whether a Sensor Has Failed

Assume we have a sensor, S, that provides observations, O_t, for some safety-critical process, from which we wish to determine the hidden, unobservable, position, and velocity values (V_t, P_t). The sensor can either

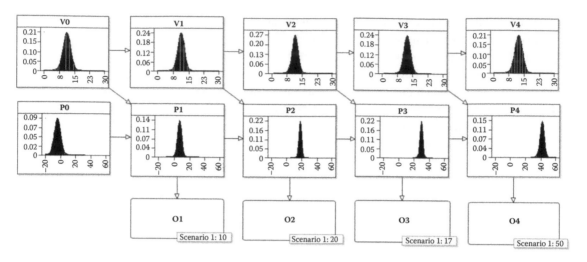

Figure 14.8 DBN for inferring hidden (V_t, P_t) states from observations, O_t.

be working properly or it might fail. When it fails it still gives a reading but the reading is much noisier than when it operates normally:

$$\sigma_S^2 = \begin{cases} 1000 & \text{if } S = Faulty \\ 10 & \text{if } S = OK \end{cases}$$

We can represent this in a DBN by supplementing the model in Figure 14.8 with a node for σ_S^2 that depends on a discrete node representing the state of the sensor, $S=\{Faulty,OK\}$. This extended model is shown in Figure 14.9. The observations, O_t, are now dependent on the sensor noise as well as the true but unknown position, P_t:

$$p(O_t \mid P_t, \sigma_S^2) = Normal(P_t, \sigma_S^2)$$

Assuming a 50:50 prior for $S=\{Faulty,OK\}$ and the same observations as shown in Figure 14.8, we can now execute this model and determine what the posterior probability is for the sensor. This is shown in

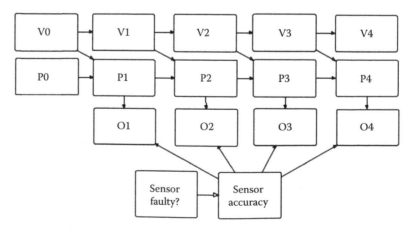

Figure 14.9 DBN for inferring whether the sensor is faulty states from observations, O_t.

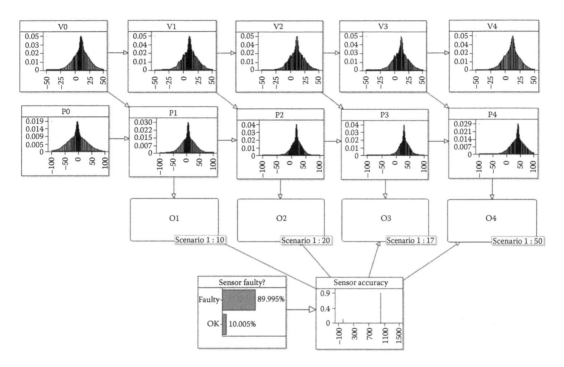

Figure 14.10 DBN showing diagnosis of sensor state from observations, O_t.

Figure 14.10 and the percentage probability of the sensor being faulty is approximately 90%.

The last observation, $O_4 = 50$, is very surprising given the positions estimated previously, the starting conditions, and the normal operating specifications for the sensor. Also, notice that the posterior distributions for the position and velocity variables are all mixtures of the probability distributions for the two possible states of the sensor, showing that our uncertainties about the sensor have propagated throughout the whole model.

14.5 Dynamic Fault Trees (DFTs)

Most reliability analysis methods are based on parametric and nonparametric statistical models of TTF data and associated metrics. The underlying assumption of these methods is that a coherent, statistical model of system failure time can be developed that will prove stable enough to accurately predict a system's behavior over its lifetime. However, given the increasing complexity of the component dependencies and failure behaviors of today's real-time safety-critical systems, the statistical models may not be feasible to build or computationally tractable. This has led to an increasing interest in more flexible modeling frameworks for reliability analysis. The most notable such framework, dynamic fault trees (DFTs), extends FTs by incorporating event dependent behaviors (sequence-dependent failures, functional dependencies, and stand-by spares) of fault-tolerant systems.

Here a DFT is represented by an equivalent "event-based" BN with continuous random variables representing the TTF of the components of the system. These can be either the TTF of elementary components of the system, or the TTF of the fault tree constructs. In the latter case, the nodes in the BN are connected by means of incoming arcs to several components' TTFs and are defined as deterministic functions of the corresponding input components' TTF as shown in the example in Figure 14.11.

In order to specify the probability distribution of the BN, we must give the marginal probability density functions of all root nodes and the NPTs and functions of all nonroot nodes. If the TTF nodes corresponding to elementary components of the system (or some subsystem) are assumed statistically independent (as is the case in standard static FT analysis, or dynamic gates with independent inputs), these are characterized by their marginal probability distributions. The marginal TTF distributions of the root nodes are generally given by standard probability density functions. For example, in Figure 14.11 there are two root nodes corresponding to components A, and B. The TTF distributions for component A is defined as an exponential distribution with parameter λ_A, while the TTF distribution for component B is defined as a Weibull distribution with shape parameter β_C and scale parameter μ_C. The values of the parameters of these density functions can be either obtained as prior information according to expert knowledge or estimated in a previous reliability data analysis step if some failure data is available.

The NPTs for both static and dynamic gates are probability distributions of variables that are a deterministic function of its parents and are

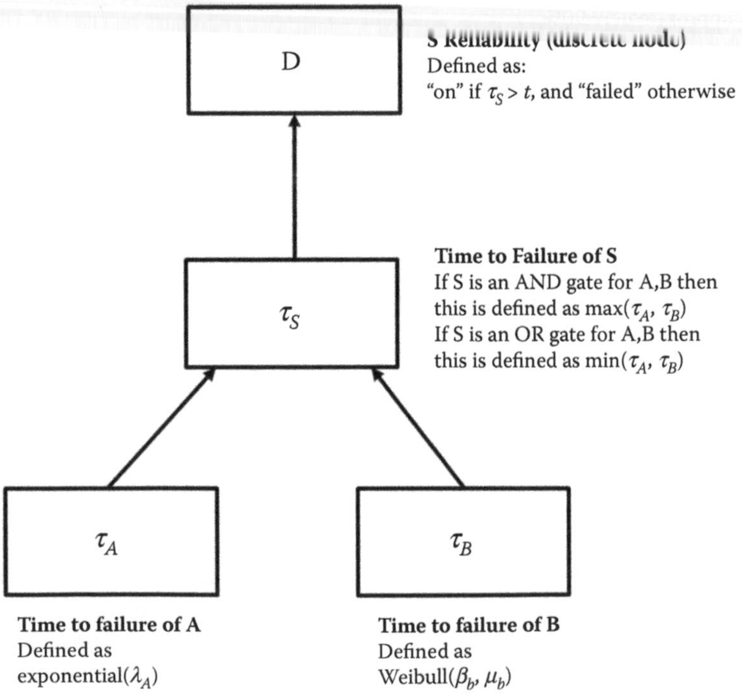

> Dynamic fault trees are an extension of the static fault trees we encountered in Chapter 13.

Figure 14.11 An event-based reliability BN example with statistical and deterministic functions with two components, A, and B, where A has Exponential failure distribution and B has a Weibull failure distribution.

determined according to the types of constructs used in the corresponding DFT. For example, in Figure 14.11 the system S could be defined as an AND gate construct of components A and B and so (as explained in Box 14.2) the TTF of S is defined as the maximum of the components' TTFs. If S is defined as an OR gate construct then the TTF of S is defined as the minimum of the components' TTFs.

For general components' failure distributions, a closed-form expression for the NPTs of dynamic gates may not be feasible, so we approximate them using the dynamic discretization algorithm. Specifically, if the continuous random variable τ_S represents the TTF of a system, S, then a discrete child node, D, of parent τ_S, may be included in the model to represent the reliability of the system.

The NPT for the discrete node D, which defines the probability distribution of the system states at a given time, t, can be automatically computed from the system TTF distribution e.g., $P_t(S = fail) = P(\tau_S \leq t)$. The resulting model (for example, Figure 14.11) is then a hybrid BN containing both continuous as well as discrete variables, with general static and time-dependent failure distributions.

The BN fragment that implements the DFT TTF equivalent of the basic Boolean FT functions is described in Box 14.2.

Box 14.2   BN Modeling of the DFT Equivalent to FT Boolean Functions

We now define the basic BN constructs (OR, AND) as used in static FT analysis. Let us denote it by τ_i, $i = 1, \ldots, n$, the TTF of the ith input component of the construct.

The AND gate. In order for the output of an AND gate to fail, all input components of the gate must fail. So, if τ_{AND} represents the TTF of the output event, then the probability of failure of the output of the AND gate in the time interval $(0, t]$ is given by

$$P(\tau_{AND} \leq t) = P(\tau_1 \leq t, \ldots, \tau_n \leq t)$$
$$= P\left(\max_i \{\tau_i\} \leq t\right)$$

That is, the TTF of the AND gate is a random variable defined as a function of its parents by $\tau_{AND} = \max_i \{\tau_i\}$.

The OR gate. The output of the OR gate will fail if at least one of the input components of the gate fail, as given by

$$P(\tau_{OR} \leq t) = 1 - P(\tau_1 > t, \ldots, \tau_n > t)$$
$$= P\left(\min_i \tau_i \leq t\right)$$

We then define the TTF of the OR gate by $\tau_{OR} = \min_i \{\tau_i\}$.

Example 14.7   Dynamic AND and OR Gates

In Figure 14.12 we use the BN structure of Figure 14.11 with the following specific parameter distribution values:

- For the exponential distribution (TTF of A) the rate parameter is $\lambda_A = 1/10,000$

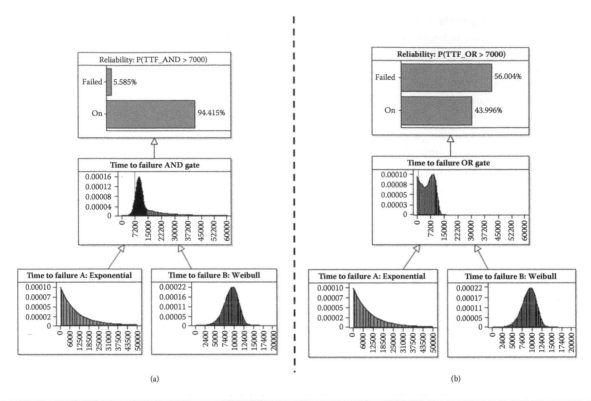

(a)   (b)

Figure 14.12   BNs for DFT (a) AND and (b) OR gates

- For the Weibull distribution (TTF of B) the shape parameter $\beta_C = 6$ and the scale parameter $\mu_C = 1/10,000$

We also define $t = 7000$ hours, so the discrete node representing the reliability of S is defined as "on" if $\tau_S > t$ and "failed" otherwise. With these values we get the following results:

- OR gate: the MTTF of S is 5991 and the reliability ("on") probability is 44%
- AND gate: the MTTF of S is 13317 and the reliability ("on") probability is 94%

In the DFT formalism, new special purpose constructs (called dynamic gates) have been added to the traditional Boolean gates to cope with many complex sequence dependencies among failure events or component states. These include:

- The *Functional and Probability Dependency* gates (FDEP and PDEP) model dependencies that propagate failure in one component to others and can be modeled by DFT OR gates.
- The next simplest case is the *Priority AND* gate (PAND), which models situations where failures can occur only in a predefined order.
- The *Spare* gate models dynamic replacement of a failed main component from a pool of (independent) spares. The Spare

gate can be quite complex and is discussed in Box 14.3 and in Examples 14.9 and 14.10.
- The *Sequence Enforcing* gate (SEQ) models failures that occur only if others occur in certain order. The SEQ gate is a special case of the Cold spare gate, so the same BN model can be used for both types.

Box 14.3   Hot, Cold, and Warm Spare Gates

In a *spare* or *standby* redundancy configuration, the spare components have two operation modes: a *standby* mode and an *active* mode. Each operation mode is represented by its failure distribution. A standby spare component becomes active (is called into service) when the current active component fails. A system in a spare configuration fails if all, main and spare, components fail. According to the standby mode failure distribution of the spare component, Spare gates are classified as Hot, Warm, and Cold spares.

- *Hot spare (HSP) gate*—Both standby and active failure distributions of the spare are the same as the failure distribution of the main component. This is equivalent to a static AND gate.
- *Cold spare (CSP) gate*—The spare components never fail when in standby mode. If τ_{main} represents the TTF of the main component and τ_i is the TTF of the ith spare component when in active mode, then the probability that a system in a cold standby configuration fails in the time interval $(0,t]$ is given by the equation

$$P\left(\tau_{CSP} \leq t\right) = P\left(\tau_{main} + \tau_1 + \cdots \tau_n \leq t\right)$$

That is, the TTF of the CSP gate with n spare components is a conditionally deterministic function $\tau_{CSP} = \tau_{main} + \tau_1 + \cdots \tau_n$, and the failure distribution of the CSP gate is calculated by dynamic discretization.

- *Warm spare (WSP) gate*—The reliability of the spare components is different (usually less) in standby mode than in active mode. Consider a warm standby system consisting of one main component and one spare. Let us denote by τ_{main} the TTF of the main component, τ_{spare}^{sb} the TTF of the spare component when in standby mode, and τ_{spare}^{act} the TTF of the spare component when in active mode. Then, the mutually exclusive events leading to the warm standby system failing in the time interval $(0,t]$ are
    - The spare component fails (when in standby mode) before the main component and the main component fails at time $t_1 < t$.
    - The main component fails at time $t_1 < t$, the spare component has not failed at time t_1 (when in standby mode), and the spare component fails in the active mode in the time interval (t_1,t).

For spare components that wear out, the preceding statements can be directly written in terms of the components TTF by equations

$$\tau_{WSP} = \begin{cases} \tau_{main} & \text{if } \tau_{spare}^{sb} < \tau_{main} \\ \left(\tau_{main} + \tau_{spare}^{act} - t_e \,\middle|\, \tau_{spare}^{act} > t_e\right) & \text{if } \tau_{spare}^{sb} > \tau_{main} \end{cases}$$

This simply means that the WSP has a TTF gate as good as the main component if the spare has failed by the time it is needed (i.e., when the main fails). If, however, the spare component is called upon, its residual TTF will differ depending on how long it has been inactive, which is the equivalent operating time, t_e. This expression can then be built into the BN model using discrete comparative nodes.

Let's look at the PAND gate to get a flavor of how dynamic gates work and how they can be implemented using a hybrid BN. The output of the PAND gate will fail if all of its input components fail in a

The NPT for the discrete node "$T_1 < T_2$" is defined as:

$$if(T_1 < T_2,\text{"Yes"},\text{"No"})$$

The NPT for the discrete node "System Reliability" is defined as:

$$if(\text{Tao\_system} < 800,\text{"Fail"},\text{"On"})$$

From Figure 14.13 we can see that the probability of meeting the reliability target is 75%.

Example 14.9   Cold Spare Gate

Consider a cold spare configuration with two components. Both the main and active spare failure distributions are Weibull with shape, $s = 1.5$, and inverse scale, $\beta = 1/1000$. The BN model is shown in Figure 14.14. The reliability target is a mission time $t = 1000$ hours; the probability of meeting this target is 82%.

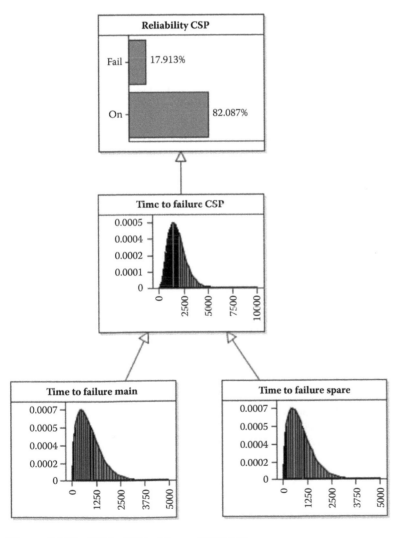

Figure 14.14   BN for DFT cold spare (CSP) gate.

Example 14.10 Warm Standby Gate

Figure 14.15 shows a system with two components in a warm standby configuration. The main component has Weibull failure distribution with shape, $s_1 = 1.5$, and inverse scale $\beta_1 = 1/1500$. When operating in the active mode, the spare component also has Weibull failure distribution with shape, s_2, and inverse scale, β_2, parameters identical to the main component. When operating in the standby mode, the spare failure distribution is Weibull with shape $s_3 = 1.5$ and inverse scale $\beta_3 = 1/2000$.

Currently, due to model size restrictions in the junction tree algorithm we cannot support *m from n* gates in DFT form.

We must account for the accumulated operation time of the spare component when it becomes active. This is achieved by simply adding a Boolean child node to spare TTF, with state values "yes" and "no" representing the statement that the component has or has not already operated for a given period of time, and entering "yes" as evidence. The resulting posterior failure density function represents the conditional active TTF distribution for the spare component, given that it has already operated during a period of time equivalent to the period of operation in standby mode. The reliability

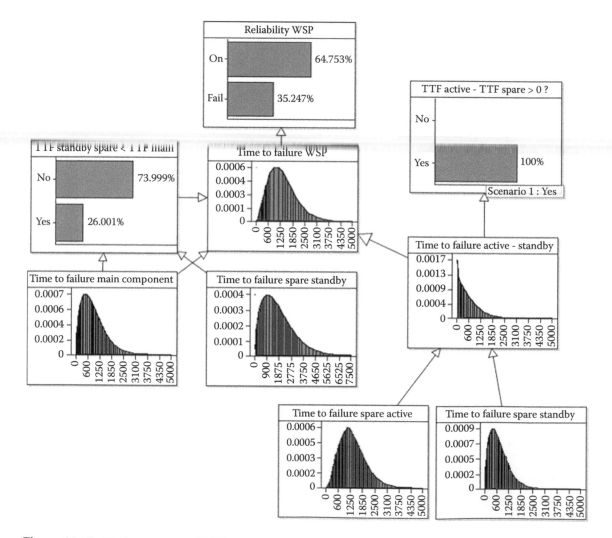

Figure 14.15   BN for warm spare (WSP) gate.

target is a mission time $t = 1000$ hours; the probability of meeting this target is 65% and the MTTF is 1358.

Box 14.4 illustrates how to use AgenaRisk to learn a system's components' reliability from data and then combine this analysis using a DFT model of the system's structure to arrive at the final system's reliability.

Box 14.4   Assessing Systems Reliability Using a DFT in AgenaRisk

Figures 14.16 and 14.7 illustrate how AgenaRisk is used to learn the system's components' reliability from data and then combine this analysis using a DFT model of the system's structure to arrive at the final system's reliability.

Here we have a system modeled as a DFT with a number of gates: OR, PAND, and a Warm spares gate. Each of the components of the system is represented in Figure 14.16, with their respective BN used to calculate their TTF distributions from failure data. These distributions are then used at the system level to calculate the system level TTF and the reliability given the target required.

In AgenaRisk the system level DFT model, shown in Figure 14.17, is declared using risk objects that are then connected together using input-output nodes. Output nodes are declared at each of the component reliability models and connected to the system level model, at input nodes, corresponding to the output nodes. When the model is calculated the marginal probabilities are then passed from the components to the system level risk objects.

One interesting and powerful way to extend this model is to do KUUUB analysis to fold in subjective assessment of reliability changes due to process, people, or environmental influences.

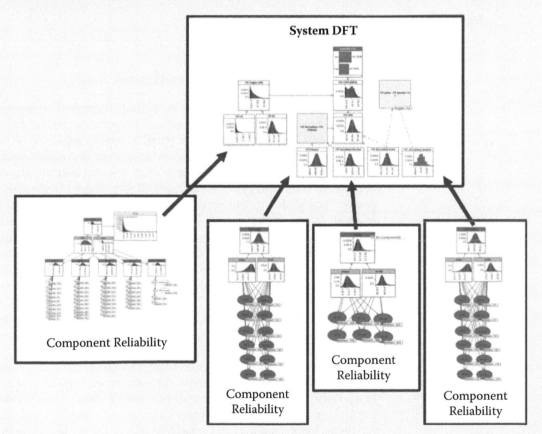

Figure 14.16   System level DFT model.

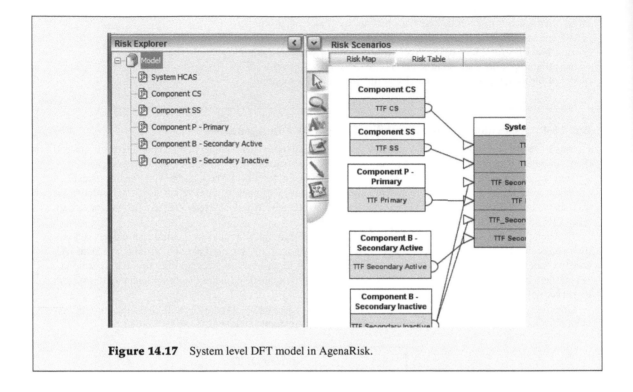

Figure 14.17   System level DFT model in AgenaRisk.

14.6   Software Defect Prediction

The reliability of software products is typically expressed in terms of software defects and we are most interested in predicting the number of defects found in operation (i.e., those found by customers).

We know this is clearly dependent on the number of *residual defects* in the product after it has been released. But it is also critically dependent on the amount of *operational usage*. If you do not use the system you will find no defects irrespective of the number there. The number of residual defects is determined by the number you introduce during development minus the number you successfully find and fix. Obviously, the number of defects found and fixed is dependent on the number introduced. The number introduced is influenced by problem complexity and design process quality. The better the design, the fewer the defects; and the less complex the problem, the fewer defects. Finally, how many defects you find is influenced not just by the number there to find but also by the amount of testing effort. A BN model that captures this information is shown in Figure 14.18.

This particular model is a simplified version of a model that has been used extensively in predicting defects of software components embedded in electronic devices.

The NPTs for the nodes with parents are provided in Table 14.1. Some are deterministic (and self-explanatory) such as the *Residual defects* being simply the numerical difference between the *Defects inserted* and the *Defects found and fixed*. In other cases, we use standard statistical functions. For example, in this simple version of the model we assume that *Defects found and fixed* is a *Binomial(n, p)* distribution where *n* is the number of defects inserted and *p* is the probability of finding and fixing a defect (which in this case is derived from the *testing quality*); in

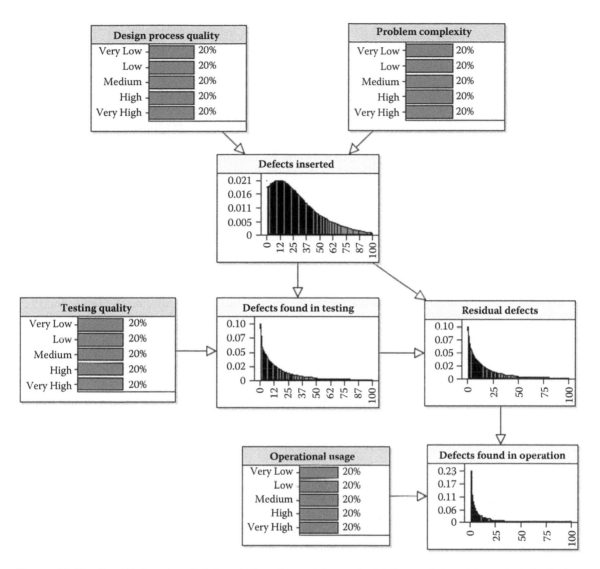

Figure 14.18 Simplified version of a BN model for software defects and reliability prediction with marginal distributions shown.

Table 14.1

NPTs for the Nodes of BN Model in Figure 14.1

Node Name	NPT
Defects found in operation	Binomial (n, p), where n = residual defects and p = operational usage
Residual defects	Defects inserted—Defects found (and fixed) in testing
Defects found in testing	Binomial (n, p), where n = defects inserted and p = testing quality
Defects inserted	Truncated Normal on range 0 to 500 with mean complexity $\times$ (1 − design) $\times$ 90 and variance 300.

The mean and variance values for the defects inserted node were based on empirical data about components in a particular environment and would obviously need to be tailored for different environments.

more sophisticated versions of the model the p variable is also conditioned on n to reflect the increasing relative difficulty of finding defects as n decreases. The nodes *design quality*, *complexity*, *testing quality,* and *operational usage* are all ranked nodes. The nodes without parents are all assumed to have a prior Uniform distribution, that is, one in which any state is equally as likely as any other state (in the "real" models the distributions for such nodes would normally not be defined as Uniform but would reflect the historical distribution of the organization either from data or expert judgment).

Figure 14.18 shows the marginal distributions of the simple model before any evidence has been entered. So this represents our uncertainty before we enter any specific information about this product. Since we assumed Uniform distributions for nodes without parents we see, for example, that the product is just as likely to have very high complexity as very low, and that the number of defects found and fixed in testing is in a wide range where the median value is in the region of 18–20 (the prior distributions here were for the particular environment). Figure 14.19 shows the result of entering two observations about this product:

- It had zero defects found and fixed in testing.
- The problem complexity is high.

Note that all the other probability distributions updated. The model is doing both forward inference to predict defects in operation and backward inference about, say, design process quality. Although the fewer than expected defects found does indeed lead to a belief that the postrelease faults will drop, the model shows that the most likely explanation is *inadequate testing*.

So far we have made no observation about *operational usage*. If, in fact, the operational usage is very high, then entering this as an observation in the model results in an increased expected number of defects found in operation. In fact, what we have done is replicate the apparently counterintuitive empirical observations whereby a product with no defects found in testing has a high number of defects postrelease.

The ability to do the kind of prediction and what-if analysis described has proved to be very attractive to organizations that need to monitor and predict software defects and reliability, and that already collect defect-type metrics.

But suppose that additionally we find out that the test quality was "very high" (Figure 14.20). Then we completely revise our beliefs. We are now fairly certain that the product will be fault free in operation. Note also that the explanation is that the design process is likely to be "very high" quality. This type of reasoning is unique to BNs. It provides a means for decision-makers (such as quality assurance managers in this case) to make decisions and interventions dynamically as new information is observed.

It is beyond the scope of this book to describe the details of these models and how they were constructed and validated, but what typifies the approaches is that they are based around a sequence of testing phases, by which we mean those testing activities such as system testing, integration testing, and acceptance testing that are defined as part of the organization's software processes (and, hence, for which relevant defect and effort data is formally recorded). In some cases a testing phase is one that does not involve code execution, such as design review. The final testing phase is generally assumed to be *operational testing*, by which we normally mean some fixed period of postrelease testing; it is information on this final phase that enables the organization to monitor and predict reliability.

To link the phases together we use the object-oriented approach described in Chapter 8. So, corresponding to each phase is a risk object like that in Figure 14.21. For the final operational phase, there is, of course, no need to include the nodes associated with defect fixing and insertion.

The distributions for nodes such as *probability of finding defect* derive from other risk objects such as that shown in Figure 14.22. The particular nodes and distributions will, of course, vary according to the type of testing phase.

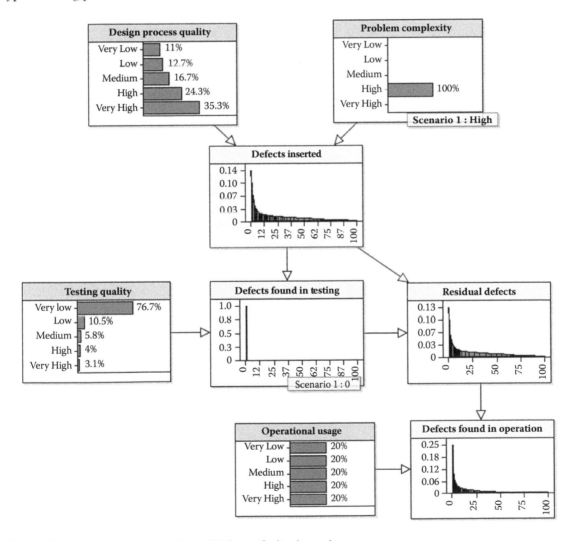

Figure 14.19   Zero defects in testing and high complexity observed.

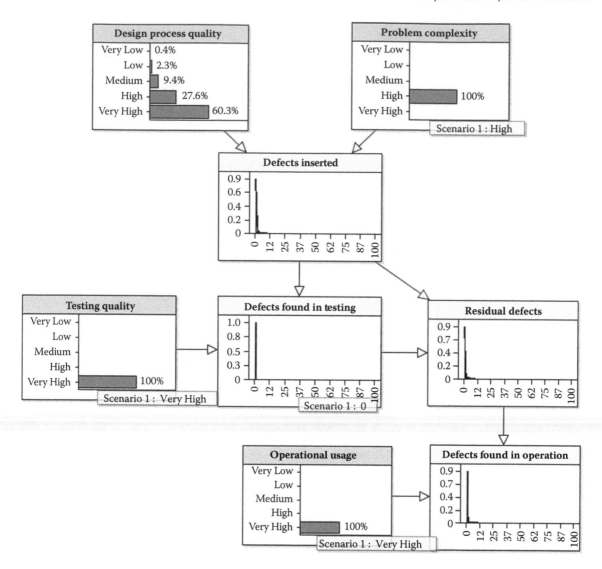

Figure 14.20 Testing quality is very high.

To give a feel for the kind of expert elicitation and data that was required to complete the NPTs in these kinds of models, we look at two examples, namely, the nodes *probability of finding a defect* and *testing process overall effectiveness*:

1. *The NPT for the node probability of finding a defect.* This node is a continuous node in the range [0,1] that has a single parent *testing process overall effectiveness* that is a ranked node. For a specific type of testing phase (such as integration testing) the organization had both data and expert judgment that enabled it to make the following kinds of assessment:

   ■ Typically (i.e., for our "average" level of test quality) this type of testing will find approximately 20% of the residual defects in the system.

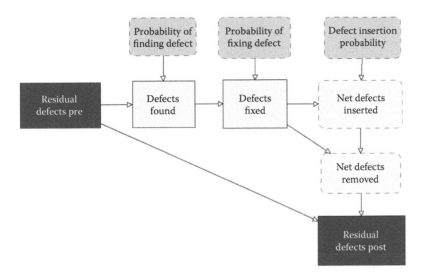

Figure 14.21   Defects phase risk object.

- At its best (i.e., when our level of testing is at its best) this type of testing will find 50% of the residual defects in the system; at its worst it will only find 1%.

Based on this kind of information the NPT for the node *probability of finding a defect* is a partitioned expression like the one in Table 14.2. Thus, for example, when overall testing process effectiveness is "average," the probability of finding a defect is a truncated Normal distribution over the range [0,1] with mean 0.2 and variance 0.001.

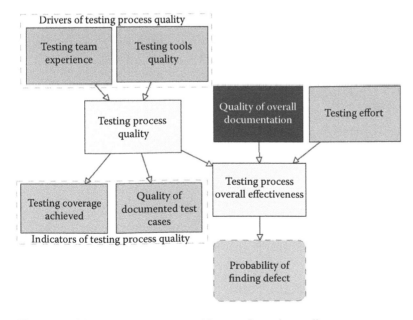

Figure 14.22   Typical risk object architecture for testing quality.

Table 14.2
NPT for Node *Probability of Finding a Defect*

Parent (Overall Testing Process Effectiveness) State	Probability of Finding a Defect
Very low	*TNormal* (0.01, 0.001, 0, 1)
Low	*TNormal* (0.1, 0.001, 0, 1)
Average	*TNormal* (0.2, 0.001, 0, 1)
High	*TNormal* (0.35, 0.001, 0, 1)
Very high	*TNormal* (0.5, 0.001, 0, 1)

The assumptions about the probability of finding a defect are especially acute in the case of the operational testing phase because, for example, in this phase the various levels of operational usage will be much harder to standardize on. What we are doing here is effectively predicting reliability and to do this accurately may require the operational usage node to be conditioned on a formally defined operational profile such as described in the literature on statistical testing.

2. *The NPT for the node testing process overall effectiveness.* This node is a ranked node on a 5-point ranked scale from "very low" to "very high." It has three parents: *testing process quality*, *testing effort*, and *quality of overall documentation*, which are all also ranked nodes on the same 5-point ranked scale from very low to very high. To avoid having to specify the 625 entries in the NPT we were able to use a weighted minimum expression (as described in Chapter 9) to capture expert judgment like "documentation quality cannot compensate for lack of testing effort, although a good testing process is important." As an illustration, Figure 14.23 shows the resulting distribution for overall testing process effectiveness when *testing process quality* is "average," *quality of documentation* is "very high," but *testing effort* is "very low."

Using AgenaRisk the various risk object models are joined as shown in Figure 14.24. Here each risk object represents a BN where only the input and output nodes are shown. For example, for the BN representing the defects in phase 2 the input node *residual defects pre* is defined by the marginal distribution of the output node *residual defects post* of the BN representing the defects in phase 1.

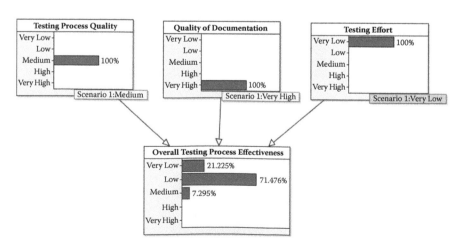

Figure 14.23   Scenario for overall testing effectiveness.

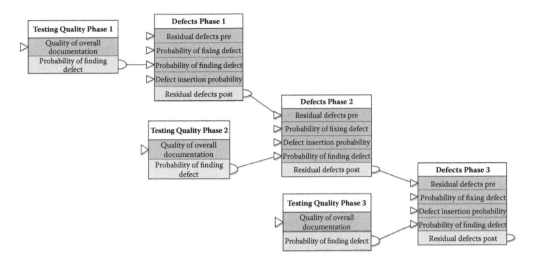

Figure 14.24 Sequence of software testing phases as linked risk objects.

The general structure of the BN model proposed here is relevant for any software development organization whose level of maturity includes defined testing phases in which defect and effort data are recorded. However, it is important to note that a number of the key probability distributions will inevitably be organization- or project-specific. In particular, there is no way of producing a generic distribution for the probability of finding a defect in any given phase (and, as we discuss in the sidebar, this is especially true of the operational testing phase); indeed, even within a single organization this distribution will be conditioned on many factors (such as ones that are unique to a particular project) that may be beyond the scope of a workable BN model. At best we assume that there is sufficient maturity and knowledge within an organization to produce a benchmark distribution in a given phase. Where necessary this distribution can then still be tailored to take account of specific factors that are not incorporated in the BN model. It is extremely unlikely that such tailoring will always be able to take account of extensive relevant empirical data; hence, as in most practically usable BN models, there will be a dependence on subjective judgments. But at least the subjective judgments and assumptions are made explicit and visible.

For organizations that already collect software defect data there is a compelling argument for using BNs to predict software defects and reliability. The causal models enable us to incorporate simple explanatory factors, such as testing effort, which can have a major impact on the resulting predictions.

14.7 Summary

In this chapter we have introduced the concept of reliability modeling and applied it to a number of complex systems examples. You have seen how we can learn the reliability parameters for a system or component from test and operational data using a number of Bayesian learning models, including complex situations involving censored data. Furthermore, we have covered the latest ideas on dynamic fault trees, which allow us to predict reliability of complex systems involving logic gates and to do so by producing TTF distributions at any level of granularity in the system.

Finally we discussed the role of software defect modeling, with an emphasis on defects in software, but given that software is the embodiment of design the ideas covered here are just as relevant to all design artifacts.

Further Reading

Fenton, N., and Neil, M. (1999). A critique of software defect prediction models. *IEEE Transactions on Software Engineering* 25(5), 675–689.

Fenton, N. E., Neil, M., and Krause, P. (2002). Software measurement: Uncertainty and causal modelling. *IEEE Software* 10(4), 116–122.

Fenton, N. E., Neil, M., and Marquez, D. (2008). Using Bayesian networks to predict software defects and reliability. *Proceedings of the Institution of Mechanical Engineers, Part O: Journal of Risk and Reliability* 222(4), 701–712.

Fenton, N. E., Neil, M., Marsh, W., Hearty, P., Radlinski, L., and Krause, P. (2008). On the effectiveness of early life cycle defect prediction with Bayesian nets. *Empirical Software Engineering* 13, 499–537.

Kang, C. W., and Golay, M. W. (1999). A Bayesian belief network-based advisory system for operational availability focused diagnosis of complex nuclear power systems. *Expert Systems with Applications* 17(1), 21–32.

Neil, M., Fenton, N., Forey, S., and Harris, R. (2001). Using Bayesian belief networks to predict the reliability of military vehicles. *IEE Computing and Control Engineering* 12(1), 11–20.

Neil, M., Tailor, M., Marquez, D., Fenton, N., and Hearty P. (2008). Modelling dependable systems using hybrid Bayesian networks. *Reliability Engineering & System Safety* 93(7), 933–939.

Rausand, M., and Hoyland, A. (2004). *System Reliability Theory: Models and Statistical Methods*, Wiley.

15

The Role of Bayes in Forensic and Legal Evidence Presentation

15.1 Introduction

The aim of this chapter is to explain the context and application of Bayes in forensic and legal reasoning. Proper use of Bayesian reasoning has the potential to improve dramatically the efficiency, transparency, and fairness of the criminal justice system and the accuracy of its verdicts by enabling the relevance of evidence to be meaningfully evaluated and communicated. Bayesian reasoning can help experts formulate accurate and informative opinions and help the court in determining the admissibility of evidence. When there are multiple dependent and conflicting pieces of evidence (as is inevitable in all but the simplest cases), BNs provide a rigorous method for combining such evidence, helping prosecutors determine which cases should and should not be pursued, and helping lawyers explain, and jurors evaluate, the weight of evidence during a trial.

In Section 15.2, we provide a brief history of the use of probability in legal reasoning, which allows us to introduce some of the well-known cases that have been of special interest for Bayes. In Section 15.3, we explain why Bayes is a natural method for reasoning about legal evidence, especially forensic evidence such as DNA (Deoxyribonucleic acid). It turns out that much of the focus of Bayes in the law has been on the presentation of forensic evidence, and in this context, the *likelihood ratio* (LR), which we introduced in Chapter 6, has been widely promoted. We will explain the special attraction of using the LR in Section 15.4, and in Section 15.5, we will explain how, in practice, BNs are needed for its computation. In Section 15.6, we discuss some practical limitations of focusing on the LR as a means of presenting the impact of forensic evidence. In particular, it is commonly assumed that with the LR approach, Bayes—and BNs—can somehow be avoided. But one of the key lessons of this chapter is that BNs are required to properly model the uncertainty of the causal context for any evidence. The challenging task of actually *building* appropriate BNs to model specific legal cases and arguments is deferred to Chapter 16.

Misuse of probability in the law is a continuing concern. There have been widely documented cases (described in this chapter) in which verdicts have been influenced by incorrect probabilistic reasoning. However, such cases are only the tip of the iceberg. Most common fallacies of probabilistic reasoning in the law are easily avoided by applying Bayes' theorem.

The central idea of Bayes' theorem—that we start with a prior belief about the probability of an unknown hypothesis and revise our belief about it once we see evidence—is also a central concept of the law.

15.2   Context and Review of Bayes in Legal Proceedings

The use of statistics in legal proceedings (both criminal and civil) has a long, but not terribly distinguished, history. This is well documented in the references at the end of this chapter. The earliest reported case of a detailed statistical analysis being presented as evidence was in the Howland case in 1867. In this case, Benjamin Peirce attempted to show that a contested signature on a will had been traced from the genuine signature by arguing that their agreement in all 30 downstrokes was extremely improbable under a binomial model. However, there were misuses of the "product rule" in Peirce's evidence for multiplying probabilities of independent events. In any case, the court found a technical excuse not to use the evidence.

The historical reticence to accept statistical analysis as valid evidence is, sadly, not without good reason. When, in 1894, a statistical analysis was used in the Dreyfus case, it turned out to be fundamentally flawed. Not until 1968 was there another well-documented case in which statistical analysis played a key role. In that case, another flawed statistical argument further set back the cause of statistics in court. The Collins case (see Box 15.1) was characterized by two errors:

1. It underestimated the probability that some evidence would be observed if the defendants were innocent by failing to consider dependence between components of the evidence.

2. It implied that the low probability from the calculation in (1) was synonymous with innocence (the so-called "prosecutor's fallacy" that we introduced in Chapter 6 and will discuss further in Section 15.3).

Box 15.1   *People v Collins* (1964–68)

In 1964, Malcolm and Janet Collins were convicted of mugging and robbing an elderly woman in an alleyway in Los Angeles. The victim had described her assailant as a young blonde woman, and another witness saw a blonde woman with a ponytail run out of the alley and jump into a waiting yellow car driven by a black man with a moustache and beard. The Collinses were a local couple who "matched" these various characteristics. The prosecution's case was largely based on using probability to show that the couple were guilty. In particular, a maths instructor assigned the following approximate estimates to the probability of the particular characteristics

- Yellow car: 1/10
- Man with moustache: 1/4
- Woman with ponytail: 1/10
- Woman with blonde hair: 1/3
- Black man with beard: 1/10
- Interracial couple in car: 1/1000

He claimed that we could assume these were "independent" events, and hence use the "product rule" to multiply them together to get an estimate of the probability that *another* couple would have all six characteristics. The prosecutor used the resulting probability—1 in 12 million—to conclude (wrongly) that there was only a 1 in 12 million chance that the couple were innocent, and thus were guilty beyond reasonable doubt.

After their conviction, the case was appealed. On the basis of the probability errors, the conviction was reversed in 1968 (after Malcolm Collins had served four years in prison).

Since then, the same errors (either in combination or individually) have occurred in other cases. One of the most notorious was that of *R v Sally Clark* (Box 15.2).

The Lucia de Berk case (see Box 15.3) is another example whereby a prosecution case was fitted around the fact that one person was connected to a number of related deaths. People die in hospitals, and it is inevitable that there will be instances of individual nurses associated with much higher than normal death rates in any given period of time. The extremely low probability P(evidence | not guilty)—which, without any real evidence that the defendant has committed a crime, tells us very little—was used to drive the prosecution case. Another very similar case (currently subject to an appeal based on precisely these concerns) is that of Ben Geen, who was also a nurse convicted of multiple murders and attempted murder.

The problem of poor use of probability in the courtroom is compounded when one considers that the most common use of statistics in the courtroom over the last 50 years has been classical methods of hypothesis testing using *p*-values and confidence intervals. The limitations of such an approach, which we discussed extensively in Chapter 12, are especially critical when used in legal proceedings.

Box 15.2   The Case of *R v Sally Clark* (1998–2003)

In 1999, Sally Clark was convicted of the murder of her two young children who had died one year apart. The prosecution case relied partly on flawed statistical evidence presented by paediatrician Professor Sir Roy Meadow to counter the hypothesis that the children had died as a result of Sudden Infant Death Syndrome (SIDS) rather than murder. He asserted that there was a 1 in 73 million probability of two SIDS deaths in the same family. But this was based on assuming that two deaths would be independent events, and hence that the assumed probability of 1/8500 for a single SIDS death could be multiplied by 1/8500. The error is compounded when the resulting (very low) probability is assumed to be equivalent to the probability of Sally Clark's innocence (this is an example of the prosecutor's fallacy).

Moreover, the (prior) probability of a SIDS death was considered in isolation, that is, without comparing it with the (prior) probability of the proposed alternative, namely of a child being murdered by a parent. The convictions were upheld on appeal in October 2000, but overturned in a second appeal in 2003 (when additional medical evidence suggesting the second child had died of natural causes was presented), and Clark was released.

Box 15.3   The Case of Lucia de Berk (2003–10)

As a result of an unexpected death of a baby in a Dutch hospital in 2001, an investigation focused on nine previous incidents, which earlier had all been thought unremarkable but now were considered suspicious. Lucia de Berk had been on duty at the time of those incidents, and was charged in connection with all of them. In 2003, she was sentenced to life imprisonment for four murders and three attempted murders of patients in her care. In 2004, after an appeal, she was convicted of seven murders and three attempts. Flawed statistical arguments drove both the initial investigation into de Berk and the prosecution case at trial. Specifically, this was an example of what we described in Chapter 5 (see, e.g., Box 5.14) as a case where highly probable events are assumed to be highly improbable: the prosecution claimed that the chance of the same nurse being present at the scene of so many unexplained deaths and resuscitations was one in 342 million. In contrast, statistician Richard Gill and colleagues demonstrated that the probability a nurse could experience a sequence of events of the same type as Lucia de Berk was 1 in 25. Because of these concerns, the case was reopened in October 2008 and Lucia was exonerated in April 2010.

The most common use of probability in the law has been in using "trait" evidence to help with identification. Traits range from "forensic" physical features like DNA (coming from different parts of the body), fingerprints, or footprints, through to more basic features such as skin

The interpretation of DNA match probabilities is critically dependent on the context for the match, and, in particular, serious errors of probabilistic reasoning may occur in cases (such as *People v Puckett*, USA 2008) where the match arises from a database search. In such cases, Bayesian reasoning can again avoid errors.

color, height, hair color or even name. But it could also refer to nonhuman artifacts (and their features) related to a crime or crime scene, such as clothing and other possessions, cars, weapons, glass, soil, and so on.

Any statistical use of trait evidence requires some estimate (based on sampling or otherwise) of the trait incidence in the relevant population. Much of the resistance to the use of such evidence is due to concerns about the rigor and validity of these estimates—with the errors made in the Collins case still prominent in many lawyers' minds.

The use of Bayes in presenting trait evidence has been most extensive with respect to DNA evidence for determining paternity For example, in 2009, North Carolina's court of appeals and supreme court upheld the admissibility of genetic paternity test results using Bayes' theorem with a 50% prior, nongenetic probability of paternity. In contrast to the 50% prior recommended by North Carolina, in *Plemel v Walter* (1987), the need to present results against a range of priors was recommended; indeed, this was also recommended in the criminal case of *State of New Jersey v J.M. Spann* in 1993.

Outside of paternity testing, the primary use of Bayes in trait evidence has been in presenting DNA evidence as a likelihood ratio. In Section 15.4, we will explain this approach in detail. In a number of cases, this approach has been used to expose the prosecutor's fallacy in the original statistical presentation of a random match probability. An especially well-known case is that of *R v Adams*, described in Box 15.4 (one that we will return to in Section 15.5). Similar rape cases were *R v Denn* (1994) and *R v Alan James Doheny and Gary Adams* (1996). In each case, an appeal accepted the Bayesian argument showing that there was the potential to mislead in the way the DNA match evidence against the defendants had been presented by the prosecution at the original trial. Similar uses of the LR for presenting DNA evidence being accepted as "valid" have been reported in New Zealand and Australia.

Box 15.4   *R v Adams* (1996–98)

This was a rape case in which the only prosecution evidence was that the defendant's DNA matched that of a swab sample taken from the victim. The defense evidence included an alibi and the fact that the defendant did not match the victim's description of her attacker. At trial, the prosecution had emphasized the very low random match probability (1 in 200 million) of the DNA evidence. There was some concern that the prosecutor's fallacy was made in the presentation of this evidence. The defense argued that, if statistical evidence was to be used in connection with the DNA evidence, it should also be used in combination with the defense evidence and that Bayes' theorem was the only rational method for doing this. The defense called a Bayesian expert (Prof. Peter Donnelly), who explained how, with Bayes, the posterior probability of guilt was much lower when the defense evidence was incorporated (we will explain this is in detail in Section 15.5). The appeal rested on whether the judge misdirected the jury as to the evidence in relation to the use of Bayes and left the jury unguided as to how that theorem could be used in properly assessing the statistical and nonstatistical evidence in the case. The appeal was successful and a retrial was ordered, although the Court was scathing in its criticism of the way Bayes was presented, stating:

> The introduction of Bayes' theorem into a criminal trial plunges the jury into inappropriate and unnecessary realms of theory and complexity deflecting them from their proper task.

> The task of the jury is ... to evaluate evidence and reach a conclusion not by means of a formula, mathematical or otherwise, but by the joint application of their individual common sense and knowledge of the world to the evidence before them.
>
> At the retrial, it was agreed by both sides that the Bayesian argument should be presented in such a way that the jury could perform the calculations themselves (a mistake, in our view). The jury were given a detailed questionnaire to complete to enable them to produce their own prior likelihoods, and calculators to perform the necessary Bayesian calculations from first principles. Adams was, however, again convicted. A second appeal was launched (based on the claim that the judge had not summed up Donnelly's evidence properly and had not taken the questionnaire seriously). This appeal was also unsuccessful, with the Court not only scathing about the use of Bayes in the case but essentially ruling against its future use.

The most well-publicized success of the use of the LR in reasoning about non-DNA evidence concerns its use in relation to firearm discharge residue (FDR) in the case of *R v Barry George* that we introduced in Chapter 6. The use of the LR in the appeal—showing that the FDR evidence had "no probative value"—was the reason granted for a retrial (in which George was found not guilty), with the FDR evidence deemed inadmissible. However, as we will discuss in Sections 15.5 and 15.6, we believe there were some misunderstandings in this case about the notion of probative value of evidence that can only be properly resolved within the context of a full BN model (that was not used in the appeal).

While the *R v George* appeal judgment can be considered a major success for the use of Bayes, the 2010 UK Court of Appeal Ruling—known as *R v T* (2010)—dealt it a devastating blow (see Box 15.5).

While specific appeal rulings "against Bayes" such as those in the cases of *R v Adams* and *R v T* have contributed to the poor take-up of Bayes in the law, there are also more fundamental impediments. In particular, there

While the *R v T* ruling in 2010 dealt a devastating blow to the use of Bayes in presenting forensic (non-DNA) evidence, the ruling against the use of Bayes in *R v Adams* is actually far more damaging. This is because it rules against the very use where Bayes has the greatest potential to simplify and clarify complex legal arguments. The fact that the complex presentation of Bayes in the case was (rightly) considered to be its death knell is especially regrettable given that in 1996, the tools for avoiding this complexity were already widely available.

Box 15.5   The *R v T* Case (2010)

Defendant T had been convicted of murder in a case in which the prosecution had relied heavily on footwear matching evidence presented using Bayes and the LR. On appeal, the conviction was quashed. Specifically, even though there was recognition that Bayes and the LR had been used with footwear matching evidence in Netherlands, Slovenia, and Switzerland, points 86 and 90 of the ruling, respectively, assert:

> We are satisfied that in the area of footwear evidence, no attempt can realistically be made in the generality of cases to use a formula to calculate the probabilities. The practice has no sound basis.
> It is quite clear that outside the field of DNA (and possibly other areas where there is a firm statistical base) this court has made it clear that Bayes' theorem and likelihood ratios should not be used.

Given its potential to change the way forensic experts analyze and present evidence in court, there have been many articles criticizing the ruling. These articles recognize that there were weaknesses in the way the expert presented the probabilistic evidence (in particular, not making clear that likelihood ratios for different aspects of the evidence were multiplied together to arrive at a composite likelihood ratio), but nevertheless express deep concern about the implications for the future presentation by experts of forensic evidence. The papers recognize positive features in the ruling (notably that experts should provide full transparency in their reports and calculations), but they provide compelling arguments as to why the main recommendations stated above are problematic. In Section 15.6.3, we will look at the ramifications of the ruling.

is a persistent attitude among some members of the legal profession that probability theory has no role to play at all in the courtroom. Indeed, the role of probability—and Bayes in particular—was dealt another blow in a 2013 UK Appeal Court case ruling (Box 15.6).

Box 15.6   The Case of *Nulty & Ors v Milton Keynes Borough Council* (2013)

The case was a civil dispute about the cause of a fire. Originally, it was concluded that the fire had been started by a discarded cigarette, although this seemed an unlikely event in itself, because the other two explanations were even more implausible. The Appeal Court rejected this approach, effectively arguing against the entire Bayesian approach to measuring uncertainty by asserting essentially that there was no such thing as probability for an event that has already happened but whose outcome is unknown. Specifically, Point 37 of the ruling asserted (about the use of such probabilities):

I would reject that approach. It is not only over-formulaic but it is intrinsically unsound. The chances of something happening in the future may be expressed in terms of percentage. Epidemiological evidence may enable doctors to say that on average smokers increase their risk of lung cancer by X%. But you cannot properly say that there is a 25% chance that something has happened.... Either it has or it has not.

Although we explained in Chapter 5 why Point 37 of the ruling is provably irrational, supporters of it often point to a highly influential paper by Tribe in 1971 that was written as a criticism of the prosecutor's presentation in the Collins case. While Tribe's article did not stoop to the level of the "no such thing as probability" argument, it is especially sceptical of the potential use of Bayes' theorem since it identifies the following concerns that are especially pertinent for Bayes:

- That an accurate and/or nonoverpowering prior cannot be devised.
- That in using statistical evidence to formulate priors, jurors might use it twice in reaching a posterior.
- That not all evidence can be considered or valued in probabilistic terms.
- That no probability value can ever be reconciled with "beyond a reasonable doubt."
- That due to the complexity of cases and nonsequential nature of evidence presentation, any application of Bayes would be too cumbersome for a jury to use effectively and efficiently.
- That probabilistic reasoning is not compatible with the law, for policy reasons. In particular, that jurors are asked to formulate an opinion of the defendant's guilt during the prosecutor's case, which violates the obligation to keep an open mind until all evidence is in.

One of the objectives of this chapter, and also Chapter 16, is to show that BNs make many of Tribe's concerns redundant in practice.

Although many of Tribe's concerns have long been systematically demolished (see the reference section), the arguments against are far less well known among legal professionals than those in favor.

15.3 Basics of Bayes for Reasoning about Legal Evidence

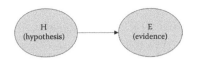

Figure 15.1 Causal view of evidence.

Probabilistic reasoning of legal evidence often boils down to the simple BN model shown in Figure 15.1. The *hypothesis H* is a statement whose truth value we seek to determine, but is generally unknown—and which may never be known with certainty.

Examples include:

■ "Defendant is guilty of the crime as charged" (this is an example of what is called an *offense-level hypothesis*, but also called the *ultimate hypothesis*, since in many criminal cases, it is ultimately the only hypothesis we are really interested in)
■ "Defendant was at the crime scene" (this is an example of what is often referred to as *an activity-level hypothesis*)
■ "Defendant was the source of DNA found at the crime scene" (this is an example of what is often referred to as *a source-level hypothesis*)

The *evidence E* is a statement that claims to support either the hypothesis *H* or the *alternative hypothesis* (which is the negation of *H*). For example, a statement by an eyewitness who claims he or she saw the defendant at the scene of the crime might be considered evidence to support any of the above types of hypotheses. Of special interest is forensic evidence, such as the example shown in the explicit BN of Figure 15.2.

In this example, we assume:

■ The evidence *E* is a DNA trace found at the scene of the crime (for simplicity, we assume the crime was committed on an island with 10,000 people who therefore represent the entire set of possible suspects).
■ The defendant was arrested and some of his DNA was sampled and analyzed (see Box 15.7 for the necessary background about DNA profiles).

The direction of the causal structure makes sense here because *H* being true (resp. false) can *cause E* to be true (resp. false), while *E* cannot "cause" *H*. However, inference can go in *both* directions. If we

One of the great challenges in practice in determining the necessary conditional probability values in Figure 15.2 is that we have to factor into the hypothesis *not H* (defendant is not the source of the DNA trace) every person other than the defendant who could have been the source (potentially every other person in the world). For example, $P(E \mid Hr)$ is much higher than $P(E \mid Hu)$, where *Hr* is the hypothesis "a close relative of the defendant is the source of the trace" and *Hu* is the hypothesis "a totally unrelated person is the source." This means that, in reality, *not H* is made up of multiple hypotheses that are difficult to articulate and quantify. The standard pragmatic solution is to assume that *not H* represents a "random person unrelated to the defendant." But this raises concerns about the homogeneity of the population used for the random match probabilities, and also requires separate assumptions about the extent to which relatives can be ruled out as suspects.

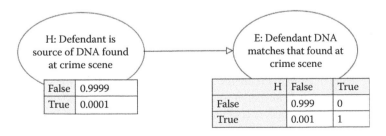

H: Defendant is source of DNA found at crime scene	
False	0.9999
True	0.0001

E: Defendant DNA matches that found at crime scene		
H	False	True
False	0.999	0
True	0.001	1

Figure 15.2 Causal view of forensic evidence, with prior probabilities shown in tables.

Box 15.7   DNA Profiles

DNA is formed from four chemical "bases" that bind together in pairs, called "base pairs." Each person's DNA contains millions of such base pairs. Typically, a swab of a person's saliva (or other tissue) is analyzed using special equipment. Instead of analyzing all of the millions of base pairs, a person's DNA profile is determined by analyzing just a small number of regions of the DNA, known as *loci* or *markers*. These are specific fixed regions that have informative genetic properties. For example, the locus named "FGA" is at a region that provides instructions for making a protein that is important for blood clot formation. At each locus, there are short sequences of base pairs repeated multiple times—these are called *short tandem repeats* (STRs). The number of times that the sequences are repeated varies between individuals. The length of each repeated sequence can be measured and expressed as the number of repeats in the sequence. This is called an *allele*. At each locus there are two alleles—one inherited from the father and one from the mother. So a person's DNA profile is just a sequence of number pairs like:

...	D3	D8	D18	D19	D21	vW1	TH01	FGA
...	14,15	13,13	12,15	14,16	29,29	16,16	6,7	24,25

The pair at a locus may be the same (in which case the person is *homozygous* at that locus rather than *heterozygous*). So, in the above example, D8 and vW1 are homozygous, while the others are all heterozygous. The pair of alleles is called the *genotype* of that locus. Typically, a genotype will be shared by around 5–15% of a particular population (this information is collected in DNA databases). For example, at the TH01 locus, the genotype (6,7) is shared by approximately 7% of the population, while at the vW1 locus, the genotype (16, 16) is shared by approximately 5%.

In general, a DNA profiling system consists of a predefined set of loci. In North America, the CODIS 20 system (consisting of 20 core loci) is generally used, whereas in the UK, the DNA-17 system (17 loci) is used. A person's DNA profile is typically analyzed from a sample of saliva, and a machine provides a printout in which the alleles are identified by "peaks" such as those shown in Figure 15.3.

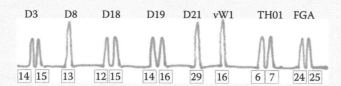

Figure 15.3   DNA profile.

observe E to be true (resp. false), then our belief in H being true (resp. false) increases. It is this latter type of inference that is central to all legal reasoning, since, informally, lawyers and jurors normally use the following widely accepted procedure for reasoning about evidence:

- Start with some (unconditional) prior assumption about the ultimate hypothesis H (e.g., the "innocent until proven guilty" assumption might equate to a belief that "the defendant is no more likely to be guilty than any other member of the population").
- Update our prior belief about H once we observe evidence E. This updating takes account of the *likelihood* of the evidence.

This informal reasoning is a perfect match for Bayesian inference, where the prior assumption about H and the likelihood of the evidence E

are captured formally by the probability tables shown in Figure 15.2. Specifically, these are the tables for the *prior probability* of H, written P(H), and the conditional probability of E given H, which we write as P(E | H). Bayes' theorem provides the formula for updating our prior belief about H in the light of observing E.

The first table in Figure 15.2 (the probability table for H) captures our knowledge that the defendant is one of 10,000 people who could have been the source of the DNA. The second table (the probability table for E given H) captures the assumptions that:

- The probability of correctly matching a DNA trace is one (so, for simplicity here, we are assuming there is no chance of a false negative DNA match). This probability P(E | H) is called the *prosecution likelihood* for the evidence E.
- The probability of a match in a person who did not leave their DNA at the scene (this is often called the "random DNA match probability"—see Box 15.8 for how this is computed) is 1 in 1000. This probability P(E | not H) is called the defense *likelihood* for the evidence E.

Box 15.8   Calculating the DNA Random Match Probability

Suppose that what we know of Fred's DNA profile is as shown in the first two rows of Table 15.1.

Table 15.1
Fred's DNA profile (with population frequencies shown in row 3)

Locus	D21	vW1	TH01	FGA
Genotype	29,29	16,16	6,7	24,25
Genotype Frequency (from appropriate database)	9%	14%	7%	5%

The genotype frequency information typically comes from a database that is supposed to be representative of Fred (e.g., if Fred is a Caucasian male from the United Kingdom, then the frequencies will be based on a database of Caucasian males from the United Kingdom). The frequencies are therefore an estimate of the probability that a person has that particular genotype. Assuming that genotype frequencies of the loci are independent, we estimate the probability that another unrelated person will share the same profile by multiplying together the probabilities. In this case, we therefore estimate that the probability an arbitrarily selected unrelated person has the same profile as Fred (on these four loci) is:

$$0.09 \times 0.14 \times 0.07 \times 0.05 = 0.0000441 = 0.00441\%$$

In other words, in a group of 100,000 UK Caucasian males unrelated to Fred we expect between 4 and 5 will share the same profile on these four loci. The probability is what is often referred to as the *random match probability* for that set of genotypes.

If we assume that a 17-loci system is used and that we have a person's full 17-loci profile, then the random match probability for the DNA profile is incredibly low—if we assume each of the 17 genotypes has a frequency of approximately 10%, then the random match probability is 10^{-17}. Since there are only 8 billion people living on our planet, that suggests that DNA profiles are "unique" and that finding some DNA at a crime scene that matches the profile of a particular suspect means the DNA must actually come from the suspect. However, there are a number of problems with this assumption:

1. The more closely two people are related, the more likely they are to share genotypes. At the extreme, identical twins will generally share the same genotypes at all loci and therefore have identical DNA profiles. Even distant relations have higher match frequencies on most genotypes than unrelated people.

2. In practice, it is generally straightforward to get a full DNA profile from a suspect by, for example, analyzing a sample of their saliva (this is the so-called *reference sample*). However, DNA found at a crime scene—called the *questioned sample*—may be so tiny and/or degraded that, in practice, very few clear "peaks" (such as those shown in the printout of Figure 15.3) can be identified. This is what is called a "low template" sample. It may only provide information on a small number of loci (a so-called *partial DNA profile*). For low-template samples, it is often difficult to be confident even about what tissue type it came from (semen, hair, blood, etc.). It is not uncommon to report a "DNA match" when the questioned sample is from unknown human tissue that provides clear genotype information on only two or three loci. Suppose, for example, that we have a full 17-loci profile for Fred (so the random match probability for this profile is about 10^{-17}). Fred is the defendant in a rape case and the defense claims to have found DNA on the victim that "matches" Fred's DNA. Can we conclude that the probability of finding this evidence from an unrelated UK male is 10^{-17}? No, because it depends on how much of the profile of the reference DNA sample can be conclusively determined. Suppose that only two of the loci, say, D21 and vW1, can be identified in this sample. Then all we know is that the reference sample matches Fred's profile on these two loci. In this case, the random match probability is $0.09 \times 0.14 = 0.0126 = 1.26\%$, that is, a random match probability of greater than 1 in 100. We have been involved as expert witnesses in cases where exactly this type of DNA match evidence has been presented. The claim of a "DNA match" in such cases is far less probative than assumed by judge and jury.

3. Things become very complex and uncertain when the questioned sample is a *mixed profile*, that is, involving DNA from more than two individuals. For example, suppose three different alleles—say, 7,8,9—are found at locus TH01; then we know it must be a mixed profile. Suppose the suspect Jane's genotype is (7,7) on this locus. The problem is that even if we assume the mixture is from only two people, there are multiple combinations of possible sources of the DNA. The two contributors could have the respective genotypes:

- (7,7), (8,9)
- (7,8), (8,9)
- (7,8), (7,9)
- (7,8), (9,9)
- (8,8), (7,9)

All but the first of these five possibilities would actually *rule out* Jane as a possible donor of the mixture. So, if we assume that each of the pairs is equally likely, then there is a 4/5 probability that Jane *cannot be a donor*. To calculate the impact of the evidence in this case [i.e., knowing that Jane's genotype is (7,7) on this locus and that there is a (7,8,9) mixture with two contributors], we need to build a BN model such as the one in Figure 15.4. This assumes that each of the six named genotypes at the locus has a 10% frequency and that there is a 50:50 prior that Jane is one of the contributors of the DNA found (note that this is to make the example as simple as possible and does not reflect reality; in practice, heterozygote genotypes at a locus have higher likelihoods than homozygote genotypes at the same locus).

Comparing the prior (Figure 15.4a) and posterior (Figure 15.4b) probabilities after entering the DNA mixture evidence, we find only a small increase (from 50% to 62.5%) in the probability that Jane is a contributor. If similar evidence was found on, say, two other loci, then, assuming independence between the loci, we could conclude that the probability Jane is one of the donors of the DNA is

$$1 - (0.325^3) = 0.966 = 96.6\%$$

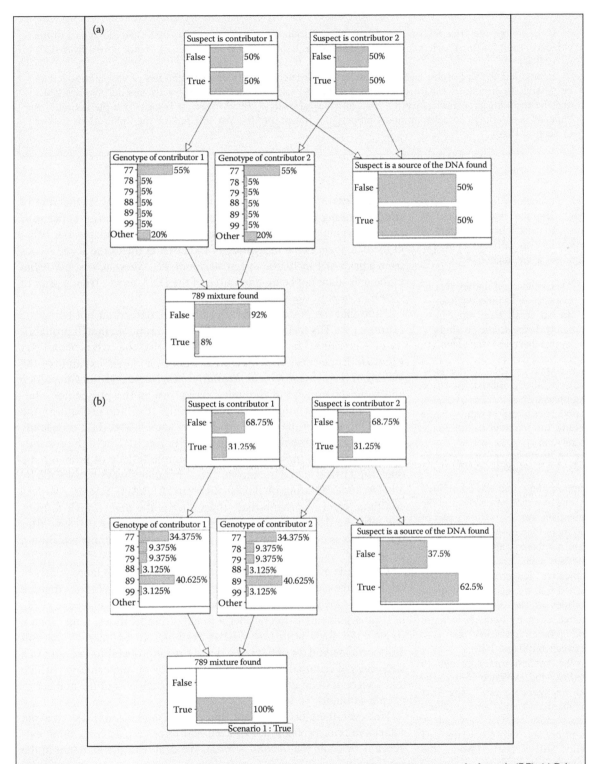

Figure 15.4 Impact of DNA mixture evidence at single locus—assumes suspect genotype at the locus is (7,7). (a) Prior probabilities. (b) Posterior probabilities after finding 789 mixture.

If we change the prior probability of Jane being a source of the DNA to 1 in 1000, then the impact of the evidence, even for three loci, is minimal (the posterior probability that Jane is a contributor moves from 0.1% to 0.249%).

In practice, the size of the "peaks" at the locus will determine different probabilities of the pairs, so it may be possible to discount some (there are many dedicated software packages for exactly this purpose available, and the resulting probabilities can be used in a Bayesian network). However, it is clear that the notion of a "match" is extremely complex to model properly for mixed profiles. We will explore the implications further in Section 15.6.

The prosecutor's fallacy is extremely common, because you do not need an explicit statement of probability to fall foul of it. For example, a statement like:

"The chances of finding this evidence in an innocent man are so small that you can safely disregard the possibility that this man is innocent."

is a classic instance of the prosecutor's fallacy. Indeed, based on examples such as these and our own experiences as expert witnesses, we believe the reported instances are merely the tip of the iceberg.

Suppose the crime was committed by one of a mob of people and the defendant was known to be part of the mob. One eyewitness says the mob numbered "about 60," while another says "about 80," These subjective judgements could be turned into a range of prior probabilities of the defendant's guilt agreed upon by both the defense and prosecution (such as, e.g., between 1/100 and 1/50).

We applied this approach to evaluate the evidence in a medical negligence case, using first the claimant's and then the defendant's most extreme assumptions as priors. In both cases, the evidence still strongly favored the claimant's hypothesis.

In the example of Figure 15.2, it follows (from Bayes' theorem or simply running the model in AgenaRisk) that the posterior belief in H after observing the evidence E being true is about 9%; that is, our belief in the defendant being the source of the DNA at the crime scene moves from a prior of 1 in 10,000 to a posterior of 9%. Alternatively, our belief in the defendant not being the source of the DNA moves from a prior of 99.99% to a posterior of 91%.

Note that the posterior probability of the defendant not being the source of the DNA is very different from the random match probability of 1 in 1000. The incorrect assumption that the two probabilities $P(\text{not } H \mid E)$ and $P(E \mid \text{not } H)$ are the same is a classic example of the *fallacy of the transposed conditional* that we discussed in Chapters 2 and 6. In the context of the law, it is known as the prosecutor's fallacy. A prosecutor might state, for example, that "the probability the defendant was not the source of this evidence is one in a thousand" (i.e., a very low probability) when actually it is 91% (a high probability). As discussed in Section 15.2, this simple fallacy of probabilistic reasoning has affected numerous cases, but can always be avoided by a basic understanding of Bayes' theorem. A closely related, but less common, error of probabilistic reasoning is the *defendant's fallacy*, whereby the defense argues that since $P(\text{not } H \mid E)$ is still low after taking into account the prior and the evidence, the evidence should be ignored.

One of the major barriers to using Bayesian reasoning more widely in the law is the extreme reluctance to consider prior probabilities that are subjective. One way to minimize the objection to using subjective priors in legal arguments is to consider a range of probabilities, rather than a single value for a prior (see sidebar example). In Chapter 16, we will describe a method for obtaining objective prior probabilities of guilt in a large class of criminal cases based on the notion of "opportunity" (which takes account of an extended "crime scene location" and the number of people within it).

However, the most important (but also misunderstood) way to avoid subjective prior probabilities in legal and forensic reasoning about evidence is by using the *likelihood ratio*. We deal with this in detail in the next section.

15.4 The Problems with Trying to Avoid Using Prior Probabilities with the Likelihood Ratio

It is possible to avoid the delicate and controversial issue of assigning a subjective prior probability to the ultimate hypothesis (or, indeed, to any specific hypothesis) if we instead are prepared to focus only on the "probabilistic value" of the evidence. Recall from Chapter 6 and the sidebar that the impact of any single piece of evidence E on a hypothesis H can be determined by considering only the likelihood ratio of E, which is the probability of seeing that evidence if H is true (e.g., "defendant is guilty") divided by the probability of seeing that evidence if H is not true (e.g., "defendant is not guilty"), that is, $P(E \mid H)$ divided by $P(E \mid \text{not } H)$.

In the DNA example of Figure 15.2, the likelihood ratio is 1000 (since the prosecution likelihood is 1 and the defense likelihood is 1/1000). So this particular evidence is clearly very probative in favor of the prosecution hypothesis (that the suspect is the source of the DNA found at the crime scene). In court, the DNA expert would be able to present this likelihood ratio without making any assertions about the prior probability of the prosecution hypothesis by simply saying:

> We are 1000 more times likely to find this evidence under the assumption that the prosecution hypothesis is true than under the assumption that the defense hypothesis is true.

While this is helpful, the extent to which it can lead a judge or jury to "believe" the prosecution hypothesis still depends on the prior probability; as explained in Box 15.9, we use the "odds" version of Bayes (that we introduced in Chapter 6) to make the necessary conclusions.

The likelihood ratio is very well suited to the legal context because it enables us to evaluate the impact of the evidence without having to specify what our prior belief is in the prosecution or defense hypothesis. Values above one favor the prosecution (the higher the better), while values below one favor the defense (the lower the better). A value of one means the evidence is worthless. *However, recall from Chapter 6 that this only applies when the defense hypothesis is the exact negation of the prosecution hypothesis.*

Box 15.9 Using the Likelihood Ratio to Update the Posterior Odds

The "odds" version of Bayes' theorem tells us that, whatever our prior odds were for the prosecution hypothesis, the result of seeing the evidence is such that those odds are multiplied by the likelihood ratio:

Posterior odds = Likelihood ratio × Prior odds

So, suppose the likelihood ratio is 1000 (as in the DNA example of Figure 15.2). Then, according to Bayes, if we started off assuming that the odds in favor of the defense hypothesis were 10,000 to 1, then the "correct" revised belief once we see the evidence is that the odds *still* favor the defense, but only by a factor of 10 to 1:

	Prior odds		Likelihood ratio		Posterior odds
Prosecutor	1	×	1000	=	1
Defense	10,000		1		10

But, if we started off assuming that the odds in favor of the defense hypothesis were 4 to 1, then the "correct" revised belief once we see the evidence is that the odds now favor the prosecution by a factor of 250 to 1:

	Prior odds		Likelihood ratio		Posterior odds
Prosecutor	1		$\dfrac{1000}{1}$		250
Defense	4	$\times$		$=$	1

Note that if we assume that the prior odds are "evens," that is, 50:50, then the posterior odds will be the same as the likelihood ratio; that is, they favor the prosecution by a factor of 1000 to 1.

Can the Likelihood Ratio Be Used without Bayes' Theorem?

The fact that the likelihood ratio has become an extremely popular method for presenting the impact of evidence (especially forensic trace evidence) is not just because it can be presented without assuming prior probabilities. Many have argued that it also "avoids using Bayes' theorem," which is considered a good thing in the law because of controversies that we described in Section 15.2. However, there is a fundamental "circular" problem with this claim. As we explained in Chapter 6, the *only reason* the LR can be meaningfully used as a measure of probative value of evidence *is because of Bayes' theorem*. In other words, if a forensic expert were asked in court to explain why a high likelihood ratio for a piece of evidence was probative in favor of the prosecution hypothesis, he or she would end up having to use Bayes' theorem to prove it, as we did in Chapter 6.

We also show in Section 15.5 that—except in simple cases—we actually require Bayes' theorem simply to compute the LR.

In summary, whether the likelihood ratio of 1000—or even 10 million—"swings" the odds sufficiently in favor of the prosecution hypothesis depends entirely on the prior odds. So, while the likelihood ratio enables us to assess the impact of evidence on H without having to consider the prior probability of H, it is clear from the above DNA example that the prior probability must ultimately be considered before returning a verdict. With or without a Bayesian approach, jurors and judges intuitively make assumptions about the prior probability and adjust this in light of the evidence. A key benefit of the Bayesian approach is to make explicit the ramifications of different prior assumptions. So, a judge could state something like the following where, say, the evidence has a likelihood ratio of one million:

Whatever you believed before about the possible guilt of the defendant, the evidence is one million times more likely if the defendant is guilty than if he is innocent. So, if you believed at the beginning that there was a 50:50 chance that the defendant was innocent, then it is only rational for you to conclude with the evidence that there is only a million to one chance the defendant really is innocent. On this basis you should return a guilty verdict. But if you believed at the beginning that the defendant is no more likely to be guilty than a million other people in the area, then it is only rational for you to conclude with the evidence that there is a 50:50 chance the defendant really is innocent. On that basis you should return a not guilty verdict.

Note that such an approach does not attempt to force particular prior probabilities on the jury (the judiciary would always reject such an attempt)—it simply ensures that the correct conclusions are drawn from what may be very different subjective priors.

It is also important to note that the idea the LR avoids having to consider priors is something of a misconception because it is impossible to specify the likelihoods meaningfully without knowing something about the priors. This is because in strict Bayesian terms (see sidebar), we say the likelihoods and the priors are all *conditioned on some background knowledge K* and it follows that, without an agreed common understanding about this background knowledge, we can end up with vastly different LRs associated with the same hypotheses and evidence. This is illustrated in the following example.

Example 15.1

Suppose the evidence E in a murder case is: "DNA matching the defendant is found on victim." While the prosecution likelihood $P(E \mid H_p)$ might be agreed to be close to one, there is a problem with the defense likelihood, $P(E \mid H_d)$. For DNA evidence such as this, the defense likelihood is usually assumed to be the random match probability (RMP) of the DNA type, which as we have seen, can typically be as low as one in several billion. But consider two extreme values that may be considered appropriate for the prior $P(H_p)$, derived from different scenarios used to determine K:

a. $P(H_p) = 0.5$, where the defendant is one of two people seen grappling with the victim before one of them killed the victim
b. $P(H_p) = 1/40$ million, where nothing is known about the defendant other than he is one of 40 million adults in the UK who could have potentially committed the crime

Whereas a value for $P(E \mid H_d) = $ RMP seems reasonable in case (b), it is clearly not in case (a). In case (a), the defendant's DNA is very likely to be on the victim regardless of whether he is the one who killed the victim. This suggests a value of $P(E \mid H_d)$ close to 1. It follows that, without an understanding about the priors and background knowledge, we can end up with vastly different LRs associated with the same hypotheses and evidence.

Background Knowledge

The prosecution prior $P(H_p)$ and the defense prior $P(H_d)$, really refer to $P(H_p \mid K)$ and $P(H_d \mid K)$, respectively. The likelihoods must take account of the same background knowledge K that is implicit in these priors. So the "real" likelihoods we need are $P(E \mid H_p, K)$ and $P(E \mid H_d, K)$.

15.5 Computing the Likelihood Ratio in Practice: When It Is Easy and When It Requires Full BN Inference

In addition to the reasons explained in the previous section, the LR has become an extremely popular method for presenting the impact of forensic evidence primarily because—in the simplest cases—it is easily calculated without any need for Bayes' theorem at all.

An example of one such "simplest case" is that of Figure 15.2, where there is a single piece of evidence and a single hypothesis. Assuming that the two likelihoods $P(E \mid H)$ and $P(E \mid \text{not } H)$ are provided by the forensic expert, we know that the LR is simply the ratio of these two probabilities. Moreover, even if there are multiple *independent* pieces of evidence $E_1, E_2, \ldots, E_n$ and a single hypothesis H, then the LR for the combined evidence is simply calculated by multiplying the individual LRs; that is,

$$\frac{P(E_1,\ldots,E_n \mid H)}{P(E_1,\ldots,E_n \mid \text{not } H)} = \frac{P(E_1 \mid H)}{P(E_1 \mid \text{not } H)} \times \frac{P(E_2 \mid H)}{P(E_2 \mid \text{not } H)}$$
$$\times \cdots \times \frac{P(E_n \mid H)}{P(E_n \mid \text{not } H)}$$

For example, Box 15.10 describes how three pieces of independent evidence—one supporting the prosecution case and two supporting the defense—were used to reason about probability of guilt in the case of *R v Adams* that we introduced in Section 15.2.

Box 15.10   Combining Independent Evidence

In Section 15.2, we described the case of *R v Adams* (1996). The Bayesian argument used in the case was actually the calculation of the LR for the three pieces of evidence, which were assumed to be independent, namely:

- The defendant's DNA matched that of a swab sample taken from the victim.
- Defendant alibi.
- Defendant did not match the victim's description of her attacker.

The DNA match probability was disputed (the prosecution assumed 1 in 200 million, but the defense argued that the correct figure was closer to 1 in 2 million).

Figure 15.5 shows a BN that incorporates both the prosecution evidence and the defense evidence, with the priors and NPTs based on discussions during the case.

Using the assumption of a 1 in 200 million random match probability for the DNA, the LR for the DNA evidence is 200,000,000, which is clearly highly probative in favor of the prosecution. However, the LR of the identification evidence is 1/9 and the LR of the alibi evidence is 1/2 (both probative in favor of the defense). Since we are assuming the pieces of evidence are independent, the LR of the combined evidence is simply

$$200,000,000 \times \frac{1}{9} \times \frac{1}{2} = 1,111,111$$

It was proposed that a reasonable prior probability of guilt was approximately 1 in 200,000 prior based on there being 150,000 males aged between 15 and 60 in the area plus 50,000 who may have come from outside the area. When we consider the prior odds of 200,000 to 1 in favor of the defense, we calculate the posterior odds as:

	Prior odds		Likelihood ratio		Posterior odds
Prosecutor	1	$\times$	200,000,000	=	1000
Defense	200,000		18		18

So, taking account of the prior odds and all of the evidence, the posterior odds are 18 to 1000 in favor of not guilty—that means a probability of 1.8% for not guilty. Although this probability is low, it is clearly nothing like the 1 in 200 million probability that may have been implied by considering the DNA evidence alone and

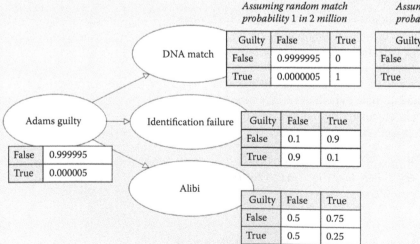

Figure 15.5   Hypothesis and evidence in the case of *R v Adams*.

also making the prosecutor's fallacy. Moreover, repeating the same calculations under the assumption of a 1 in 2 million random match probability results in posterior odds of 18 to 10 in favor of not guilty—that means a probability of 64% for not guilty.

Of course, it would be up to the jury to decide which of the assumptions in the model are reasonable and whether the resulting probability of guilt leaves room for doubt.

Note that because the independent evidence assumptions make calculating the overall LR so simple, there is no need at all to run the BN model in order to calculate the posterior probability of guilt. Nevertheless, because the defense expert attempted to both explain Bayes' theorem in detail and get the jury and judge to perform all the calculations themselves (with the aid of calculators), the judge ruled this approach to be overly complex.

Unfortunately, as soon as a case involves dependent pieces of evidence and/or multiple related hypotheses, the LR calculation cannot be done in this simple way. We need to use a BN model and full BN inference. Indeed, even the case of a "single piece of DNA evidence" (as in Figure 15.2) could more realistically be modeled as the five-node BN model shown in Figure 15.6.

This incorporates the possibility that testing is not perfectly accurate and hence that we do not know the true DNA profile of either the defendant or the source trace. This means it may be possible that both profiles are tested as having the same profile X, even though at least one has a different profile, and that it may even be possible that the defendant and source trace have a *different* matching profile than the one they are tested to have. The direct calculation of the LR for this model is extremely complex. Yet, by simply running the BN model, we can get the posterior probabilities as well as the LR.

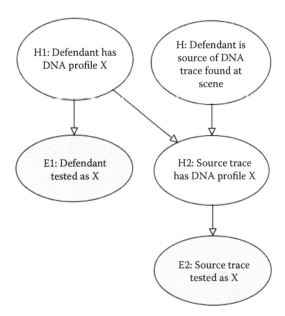

Figure 15.6 Full BN for simple DNA match evidence.

This example is very simplified. In practice, DNA profiling errors are not easily summarized as simple false positive and false negative probabilities. For example, it is common to take at least four swabs as a reference sample from a suspect. For a false negative/positive, it is necessary that all four provide the "wrong" result at one of the loci. Also, in practice, determining profiles from a graphical printout may involve complex decisions relating to the observed peaks and their heights, and in such cases, the notion of false positive and false negative is insufficient to capture the range of possible errors that may occur.

Example 15.2

Assume (as in Figure 15.2) that the profile X is of a "1 in 1000" type (as might be the case if the source trace has three identifiable loci) and that, additionally, the DNA testing has a false positive rate of 0.1% and a false negative rate of 1%. Using these values to define the NPTs of the BN in Figure 15.6 and assuming a prior probability of 1 in 10,000 (i.e., 0.01%) for the hypothesis H results in a posterior probability of 2.42% when we enter "true" for both pieces of evidence E1 and E2 (see Figure 15.7). So, although the evidence is probative in favor of H, it is still extremely unlikely that H is true after observing the evidence. We can use the BN to easily calculate the LR (see Box 15.11) as 248.

Box 15.11 explains how it is easy to calculate the LR from a BN model in general.

Even the model of Figure 15.7 is unrealistically simple when it comes to forensic evidence analysis in practice in the full context of a case. Taking account of the various errors and poor practice (which can lead to contaminated or wrong samples being tested), a more realistic model is the one shown in Figure 15.8. The real interest would be the LR of the match evidence with respect to either the activity-level hypothesis (defendant was at the scene of the crime) or the offense-level hypothesis (defendant committed the crime).

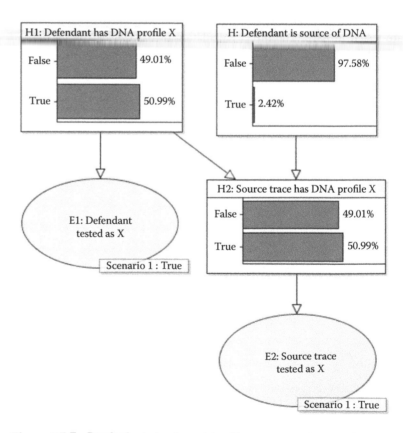

Figure 15.7 Result of entering the match evidence.

Box 15.11   Calculating the Likelihood Ratio from a Bayesian Network Model

No matter how many nodes there are in a BN, we can, for any node H and any set of nodes that have evidence, calculate the LR of the combined evidence with respect to H. To do this, we simply rearrange the odds version of Bayes' theorem to get:

$$LR = \frac{\text{posterior odds}}{\text{prior odds}}$$

We can read off the posterior odds from the BN posterior values of H. For example, in Figure 15.7, that is 2.42/97.58. The prior odds are 1/999. So:

$$LR = \frac{2.42}{97.58} \times \frac{999}{1} = 248$$

Even more simply, if we set the prior for H to 50:50, then the LR for H is simply the posterior odds. Making this change to the model in Figure 15.7 results in posterior odds for H of 99.59/0.402 = 248.

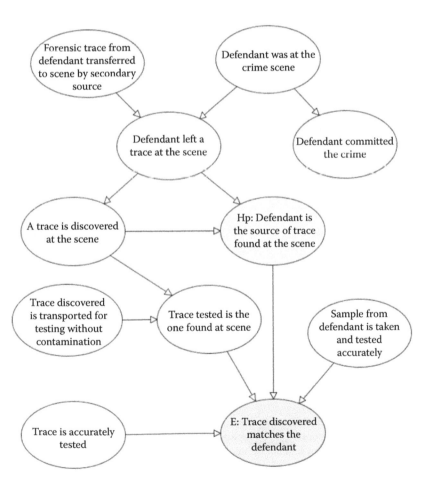

Figure 15.8 Model of the impact of forensic evidence that takes account of additional sources of uncertainty.

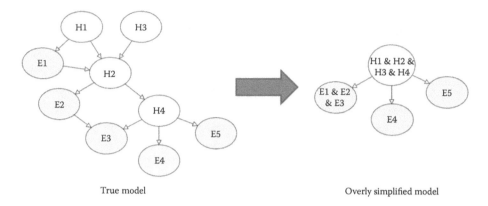

Figure 15.9   Artificially collapsing a model into one where the LR is easy to calculate.

The desire to achieve a "simple-to-calculate" LR in forensic practice often leads to an oversimplistic model in which multiple uncertain hypotheses must be collapsed into a single hypothesis and all evidence must be collapsed in such a way that it can be reduced to pieces of independent evidence (as shown in Figure 15.9). As reported in the case of the Barry George appeal, this can result in confusion and misleading conclusions. The trade-off is too great. In achieving a "simple" LR calculation, the necessary likelihoods are generally too difficult to define.

It is not just the hypotheses that may need to be "decomposed." In practice, even an apparently "single" piece of evidence E may comprise multiple separate pieces of evidence, and it is only when the likelihoods of these separate pieces of evidence are considered that correct conclusions about probative value of the evidence can be made. This is illustrated in Example 15.3.

Example 15.3

The authors were asked to provide expert advice on the statistical issues in a rape case conviction in preparation for an appeal. The key piece of prosecution evidence was a tiny trace of DNA on the back of the alleged victim's knickers that "matched" the defendant. The DNA was a mixture of such poor quality that the "match" to the defendant was identifiable on just two loci (which meant there was a probability of about 1 in 100 of finding such a "match" if the suspect was not a donor). Yet, the forensic experts argued that the "matching" DNA trace provided "strong support" for the prosecution case (and there was no clear separation made between the two hypotheses "defendant was source of DNA" and "defendant raped the victim"). So, essentially, the "tiny piece of matching DNA" was used as a single piece of evidence to support the hypothesis "defendant raped victim."

However, the victim had been sitting in the defendant's car, which is known to have contained multiple traces of the defendant's DNA. Moreover, if the defendant had committed the rape as charged, there should have been more than just a tiny trace of the defendant's DNA on the victim. Instead of the prosecution's two-node model shown in Figure 15.10(a), the more appropriate model is the one shown in Figure 15.10(b).

(a)

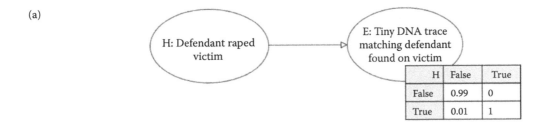

(b)

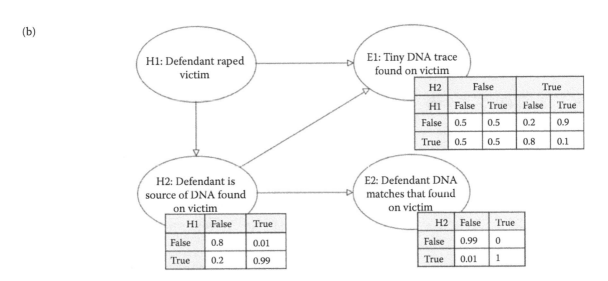

Figure 15.10 Transforming oversimplistic argument into a more realistic one. (a) Oversimplistic prosecution argument. (b) More realistic argument.

The LR for the evidence E in model (a) is 100 (in support of H1).

Assuming a 50:50 prior for H1, when we enter only the evidence E2 into model (b), the posterior for H1 is 83%, so for this evidence alone, the LR is just under 5. While this still supports the prosecution hypothesis H1, when we also enter the evidence E1, we see that the posterior for H1 drops to 38%, meaning that the combined evidence has a LR of 0.6, so the combined evidence actually provides more support for the defense hypothesis.

Example 15.3 also illustrates the importance of taking account of "absence of evidence." In this case, lack of any substantial DNA evidence from the defendant supports the defense hypothesis.

15.6 Practical Limitations of the Use of the Likelihood Ratio in Legal Reasoning

In the previous section, we explained why, in most realistic situations, a nontrivial BN model is required to compute the LR. Until forensic experts are routinely trained in both the use of BNs and their software tools, this creates a practical barrier to the effective use of the LR in practice. However, there are even more fundamental limitations that we discuss in this section.

We believe that much of the recent legal resistance to Bayes is due to confusion, misunderstanding, oversimplification, and overemphasis on the role of the LR, as explained in this section.

15.6.1 Exhaustiveness and Mutual Exclusivity of Hypotheses Is Not Always Possible in Practical Likelihood Ratio Uses

The most powerful benefit of the LR is that it provides a valid measure of the probative value of the evidence E in relation to a pair of hypotheses H_p (prosecution hypothesis) and H_d (defense hypothesis). But, when we say the evidence E "supports the hypothesis H_p," what we mean is that the "posterior probability of H_p after observing E is greater than the prior probability of H_p," and "no probative value" means the "posterior probability of H_p after observing E is equal to the prior probability of H_p." As we explained in Chapter 6, the proof of the meaning of "probative value" in this sense relies both on Bayes' theorem and the fact that H_p and H_d are mutually exclusive and exhaustive—that is, are negations of each other (H_d is the same as "not H_p"). If H_p and H_d are mutually exclusive but not exhaustive, then the LR might tell us nothing about the relationship between the posterior and the priors of the individual hypotheses.

Example 15.4

In the game of Cluedo, there are six people at the scene of a crime (three men and three women). One woman, Mrs. Peacock, is charged with the crime (so H_p is "Mrs. Peacock guilty"). The prosecution provides evidence E that the crime must have been committed by a woman. Since $P(E \mid H_p) = 1$ and $P(E \mid \text{not } H_p) = 2/5$, the LR is 2.5 and so clearly the evidence is probative in favor of the prosecution. However, suppose the defense is allowed to focus on an alternative hypothesis H_d that is not "not H_p"—namely "Miss Scarlet guilty"—then $P(E \mid H_d) = 1$. In this case, the defense can argue that the LR is 1 and so has no probative value. But while the evidence does not distinguish between the (nonexhaustive) H_p and H_d, it certainly is probative for H_p.

Example 15.5

There is a well-known DNA example called the "twins paradox" that is similar to Example 15.4. In this example, DNA matching the defendant is found at the scene of a crime. The prosecution argues that this is highly probative in support of the hypothesis that the defendant was the source of the DNA found at the scene. The defense, however, notes that the defendant has a twin brother and that the DNA match evidence has LR = 1 when the prosecution hypothesis is considered against the alternative that the twin was at the scene. As the evidence has "no probative value," the defense argues it should not be considered. While some legal scholars have suggested that this example demonstrates the LR is incompatible with legal reasoning, all it actually demonstrates is the importance of being clear about which hypotheses to use, and to talk about probative value only in the context of mutually exhaustive and exclusive hypotheses. Assuming a 1 in 1000 random match probability for the DNA evidence, an appropriate simple BN model for this problem is the one shown in Figure 15.11. This shows that the DNA evidence *is* probative; even though it does not distinguish between whether the defendant or twin is the source, the probability that the defendant is the source increases from 33.33% to 49.975% after observing the evidence.

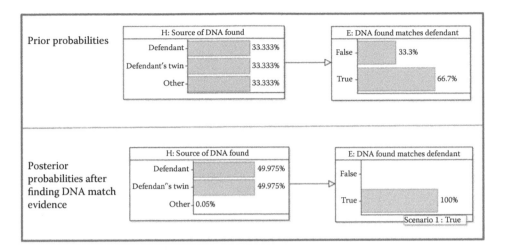

Figure 15.11 Resolving the "twins paradox" by including possibility that person other than defendant or twin is the source of DNA found.

Example 15.6

This problem is especially pertinent in the case of "DNA mixture" evidence, as introduced in Box 15.8; that is, when some DNA sample relevant to a case comes from more than one person. It is common to find DNA samples with multiple (but an unknown number of) contributors. In such cases, there is no obvious "pair" of hypotheses that are mutually exclusive and exhaustive, since we have individual hypotheses such as:

H1:   suspect + one unknown
H2:   suspect + one known other
H3:   two unknowns
H4:   suspect + two unknowns
H5:   suspect + one known other + one unknown
H6:   suspect + two known others
H7:   three unknowns
H8:   one known other + two unknowns
H9:   two known others + one unknown
H10: three known others
H11: suspect + three unknowns
and so on.

It is typical in such situations to focus on the "most likely" number of contributors (say, n) and then compare the hypothesis "suspect + $(n - 1)$ unknowns" with the hypothesis "n unknowns." For example, if there are likely to be three contributors, then typically the following hypotheses are compared:

H1: suspect + two unknowns
H2: three unknowns

Now, to compute the LR, we have to compute the likelihood of the particular DNA trace evidence E under each of the hypotheses. Generally,

In practice, for DNA mixture profiles, calculating the likelihood $P(E \mid H)$ is extremely complex, and many researchers argue that complex BN models are required for this purpose alone. As the example in Box 15.8 made clear, even for a single locus, the necessary BN model is quite complex. Describing the necessary BN models for realistic cases is beyond the scope of this book, but examples are provided in the "Further Reading" section.

both of these are extremely small numbers; that is, both the probability values $P(E \mid H1)$ and $P(E \mid H2)$ are very small numbers. For example, we might get something like

$$P(E \mid H1) = 10^{-20}$$
$$P(E \mid H2) = 10^{-26}$$

For a statistician, the size of these numbers does not matter—we are only interested in the ratio (i.e., precisely what the LR is), and in the above example, the LR is very large (one million), meaning that the evidence is a million times more likely to have been observed if $H1$ is true compared to $H2$. This seems to be overwhelming evidence that the suspect was a contributor.

Case closed? Apart from the communication problem in court of getting across what this all means (defense lawyers can, and do, exploit the very low probability of E given $H1$) and how it is computed, there is an underlying statistical problem with small likelihoods for nonexhaustive hypotheses, and we can highlight the problem with two scenarios involving a simple urn example. Superficially, the scenarios seem identical. The first scenario causes no problem, but the second one does. The concern is that it is not at all obvious that the DNA mixture problem always corresponds more closely to the first scenario than the second.

In both scenarios, we assume the following:

There is an urn with 1,000 balls, some of which are white. Suppose W is the (unknown) number of white balls. We have two hypotheses:

H1: W = 100
H2: W = 90

We can draw a ball as many times as we like, note its color and replace it (i.e., sample with replacement). We wish to use the evidence of 10,000 such samples.

Scenario 1: We draw 1001 white balls. In this case, using standard statistical assumptions, we calculate $P(E \mid H1) = 0.013$, $P(E \mid H2) = 0.0000036$. Both values are small, but the LR is large, 3611, strongly favoring H1 over H2.

Scenario 2: We draw 1100 white balls. In this case, $P(E \mid H1) = 0.000057$, $P(E \mid H2) < 0.00000001$. Again, both values are very small, but the LR is very large, strongly favoring H1 over H2.

(Note: In both cases, we could have chosen a much larger sample and gotten truly tiny likelihoods, but these values are sufficient to make the point.)

So in what sense are these two scenarios fundamentally different, and why is there a problem?

In scenario 1, not only does the conclusion favoring H1 make sense, but the actual number of balls drawn is very close to the expected number we would get if H1 were true (in fact, W = 100 is the maximum likelihood estimate for number of balls). So not only does the evidence point to H1 over H2, but also to H1 over any other hypothesis (and there are 1000 different hypotheses: W = 0, W = 1, W = 2, etc.).

In scenario 2, the evidence is actually even much more supportive of H1 over H2 than in scenario 1. But it is essentially meaningless because it is virtually certain that *both* hypotheses are false.

So, returning to the DNA mixture example, it is certainly not sufficient to compare just two hypotheses. The LR of one million in favor of H1 over H2 may be hiding the fact that neither of these hypotheses is true. It is far better to identify as exhaustive a set of hypotheses as is realistically possible and then determine the individual likelihood value of each hypothesis. We can then identify the hypothesis with the highest likelihood value and consider its LR compared to each of the other hypotheses. Indeed, there are now dedicated software packages for DNA mixture profiles (typically based on inclusion probabilities instead of random match probabilities) that can return the full set of possible likelihoods based on a maximum number of donors and possible known donor profiles.

We already noted that it may be very difficult in practice to elicit $P(E \mid H_d)$, where H_d is simply the negation of H_p, because H_d may encompass too many options. The inevitable temptation is to "cherry pick" an H_d that seems a reasonable alternative to H_p for which $P(E \mid H_d)$ can be elicited. The above examples show the dangers in such an approach. In the case of *R v Barry George* that we discussed previously, the argument that the key forensic evidence had a LR of 1 (and so had no "probative value") was key to the judge declaring the evidence was inadmissible. However, the LR argument may have been compromised by the problem of using nonexhaustive hypotheses in this case.

In the Sally Clark case that we described in Section 15.2, similar problems—of failing to choose appropriate (mutually exclusive and exhaustive) hypotheses—compromised both the original statistical arguments and the arguments used by Bayesians to expose the original flaws. A simple model that resolves both this problem and the original failure to consider dependence between the two deaths is shown in Figure 15.12.

In this model, "SIDS" is synonymous with "death due to unknown medical cause." Crucially, in addition to modeling the dependence between the two child deaths, we incorporate the possibility that exactly one of the children may have been murdered (this possibility—which would still have meant Clark was guilty—was ignored in the probabilistic arguments). The prior for Child A death makes explicit that, although there is a low prior for SIDS (we accept the 1 in 8500 stated at trial), there is a much lower prior for murder. Given that we know the two children are dead and that either SIDS or murder is the "cause" in both cases, we see that the prior marginal for "Clark guilty" is 7.89%. As we add the evidence presented in the original case, we get the revised probability of guilt shown in Table 15.2.

However, when we add the evidence presented in the successful appeal (i.e., signs of disease "True" rather than "False" for Child A), the probability of guilt drops to 4.59% and to 0.09% if there are also signs of disease in Child B.

An additional danger in allowing H_d to be something different from "not H_p" is that, in practice, forensic experts may come up with an H_d that is not even mutually exclusive to H_p. This was also shown to be a problem in the transcript of the *R v Barry George* appeal. Two hypotheses that are not mutually exclusive may both be true, and in such circumstances, the LR is meaningless. The details of all the probabilistic problems in the Barry George case appeal are discussed in Fenton et al. (2014a).

The details of the probabilistic problems in the Sally Clark case are discussed in Fenton (2014).

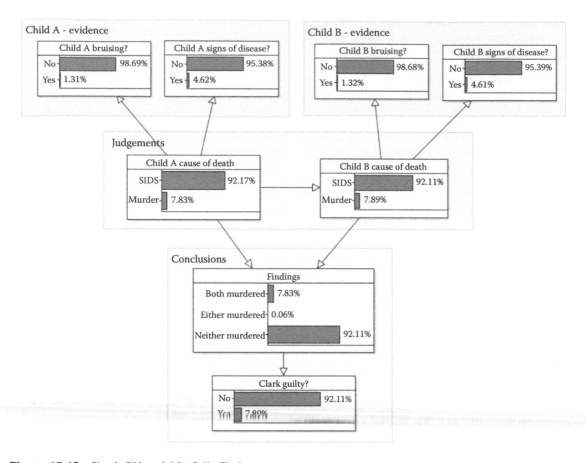

Figure 15.12 Simple BN model for Sally Clark case.

Table 15.2
Impact of evidence in original trial

Evidence Added	Probability Clark Guilty (%)
None	7.89
Child A bruising True	28.87
Child A signs of disease False	30.93
Child B bruising True	69.13
Child B signs of disease False	70.19

15.6.2 The Likelihood Ratio for Source-Level Hypotheses May Tell Us Nothing about Activity and Offense-Level Hypotheses

Even when hypotheses are mutually exclusive and exhaustive, there remains the potential during a case to confuse different, but related, hypotheses. Forensic experts normally report only on the probative value of their evidence on source-level hypotheses (such as "the defendant is the source of the DNA found at the crime scene"), and judges and juries may wrongly assume that the probative value is the same for activity-level hypotheses ("the defendant was at the crime scene") or even offense-level hypotheses ("the defendant committed the crime as charged").

The danger of these assumptions was highlighted in Example 15.3 in Section 15.5; we demonstrated there how some particular DNA match evidence was probative in favor of the prosecution's source-level hypothesis ("the defendant is the source of the DNA found on the victim") but was actually probative in favor of the defense's offense-level hypothesis ("defendant did not commit the crime against the victim").

Similarly, an LR that is neutral under the source-level hypotheses may actually be significantly non-neutral under the associated offence-level hypotheses. This was exactly one of the problems that we identified when we analyzed the transcript of the *R v Barry George* appeal.

15.6.3 Confusion about Likelihood Ratio Being Expressed on a Verbal Scale

Many European forensic experts and organizations recommend that, to help lawyers and jurors understand the significance of an LR, it should be presented on an equivalent verbal scale such as that shown in Table 15.3.

This recommendation contrasts with US courts that have advised against verbal scales and instead recommended that posterior probabilities should be provided based on a range of priors for the given LR. Unfortunately, in the United Kingdom, explicit use of numerical LRs are increasingly snubbed following the *R v T* judgment.

As part of our own legal advisory/expert witness work, we have examined numerous expert reports in the last few years (primarily, but not exclusively, from forensic scientists). These reports considered different types of match evidence in murder, rape, assault, and robbery cases. The match evidence includes not just DNA, but also handprints, fiber matching, footwear matching, soil and particle matching, matching specific articles of clothing, and matching cars and their number plates. In all cases, there was some kind of database or expert judgment on which to estimate frequencies and "random match" probabilities, and in most cases there appears to have been some attempt to compute the LR. However, in all but the DNA cases, the explicit statistics and probabilities were not revealed in court—in several cases, this was as a direct result of the *R v T* ruling, which has effectively pushed explicit

While legal professionals generally insist that forensic experts restrict their conclusions to source-level hypotheses, efforts are currently being made to enable experts to report on activity level in a range of fields.

In the Barry George case appeal, the forensic evidence was argued to have an LR of 1 with respect to the source-level hypotheses (the FDR did/did not come from a gun owned by Barry George). However, while the evidence may not have been probative for that pair of hypotheses, we believe it may have provided some support for the offense-level hypothesis.

Table 15.3
Verbal scale for likelihood ratio

LR	Verbal Scale
1	No support
1–10	Weak support
10–100	Moderate support
100–10,000	Moderately strong support
10,000–1,000,000	Very strong support
>1,000,000	Extremely strong support

use of numerical LR "underground." Indeed, we have seen expert reports that contained explicit data being formally withdrawn as a result of *R v T.* This is one of the key negative impacts of *R v T*—we feel it is extremely unhelpful that experts are forced to suppress explicit probabilistic information.

15.7 Summary

We have explained why Bayes is a natural method for reasoning about legal evidence—especially forensic evidence—and how its proper use can improve the transparency and fairness of the justice system by enabling the relevance of evidence to be meaningfully evaluated and communicated. Yet, the acceptance level of Bayes by legal professionals has been disappointingly low for reasons we highlighted in our historical overview.

We described the widespread use of the LR as a measure of probative value for legal and forensic evidence. However, we also highlighted the some practical limitations of focusing on the LR and described how BNs avoided these. The proposal to use BNs for legal arguments is by no means new. Indeed, in the paper

> Edwards, W. (1991). "Influence diagrams, Bayesian imperialism, and the Collins case: an appeal to reason." *Cardozo Law Review* 13: 1025–1079.

Edwards provided an outstanding argument for the use of BNs, about which he said:

> I assert that we now have a technology that is ready for use, not just by the scholars of evidence, but by trial lawyers.

He predicted such use would become routine within "two to three years." Unfortunately, he was grossly optimistic for reasons that we have partly explained in this chapter. A major reason that we did *not* discuss in this chapter was the lack of any systematic method for *building* appropriate BNs to model specific legal cases and arguments. We address that important challenge in the next chapter.

Further Reading

An extensive set of relevant references (including to all of the cases discussed in this chapter and many more) can be found in:

Fenton, N., Neil, M., and Berger, D. Bayes and the law. (2016). *Annual Review of Statistics and Its Application*, 3.

The probabilistic issues arising in the Barry George case are described in:

Fenton, N.E., Berger, D., Lagnado, D., Neil, M., and Hsu, A. (2014a). When "neutral" evidence still has probative value (with implications from the Barry George Case). *Science and Justice*, 54(4), 274–287.

The probabilistic issues arising in the Sally Clark case are described in:

Fenton, N.E. (2014). Assessing evidence and testing appropriate hypotheses. *Science and Justice*, 54(6), 502–504.

The issue of incorporating errors in evaluating DNA and other forensic evidence is described in:

Fenton, N.E., Neil, M., and Hsu, A. (2014b). Calculating and understanding the value of any type of match evidence when there are potential testing errors. *Artificial Intelligence and Law*, 22, 1–2.

Presentations and videos by many of the world's leading forensic and legal scholars covering the issues in this chapter can be found here:

Isaac Newton Institute, Probability and Statistics in Forensic Science. (2016). https://www.newton.ac.uk/event/fos.

More detailed coverage of BNs for forensic analysis can be found in:

Taroni, F., Aitken, C., Garbolino, P., and Biedermann, A. (2014). *Bayesian Networks and Probabilistic Inference in Forensic Science*, 2nd ed. John Wiley & Sons, Chichester, UK.
Tribe, L.H. (1971). Trial by Mathematics. *Harvard Law Review*, 84, 1329.

Detailed coverage of the probabilistic and computational issues associated with DNA mixtures can be found in:

Cowell, R.G., Graversen, T., Lauritzen, S., and Mortera, J. (2015). *Journal of the Royal Statistical Society*, series C, 64(1), 1–48.
Graversen T., and Lauritzen, S. (2015). Computational aspects of DNA mixture analysis. *Statistics and Computing*, 25(3), 527–541.

The medical negligence case where we used Bayes to compare the risk of alternative treatments is described in:

Fenton, N., and Neil, M. (2010). Comparing risks of alternative medical diagnosis using Bayesian arguments. *Journal of Biomedical Informatics*, 43, 485–495.

16

Building and Using Bayesian Networks for Legal Reasoning

Of all of the application domains where BNs have been used (or proposed), the domain of legal reasoning is possibly the most natural, because we are primarily concerned with updating prior hypotheses as new evidence is presented. Moreover, because people's lives depend on reaching "correct" decisions from the evidence, it could be argued that the domain of legal reasoning is also the most critical for BNs. However, as discussed in Chapter 15, to date, the impact of BNs in the law has been low. One of the reasons for the minimal take-up is that traditionally the process of building BNs for legal arguments has been ad hoc, with little possibility for learning and process improvement.

Hence, the objective of this chapter is to show that there is a more systematic method for building legal BNs that can be practiced and learned by nonmathematicians. We show how the idioms approach of Chapter 8 enables us to build useful (and even complex) legal arguments in a consistent and repeatable way. We introduce a number of specially adapted idioms (Sections 16.1 to 16.6) and present examples throughout of their use. In Section 16.7, we provide a comprehensive case example that puts together all of the idioms. Finally, in Section 16.8, we describe how our approach helps expose and avoid common probabilistic fallacies made in legal arguments, and also dispels the myth that Bayesian reasoning is incompatible with legal reasoning.

Despite Bayes' elegant simplicity and natural match to intuitive reasoning about evidence, practical legal arguments normally involve multiple pieces of evidence (and other issues) with complex causal dependencies. Fortunately, there are unifying underlying concepts that mean we can build relevant BN models, no matter how large, that are still conceptually simple because they are based on a very small number of special cases of the idioms that were described in Chapter 8.

16.1 The Evidence Idiom

We can think of the simple two-node BN (consisting of a hypothesis node H, i.e., a parent of an evidence node E)—and its extension to multiple pieces of evidence—as the most basic BN idiom for legal reasoning. This basic idiom, which we call the evidence idiom, is an instantiation of the cause–consequence idiom and has the generic structure shown in Figure 16.1.

We do not distinguish between evidence that supports the prosecution (H true) and evidence that supports the defense (H false), since the

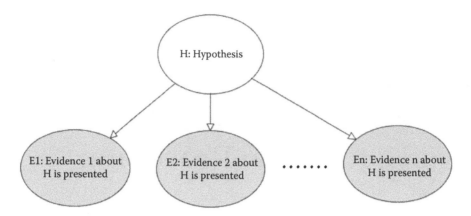

Figure 16.1   Evidence idiom.

The Adams case discussed in Chapter 15 was a good example of the evidence idiom with a conflict pattern. Specifically, there was one piece of evidence (DNA match) that supported the prosecution hypothesis and two pieces of evidence (defendant alibi and failure of victim to identify defendant in ID parade) that supported the defense hypothesis. As we showed in Chapter 15, Bayes enables us to seamlessly incorporate the conflicting evidence.

BN model handles both types of evidence seamlessly. Hence, this idiom subsumes two basic patterns:

1. *Corroboration pattern*: this is simply the case where there are two pieces of evidence, E1 and E2, that both support one side of the argument.
2. *Conflict pattern*: this is simply the case where there are two pieces of evidence, E1 and E2, with one supporting the prosecution and the other supporting the defense.

The evidence idiom assumes that the various pieces of evidence are independent, and it is this assumption that makes it possible to compute the overall likelihood ratio for the combined evidence by simply multiplying the individual LRs (as we explained in Chapter 15). In other words, we do not need BN computations to calculate the LR. It is not surprising, therefore, that in practice, forensic experts try to force their evidence to fit this idiom. However, as we also discussed in Chapter 15, in real cases, the assumption is oversimplistic, meaning that the evidence idiom has a number of limitations. The following idioms identify and address these various limitations in turn.

16.2   The Evidence Accuracy Idiom

The evidence accuracy idiom (Figure 16.2) is the legal equivalent of the measurement idiom. Just as the measurement idiom is crucial for distinguishing between a potentially inaccurate "measured value" of some attribute and the true but unknown value itself (such as an IQ test to measure the intelligence of a person or a mobile phone app to measure the temperature outside), any piece of evidence in a legal case can be considered a potentially inaccurate assertion about the true value of some unknown hypothesis. So, when an expert witness testifies that a defendant had an excess amount of alcohol in his or her blood, we cannot treat this as being exactly the same as the *fact* that the defendant had an

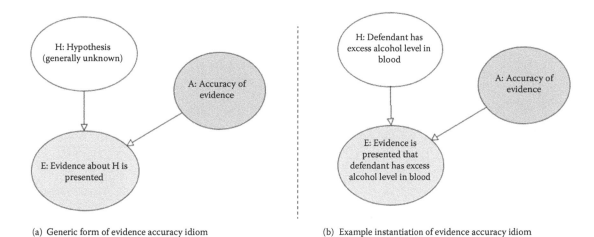

(a) Generic form of evidence accuracy idiom   (b) Example instantiation of evidence accuracy idiom

Figure 16.2   Evidence accuracy idiom.

excess amount of alcohol in his or her blood. The expert could be using inaccurate test equipment or might make errors in reporting.

For simplicity, we have lumped together all possible sources of inaccuracy into a single node (see sidebar "Modeling Different Notions of 'Accuracy'").

In general, what is often assumed to be a "single piece of evidence for a single hypothesis" turns out to be more realistically modeled by multiple instances of the evidence accuracy idiom.

Example 16.1

Let us return to the two-node example of Chapter 15, Figure 15.2, where H represents the hypothesis "Defendant is source of DNA found at crime scene" and E represents the evidence "Defendant DNA matches that found at the crime scene." The two-node model presented made all of the following unstated assumptions:

- The DNA trace tested really was that found at the scene of the crime (i.e., the crime-scene procedures were good to ensure the sample was taken to the lab, and the lab procedures were good to ensure the same sample was tested).
- The DNA sample obtained from the defendant is the one that really was tested (so the lab procedures for reference samples were good).
- Neither of the two DNA samples became contaminated at any time (i.e., good lab procedures generally).
- The DNA testing process and equipment are perfect, with no possibility of an incorrect genotype being recorded at any locus for either sample (i.e., perfectly accurate DNA testing and reporting equipment).

If any of the above is uncertain, then the necessary BN model required to understand the value of the DNA match evidence is something like the one shown in Figure 16.3.

Modeling Different Notions of "Accuracy"

There are a number of ways in which evidence accuracy can be tailored. For example, there is no need to restrict the node *accuracy of evidence* to being a Boolean (false, true). In general, it may be measured on a more refined scale, for example, a ranked scale like {very low, low, medium, high, very high}, where very low means "completely inaccurate" and very high means "completely accurate," or even a continuous scale. Also, it is possible to extend the idiom by decomposing "accuracy" into three components: *competence*, *objectivity*, and *veracity*. This is especially relevant for eyewitness evidence.

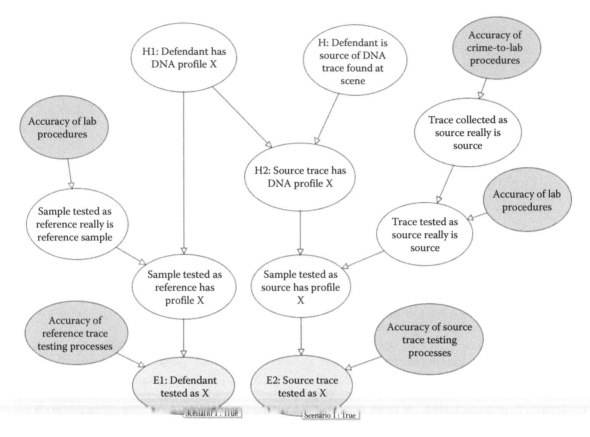

Figure 16.3 BN model of DNA match evidence that incorporates two instances of the evidence accuracy idiom and three instances of the standard measurement idiom.

In this model, we have used the same prior for H true as in Figure 15.2 of Chapter 15 (i.e., 1 in 10,000) and the same random match probability for the particular profile X (namely 1 in 1000). We assume 90% prior probability for each of the evidence accuracy nodes. With these assumptions, when we run the model with the two pieces of evidence (corresponding to the assertion of a match), we get a posterior probability of just over 4% for H true. This compares with a posterior of just over 9% using the two-node model assumptions. When all the accuracy nodes are set to true, the posterior is, of course, the same as the two-node model.

16.3 Idioms to Deal with the Key Notions of "Opportunity" and "Motive"

In Chapter 15, we explained the importance of including all relevant context when building a BN model to evaluate the probative value of legal evidence. In particular, our ultimate goal is to determine the impact of the evidence on the ultimate hypothesis (defendant is guilty). Hence, if we wish to build BNs for complete legal cases, we need to include the ultimate hypothesis as a node in the model. It may seem "correct" that such a node should have no parents, as this fits naturally with the intuitive approach to legal reasoning whereby it is the hypothesis about which we start with an unconditional prior belief before observing evidence to update that belief. However, there are two very common types of evidence that, unlike the examples seen so far, support hypotheses that are *causes*, rather than *consequences*, of guilt. These hypotheses are concerned with "opportunity" and "motive," and they inevitably change the fundamental structure of the underlying causal model.

16.3.1 Opportunity

When lawyers refer to "opportunity" for a crime, they actually mean a necessary requirement for the defendant's guilt. By far the most common example of opportunity is "being present at the scene of the crime." So, for example, if Joe Bloggs is the defendant charged with slashing the throat of Fred Smith at 4 Highlands Gardens on 1 January 2018, then Joe Bloggs had to be present at 4 Highlands Gardens on 1 January 2018 in order to be guilty of the crime. The correct causal BN model to represent this situation (incorporating the evidence accuracy idiom) is shown in Figure 16.4(a).

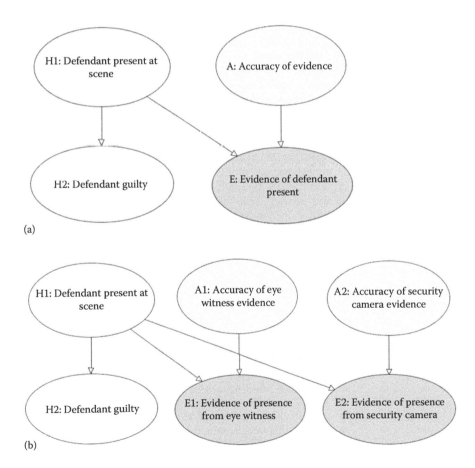

Figure 16.4 Opportunity idioms. (a) Idiom for incorporating "opportunity" (defendant present at scene of crime). (b) Multiple types of evidence for opportunity hypothesis.

Note that, just as the hypothesis "defendant is guilty" is unknowable to a jury, the same is true of the opportunity hypothesis. Just like any hypothesis in a trial, its truth value must be determined on the basis of evidence. In this particular example, there might be multiple types of evidence for the opportunity hypothesis, each with different levels of accuracy, as shown in Figure 16.4(b).

Given the natural reluctance of lawyers to consider prior probabilities for the ultimate hypothesis of "guilty," there is a clear benefit in the case where the notion of opportunity is relevant. Box 16.1 explains how, in many cases, it may be possible to arrive at a noncontroversial probability of guilt conditioned on opportunity.

Box 16.1   The "Opportunity Prior"

Suppose it is not disputed that the defendant was one of a group of 100 people at the crime scene, among whom only one committed the crime. Then there is no need to consider an unconditional prior for "guilty"; given that Opportunity (being at the scene of the crime) is true in this case, there is a noncontroversial probability of guilt equal to 1%, and this is also as consistent as possible with the legal notion of "innocent until proven guilty." This idea can be extended to arrive at a realistic prior even when it is not known that the defendant was "at the crime scene." The approach applies to all cases where we know that a crime has taken place and that it was committed by one person against one other person, for example, murder, assault, robbery.

The approach is based on simple location and time parameters that determine both (a) the crime scene/time (within which it is certain the crime took place) and (b) the extended crime scene/time that is the "smallest" window within which it is certain the suspect was known to have been "closest" in location/time to the crime scene. In each case, we rely on an estimate of the number of people who were actually inside the location/time.

For example, if a person was mugged while standing next to a particular lamppost by Piccadilly Circus tube station, then the Crime Scene would be an area about one meter around the lamppost. If the time of the attack was between 20:30–20:33 on a Thursday, then the number of people n who were "inside" that location/time could be estimated at 50. If the suspect is known to have been inside that location/time (for example, from CCTV images), then the "opportunity prior" is 1/50. However, suppose the only CCTV images available show the suspect was 100 meters from the crime scene at 20:40. Then the "extended" crime scene is the area whose center is the lamppost and whose perimeter is 100 meters from the lamppost. We now have to estimate N, the number of people who were in this area between 20:30 and 20:40. Exactly one of these people—which we know includes the suspect—must have committed the crime. So, suppose $N = 500$. Then the opportunity prior is 1/500. Full details of this approach are provided in "Further Reading."

The following example brings together all of the idioms discussed so far.

Example 16.2

A serious assault has taken place in a pub at a time when approximately 100 people were present. One witness tells police that the offender looked like a man who lives close to the pub, whom he believes is called Jack Smith. A second witness tells police several days later that Jack Smith told her that he committed the assault in the pub. After Jack Smith is arrested, his DNA is sampled and found to match a trace of DNA found on an ashtray in the pub; the trace is low quality with identifiable genotypes at only four loci, which a DNA expert asserts has a 1 in 1000 random match probability. The prosecution case is based on the two witness statements and the DNA match evidence. The defense case is based on an

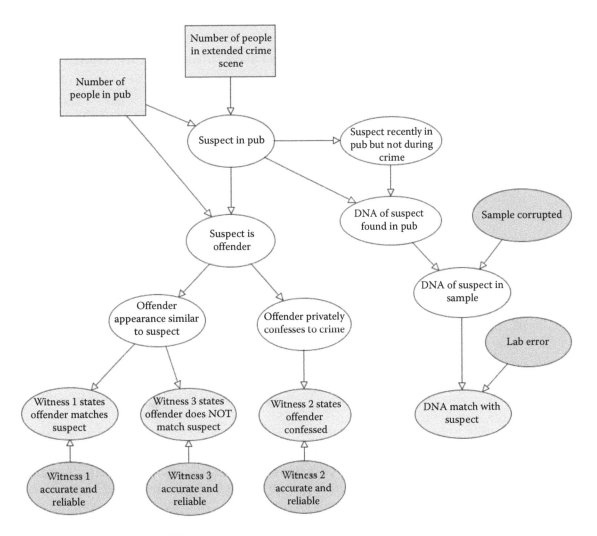

Figure 16.5 Pub offender model.

eyewitness to the assault who asserts that Jack Smith does *not* match the offender. A BN model for the case is shown in Figure 16.5.

The three witness statements are examples of the evidence accuracy idiom. For example, the NPT for the node "witness 1 states offender matches suspect" is shown in Table 16.1.

So, we are assuming that if the witness is not accurate and reliable, there is still a 10% chance that he or she would make the assertion irrespective of whether it is true. The other witness nodes are defined similarly, and we also assume a prior accuracy of 90% for each witness.

For the DNA evidence, we assume there are small probabilities of both lab error and sample corruption, and also that an "innocent" possible explanation for the suspect's DNA being in the pub is that he had been there recently but not at the time of the crime. We use the "opportunity prior" approach (Box 16.1) to determine that, even if the suspect was not at the crime scene (which has 100 people), he was in an "extended crime scene area" (a short walk away from the pub), for which it is estimated there were at most 1000 people at the time of the crime. Consequently,

Table 16.1

NPT for Node "Witness 1 States Offender Matches Suspect"

Witness 1 Accurate and Reliable	False		True	
Offender Appearance Similar to Suspect	False	True	False	True
False	0.9	0.9	1	0
True	0.1	0.1	0	1

Table 16.2

Changing Probabilities as Evidence Is Entered

	Suspect in Pub at Time of Crime (%)	Suspect Is Offender (%)
No evidence	10	0.1
DNA evidence	84	0.84
Witness 1 evidence	89	29
Witness 2 evidence	98	88
Witness 3 evidence	86	13

we get the prior probabilities shown in row 1 of Table 16.2. As we enter evidence in order, we get the changing probabilities shown in successive rows of Table 16.2.

16.3.2 Motive

There is a widespread acceptance within the police and legal community that a crime normally requires a motive (this covers the notions of "intention" and "premeditation"). The existence of a motive increases the chances of a crime happening. This means that, as with opportunity, the correct causal BN model to represent motive in a legal argument is shown in Figure 16.6.

If we wish to include both opportunity and motive in the argument, then the appropriate BN idiom is shown in Figure 16.7.

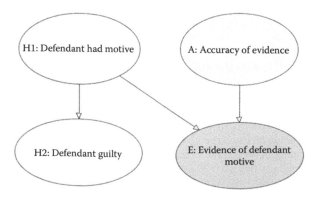

Figure 16.6　Idiom for incorporating motive.

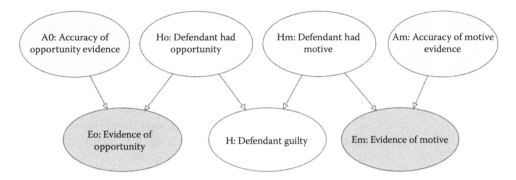

Figure 16.7 Incorporating both opportunity and motive.

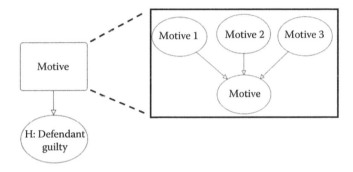

Figure 16.8 Appropriate model for multiple motives (using object-oriented notation).

This makes the task of defining the NPT for the ultimate hypothesis *H* a bit harder, since we must consider the probability of guilt conditioned on both opportunity and motive, but, again, these specific conditional priors are inevitably made implicitly anyway. What we *do* need to avoid is conditioning *H* directly on *multiple* motives, that is, having multiple motive parents of *H*. Instead, if there are multiple motives, we simply model what the lawyers do in practice in such cases: specifically, they consider the accuracy of each motive separately but jointly think in terms of the strength of overall motive. The appropriate model for this is shown in Figure 16.8 (using object-oriented notation).

16.4 Idiom for Modeling Dependency between Different Pieces of Evidence

In the case of a hypothesis with multiple pieces of evidence (such as in Figure 16.1), we have so far assumed that the pieces of evidence were *independent* (conditional on *H*). But, in general, we cannot make this assumption. Suppose, for example, that the two pieces of evidence for "defendant present at scene" were images from two video cameras. If the cameras were of the same make and were pointing at the same spot, then there is clear dependency between the two pieces of evidence: if we know that one of the cameras captured an image of a person matching the defendant, there is clearly a very high chance that the same will be true of the other camera, irrespective of whether the defendant really was or was not present. Conversely, if one of the cameras did not capture such an image, there is clearly a very high chance that the same will be true of the other camera, irrespective of whether the defendant really was not present.

The appropriate way to model this would be as shown in Figure 16.9(a) (for simplicity, we are ignoring the issue of accuracy here), with a direct dependency between the two pieces of evidence. Also, for simplicity, note from the NPTs that "dependence" here means the cameras will produce identical results; we could easily adjust the NPT to reflect partial dependence by, for example, making the probability 0.9 (as opposed to 1) that camera 2 will return "true" when H is true and camera 1 is true.

If we assume that the prior for H being true is 0.1 and the prior for the cameras being dependent is 0.5, then the initial marginal probabilities are shown in Figure 16.9(b).

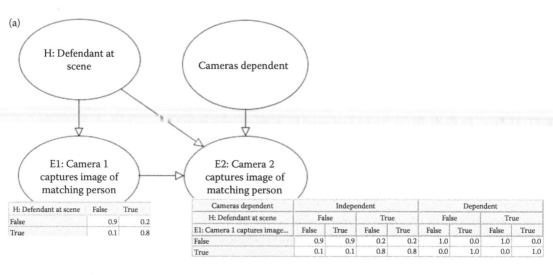

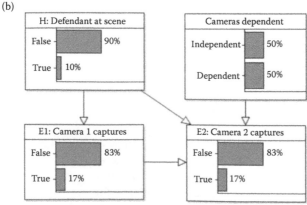

Figure 16.9 Idiom for modeling dependency between different pieces of evidence. (a) Idiom instantiation. (b) Running model with initial probabilities.

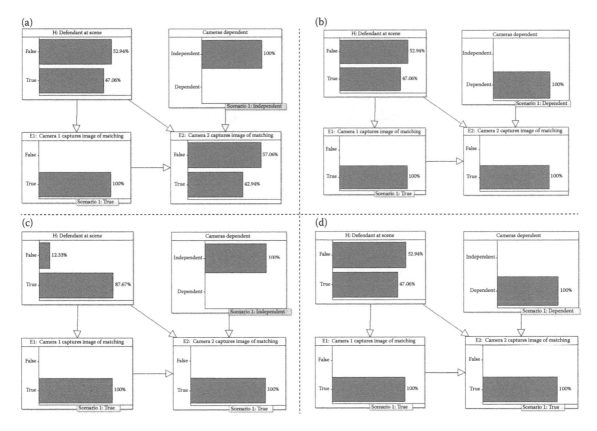

Figure 16.10 Running the dependent evidence model with different evidence observed. (a) E1 true, the two cameras independent. (b) E1 true, the two cameras dependent. (c) E1 and E2 true, the two cameras independent. (d) E1 and E2 true, the two cameras dependent.

It is instructive to compare the results between the two models: (a) where there is no direct dependence between E1 and E2, and (b) where there is. Hence, in Figure 16.10(a),(b), we show both these cases where evidence E1 is true. Although both models result in the same (increased) revised belief in H, the increased probability that E2 will also be true is different. In (a), the probability increases to 43%, but in (b), the probability is 100%, since here we know E2 will replicate the result of E1.

Figure 16.10(c) and (d) show Romaine's suspicions about the value as true in both cases. When they are dependent, the additional E2 evidence adds no extra value. However, when they are independent, our belief in *H* increases to 88%.

There are other types of dependent evidence that require slightly different BN idioms that are beyond the scope of this chapter. These include:

■ Dependent evidence through confirmation bias: in this case, there are two experts determining whether there is a forensic match (the type of forensics could even be different, such as DNA and fingerprinting). It has been shown that the second

The benefits of making explicit the direct dependence between pieces of evidence are enormous. For example, in a murder case in which we acted as expert witnesses, the prosecution presented various pieces of directly dependent evidence in such a way as to lead the jury to believe that they were independent, hence drastically overstating the impact on the hypothesis being true. A simple model of the evidence, based on the structure above, showed that once the first piece of evidence was presented, the subsequent evidence was almost useless, in the sense that it provided almost no further shift in the hypothesis probability. A similar problem of treating dependent evidence as independent was a key issue in some of the famous cases discussed in Chapter 15 (notably the Collins case and Sally Clark).

expert's conclusion will be biased if he/she knows the conclusion of the first expert.

■ Dependent evidence through common biases, assumptions, and sources of inaccuracies.

16.5   Alibi Evidence Idiom

A special case of additional direct dependency within the model occurs with so-called *alibi* evidence. In its most general form, alibi evidence is simply evidence that directly contradicts a prosecution hypothesis. The classic example of alibi evidence is an eyewitness statement contradicting the hypothesis that the defendant was present at the scene of the crime, normally by asserting that the defendant was in a different specific location. What makes this type of evidence special is that the hypothesis itself may directly influence the accuracy of the evidence, such as when the eyewitness is either the defendant him- or herself or a person known to the defendant. Figure 16.11 shows the appropriate model with the revised dependency in the case where the witness is known to the defendant. A possible NPT for the node A1 (accuracy of alibi witness) is also shown in Figure 16.11.

Imagine that the witness is the partner of the defendant. Then what the NPT is saying is that if the defendant is not guilty, there is a very good chance the partner will provide an accurate alibi statement. But if the defendant *is* guilty, there is a very good chance the partner's statement will be inaccurate. Of course, if the witness is an enemy of the defendant the NPT will be somewhat inverted. But with the NPT of Figure 16.11, we can run the model and see the impact of the evidence in Figure 16.12.

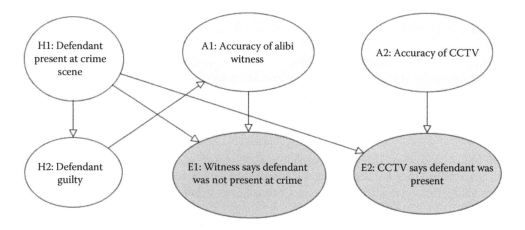

H2: Defendant guilty	False	True
False	0.1	0.9
True	0.9	0.1

Figure 16.11   Alibi evidence idiom, with NPT for A1 (accuracy of alibi witness).

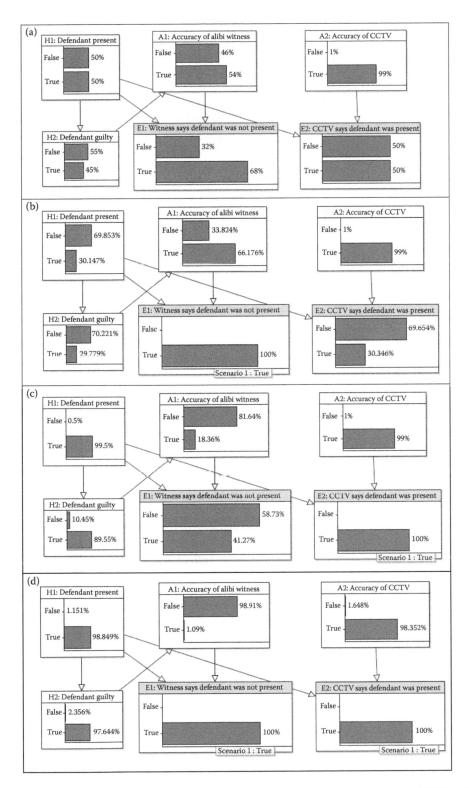

Figure 16.12 Impact of alibi evidence. (a) Prior probabilities. (b) Alibi evidence only. (c) CCTV evidence only. (d) Conflicting evidence (both CCTV and alibi).

The model provides some very powerful analysis, notably in the case of conflicting evidence (i.e., where one piece of evidence supports the prosecution hypothesis and one piece supports the defense hypothesis). The most interesting points to note are:

- When presented on their own [(b) and (c), respectively], both pieces of evidence lead to an increase in belief in their respective hypotheses. Hence, the alibi evidence leads to an increased belief in the defense hypothesis (not guilty), and the CCTV evidence leads to an increased belief in the prosecution hypothesis (guilty). Obviously, the latter is much stronger than the former because of the relative priors for accuracy, but nevertheless, on their own, they both provide support for their respective lawyers' arguments.
- When both pieces of evidence are presented (d), we obviously have a case of conflicting evidence. If the pieces of evidence were genuinely independent, the net effect would be to decrease the impact of both pieces of evidence on their respective hypotheses compared to the single evidence case. However, because the alibi evidence is dependent on H2, the result is that the conflicting evidence actually strengthens the prosecution case even more than if the CCTV evidence was presented on its own. Specifically, because of the prior accuracy of the CCTV evidence, when this is presented together with the alibi evidence, it leads us to doubt the accuracy of the latter (we tend to believe the witness is lying) and hence, by backward inference, to increase the probability of guilt.

16.6 Explaining Away Idiom

As we saw in Chapter 7, one of the most powerful features of BN reasoning is the concept of "explaining away." An example of this can be seen in the evidence accuracy idiom of Figure 16.2(b) above:

> When the expert asserts that the defendant had an excess alcohol level (i.e., when the evidence E is observed), then there are two possible explanations for this: either the defendant really did have an excess level (i.e., the hypothesis H is true) or the expert is not accurate. If we have a high prior for accuracy and a 50:50 prior for H, then after observing E, the hypothesis H becomes very likely. However, if we discover that H is false, then the evidence is explained away by inaccuracy.

We could consider "explaining away" an explicit idiom, as shown in Figure 16.13.

However, it turns out that traditional "explaining away" does not work in a very important class of situations that are especially relevant

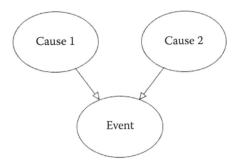

Figure 16.13 Explaining away idiom.

for legal reasoning. These are the situations where the causes are *mutually exclusive* (as discussed in Chapter 8); that is, if one of them is true, then the others must be false.

Suppose, for example, that we have evidence E that blood found on the defendant's shirt matches the victim's blood. Since there is a small chance (let us assume 1%) that the defendant's blood is the same type as the victim's, there are two possible causes of this:

- Cause 1: the blood on the shirt is from the victim.
- Cause 2: the blood on the shirt is the defendant's own blood.

In this case, only one of the causes can be true. But the standard approach to BN modeling will not produce the correct reasoning in this example, as we explained in Chapter 8. So we have to use the special solution described in Chapter 8 whereby we have to introduce an auxiliary node and a constraint node.

Example 16.3

In the BN in Figure 16.14, the blood on shirt example is put into a fuller context and the necessary auxiliary and constraint nodes are shown along with their NPTs.

Table 16.3 shows how the respective probabilities of the two mutually exclusive causes change as we enter the various pieces of evidence. With only the blood match evidence, it is far more likely the blood is from the victim, but this changes completely when the other evidence is observed.

Note that that the values 0.09 and 0.5 in the top row of the NPT of the constraint node in Figure 16.14 correspond to the respective priors for the blood on the shirt from victim and defendant. These in turn come from setting the prior for "defendant guilty" to 0.1 and the prior for "defendant cut himself" to 0.5.

The key point about the special "explaining away" idiom is that it should be used when there are two or more mutually exclusive causes of some evidence or observations, each with separate causal pathways. The auxiliary node has the benefit of (a) revealing assumptions about the state space that would be otherwise tacit or implicit and (b) helping to keep causal pathways cleaner at a semantic level.

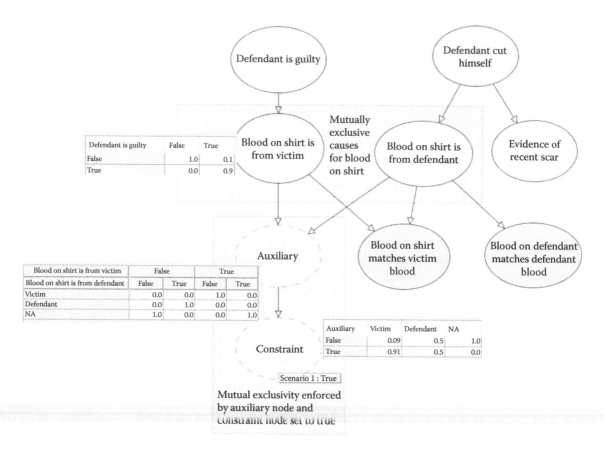

Figure 16.14 Full example of mutually exclusive causes.

Table 16.3

Changing Probability Values as Evidence Is Entered

		Blood on the Shirt from Victim (%)	Blood on the Shirt from Defendant (%)
1	No evidence	9	50
2	With constraint true (normalizes the priors, assuming one must be true)	15.3	84.7
3	Blood match evidence	94.7	5.3
4	Blood on defendant matches defendant blood	15.3	84.7
5	Evidence of recent scar	2	98

16.7   Putting It All Together: Vole Example

Lagnado describes the fictional case based on Agatha Christie's play *Witness for the Prosecution*, as shown in Box 16.2.

Box 16.2   The Case of Vole

Leonard Vole is charged with murdering a rich elderly lady, Miss French. He had befriended her and visited her regularly at her home, including the night of her death. Miss French had recently changed her will, leaving Vole all her money. She died from a blow to the back of the head. There were various pieces of incriminating evidence: Vole was poor and looking for work; he had visited a travel agent to enquire about luxury cruises soon after Miss French had changed her will; the maid claimed that Vole was with Miss French shortly before she was killed; the murderer did not force entry into the house; Vole had blood stains on his cuffs that matched Miss French's blood type.

As befits a good crime story, there were also several pieces of exonerating evidence: the maid admitted that she disliked Vole; the maid was previously the sole benefactor in Miss French's will; Vole's blood type was the same as Miss French's, and thus also matched the blood found on his cuffs; Vole claimed that he had cut his wrist slicing ham; Vole had a scar on his wrist to back this claim. There was one other critical piece of defense evidence: Vole's wife, Romaine, was to testify that Vole had returned home at 9:30 p.m. This would place him far away from the crime scene at the time of Miss French's death. However, during the trial, Romaine was called as a witness for the prosecution. Dramatically, she changed her story and testified that Vole had returned home at 10:10 p.m., with blood on his cuffs, and had proclaimed: "I've killed her." Just as the case looked hopeless for Vole, a mystery woman supplied the defense lawyer with a bundle of letters. Allegedly, these were written by Romaine to her overseas lover (who was a communist!). In one letter, she planned to fabricate her testimony in order to incriminate Vole and rejoin her lover. This new evidence had a powerful impact on the judge and jury. The key witness for the prosecution was discredited, and Vole was acquitted.

After the court case, Romaine revealed to the defense lawyer that she had forged the letters herself. There was no lover overseas. She reasoned that the jury would have dismissed a simple alibi from a devoted wife; instead, they could be swung by the striking discredit of the prosecution's key witness.

What we will now do is build the model from scratch using only the idioms introduced. Most importantly, we are able to run the model to demonstrate the changes in posterior guilt that result from presenting evidence in the order discussed in the example.

Step 1: Identify the key prosecution hypotheses (including opportunity and motive):

- The ultimate hypothesis "H0: Vole guilty"
- Opportunity: "Vole present"
- Motive: There are actually two possible motives: "Vole poor" and "Vole in will"

Step 2: Consider what evidence is available for each of the above and what the accuracy is of the evidence:

Evidence for H0. There is no direct evidence at all for H0, since no witness testifies to observing the murder. But we have evidence for two hypotheses that depend on H0:

Of course, neither of these hypotheses is guaranteed to be true if H0 is true, but this uncertainty is modeled in the respective NPTs.

- H1: Vole admits guilt to Romaine
- H2: Blood on Vole's shirt is from French

The (prosecution) evidence to support H1 is the witness statement by Romaine. Note that Romaine's evidence of Vole's guilt makes her evidence of "Vole present" redundant (so there is no need for the link from "Vole present" to Romaine's testimony in the original model).

The issue of accuracy of evidence is especially important for Romaine's evidence. Because of her relationship with Vole, the H1 hypothesis influences her accuracy.

Evidence to support H2 is that the blood matches French's.

The evidence to support the opportunity "Vole present" is a witness statement from the maid.

Step 3: Consider what defense evidence is available to challenge the above hypotheses.

- The evidence to challenge H1 is the (eventual) presentation of the love letters and the introduction of a new (defense) hypothesis "H4: Romaine has lover."
- The evidence to challenge the opportunity "Vole present" is (a) to explicitly challenge the *accuracy* of the maid's evidence and (b) Vole's own alibi evidence.
- The evidence to challenge H2 is that the blood matches Vole's (i.e., Vole and French have the same blood type). Additionally, the defense provides an additional hypothesis, "H3: Blood on Vole is from previous cut," that depends on H2.

Finally, for simplicity, we shall assume that some evidence (such as the blood match evidence) is perfectly accurate and that the motives are stated (and accepted) without evidence.

From this analysis, we get the BN shown in Figure 16.15.

Note how the model is made up only from the idioms we have introduced (the blood match component is exactly the special "explaining away" idiom example described in Section 16.6).

With the exception of node H5 (Romaine has lover), the priors for all parentless nodes are uniform. The node H5 has prior set to True = 10%. What matters when we run the model is not so much whether the probabilities are realistic but rather the way the model responds to evidence. Hence, Table 16.4 shows the effect (on probability of guilt) of the evidence as it is presented sequentially, starting with the prosecution evidence.

The key points to note here are that:

- The really big jump in belief in guilt comes from the introduction of the blood match evidence (at this point, it jumps from 56.7% to 86.7%). However, if Romaine's evidence had been presented before the blood match evidence, that jump would have been almost as great (52.6–81%).
- Once all the prosecution evidence is presented (and bear in mind that at this point, the defense evidence is set to the prior values) the probability of guilt seems overwhelming (96.7%).

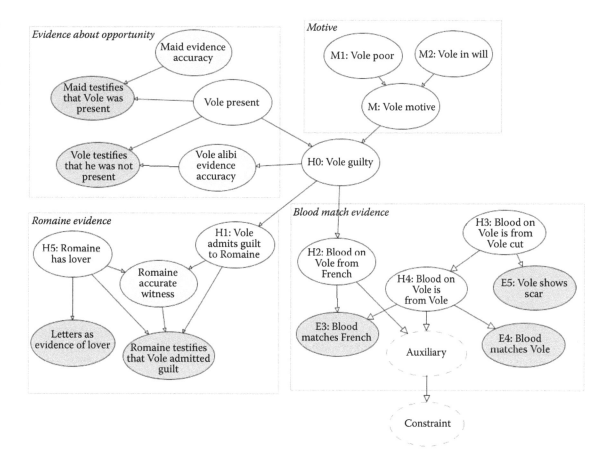

Figure 16.15 Full Vole model.

Table 16.4

Effect on Probability of Guilt of Evidence Presented Sequentially

Sequential Presentation of Evidence	H0 Vole Guilty Probability (%)
Prosecution evidence presented	
1. Prior (no observations)	37
2. Motive evidence added (M1 and M2 = true)	39.7
3. Maid testifies Vole was present = true	56.7
4. E3 blood matches French evidence = true	86.7
5. Romaine testifies Vole admitted guilt = true	96.7
Defense evidence presented	
6. Vole testifies he was not present = true	97
7. Maid evidence accuracy = false	91.4
8. E4 blood matches Vole = true	68.2
9. E5 Vole shows scar = true	41.9
10. Letters as evidence = true	15.7

- If, as shown, the first piece of defense evidence is Vole's own testimony that he was not present, then the impact on guilt is negligible (in fact, there is a very small increase). This confirms that, especially when seen against stronger conflicting evidence, an alibi that is not "independent" is very weak and may even be seen as evidence of lying and having something to hide. Although the model does not incorporate the intended (but never delivered) alibi statement by Romaine, it is easy to see that there would have been a similarly negligible effect; that is, Romaine's suspicions about the value of her evidence are borne out by the model.
- The first big drop in probability of guilt comes with the introduction of the evidence that Vole and French have the same blood type.
- However, when all but the last piece of defense evidence are presented, the probability of Vole's guilt is still 41.9%—higher than the initial probability. Only when the final evidence— Romaine's letters—is presented do we get the dramatic drop to 15.7%. Since this is considerably *less* than the prior (37%), this should certainly lead to a not guilty verdict if the jury were acting as rational Bayesians.

The overall likelihood ratio for the evidence is less than one—it supports the defense case.

It is also worth noting the way the immediate impact of different pieces of evidence is very much determined by the order in which the evidence is presented. To emphasize this point, Figure 16.16 presents a sensitivity analysis, in the form of a tornado chart, of the impact of each

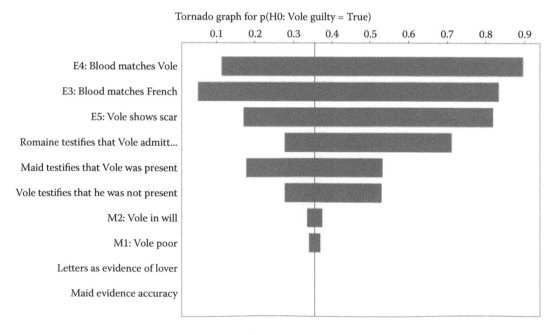

Figure 16.16 Sensitivity analysis on guilty hypothesis.

possible piece of evidence *individually* on the probability of guilt. From this graph, we can see, for example, that if all other observations are left in their prior state, the Vole blood match evidence has the largest impact on Vole guilty; when it is set to true, the probability of guilt drops to just over 10%, and when it is set to false, the probability of guilt jumps to nearly 90%. The French blood match evidence has a very similar impact. At the other extreme, the individual impact of the Romaine letters is almost negligible. This is because this piece of evidence only becomes important when the particular combination of other evidence has already been presented.

16.8 Using Bayesian Networks to Expose Further Paradoxes and Fallacies of Legal Reasoning

We have demonstrated how the BN approach is effective in avoiding the most common fallacies of legal reasoning (namely fallacies relating to the transposed conditional and to treating dependent evidence). We have also demonstrated how the BN approach resolves many so-called "paradoxes" of legal reasoning such as the "Twins paradox" in Chapter 15. We conclude with three further interesting examples of such problems resolved by the BN approach.

16.8.1 The Conjunction Paradox

In civil cases, plaintiffs only have to prove that there is a greater than 50% probability that their claim is true. Suppose that, in a particular case, the plaintiff's claim is true if two independent facts A and B are true. Suppose that evidence presented suggests that A and B each have a probability of 70% of being true. Then, while separately, each of the facts is well above a 50% threshold for "truth," the case fails to meet the 50% threshold, since

$$P(A \text{ and } B) = P(A) \times P(B) = 0.7 \times 0.7 = 0.49 = 49\%$$

While this is, of course, correct from a formal probability perspective, legal scholars regard this as being inconsistent with both legal reasoning and intuition: the plaintiff's case is rejected despite having shown strong support for both A and B. This is what Cohen originally referred to as the "conjunction paradox."

In fact, it is possible to reconcile both the formal probabilistic interpretation and the intuitive legal perspective once we properly articulate the assumptions about the evidence. Specifically, what will be observed by a judge in the case are witnesses W1 and W2 respectively asserting that A and B are true. In other words, we have an instance of the *evidence accuracy idiom*, as shown in Figure 16.17.

Rather than "observing" a 70% probability that A is true, it is more reasonable to assume that the judge observes witness W1 and

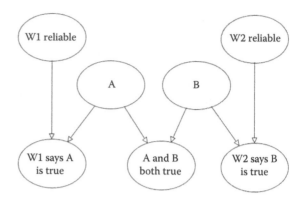

Figure 16.17   BN model for the conjunction paradox.

determines there is a 70% probability that W1 is reliable (which we assume for simplicity means telling the truth). With the same assumption about witness *B*, we assign probabilities of 70:30 for True/False for the nodes "W1 reliable" and "W2 reliable." Suppose we also assume that the priors for nodes *A*, *B* are 50:50 True/False (and so the prior probability that the plaintiff claim is true is 25%; if we wanted to assume the prior probability that the plaintiff claim is true is 50%, we would be assuming that both *A* and *B* had priors of 70.7% true). The conditional probability for the nodes representing what the witnesses say are as shown in Table 16.5.

Then the results of running the model pre- and post-observation of the witness evidence are shown in Figure 16.18(a),(b), respectively.

So, with the 70% assumptions applying to the witness accuracy about *A* and *B* (rather than directly to *A* and *B*), the probabilistic results are perfectly consistent with the legal intuition—the plaintiff case (i.e., "*A* and *B* both true") posterior is 72.25% and so easily surpasses the 50% threshold. This is despite the fact that, with the 50:50 prior assumptions for *A* and *B* individually, the prior for the plaintiff case is only 25%. If we assume the prior probability of the plaintiff case is 50% (i.e., that *A* and *B*, respectively, have priors of 70.7%), then the posterior probability for the plaintiff case goes to 86.83%.

The BN model here is essentially the argument that was presented using first principles in Dawid (1987), which for nonmathematicians is very hard to understand. Using a BN model makes things clearer and is also scalable to additional pieces of evidence.

Table 16.5
Conditional Probability Table for the Nodes Representing What the Witnesses Say

Witness Reliable		True		False	
Fact		True	False	True	False
Witness says fact is True		1	0	0.5	0.5
Witness says fact is False		0	1	0.5	0.5

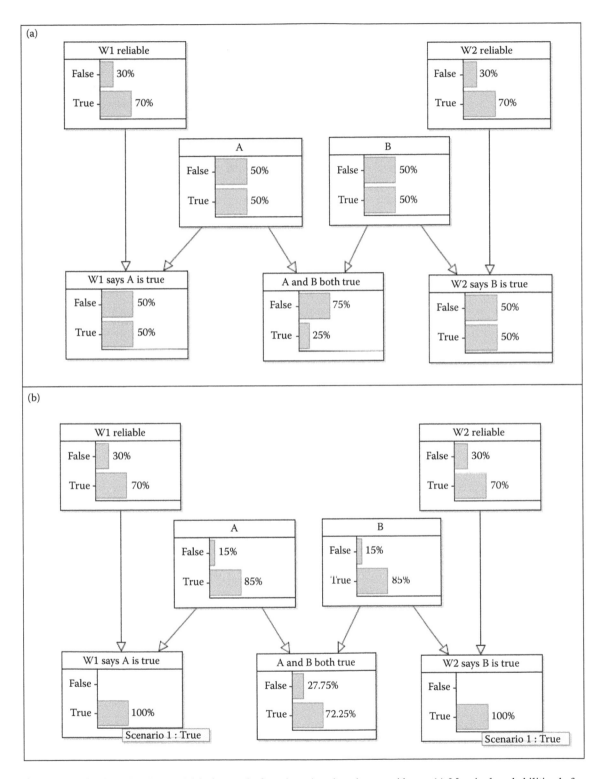

Figure 16.18 Running the model before and after observing the witness evidence. (a) Marginal probabilities before evidence is observed. Note that the probability "A and B both true" is 25%. (b) Posterior probabilities after entering the witnesses' evidence. Note that the probability "A and B both true" is well over the 50% threshold. The LR for the evidence with respect to the plaintiff claim being true rather than false is over 8.

16.8.2   The Jury Observation Fallacy

Suppose you hear that a defendant is on trial for a serious crime and that, after all the evidence is heard, the jury returns a not guilty verdict. After the defendant is set free, the press and police release information that the defendant had a previous conviction for a similar crime. When scenarios exactly such as this have happened, it is typical to hear that members of the jury are reduced to tears, with the assumption being that they got the verdict wrong. But this assumes that, whatever the probability of guilt the jury arrived at to make their not guilty verdict, the probability of guilt must *increase* once the additional evidence of the previous conviction is presented. In fact, the intuitive belief that the "jury got it wrong" (and so that the probability should drop when you find out about the previous conviction) is a fallacy.

The key point is that there is a (known) small probability that the defendant will be charged in the absence of hard evidence, purely on the basis of a previous similar conviction. Yet it is the hard evidence that is required to convince the jury.

The appropriate BN structure for this problem, with example NPTs, is shown in Figure 16.19.

So let us, as an external observer of the trial, enter the information as it becomes available:

Step 1: When we know that the defendant has been charged and brought to trial, both the probability of "Verdict" = "Guilty" and the probability of "guilty" = "Yes" increase, as shown in Figure 16.20 (this also reflects the empirical knowledge that most defendants brought to trial for a serious crimes are eventually found guilty—the UK figure in 2011 was 81%).

Step 2: We hear that the jury Verdict = "Innocent" (Figure 16.21). Note that the probability of not guilty jumps to 92.59%.

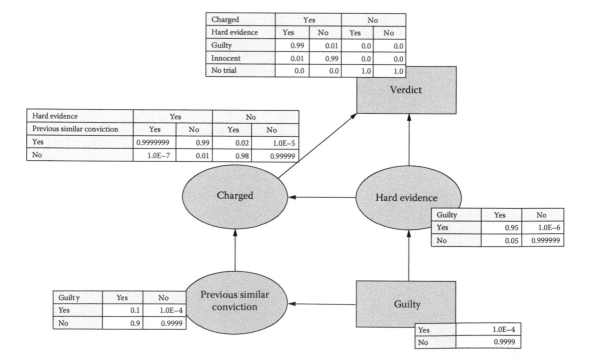

Figure 16.19   BN for jury observation fallacy.

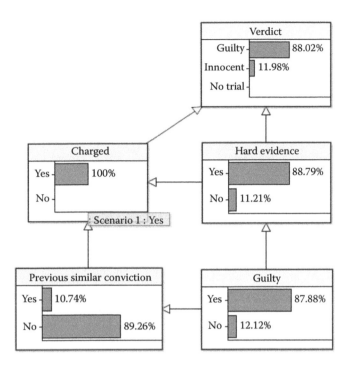

Figure 16.20   Defendant is charged.

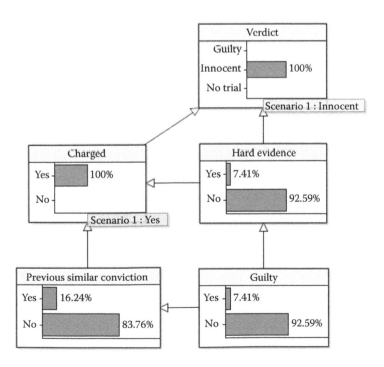

Figure 16.21   Defendant found innocent by jury.

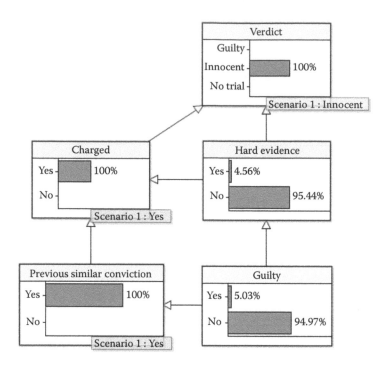

Figure 16.22 Previous conviction is revealed.

Step 3: Now we discover that the defendant has a previous conviction for a similar crime, so enter "Yes" in that node, as in Figure 16.22. The probability of "not guilty" does not decrease— rather, it increases slightly to 94.97%. There is also a small decrease in the probability of the hard evidence. Essentially, the fact that the defendant was charged in the first place is explained away more by the previous conviction than the evidence. It can be shown that the fallacy "holds" for a wide range of sensitivity.

16.8.3 The Crimewatch UK Fallacy

We discovered this fallacy as a result of our work on the case of *R v Bellfield*. Levi Bellfield was charged with two murders (Amelie Delagrange and Marsha Macdonnell) and three attempted murders. A key piece of evidence related to a car near the scene of the Marsha Macdonnell murder that the prosecution believed to be a match to a car owned by Bellfield.

This fallacy is called the "Crimewatch" fallacy after the name (Crimewatch UK) of a popular TV program in which the public is invited to provide evidence about unsolved crimes. The fallacy is stated in general terms in Box 16.3.

The prosecution opening contained an explicit example of this fallacy. The evidence X was the CCTV still image of the car that the police believed was driven by Marsha Macdonnell's attacker. Fact 2 was the prosecution hypothesis that this was a car belonging to Levi Bellfield. Fact 3 and the conclusion were the exact statements made in the prosecution opening.

Box 16.3 Generic Crimewatch UK Fallacy

Suppose we know the following:

Fact 1: item X was found at the crime scene that is almost certainly linked to the crime.
Fact 2: item X could belong to the defendant.
Fact 3: despite many public requests (including, e.g., one on the BBC Crimewatch UK program) for information for an innocent owner of item X to come forward and clear themselves, nobody has done so.

The fallacy is to conclude from these facts that:

It is therefore highly improbable that evidence X at the crime scene could have belonged to anybody other than the defendant.

The conclusion is a fallacy. Intuitively, we can explain the fallacy as follows: because fact 2 is "almost" certain (the police were already convinced that the car was driven by the attacker), we can predict fact 3 will almost certainly be true even before the Crimewatch program is screened (the attacker, whether it is the defendant or somebody else) is almost certainly *not* going to come forward as a result of the Crimewatch program. Hence, when fact 3 is confirmed (nobody does come forward), it has negligible impact on the probability of guilt. In other words, the Crimewatch UK evidence, far from proving that it is "highly improbable that the car could have belonged to anybody other than the defendant," actually tells us almost nothing more than we already knew or assumed.

A proper BN "solution" for this problem is shown in Figure 16.23, where we start with priors that are very favorable to the prosecution case. Thus, we assume a very high probability, 99%, that evidence X was directly linked to the crime.

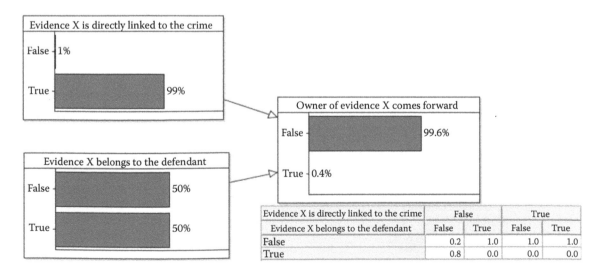

Figure 16.23 Crimewatch UK priors.

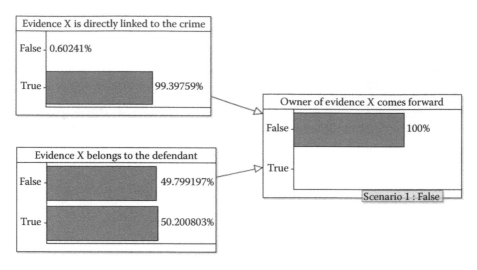

Figure 16.24   Now we enter fact 3 (nobody comes forward).

Looking at the NPT, we assume generously that if the owner of X were innocent of involvement, then there is an 80% chance he/she would come forward (the other assumptions in the NPT are not controversial).

What we are interested in is how the *prior probability of the evidence X being the defendant's* changes when we enter the fact that no owner comes forward. The prosecution claim is that it becomes almost certain. The key thing to note is that, with these priors, there is already a very low probability (0.4%) that the owner would come forward. Consequently, when we now enter the fact that nobody comes forward (Figure 16.24), we see that the impact on the probability that X belongs to the defendant is almost negligible (moving from 50% to 50.2%).

This demonstrates the Crimewatch fallacy, and that the evidence of nobody coming forward is effectively worthless despite what the prosecution claims.

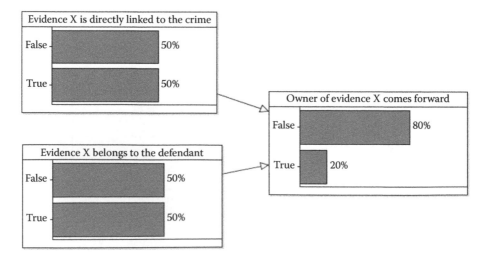

Figure 16.25   Different priors.

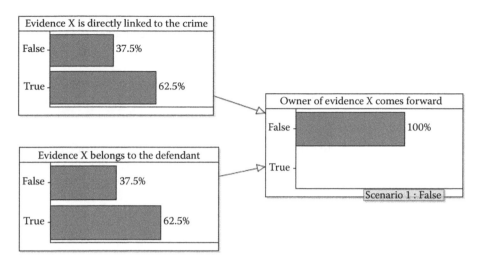

Figure 16.26   Evidence now makes a difference.

In fact, the only scenarios under which the evidence of nobody coming forward has an impact are those that contradict the heart of the prosecution claim. For example, let us assume, as in Figure 16.25, that there is only a 50% probability that the evidence X is directly linked to the crime.

Then, when we enter the fact that nobody comes forward (Figure 16.26), the impact on our belief that X is the defendant's is quite significant (though still not conclusive), moving from 50% to 62.5%. But, of course, in this case, *the priors contradict the core of the prosecution case.*

16.9 Summary

While Chapter 15 demonstrated the enormous potential benefits of using BNs in the justice system, in this chapter, we have demonstrated that it is possible to build BNs that capture entire complex legal BN arguments in a systematic and repeatable way by using a small set of BN idioms. While lawyers are not yet ready to consider BNs in the courtroom, the kind of BNs we have presented are highly suited for assessing the merits of a case pretrial when all key evidence has been logged.

Only the simplest Bayesian legal argument (a single hypothesis and a single piece of evidence) can be easily computed manually; inevitably, we need to model much richer arguments involving multiple pieces of possibly linked evidence. While humans must be responsible for determining the prior probabilities (and the causal links) for such arguments, it is simply wrong to assume that humans must also be responsible for understanding and calculating the revised probabilities that result from observing evidence. The Bayesian calculations quickly become impossible to do manually, but any BN tool enables us to do these calculations instantly.

The results from a BN tool can be presented using a range of assumptions including different priors. What the legal professionals (and perhaps even jurors if presented in court) should never have to think about is how to perform the Bayesian inference calculations. They do, of course, have to consider the prior assumptions needed for any BN model. Specifically, they have to consider:

- Whether the correct nodes and causal links have been modeled
- What suitable values (or range of values) are required for the priors and conditional priors

But these are precisely what have to be considered in weighing up any legal argument. The BN simply makes this all explicit rather than hidden, which is another clear benefit of the approach.

One of the greatest opportunities offered by BNs is to provide a much more rigorous and accurate method to determine the combined impact of multiple types of evidence in a case.

Further Reading

The conjunction paradox was described in:

Cohen, L. J. (1977). *The Probable and the Provable*. Clarendon Press, Oxford, UK.

The paper that resolved the paradox using Bayes was:

Dawid, A. P., (1987). The difficulty about conjunction. *J. R. Statistics Society Series D The Stat*, 36, 91, 97.

The following papers cover the material in this chapter in greater depth:

Fenton, N. E., Lagnado, D., and Neil, M. (2013). A general structure for legal arguments using Bayesian networks. *Cognitive Science*, 37, 61–102.

Fenton, N. E., and Neil, M. (2011). Avoiding legal fallacies in practice using Bayesian networks. *Australian Journal of Legal Philosophy*, 36, 114–151.

Fenton, N. E., Lagnado, D. A., Dahlman, C., and Neil, M. (2017). The opportunity prior: A simple and practical solution to the prior probability problem for legal cases. In *International Conference on Artificial Intelligence and the Law (ICAIL 2017)*. Published by ACM, pp 69–76, 10.1145/3086512.3086519.

The jury observation fallacy is described in:

Fenton, N. E., and Neil, M. (2000). The jury fallacy and the use of Bayesian nets to simplify probabilistic legal arguments. *Mathematics Today (Bulletin of the IMA)*, 36(6), 180–187.

The impact of confirmation bias is described in,:

Dror, I. E., and Charlton, D. (2006). Why experts make errors. *Journal of Forensic Identification*, 56(4), 600–616.

A comprehensive list of references on Bayes and the law can be found here:

Fenton, N. E., Neil, M., and Berger, D. (2016). Bayes and the law. *Annual Review of Statistics and Its Application*, 3, 51–77.

Visit www.bayesianrisk.com for exercises and worked solutions relevant to this chapter.

17

Learning from Data in Bayesian Networks

17.1 Introduction

A wide variety of different objectives are covered by learning from data in Bayesian networks. First, there is parameter and hyper parameter learning, which was covered in Chapter 10. During parameter learning, we update our beliefs about the distribution and value of continuous parameters based on observed data. We can also view hypothesis testing, the subject of Chapter 12, as a form of learning from data since this process involves using data to update our beliefs about hypotheses. Furthermore, Chapter 12 discussed complex causal hypotheses, which might be considered a form of structural learning. This involves identifying which causal connections between variables are best explained by the available data. Last, in various other chapters, we touched on learning but with a different focus; for instance, in Chapter 13 on systems reliability modeling, we performed learning with censored data. This involves learning from data also, but from intervals rather than data points.

In this chapter, our focus is on learning probability tables for BNs where we have a predefined BN structure. We start in Section 17.2 with the simplest case of discrete BNs learned entirely from data sets that are "complete" in the sense that there are no missing values. For such cases, different standard learning algorithms not only produce identical results but can also be easily understood and calculated. In Section 17.3, we stick with discrete BNs and complete data sets but show how we can supplement the data by incorporating prior expert judgment in a simple—but mathematically rigorous—way; indeed, this approach is exactly the same that is used to incorporate expert judgment for the more complex cases that follow. In Section 17.4, we stay with discrete BNs but consider the important and common situation whereby the data set is not complete (although, where data are missing, we do assume they are missing at random). This is where we cover one of the most popular algorithms for learning node probability tables from incomplete data. This is the expectation-maximization (EM) algorithm.

A good example of where data may not be missing at random might occur in salary tables published by large corporations, such as the BBC. Here, senior staff, on presumably high salaries, may be underreported, for obvious reasons.

Learning from data to define statistical distributions can also be done using the EM algorithm, and in Section 17.5, we demonstrate this for a variety of hybrid BNs. The state of the art here applies to continuous

The perceptive reader will notice that this chapter does not cover techniques used to learn BN structure, of which there are many. It is our view that it is better, for risk problems, at least, for the analyst to postulate possible causal structures in advance of learning from data rather than take a "brute-force" approach where the concomitant danger is that the resulting model will make little causal sense. Hence, for now, we recommend learning from data on BNs with a predefined structure.

The background to this model is the knowledge that, while single women are more likely to be in full-time employment than single men, the opposite is true for married women and men. People who do not work full time and are married are more likely to receive state benefits than those who are single. This knowledge determines the model structure and is later also used to supplement the data, but initially we will learn the NPTs from data only.

valued BNs (where all nodes are numeric and normally distributed) and BNs containing both discrete and continuous variables. Here, where the variables are all continuous, EM can learn the multivariate Normal distribution for a group of continuous connected parent-child nodes in the BN that best explains the data. In the hybrid case, Normal mixture distributions will be learned where each mixture component is dependent on combinations of discrete parent states.

17.2 Learning Discrete Model Tables from Complete Data Sets

To explain how we learn NPTs from data, we will use the example model shown in Figure 17.1 (see the sidebar for background).

To learn the NPT values, we need a data set like that in Table 17.1. Here we have a sample of 10 people.

The two most popular methods for learning the table values—maximum likelihood estimation, which we introduced in Chapter 12) and expectation maximization, which we describe in this chapter—produce identical results in cases similar to this example where the following all hold:

- All nodes are discrete.
- There are no missing values in the data
- We assume no "prior" knowledge.

Although the statistical concepts involved (so-called Dirichlet distributions and their "conjugate" Multinomial distributions) sound daunting, the necessary calculations are actually extremely simple. All we do is count the frequencies of the observed values for each node's state combinations to arrive at the "learned" NPTs in Figure 17.2.

So, for the node "Sex," which has no parents, we observed six instances of "Female" and four instances of "Male." Hence, we have "learned" that 60% of the population is Female and 40% is Male, resulting in the NPT for Sex shown in Figure 17.2. For the node "Work," we

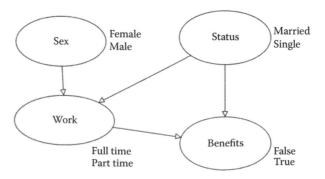

Figure 17.1 Simple model for determining whether individual is receiving state benefits.

Table 17.1
Data Set for Model in Figure 17.1

Sex	Status	Work	Benefits
Female	Single	Full time	FALSE
Female	Married	Part time	TRUE
Male	Single	Full time	FALSE
Female	Married	Part time	FALSE
Female	Married	Full time	FALSE
Male	Married	Part time	TRUE
Female	Married	Part time	FALSE
Male	Married	Full time	TRUE
Female	Married	Part time	TRUE
Male	Single	Full time	TRUE

count the number of instances of "Full time" and "Part time" for each of the four state combinations (Female, Married), (Female, Single), (Male, Married), (Male, Single). These are, respectively:

- (Female, Married): 1 "Full time" and 4 "Part time"
- (Female, Single): 1 "Full time" and 0 "Part time"
- (Male, Married): 1 "Full time" and 1 "Part time"
- (Male, Single): 2 "Full time" and 0 "Part time"

Note that, even though we only have one observation for (Female, Single), the fact that this one observation is "Full time" means that we have "learned" that 100% of single females work full time. While this

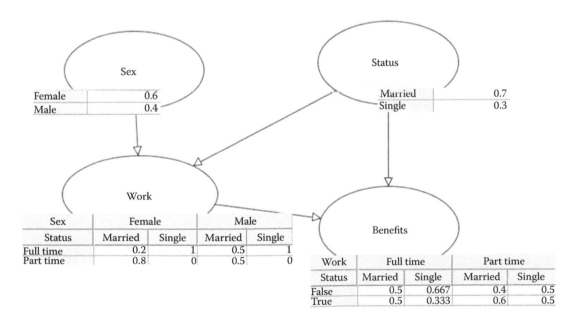

Figure 17.2 NPTs learned from data in Table 17.1.

indicates a clear weakness in learning from small data sets without using prior knowledge, an even more serious weakness is suggested by the NPT learned for the node "Benefits." Specifically, *the state combination (Part time, Single) is not observed at all in the data set.* The statistical assumptions of MLE and EM require that, in such cases, all states be assumed to be equally probable—hence the 50-50 NPT entry for this node shown in Figure 17.2.

17.3   Combing Expert Knowledge with Table Learning from Data

Fortunately, it is easy to incorporate prior expert knowledge into the learning in a mathematically rigorous way. Suppose the prior expert judgment (which may itself be based on previous data) for the node "Status" is that 60% of the relevant population is "Married" and 40% is "Single." Then we can express our confidence in the prior data compared to the new data set as a ratio r, where r is a number between 0 and 1. The value $r = 1$ means complete reliance on the prior knowledge value and ignores the data, while $r = 0$ ignores the prior knowledge entirely. Thus, we use a weighted average. In what follows, we assign a value $r = 0.8$, meaning that there is an 80:20 preference for the prior knowledge over the new data set.

With $r = 0.8$ and a data set of size 10 as in the above example, this is equivalent to having observed a data set of size 40 for the prior knowledge.

Since the prior knowledge for the state "Married" is 60% (i.e., probability 0.6) and the learned value from the data set is 70% (i.e., probability 0.7), the value for the combined knowledge and data using the ratio $r = 0.8$ is simply computed as

$$(0.6 \times 0.8) + (0.7 \times 0.2) = 0.62$$

Figure 17.3 shows the result of combining knowledge and the data set for the node "Benefits."

In general, if x is the prior knowledge probability for a given state of a node and y is the probability value learned from a data set, then the combined knowledge and data using the ratio r is

$$xr + y(1 - r)$$

In AgenaRisk, the r values can be set for individual nodes or the entire model and the calculations are done automatically.

We can consider the ratio r as the ratio of data size in a historical data set, perhaps notionally equivalent to our "knowledge" and the current data used in learning. In AgenaRisk, r can be set as a % value, where 100% "knowledge" would be equivalent to infinite prior data and 0% would be equivalent to the size of the data set used for learning. A value of $r = z\%$ for a current data set of size w_1 would therefore suggest a historical data set size w_2, where $w_2 = w_1((100/z) - 1)$.

17.4   Learning Table Values with Missing Data: Expectation Maximization for Discrete Models

The discrete case EM algorithm requires (a) a BN model with some prior probabilities, (b) a learning data set (with missing values) that corresponds to the nodes of the BN model, and (c) knowledge in the form of expert opinions that can be, optionally, mixed with the data. Where there is no missing data, the EM algorithm works exactly as described in the previous two sections. The challenging part, of course, is handling the missing data; this is where we have to use Bayesian inference by essentially "estimating" the missing values.

As described informally in Section 17.2, the discrete EM algorithm involves Dirichlet counts for observations that correspond to the joint probability distribution of each variable and its parents. The formal assumptions and overview of the algorithm are described in Box 17.1.

Prior expert knowledge

Work	Full time		Part time	
Status	Married	Single	Married	Single
False	0.8	0.9	0.4	0.8
True	0.2	0.1	0.6	0.2

Learned from data set in Table 17.1

Work	Full time		Part time	
Status	Married	Single	Married	Single
False	0.5	0.667	0.4	0.5
True	0.5	0.333	0.6	0.5

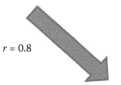 $r = 0.8$

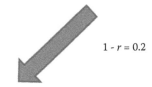 $1 - r = 0.2$

Work	Full time		Part time	
Status	Married	Single	Married	Single
False	0.74	0.853	0.4	0.74
True	0.26	0.147	0.6	0.26

Combined knowledge and data

Figure 17.3   Combined learning from knowledge and data with $r = 0.8$ (80% preference for knowledge).

Box 17.1   Expectation Maximization Algorithm Formal Assumptions and Overview

In general, we assume we are given observed data **X**, missing data **Z**, and a vector of unknown parameters θ, along with a likelihood function L(θ;**X**,**Z**) = p(**X**,**Z** | θ). The EM algorithm seeks to converge to the maximum likelihood estimate of θ by iteratively applying two steps, Expectation and Maximization, as follows:

- Expectation (E) step: Calculate the expected value of the log likelihood function log$L$(θ;**X**,**Z**) on the conditional distribution of p(**Z** | **X**,θ^t):

$$Q(\theta \mid \theta') = E_{\mathbf{Z}|\mathbf{X},\theta'}[\log L(\theta; \mathbf{X}, \mathbf{Z})]$$

- Maximization (M) step: find the parameters that maximize this quantity:

$$\theta^{t+1} = \underset{\theta}{\operatorname{argmax}} \; Q(\theta \mid \theta')$$

Let θ^t be the parameters obtained at the tth iteration. If θ^t = θ^{t+1}, then θ^t is a stationary point of the log-likelihood function L(θ;**X**,**Z**). We check the expected log-likelihood of the data with the parameters at each EM iteration. In the E-step, we use Bayesian inference to calculate the most probable value for p(**Z** | **X**,θ^t) and record the sufficient statistics for conditional distributions p(**Z** | **X**,θ^t). At the M-step, we use the recorded sufficient statistics to generate new parameters θ^{t+1}. The process is then repeated until some stopping criterion is reached, such as the number of iterations or a convergence target being exceeded.

Note that the Dirichlet counts for the observed data rely on the *Multinomial distribution*, which is the multivariate generalization of the binomial distribution:

$$P(X_1 = x_1, X_2 = x_2, \ldots, X_k = x_k) = \frac{n!}{x_1! x_2! \ldots x_k!} p_1^{x_1} p_2^{x_2} \ldots p_k^{x_k}$$

where $\sum_{i=1}^{k} x_i = n$ for non-negative integers x_i.

The method for deriving the MLE of the multinomial distribution is well known and involves using the observed frequency counts. Each component of the multinomial distribution can be treated as a binomial distribution; hence, the maxima are obtained using the sufficient statistic:

$$E(p_i) = \frac{x_k}{\sum_{i=1}^{k} x_i}$$

A sufficient statistic is a function whose values contain all the information needed to compute any estimate of a parameter of a distribution. A Normal distribution has sufficient statistics, mean, and variance. A Poisson distribution has a single sufficient statistic, the mean, from which all information about the Poisson distribution can be derived.

When using AgenaRisk to perform learning on models containing discrete nodes, only the convergence target value is automatically set to 0.000001, while the default of 0.01 is set for hybrid models. These values can be customized. Decreasing the values leads to higher accuracy but longer computation time and vice versa.

While this looks daunting, we will provide a fully worked example; moreover, the algorithm is implemented in AgenaRisk, so users simply have to import a data set to run it for a given BN model.

When the data set is complete, the EM algorithm is equivalent to MLE. Hence, it produces the results described in Sections 17.2 and 17.3. However, when we have variables with missing values, approximate inference is performed within the BN junction tree to produce "pseudo observations"; these can be considered the hypothetical observations we would wish to impute from the known data. Because of this, there is the possibility of achieving local suboptimal convergence. Therefore, checks should be done to determine if local or global convergence has been achieved, such as by performing cross validation across multiple data sets.

Box 17.2 provides an overview of the algorithm as used in AgenaRisk, including the incorporation of expert judgment using the method described in Section 17.3.

Box 17.2   Discrete Expectation Maximization Algorithm with Prior Expert Beliefs

Expert beliefs expressed as knowledge to data ratio, r:

```
Set initial conditions
While Not Converged
    Expectation Step (E-Step)
      For every data row:
        If the row has missing values
              Instantiate the model with known values in that record.
              run Junction Tree propagation
                  obtain the joint probability distribution of each variable and
                  its parents from the junction tree and infer missing value
                  distributions
            Else
                update the Dirichlet counts with the corresponding (complete) data
        End if
      End for
    Maximization Step (M-Step)
    the new parameters are calculated from the final counts obtained in the E-step.
End while
Apply knowledge to data ratio r (0 ≤ r ≤ 1)
    New parameters θ̃ = r × θ_knowledge + (1 − r) × θ_data
End If
```

Example 17.1

To demonstrate the EM learning process, we consider the very simple BN model, consisting of three variables, A, B, and C, as presented in Figure 17.4. The prior probabilities for this model are shown in Table 17.2 and form the initial conditions. These prior probabilities are all 0.5 and reflect an uninformative "no information" state at the start of learning. The data set, consisting of three records, is shown in Table 17.3. As you can see, there is a missing value in the last record, denoted by "NA."

The EM algorithm learns the parameters of the variables for Example 17.1 based on the data shown in Table 17.3. Whenever a row of data has a missing value, the EM algorithm calculates the expected distribution, or sufficient statistics, of the missing value by using the observed data for that record and the BN model learned up to that iteration. Afterward, the parameters of the Example 17.1 model are updated based on these expectations.

Let's go through the process of applying EM to Example 17.1 for the data set.

Note that the initial conditions for EM determine the speed of convergence: choosing starting values close to the true values will lead to faster convergence; however, this presumes one knows what the true values are in the first place. Therefore, here we use noninformative priors for EM rather than randomized prior values.

Iteration 1

At the expectation step, the counts of joint states are generated. The respective counts for the corresponding NPT entries' data row are shown in Table 17.4. Specifically, based on the first record {a1,b2,c2}, we increase the counts for "a1," "a1,b2," and "b2,c2," by 1. For the second record in our data {a2,b2,c2}, we increase "a2," "a2,b2," and "b2,c2."

Notice that we list the records such that they map onto the BN structure, so the first data record in our data {a1,b2,c2} maps into "a1," "a1,b1,"

Figure 17.4   Example 17.1 BN model.

Table 17.2
Initial Parameters of the Example 17.1 Model

A		B		C	
$P(a1)$	0.5	$P(b1\mid a1)$	0.5	$P(c1\mid b1)$	0.5
$P(a2)$	0.5	$P(b2\mid a1)$	0.5	$P(c2\mid b1)$	0.5
		$P(b1\mid a2)$	0.5	$P(c1\mid b2)$	0.5
		$P(b2\mid a2)$	0.5	$P(c2\mid b2)$	0.5

Table 17.3
Data Used for Learning Example 17.1

Record#	A	B	C
1	a1	b2	c2
2	a2	b2	c2
3	a1	NA	c1

Table 17.4
The Counts of Joint States after First and Second Records Are Read
(Iteration 1)

a1	a2	a1,b1	a1,b2	a2,b1	a2,b2	b1,c1	b1,c2	b2,c1	b2,c2
1	1	0	1	0	1	0	0	0	2

Table 17.5
Joint Probabilities for Record 3 with
Missing Values (Iteration 1)

A		B		C	
P(a1)	1	P(a1,b1)	0.5	P(b1,c1)	0.5
P(a2)	0	P(a1,b2)	0.5	P(b1,c2)	0
		P(a2,b1)	0	P(c1,b2)	0.5
		P(a2,b2)	0	P(b2,c2)	0

and "b2,c2," corresponding to the BN structure and the NPTs defined therein.

Now, for the third record, we need to estimate the parameters using the current BN parameterization. This is simply the initial parameters from Table 17.2. Looking at the algorithm, we can calculate this by entering the observations into the BN and determining the joint probability values for each NPT. In this case, they are uninformative given the initial parameterization is noninformative (notice they are simply entries for Table 17.2 multiplied by 0 or 1 if an entry exists in the record or not). These joint probabilities are shown in Table 17.5.

We then use the values from Table 17.3 as "pseudo observations" and add them to the counts we obtained in Table 17.3, as shown in Table 17.6. This completes the Expectation step for iteration 1.

Now that we have completed the E-step, we need to perform the Maximization step. This is simply achieved by entering the above values into the NPTs in the BN and propagating, which is simply equivalent to normalization here, as shown in Table 17.7.

Iteration 2

Table 17.7 is now the starting parameterization for our model, and we repeat the same steps we followed in iteration 1 *using the same records*. Therefore, as before, we first apply the counts for records 1 and 2 and

Table 17.6
The Counts of Joint States after First, Second, and Third Records
Are Read (Iteration 1)

a1	a2	a1,b1	a1,b2	a2,b1	a2,b2	b1,c1	b1,c2	b2,c1	b2,c2
2.0	1.0	0.5	1.5	0	1.0	0.5	0	0.5	2.0

Table 17.7
M-Step (Iteration 1)

A		B		C	
$P(a1)$	0.67	$P(b1\|a1)$	0.25	$P(c1\|b1)$	1
$P(a2)$	0.33	$P(b2\|a1)$	0.75	$P(c2\|b1)$	0
		$P(b1\|a2)$	0	$P(c1\|b2)$	0.2
		$P(b2\|a2)$	1	$P(c2\|b2)$	0.8

then infer the pseudo-counts for record 3 with missing data. The only difference is the starting parameterization; therefore, Table 17.3 is simply reused in iteration 2 to calculate the joint probabilities again, using the new parameterization in Table 17.7.

The results are as shown in Table 17.8.

We use the values from Table 17.8 as "pseudo observations" and add them to the counts we obtained in Table 17.2, as shown in Table 17.9.

Now that we have completed the E-step in iteration 2, we need to perform the Maximization step, as shown in Table 17.10.

Table 17.8
Joint Probabilities for Record 3 with Missing Values (Iteration 2)

A		B		C	
$P(a1)$	1	$P(a1,b1)$	0.625	$P(b1,c1)$	0.625
$P(a2)$	0	$P(a1,b2)$	0.375	$P(b1,c2)$	0
		$P(a2,b1)$	0	$P(b2,c1)$	0.375
		$P(a2,b2)$	0	$P(b2,c2)$	0

Table 17.9
The Counts of Joint States after First, Second, and Third Records Are Read (Iteration 2)

a1	a2	a1,b1	a1,b2	a2,b1	a2,b2	b1,c1	b1,c2	b2,c1	b2,c2
2.0	1.0	0.625	1.375	0	1.0	0.625	0	0.375	2.0

Table 17.10
M-Step (Iteration 2)

A		B		C	
$P(a1)$	0.67	$P(b1\|a1)$	0.3125	$P(c1\|b1)$	1
$P(a2)$	0.33	$P(b2\|a1)$	0.6875	$P(c2\|b1)$	0
		$P(b1\|a2)$	0	$P(c1\|b2)$	0.1578
		$P(b2\|a2)$	1	$P(c2\|b2)$	0.8421

Iterations 3–10

This iterative process continues and is guaranteed to converge. For example, the parameter $P(c2 \mid b2)$ changes over 10 iterations, as shown in Table 17.11, and convergence is clear.

After the 10th iteration, we achieve the posterior probabilities for records 1–3, as shown in Table 17.12. These are the final learned conditional probabilities, θ_{data}, for this simple model using the data.

Notice that from our data, many of the values are zero simply because these combinations did not occur in the data set. But recall Cromwell's rule: any probabilities of zero or one indicate certitude, and as a result they are immutable. So, in the presence of missing data, we can overcome this limitation in the data by mixing the likelihoods derived from data with expertise about the data in order to produce a "reasonable" result. We do this using the approach described in Section 17.3 using the ratio r ($0 \leq r \leq 1$) that represents the relative weight we wish to assign to expert knowledge and the data, respectively.

Clearly, judgment is an integral part of what is, after all, a Bayesian process, but as is the case in all Bayesian models, an increase in the amount of data will tend to "wash out" the prior judgments about the ratio value used. Of course, $r = 1$ means complete reliance on the knowledge value and ignores the data, while $r = 0$ ignores the knowledge entirely and essentially replaces the NPTs with the frequencies from the data set or imputed values when the data is missing.

In AgenaRisk, the expert knowledge is performed using the "Incorporate Expert Judgment" feature that allows the setting of data to knowledge weight for all of the variables in the model or for a subset of these. This gives added flexibility when learning from data and/or using expert opinion together in creative ways.

Table 17.11
$P(c2 \mid b2)$ Estimate Convergences

Iteration (t)	$P(c2 \mid b2)^t$	$P(c2 \mid b2)^{t+1}/P(c2 \mid b2)^t$
1	0.5	–
2	0.8	1.6
3	0.842105	1.052632
4	0.885813	1.051903
5	0.92501	1.044249
6	0.955012	1.032435
7	0.974875	1.020799
8	0.986631	1.012059
9	0.993089	1.006546
10	0.996485	1.003419

Table 17.12
Final Conditional Learned Probabilities (θ_{data})

A		B		C	
$P(a1)$	0.67	$P(b1 \mid a1)$	0.4964	$P(c1 \mid b1)$	1
$P(a2)$	0.33	$P(b2 \mid a1)$	0.5035	$P(c2 \mid b1)$	0
		$P(b1 \mid a2)$	0	$P(c1 \mid b2)$	0.0035
		$P(b2 \mid a2)$	1	$P(c2 \mid b2)$	0.9964

Table 17.13

Knowledge Parameters, $\theta_{knowledge}$, for Example 17.2

A		B		C	
$P(a1)$	0.3	$P(b1 \mid a1)$	0.8	$P(c1 \mid b1)$	0.7
$P(a2)$	0.7	$P(b2 \mid a1)$	0.2	$P(c2 \mid b1)$	0.3
		$P(b1 \mid a2)$	0.1	$P(c1 \mid b2)$	0.2
		$P(b2 \mid a2)$	0.9	$P(c2 \mid b2)$	0.8

Example 17.2

To demonstrate the EM learning process with knowledge, we consider the same very simple BN model as used in Example 17.1. The data set, consisting of three entries, is as shown in Table 17.3, and the knowledge parameters are shown in Table 17.13. The ratio of knowledge to data is set at $r = 0.5$, suggesting the data and knowledge are equally weighted.

The resulting posterior probabilities $\tilde{\theta}$ resulting from mixing θ_{data} and $\theta_{knowledge}$ are simply calculated as shown in Table 17.14. Of course, this calculation step is carried out automatically in AgenaRisk.

Example 17.3

To demonstrate the EM learning process with knowledge and a more realistically sized data set, let's consider the Asia example, originally introduced in Chapter 9. The model, with marginal probabilities prior to learning, is as shown in Figure 17.5.

We have some data on a subset of the variables, as listed in Table 17.15.

Let's assume that there is some belief that the ratio between the probabilities in the model and the data is $r = 0.1$ in favor of the data (i.e., 90% data and 10% prior). When we execute the learning function in AgenaRisk, this results in new Marginal distributions, as shown in Figure 17.6. Notice that nodes that have been learned from data and knowledge are denoted by an "M" icon, and where the original knowledge has been retained, the node shows a "K" icon (TB or Cancer is a logical node, so it makes no sense to learn it). As you can see, the marginal probabilities have changed significantly from the original model using the learned data.

In AgenaRisk, the prior probabilities used to initialize the EM process are the same as those entered into the NPTs for the model and could in practice be the same as those used to represent the expert knowledge. While this might seem unusual, this is fine because we hope that the EM will converge to a result no matter what the starting point might be, so using the expert knowledge parameters is as good a starting point as any other.

In AgenaRisk, when learning is completed, icons are temporarily added to nodes to indicate:

- "D"—the node's probability table has been learned solely from *data*, that is, 0% for knowledge to data ratio.
- "M"—the node's probability table has been learned as a *mixture*—partially from data and partially from previous NPTs, that is, knowledge to data ratio higher than 0% and lower than 100%.
- "K"—the node's probability table has been based solely on *knowledge* (previous NPTs), that is, 100% for knowledge to data ratio.

Table 17.14

Final Parameters, $\tilde{\theta}$, for Example 17.2

A		B		C	
$P(a1)$	$0.5(0.3) + 0.67(0.5)$ $= 0.485$	$P(b1 \mid a1)$	$(0.5)0.8 + (0.5)0.4964$ $= 0.6482$	$P(c1 \mid b1)$	$(0.5)0.7 + (0.5)1$ $= 0.85$
$P(a2)$	$0.5(0.7) + 0.33(0.5)$ $= 0.515$	$P(b2 \mid a1)$	$(0.5)0.2 + (0.5)0.5035$ $= 0.35175$	$P(c2 \mid b1)$	$(0.5)0.3 + (0.5)0$ $= 0.15$
		$P(b1 \mid a2)$	$(0.5)0.1 + (0.5)0$ $= 0.05$	$P(c1 \mid b2)$	$(0.5)0.2 + (0.5)0.0035$ $= 0.10175$
		$P(b2 \mid a2)$	$(0.5)0.9 + (0.5)1$ $= 0.95$	$P(c2 \mid b2)$	$(0.5)0.8 + (0.5)0.9964$ $= 0.8982$

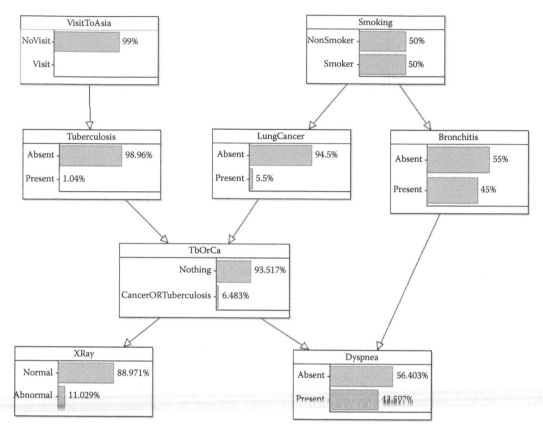

Figure 17.5 Example 17.3 Asia model prior to learning.

Table 17.15

Data Set for Example 17.3 (Missing Values Are Blank)

VisitToAsia	Tuberculosis	Smoking	LungCancer	XRay	Bronchitis	Dysponea
NoVisit	Absent	NonSmoker	Absent	Normal	Absent	Absent
NoVisit	Absent	Smoker	Present	Abnormal	Present	Present
NoVisit	Absent	–	Absent	Normal	Present	Absent
NoVisit	Absent	Smoker	Absent	Normal	Present	Absent
NoVisit	Absent	Smoker	Absent	Normal	Present	Absent
NoVisit	Absent	Smoker	Absent	Normal	Present	Present
–	Absent	Smoker	Absent	Normal	Absent	–
NoVisit	Absent	NonSmoker	Absent	Normal	Present	Present
NoVisit	Absent	Smoker	Absent	Normal	Absent	Absent
NoVisit	Absent	NonSmoker	Absent	Normal	Absent	Absent
NoVisit	Absent	NonSmoker	Absent	Normal	Present	Present
NoVisit	–	NonSmoker	Absent	Normal	Absent	Absent
NoVisit	Absent	Smoker	Absent	–	Present	Present
NoVisit	Absent	NonSmoker	Absent	Normal	Absent	Absent
NoVisit	Absent	Smoker	Present	Abnormal	Present	Present
NoVisit	Absent	Smoker	Absent	Normal	Absent	Absent
NoVisit	Absent	NonSmoker	Absent	Normal	Present	Present
NoVisit	Absent	NonSmoker	Absent	Normal	Present	Present
NoVisit	–	NonSmoker	Absent	–	Present	Present

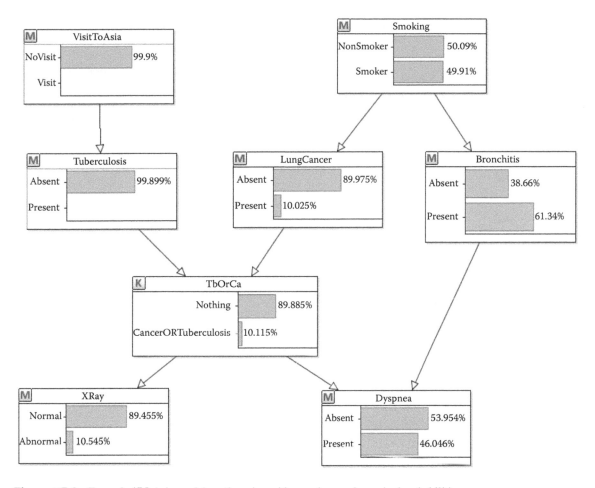

Figure 17.6   Example 17.3 Asia model postlearning with superimposed marginal probabilities.

17.5   Expectation Maximization for Hybrid Models

By hybrid models, we mean models that contain discrete and continuous variables together, where a continuous child node can have:

- Zero, one, or more continuous parents (Type I model)
- A mixture of discrete and continuous parents (Type II model)

As Type I models contain continuous parent nodes only, they will be of the general linear type (i.e., a regression model). In Type II models, the learned models will be mixtures of general linear models (Normal mixtures). In both Type I and Type II cases, we are restricted to learning parameters of multivariate Normal models.

For Type II models, where we have discrete and continuous parents, a linear model will be learned, using the continuous parent nodes, where each linear model will be indexed by the discrete parent node states, thus creating a mixture model.

The material in this section is mathematically challenging, but is all implemented in AgenaRisk so users do not have to understand the underlying mathematics in order to use the algorithm.

There are four key differences between our approach and traditional methods:

1. We incorporate a user-specified prior for discrete nodes into the learning process.
2. When we have missing data, we use dynamic discretization to perform inference and query the sufficient statistics of latent variables in the model.
3. The approach for learning Normal mixtures adopted is much more general than that available in other EM implementations.
4. Compared to other matrix inversion–based numerical methods, this approach is robust because it does not suffer from numerical overflows.

For continuous variables, the Normal probability distributions declared in the model at the beginning of EM learning will be overwritten once completed and do not play an active role in the learning process. Should there be a requirement to learn a partial subset of the variables in a model, those that are exempted from learning can be easily identified as such in the AgenaRisk learning tool.

The EM algorithm for hybrid models is specified in Box 17.3.

Box 17.3   Hybrid Expectation Maximization Algorithm with Prior Expert Beliefs

Expert beliefs expressed as knowledge to data ratio r.

```
Set initial conditions
While Not Converged
    Expectation Step (E-Step)
        For every data row:
            If the row has missing values
                Instantiate the model with known values in that record.
                Run propagation

                    obtain the joint probability distribution Γ of each continuous
                    variable C's discrete parent and discrete ancestors nodes from the JT

                    update each continuous variable C's sufficient statistics based on
                    the weighted posterior mean and squares of mean, where the weights
                    are specified by Γ

                    update each Gaussian component's mean vector and covariance matrix in
                    C based on the sufficient statistics queried
            Else
                update the sufficient statistics for each continuous node's Gaussian
            component with the corresponding (complete) data
                update each Gaussian component's mean vector and covariance matrix in C
            based on the sufficient statistics queried
            End if
        End for
    Maximization Step (M-Step)
        The new mean vector and covariance matrix parameters are calculated
End while
Apply knowledge to data ratio r (0 ≤ r ≤ 1)
        New parameters θ̃ = r × θ_knowledge + (1 − r) × θ_data
End If
```

More details on parameter estimation involved for these model types can be found in Box 17.4.

The maximum number of continuous parent variables that can be handled is four because numbers greater than this lead to combinatorial explosion in the size of clusters in the junction tree.

Box 17.4 Types of Hybrid Model Learned Using Expectation Maximization in AgenaRisk

Type I Model

A Type I model contains only continuous variables and is characterized by a multivariate Gaussian vector $\mathbf{X} = \{X_1, \ldots, X_d\}$ with mean vector:

$$\mu_{\mathbf{X}} = \begin{pmatrix} \mu_1 \\ \mu_2 \\ \ldots \\ \mu_d \end{pmatrix}$$

and covariance matrix

$$\sum\nolimits_{\mathbf{X}} = \begin{pmatrix} \sigma_1^2 & \ldots & \mathrm{cov}_{X_1 X_d} \\ \ldots & \ldots & \ldots \\ \mathrm{cov}_{X_d X_1} & \ldots & \sigma_d^2 \end{pmatrix}$$

If the number of samples is n, the sufficient statistics of a multivariate Gaussian vector are defined by:

$$\tilde{\mu}_{\mathbf{X}} = \begin{pmatrix} s_1/n \\ s_2/n \\ \ldots \\ s_d/n \end{pmatrix} \tag{17.1}$$

$$\tilde{\sum}_{\mathbf{X}} = \begin{pmatrix} \sum_{11} & \ldots & \sum_{1d} \\ \ldots & \ldots & \ldots \\ \sum_{d1} & \ldots & \sum_{dd} \end{pmatrix} = \begin{pmatrix} s_{11}/n - (s_1/n)^2 & \ldots & s_{1d}/n - s_1 s_d/n^2 \\ \ldots & \ldots & \ldots \\ s_{d1}/n - s_d s_1/n^2 & \ldots & s_{dd}/n - (s_d/n)^2 \end{pmatrix} \tag{17.2}$$

where $s_i = \sum_{j=1}^{n} x_{ji}, (i = 1, \ldots, d)$, x_{ji} is the jth variable at the ith variable and:

$$s_{ii} = \sum_{j=1}^{n} \begin{cases} x_{ji}^2, & \text{if} \quad j\text{th sample not missing} \\ \hat{\mu}_i^2 + \hat{\sigma}_i^2 & \text{if} \quad j\text{th sample missing} \end{cases}$$

where $\hat{\mu}_i$ and $\hat{\sigma}_i$ correspond to the posterior mean and standard deviation of the ith variable in $\mathbf{X}$.

Also, $s_{pq} = \sum_{j=1}^{n} x_{jp} x_{jq}, (p = 1, \ldots, d, q = 1, \ldots, d)$ if the jth sample is missing at the pth variable, x_{jp} is replaced by $\hat{\mu}_p$, the same for the qth variable in $\mathbf{X}$.

If we have zero parents, a Type I model will simply learn the parameters for a single node from the data available.

Armed with these sufficient statistics, we can then use an EM algorithm that should converge to a local maximum of the likelihood function.

Type II Model

A Type II model is a hybrid model and is characterized by a discrete variable vector $\mathbf{Y} = \{Y_1, ..., Y_m\}$, with the number of discrete states k_t, $t = 1, ..., m$ for each variable Y_t. The number of all these discrete states $\Gamma = \prod_{t=1}^{m} k_t$. Each discrete state corresponds to a Type I model.

We can construct any Type II model by partitioning expressions in a BN by a V-structure:

$$\mathbf{Y} \rightarrow C \leftarrow \mathbf{X}_{\backslash C} \leftarrow \mathbf{Z}$$

where the variable C is a single continuous child node with discrete parent vector $\mathbf{Y}$ and continuous parent vector $\mathbf{X}_{\backslash C}$ (all continuous variables without variable C) and where $\mathbf{X}$ has discrete parents node vector $\mathbf{Z} = \{Z_1, ..., Z_n\}$.

If $\mathbf{Z} = \varnothing$, we have a Type II model with discrete parents only, $\mathbf{Y} \rightarrow C$. Here, to consider each multivariate Normal component in C, we need to learn the multivariate Normal joint distribution $P(C, \mathbf{X}_{\backslash C})$, and there are Γ multivariate joint distribution components in a Type II model. Learning $P(C, \mathbf{X}_{\backslash C})$ is straightforward when we have complete data for $C \mid \mathbf{Y}$, since we simply need to calculate the sufficient statistics for all combinations. However, when there are missing data, we need to query the BN to infer what the missing data values might be across the discrete states in $\mathbf{Y}$ corresponding to the missing values. This is equal to Γ in the worst case and equivalent to calculating the probability p_i, where ($i = 1, ..., \Gamma$), from the joint distribution $p(\mathbf{Y})$ for each Normal mixture component and multiplying the sufficient statistics contained for each data sample by the relevant p_i.

If $\mathbf{Z} \neq \varnothing$, we have a Type II model with discrete parents $\mathbf{Y}$ and $\mathbf{Z}$: $\mathbf{Y} \rightarrow C \leftarrow \mathbf{X}_{\backslash C} \leftarrow \mathbf{Z}$. This makes things more complicated, since $\mathbf{Z}$ may not be fully observed and we need to learn the mixture components of $\mathbf{X}$ in the presence of missing data for $(\mathbf{Y}, \mathbf{Z})$. Hence, as before for $\mathbf{Z}$ alone, we query the BN to determine the probabilities of each mixture component and its sufficient statistics, calculated from the data available for discrete variables $(\mathbf{Y}, \mathbf{Z})$.

In both Type I and Type II models, we can split each multivariate Normal $\mathbf{X}$ into a bivariate Normal with variable C and $\mathbf{X}_{\backslash C}$. So μ_1 corresponds to the mean vector of $\mathbf{X}_{\backslash C}$ and μ_2 corresponds to the mean of C. Similarly, if Σ_{11} is the covariance matrix, then Σ_{22} is the variance and Σ_{21} are vectors except Σ_{11} and Σ_{22} from the covariance matrix of $\mathbf{X}$. The conditional mean $\mu_{2|1} = E(C \mid \mathbf{X}_{\backslash C})$ and conditional variance $\Sigma_{22|1} = Var(C \mid \mathbf{X}_{\backslash C})$ can then be obtained in conditional form. Because Σ_{22} is a constant, $\Sigma_{22|1}$ will be a constant, but $\mu_{2|1}$ will be a constant plus a linear expression of variable C's parent variables $\mathbf{X}_{\backslash C}$, since μ_1 is a vector.

Since the BN is expressed in conditional form, when learning multivariate Normal distributions, we use the fact that any multivariate Gaussian distribution can be converted to its conditional form. For example, a bivariate Gaussian (X_1, X_2) with mean vector (μ_1, μ_2) and covariance matrix $\begin{pmatrix} \Sigma_{11} & \Sigma_{12} \\ \Sigma_{21} & \Sigma_{22} \end{pmatrix}$ can be converted to a conditional Gaussian by:

$$\mu_{2|1} = \mu_2 + \sum_{21} \sum_{11}^{-1} (X_1 - \mu_1)$$

$$\Sigma_{22|1} = \Sigma_{22} - \Sigma_{21}\Sigma_{11}^{-1}\Sigma_{12}$$

which yields the conditional model in the BN.

To demonstrate the EM learning process, from data on a hybrid model with conditional distributions prior to learning, we consider the BN shown in Figure 17.7.

The node C conforms to a Type II model structure, since it is a mixture of four Normal components specified by the combinations of states for parent nodes B and D. The joint distribution $P(B, D)$ will therefore be used to calculate the weightings for each Normal mixture component for continuous node C. Node E is a single independent Normal distribution with no parents, and this conforms to the Type I model. Node F has hybrid parents, a discrete node and two continuous parent nodes, such that node G is node F's mixture variable for the two multivariate Normal mixture components with continuous variables C and E. To calculate the sufficient statistics for F, the EM algorithm queries the joint distribution $P(B, D, G)$ from the JT. Since node C has four Normal components, EM learning will be done using eight multivariate linear models in the following form:

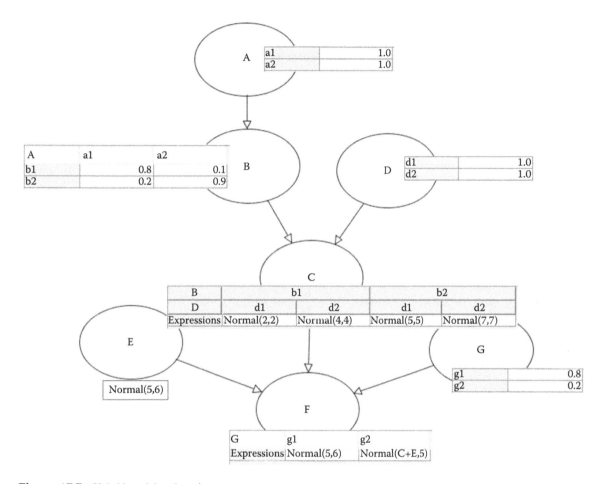

Figure 17.7 Hybrid model prelearning.

When *G* = *False*:

$$[C_{B=b1,D=F}, E, F_{G=F}], [C_{B=b1,D=T}, E, F_{G=F}], [C_{B=b2,D=F}, E, F_{G=F}],$$
$$[C_{B=b2,D=T}, E, F_{G=F}]$$

When *G* = *True*:

$$[C_{B=b1,D=T}, E, F_{G=T}][C_{B=b2,D=T}, E, F_{G=T}][C_{B=b1,D=F}, E, F_{G=T}]$$
$$[C_{B=b2,D=F}, E, F_{G=T}]$$

When any node in (B, D, G) is unobserved, then node F will be a weighted mixture of some of the above multivariate Normal distributions, and the weight of each distribution is obtained by the joint distribution of (B, D, G).

Let's assume we have some data, as shown in Table 17.16, and wish to learn the single and mixture distributions for the continuous nodes and the multinomial distributions for the discrete nodes.

Table 17.16
Data Set for Example 17.4 (Missing Values Are NA)

A	B	C	D	E	F	G
a1	NA	−0.54	d1	0.55	0.79	g2
a1	b2	0.74	d1	0.02	0.32	g2
NA	b2	0.04	d2	0	−0.97	g2
a2	b1	−0.3	NA	0.42	0.54	g1
a2	b2	−0.09	d1	−0.21	−0.98	g1
a2	b2	0.68	d1	0.55	−0.84	NA
a1	b1	−0.05	d2	−0.57	0.84	g2
a1	b2	−0.65	d2	0.88	−0.53	g1
a2	b1	0.11	NA	0.45	−0.96	g1
a2	NA	−0.6	d1	−0.84	0	g1
a1	NA	−0.54	d1	0.55	0.79	g2
a1	b2	0.74	d1	0.02	0.32	g2
NA	b2	0.04	d2	0	−0.97	g2
a2	b1	−0.3	d1	0.42	0.54	g1
a2	b2	−0.09	d1	−0.21	−0.98	g1
a2	b2	0.68	d1	0.55	−0.84	g1
a1	b1	−0.05	d2	−0.57	0.84	g2
a1	b2	−0.65	d2	0.88	−0.53	g1
a2	b1	0.11	NA	0.45	−0.96	g1
a2	NA	−0.6	d1	−0.84	0	g1
a1	NA	−0.54	d1	0.55	0.79	g2
a1	b2	0.74	d1	0.02	0.32	g2
a1	b2	0.04	d2	0	−0.97	g2
a2	b1	−0.3	NA	0.42	0.54	g1
a2	b2	−0.09	d1	−0.21	−0.98	g1
a2	b2	0.68	d1	0.55	−0.84	NA
a1	b1	−0.05	d2	−0.57	0.84	g2
a1	b2	−0.65	d2	0.88	−0.53	g1
a2	b1	0.11	d1	0.45	−0.96	g1
a2	NA	−0.6	d1	−0.84	0	g1

When we learn from this data, we achieve the revised BN NPTs and expressions as shown in Figure 17.8. Notice that the mixture model for node G is a simple independent Normal distribution and a linear regression equation dependent on nodes C and E.

The resulting marginal distributions are shown in Figure 17.9.

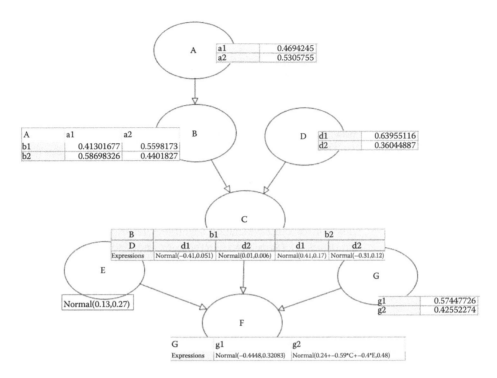

Figure 17.8 Hybrid model postlearning (the "D" icon denotes learned from data).

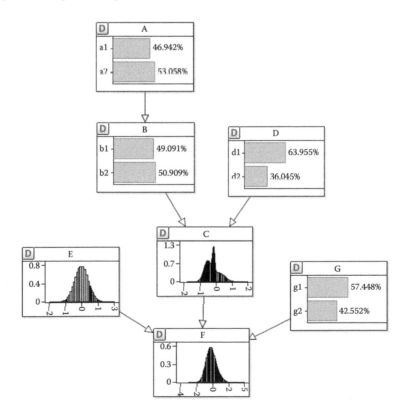

Figure 17.9 Hybrid model postlearning with superimposed marginal distributions.

17.6   Summary

In previous chapters, we covered various forms of parameter learning and parameter updating in BNs, but this was typically to learn single parameters. In contrast, in this chapter, we have shown how to learn many parameters from larger data sets and over the whole BN. Importantly, because in practice data sets are often incomplete, we have demonstrated how the learning can be done even in the presence of missing data.

The algorithm at the center of this chapter, expectation maximization, has been presented in two simple forms: for multinomial variables and for Normally distributed variables. Indeed, we have gone farther than this and presented these algorithms such that they can be applied to the mixed case where we have truly hybrid BNs containing continuous and discrete variables together. Naturally, this involves learning mixtures of Gaussian distributions indexed by the relevant discrete node states.

Finally, we have shown how expert opinions can be seamlessly added to the analysis and so can be combined with data. This is crucial given that even very large data sets on their own may be insufficient and are likely to include no observations for many state combinations. Moreover, this is of special benefit when coupled with the flexibility of being able to conduct this analysis using hybrid models expressed as a meaningful BN structure.

Further Reading

Dempster, A. P., Laird, N. M., and Rubin, D. B. (1977). Maximum likelihood from incomplete data via the EM algorithm. *Journal of the Royal Statistical Society. Series B (Methodological)*, 39(1), 1–38.

Appendix A: The Basics of Counting

The branch of mathematics that deals with counting is called combinatorics and an understanding of it is needed for probability and building risk models. We begin with two basic rules, which you will already know intuitively. Many counting problems may ultimately be decomposed into problems that are solved by these rules.

A.1 The Rule of Sum

Suppose a library has 4 different books on learning French and 5 on learning Spanish. Then in order to learn French or Spanish a student can select among these $4 + 5 = 9$ books. That is a very simple example of the following rule, called the rule of sum.

Rule of Sum

If a task can be performed in n distinct ways, while a second task can be performed in m distinct ways, and the tasks cannot be performed simultaneously, then performing either task can be accomplished in any of $n + m$ distinct ways.

The rule can be extended beyond two tasks as long as no pair of them can occur simultaneously. So, if the other language books in the library consist of 10 different books on German and 3 on Portuguese, then a student can select any one of the $5 + 4 + 10 + 3 = 22$ books to learn a language.

A.2 Rule of Product

Suppose now that the CEO of a large corporation is considering a takeover of another company. The CEO assigns 12 employees to 2 committees:

- Committee A (5 members) is to investigate possible favorable results (i.e., opportunities) from the takeover.
- Committee B (7 members) will scrutinize possible unfavorable repercussions (i.e., risks) from the takeover.

If the CEO decides to speak to just one committee member before making a decision, then by rule of sum there are 12 members who can be called for opinion. However, in order to be unbiased the CEO decides to speak with a member of committee A on Thursday and a member of committee B on Friday. There are 5×7 ways in which this can be done.

This is an example of the following rule of product.

Rule of Product

If a procedure can be broken down into first and second stages, and if there are n possible outcomes for the first stage and m for the second, then the total procedure can be carried out, in the designated order, in $n \times m$ ways.

All models in this chapter are available to run in AgenaRisk, downloadable from www.agenarisk.com.

The rule of product extends to any number of stages. So, if in a class of 30 students prizes are awarded for the best, second best, and third best student, then the number of different possible selections for the first two prizes is 30×29, and the number of different possible selections for the three prizes is $30 \times 29 \times 28$.

Example A.1

In the UK system for assigning postal codes the first two characters are predetermined by the geographical district. So, for example, every postal code in the town of Ilford begins with the characters IG, whereas every postal code in the town of Romford begins with the characters RM. The rest of a postal code is made up of the sequence of

<Number (1 to 9), Number (1 to 9), Letter (A to Z), Letter (A to Z)>

By the rule of product, the total number of postal codes that can be assigned to a particular geographical district is

$$9 \times 9 \times 26 \times 26 = 54{,}756$$

Since this is the absolute maximum number of different postal codes it follows that in any district with more than 54,756 buildings, it is inevitable that some buildings will share the same postal code. The town of Ilford has approximately 200,000 buildings and so most buildings share their postal code with three others.

At times we need both rules. Suppose, for example, we wish to assign an anonymous code to each employee in an organization, where the code consists either of a single letter or a single letter followed by a single decimal digit. By the product rule there are $26 \times 10 = 260$ two-character codes. Since there are also 26 one-character codes the rule of sum implies that there are $26 + 260 = 286$ possible codes in all. So if the organization has more than 286 employees we would need a different coding scheme.

Example A.2

Variables in a certain programming language may be up to three characters in length. The first character must be a letter of the alphabet, while the last (if there are more than one) must be a decimal digit. What is the total number of distinct variables allowed?

Solution. The total number of variables is equal to the number of 3-character variables plus the number of 2-character variables plus the number of 1-character variables. These are:

- Number of 3-character variables is $26 \times (26 + 10) \times 10 = 9360$ (by rules of sum and product).
- Number of 2-character variables is $26 \times 10 = 260$.
- Number of 1-character variables is 26.

Thus the total number is 9646.

A.3   Permutations

Suppose that in a class of 10 students, 5 are to be chosen and seated in a row for a photograph. How many such arrangements are possible? The key word is *arrangement*, emphasizing the importance of *order*. If the students are called A, B, C, D, E, F, G, H, I, and J then (B, C, E, F, I) and (C, E, F, I, B) are two different arrangements, even though they involve exactly the same five students. The question is answered by the rule of product applied to the number of ways of filling positions:

Position 1	Position 2	Position 3	Position 4	Position 5
10	9	8	7	6

So the total is $10 \times 9 \times 8 \times 7 \times 6 = 30{,}240$ possible arrangements.

If there were to be a full class photograph (with 10 students in a row), then there are $10 \times 9 \times 8 \times 7 \times 6 \times 5 \times 4 \times 3 \times 2 \times 1$ possible arrangements. Mathematicians write this more simply as 10!

The symbol ! here is called the *factorial* symbol. In general $n!$ is the product of all the numbers from n down to 1. For completeness we define 0! as being equal to 1. This is not really contentious when you consider the following observations:

$$3! = 4!/4$$
$$2! = 3!/3$$
$$1! = 2!/2$$

In fact, for any number n bigger than 0, it is easy to see that $n!$ is equal to $(n + 1)!$ divided by $n + 1$. So it is reasonable to define 0! $= 1!/1$, which is equal to 1.

Example A.3

In a normal pack of 52 playing cards the number of different possible sequences in which the cards can be ordered is 52! This is because the first card can be selected in 52 ways, the second in 51 ways, the third in 50 ways, and so on.

Example A.4

A traveling salesman has to visit 25 cities, starting and ending at the same one. Since he wishes to minimize the distance he has to drive, he is only interested in planning a route that involves visiting each other city once. How many different possible routes are there?

Solution. We label the cities as 1, 2, ..., 25. He has to start and end at city 1. There are then 24 ways to choose the second city to visit, 23 ways to visit the third city, and so on. By the rule of product there are 24! routes. In general if there are n cities there are $(n - 1)!$ routes.

Given a selection of objects, any arrangement of these objects is called a *permutation*.

Example A.5

There are 6 permutations of the letters {a, b, c}, namely, the 6 ordered triples {(a, b, c), (a, c, b), (b, c, a), (b, a, c), (c, b, a), (c, a, b)}.

If we are only interested in arranging two letters at a time then there are 6 permutations of *size* 2 from the collection, namely, the ordered pairs {(a, b), (b, a), (a, c), (c, a), (b, c), (c, b)}.

In general, if there are n objects and r is an integer between 1 and n, then the *number of permutations of size r for the n objects* is

n	$\times$	$n-1$	$\times$	$n-2$	$\times ... \times$	$n-r+1$
Position 1		Position 2		Position 3	...	Position r

The expression

$$n \times (n - 1) \times (n - 2) \times ... \times (n - r + 1)$$

is the same as

$$\frac{n!}{(n - r)!}$$

so the latter provides a formula for the number or permutations of size r for n objects.

Many problems that we consider in probability analysis rely on taking a sample of objects from some population. For example, suppose we have a set of 100 balls in an urn and wish to choose 5 of these balls one after the other for our sample. If, as earlier, we assume order is important, then there are still two relevant approaches that are different:

1. *Repetition is not allowed* (sampling without replacement). In this case the number of different samples we can choose is the number of permutations of 100 objects of size 5, that is, 100!/95!. In general for a set of n objects the number of different samples of size r (when sampling without replacement) is $n!/(n-r)!$ and clearly r must be no bigger than n.
2. *Repetition is allowed* (sampling with replacement). In this case we apply the rule of product 5 times to obtain the number of different samples by enumeration:

$$100 \times 100 \times 100 \times 100 \times 100, \text{ or } 100^5$$

In general for a set of n objects the enumerated number of different samples of size r (when sampling with replacement) is n^r, where r can be any number greater than zero.

Example A.6

A computer word consists of a string of eight 0's or 1's. How many distinct words are there? It is just the number of samples of size 8 from a population of size 2 (with replacement), that is, $2^8 = 256$.

Example A.7

We are to select three cards, noting them in order, from a pack of 52. If cards are not replaced then there are $52!/49! = 132,600$ selections, whereas with replacement there are $52^3 = 140,608$ selections.

Example A.8

The number of permutations of the letters of the word *computer* is 8! If only 4 of the letters are used, the number of permutations (of size 4) is

$$8!/(8 - 4)! = 8!/4! = 1680$$

If repetitions are allowed, the number of possible 4-letter sequences is $8^4 = 4096$. The number of 12-letter sequences is 8^{12}.

In all of the previous examples we have assumed the objects from which we are sampling are all different. But suppose now that we wish to calculate the number of permutations of the word *BALL*. The answer is not 4! (= 24) but 12. This is because there are not four distinct letters to permute; the two *L*'s are indistinguishable in the sense that if we switched the first and second *L* in the word *BALL* we still have exactly the same word. Let's label the two *L*'s as L_1 and L_2. If these two *L*'s really were distinguishable then there would be 4! permutations. However, to each permutation for which the *L*'s are indistinguishable there corresponds two permutations with distinct *L*'s. For example,

$$BALL = \frac{BAL_1L_2}{BAL_2L_1}, \qquad LABL = \frac{L_1ABL_2}{L_2ABL_1}, \qquad \text{etc.}$$

Hence the solution is $4!/2 = 12$.

As another example let us calculate how many permutations there are of the letters in *PEPPER*. Again, if we could distinguish three *P*'s and two *E*'s then there would be 6! However, to each permutation in which letters are indistinguishable there now corresponds $3! \times 2! = 12$ permutations where we can distinguish between the letters. For example,

$$PEPPER = \begin{array}{l} P_1E_1P_2P_3E_2R \\ P_2E_1P_1P_3E_2R \\ P_1E_1P_3P_2E_2R \\ \ldots \end{array}$$

Hence the solution is $6!/(3! \times 2!) = (6 \times 5 \times 4)/2 = 60$.

The two previous examples lead us to the following general rule for permutations.

General Rule for Permutations of Types of Objects

If there are n objects that are divided into r types with n_1 of first type, n_2 of second type, ..., n_r of the rth type then there are

$$\frac{n!}{n_1! \times n_2! \times \ldots n_r!}$$

permutations of the given objects.

Example A.9

There are $10!/(4! \times 3! \times 1! \times 1! \times 1!)$ permutations of the letters in *MASSASAUGA*.

A.4 Combinations

We have seen that there are two ways to select samples from a population: with and without replacement. Suppose now that from a class of 10 students we wish to select 3 as representatives for a committee. Certainly this involves sampling without replacement. However, the formula for permutations is not appropriate since we would not wish, for example, to distinguish the ordered triple B, E, F from F, E, B say. In short, order is now not important.

In this case each sample of 3 students with no reference to order corresponds to $3!$ permutations. A sample where order is not relevant is called a *combination*. So what this means is that to calculate the number of combinations of 3 students from 10 we simply divide the number of permutations by $3!$.

Since we know that the number of permutations is $10!/7!$, it follows that the number of combinations is equal to

$$\frac{10!}{3! \times 7!}$$

In general we therefore have the following rule:

The number of combinations of size r from n is

$$\frac{n!}{r! \times (n-r)!}$$

This is also written as $\binom{n}{r}$ or as ^nC_r

Example A.10

In a certain exam the students must answer four out of the seven questions. Since order is not important, the total number of different ways to answer the exam is thus

$$\frac{7!}{4! \times 3!} = \frac{7 \times 6 \times 5}{3 \times 2 \times 1} = 35$$

Example A.11

If the students must answer three questions out of five in section A and two out of four in section B, then section A can be completed in

$$\frac{5!}{3! \times 2!} = 10$$

ways and section B in

$$\frac{4!}{2! \times 2!} = 6$$

By the rule of product the exam can be completed in $10 \times 6 = 60$ ways.

Example A.12

In a lottery 49 balls numbered from 1 to 49 are placed in an urn and 6 are selected without replacement. Each ticket has 6 numbers that are chosen by the player. To win the lottery jackpot your ticket must match the 6 numbers drawn. How many tickets would you need to buy to guarantee that you win the jackpot?

Solution. Clearly to be certain of winning we would have to buy tickets with every possible combination of 6 numbers. Since this is sampling without replacement where order is not important the answer is simply the total number of combinations of size 6 from 49:

$$\frac{49!}{6! \times 43!} = \frac{49 \times 48 \times 47 \times 46 \times 45 \times 44}{6 \times 5 \times 4 \times 3 \times 2 \times 1} = 13,983,816$$

So you would need to buy nearly 14 million different tickets.

Example A.13

In a sequence of 10 coin tosses, how many different ways are there to get exactly 5 heads and 5 tails? What is the probability you would get exactly 5 heads and 5 tails?

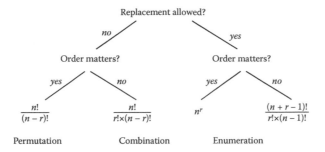

FIGURE A.1 Number of samples of size *r* from *n* objects where all *n* objects are distinct.

Solution. Once we know a sequence of 10 coin tosses has exactly 5 heads we know it must also have exactly 5 tails. So the problem reduces to counting the number of ways we can assign 5 objects to 10 positions, where order is not important. This is just the number of combinations of size 5 from 10:

$$\frac{10!}{5! \times 5!} = \frac{10 \times 9 \times 8 \times 7 \times 6}{5 \times 4 \times 3 \times 2 \times 1} = 252$$

If you toss a coin 10 times there are $2^{10} = 1024$ different sequences. So, since we know that 252 out of these 1024 sequences will result in exactly 5 heads and 5 tails, it follows that the probability of getting such a sequence is 252/1024 which is just under 25%.

A summary of the counting rules is shown in Figure A.1.

Appendix B: The Algebra of Node Probability Tables

There are a number of algebraic "tricks" that are needed when performing marginalization and using Bayes' Theorem in practice. This revolves around the manipulation of node probability tables (NPTs). You will only need to read this appendix if you are curious about the mathematical underpinnings, and if you are interested in gaining an understanding of the underlying algorithms reading this appendix is essential.

There are three operations we are interested in and we can use these along with Bayes' Theorem to compute or derive any measure of interest in the Bayesian network (BN). These are *marginalization*, *multiplication*, and *division*. Plus, instead of calculating probabilities one at a time we can use tables containing rows and columns indexed by variable state values to do so (note that these tables are not the same as matrices).

B.1 Multiplication

Consider the simple model $P(A, B, C) = P(B \mid A) P(C \mid A) P(A)$ with the following probability tables (notice that we are using x, y, z values for the probabilities and lowercase for each of the variable state values):

$P(A) =$

a_1	a_2
z_1	z_2

$P(B \mid A) =$

	a_1	a_2
b_1	x_1	x_2
b_2	x_3	x_4

$P(C \mid A) =$

	a_1	a_2
c_1	y_1	y_2
c_2	y_3	y_4

If we were to multiply the two NPTs $P(B \mid A)$ and $P(A)$ we would get a new NPT thus:

$P(A)P(B \mid A) =$

a_1	a_2
z_1	z_2

$\times$

	a_1	a_2
b_1	x_1	x_2
b_2	x_3	x_4

$=$

	a_1	a_2
b_1	$z_1 x_1$	$z_2 x_2$
b_2	$z_1 x_3$	$z_2 x_4$

Next, if we want to get $P(B \mid A)P(A)P(C \mid A)$ we can simply multiply again by $P(C \mid A)$:

$P(B \mid A)P(A)P(C \mid A) =$

	a_1	a_2
b_1	$z_1 x_1$	$z_2 x_2$
b_2	$z_1 x_3$	$z_2 x_4$

$\times$

	a_1	a_2
c_1	y_1	y_2
c_2	y_3	y_4

$=$

	a_1		a_2	
	c_1	c_2	c_1	c_2
b_1	$z_1 x_1 y_1$	$z_1 x_1 y_3$	$z_2 x_2 y_2$	$z_2 x_2 y_4$
b_2	$z_1 x_3 y_1$	$z_1 x_3 y_3$	$z_2 x_4 y_2$	$z_2 x_4 y_4$

A table's weight is the product of the states of all variables contained in that table. So the preceding table for $P(B \mid A)P(A)P(C \mid A)$ has weight 8 since it contains 8 cells.

Multiplying a table by unit values leaves the table unchanged. For example:

$$P(A) \times P(A') = \begin{array}{|cc|} a_1 & a_2 \\ \hline z_1 & z_2 \end{array} \times \begin{array}{|cc|} a_1 & a_2 \\ \hline 1 & 1 \end{array} = \begin{array}{|cc|} a_1 & a_2 \\ \hline z_1 & z_2 \end{array}$$

Note that we could normalize the table of unit values to (0.5, 0.5), but it is actually easier when carrying out calculations to leave the unit values and postpone normalization to the very end of a sequence of calculation steps.

B.2 Marginalization

Marginalization involves simply summing over terms in the table. So, if we had a table $P(A,B)$:

$$P(A,B) = \begin{array}{c|cc} & a_1 & a_2 \\ \hline b_1 & x_1 & x_2 \\ b_2 & x_3 & x_4 \end{array}$$

Then we marginalize by summing over index values:

$$P(A) = \sum_B P(A,B) = \begin{array}{|cc|} a_1 & a_2 \\ \hline x_1 + x_3 & x_2 + x_4 \end{array}$$

$$P(B) = \sum_A P(A,B) = \begin{array}{|cc|} b_1 & b_2 \\ \hline x_1 + x_2 & x_3 + x_4 \end{array}$$

Marginalization is distributive (can be broken into parts). First we can separate the parts by those variables that belong together and carry out table operations one part at a time. Thus, in this example $P(A)$ can be taken outside of a marginalization because it does not belong with the other variables:

$$\sum_C P(B \mid C)P(C)P(A) = P(A) \sum_C P(B \mid C)P(C)$$

Doing this means that we make the tables smaller, i.e., instead of calculating a big table for all the state combinations for all variables we do it for two alone and then multiply the others in. This saves time manually and for the computer.

We have to be careful to split the parts properly according to the conditional probabilities and how they are declared. For instance, for the joint probability model where each variable has two states:

$$P(A, B, C) = P(C \mid B)P(B \mid A)P(A)$$

We need to marginalize those terms that are "farther away," in terms of the chain of conditioning from the variable we are interested in so that we produce the smallest tables possible and carry out the most efficient calculations we can:

$$P(C) = \sum_{A,B} P(C \mid B)P(B \mid A)P(A) = \sum_{B} P(C \mid B)\left(\sum_{A} P(B \mid A)P(A)\right)$$

For $\sum_A P(B \mid A)P(A)$ we would need to multiply a table of weight 2 with one of weight 4 to get a table of weight 2. This is then multiplied into a table of weight 4, $\sum_B P(C \mid B)$, to get another table of weight 2. So the marginals we calculate at each stage are at most weight 2.

However, the following alternative marginalization sequence is much less efficient:

$$P(C) = \sum_{A,B} P(C \mid B)P(B \mid A)P(A) = \sum_{A} P(A)\left(\sum_{B} P(C \mid B)P(B \mid A)\right)$$

This is because, for $\sum_B P(C \mid B)P(B \mid A)$ we would multiply weight 4 by weight 4 to produce a marginal of weight 4. This is then multiplied with $\sum_B P(A)$, which is dimensions to provide a marginal result of weight 2. The largest marginal weight is 4 in this case, which is twice the size of the most efficient case.

Identifying the best marginalization sequence by eye is extremely difficult, inefficient, and tedious, and the resulting calculations are error prone. This is why we should rely on sophisticated computer algorithms to carry them out and do so efficiently by identifying from the structure of the model what the best marginalization sequence might be. This is discussed in Chapter 7.

B.3 Division

Division is equivalent to simply carrying out the fundamental rule of probability (Axiom 5.4, Chapter 5). Consider:

$$P(A \mid B) = \frac{P(A,B)}{P(B)}$$

with NPTs:

$P(B \mid A)P(A) =$

	a_1	a_2
b_1	$z_1 x_1$	$z_2 x_2$
b_2	$z_1 x_3$	$z_2 x_4$

and $P(B) =$

b_1	b_2
$x_1 + x_2$	$x_3 + x_4$

Then division is simply:

$$\frac{P(B\mid A)P(A)}{P(B)} = \frac{\begin{array}{c|cc} & a_1 & a_2 \\ \hline b_1 & z_1x_1 & z_2x_2 \\ b_2 & z_1x_3 & z_2x_4 \\ \hline & b_1 & b_2 \\ \hline & x_1+x_2 & x_3+x_4 \end{array}}{} = \begin{array}{c|cc} & a_1 & a_2 \\ \hline b_1 & \dfrac{z_1x_1}{x_1+x_2} & \dfrac{z_2x_2}{x_1+x_2} \\ b_2 & \dfrac{z_1x_3}{x_3+x_4} & \dfrac{z_2x_4}{x_3+x_4} \end{array}$$

Dividing a table by itself results in a unit table as follows:

$$\frac{P(A)}{P(A)} = \frac{\begin{array}{c|cc} & a_1 & a_2 \\ \hline & z_1 & z_2 \end{array}}{\begin{array}{c|cc} & a_1 & a_2 \\ \hline & z_1 & z_2 \end{array}} = \begin{array}{c|cc} & a_1 & a_2 \\ \hline & 1 & 1 \end{array}$$

Note that dividing a table by zeros does not result in an infinite value but the convention is to set them to zero. Similarly, for example, if a table contains a zero and is divided by any other term it also results in zero. For example, if

$$P(A') = \begin{array}{c|cc} & a_1 & a_2 \\ \hline & 0 & 0 \end{array}$$

then:

$$\frac{P(A)}{P(A')} = \frac{\begin{array}{c|cc} & a_1 & a_2 \\ \hline & z_1 & z_2 \end{array}}{\begin{array}{c|cc} & a_1 & a_2 \\ \hline & 0 & 0 \end{array}} \equiv \frac{P(A')}{p(A)} = \frac{\begin{array}{c|cc} & a_1 & a_2 \\ \hline & 0 & 0 \end{array}}{\begin{array}{c|cc} & a_1 & a_2 \\ \hline & z_1 & z_2 \end{array}} = \begin{array}{c|cc} & a_1 & a_2 \\ \hline & 0 & 0 \end{array}$$

An example of applying the algebra of NPTs is given in Box B.1.

Box B.1   An Example Using the Algebra of NPTs

Let's assume we have the model $P(A, B, C) = P(C \mid A, B)P(A)p(B)$, where

$$P(A) = \begin{array}{c|cc} & a_1 & a_2 \\ \hline & .7 & .3 \end{array} \qquad P(B) = \begin{array}{c|cc} & b_1 & b_2 \\ \hline & .6 & .4 \end{array}$$

$P(C \mid A, B) =$

	a_1		a_2	
	b_1	b_2	b_1	b_2
c_1	.7	.8	.1	.3
c_2	.3	.2	.9	.7

Let's assume we have some evidence about B and wish to calculate $P(C \mid B = b_1)$:

$$P(C \mid B = b_1) = \sum_{A,B} P(A, B = b_1, C) = \sum_{A,B} P(C \mid A, B = b_1)P(B = b_1)P(A)$$

We can use the algebra of NPTs to do this easily. First we recognize that

$P(B = b_1) =$

b_1	b_2
1	0

And then we can calculate $P(C \mid A, B = b_1)P(B = b_1)$ using multiplication:

$P(C \mid A, B = b_1)P(B = b_1) =$
b_1	b_2
1	0
$\times$

	a_1		a_2	
	b_1	b_2	b_1	b_2
c_1	.7	.8	.1	.3
c_2	.3	.2	.9	.7

$=$

	a_1		a_2	
	b_1	b_2	b_1	b_2
c_1	.7	0	.1	0
c_2	.3	0	.9	0

Next we can then multiply the result above, $P(C \mid A, B = b_1)P(B = b_1)$, by $P(A)$:

$P(A)P(B = b_1)P(C \mid A, B = b_1) =$
a_1	a_2
.7	.3
$\times$

	a_1		a_2	
	b_1	b_2	b_1	b_2
c_1	.7	0	.1	0
c_2	.3	0	.9	0

$=$

	a_1		a_2	
	b_1	b_2	b_1	b_2
c_1	.7(.7)	0	.1(.3)	0
c_2	.3(.7)	0	.9(.3)	0

We can now marginalize out A and B leaving C:

$$P(C \mid B = b_1) = \sum_{A,B} P(C \mid A, B = b_1) P(B = b_1) P(A)$$

$$= \begin{array}{cc} c_1 & c_2 \\ \hline .49 + .03 & .21 + .27 \end{array} = \begin{array}{cc} c_1 & c_2 \\ \hline .52 & .48 \end{array}$$

Visit www.bayesianrisk.com for exercises and worked solutions relevant to this chapter.

Appendix C: Junction Tree Algorithm

We break down the junction tree algorithm into two components: the algorithm for creating the tree and the algorithm for evidence propagation in the tree.

C.1 Algorithm for Creating a Junction Tree

In summary there are three steps involved in producing the junction tree of a Bayesian network (BN):

1. *Construct the moral graph*. This involves constructing an undirected graph from the BN.
2. *Triangulate the moral graph*. This involves choosing a node ordering of the moral graph, and using node elimination to identify a set of clusters (which are sets of nodes having certain properties we will explain later).
3. *Construct the junction tree*. This involves creating nodes corresponding to the clusters and inserting the appropriate separators between them.

Steps 2 and 3 are nondeterministic; consequently, many different junction trees can be built from the same BN.

We know that a BN comprises a graph G, and associated probability tables. When we speak of the junction tree associated with the BN, we are actually only interested in the underlying graph G. The algorithm for constructing a junction tree applies to any directed acyclic graph (DAG).

C.1.1 Step 1: Constructing the Moral Graph

The moral graph, G_M, corresponding to G is constructed as follows:

1. *For each node V in G identify its parents in G*. For example, in the BN graph G in Figure C.1(a), the parents of nodes F are $\{D, E\}$, the parents of H are $\{G, E\}$. The other nodes all have just one or zero parents.
2. *For any pair of such parents add an edge linking them if an arc does not already connect them.* For example, since there is no arc linking the parents $\{D, E\}$ of F we add an edge linking D and E. Similarly we need to add an edge linking G and E.
3. *Drop the direction of all the arcs.* The resulting graph G_M is the moral graph, as shown in Figure C.1(b).

The reason we call this a moral graph is because we insist on "marrying" all of the parent nodes together.

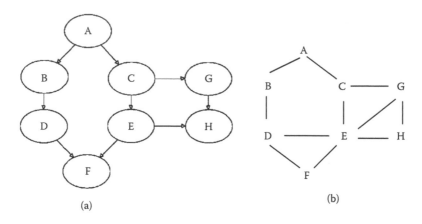

FIGURE C.1   (a) Example BN, G. (b) Moral graph, G_M

Box C.1   The Weight of a Node

For any node V the (primary) *weight* of V is the number of edges that need to be added to its set of neighbors to ensure that the node V and its neighbors form a complete subgraph (see sidebar). For example, in Figure C.1(b) the node A has two neighbors (B and C) that require one extra edge (to link B and C) to ensure that these three nodes form a complete subgraph. However, the node H, with neighbors E and G requires no extra edges since $\{G, E, H\}$ is already a complete graph.

Since we are interested in selecting nodes with the lowest weight, to reduce the number of ties we consider the (secondary) weight of a node V to be the product of the number of states in V and in neighbors. This notion is only relevant for graphs (like BNs) in which the notion of number of states associated with a node is meaningful.

C.1.2   Step 2: Triangulating the Moral Graph

The objective of this step is to identify subsets of nodes of the moral graph called *clusters*. The process involves determining at each step the node of minimum weight (see Box C.1) and eliminating it. Specifically, while there are still nodes left in the moral graph:

- Determine the node V with minimum weight (note this may not be unique; if not any node of minimum weight is selected). In determining the weight you will identify additional edges that may need to be added to neighbors of V.
- Add the edges identified in the previous step to the graph.
- Define as a cluster the node V and its neighbors.
- Remove node V from the graph.

A *complete* graph is one in which every pair of distinct nodes is connected by an edge.
A *maximal complete* subgraph is one that is not only complete but that is not contained in any larger, complete subgraph.

Example C.1

Let's apply this process to the moral graph of Figure C.1(b):

1. There are two nodes, H and F, of minimum weight zero, so we can choose either as the first for elimination. Let's choose H (Figure C.2). As the parents of H are already linked, there is no need to add any edges.
2. When we eliminate a node we must identify the cluster, so in this case the cluster formed is *GEH*.

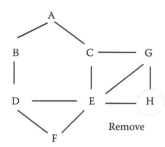

FIGURE C.2   *H* identified as node of minimum weight in the moral graph.

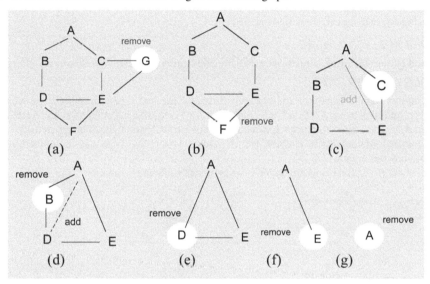

FIGURE C.3   Elimination sequence on moral graph.

3. We remove node *H* to arrive at the graph in Figure C.3(a).
4. In this graph there are two nodes, *G* and *F*, of minimum weight zero, so we can choose either as the first for elimination. Let's choose *G*. This induces the cluster *CEG* (and as the parents of *G* are already linked there is no need to add any edges).
5. We remove node *G* to arrive at the graph in Figure C.3(b).
6. Node *F* has minimum weight zero here inducing the cluster *DEF*.
7. We remove node *F*.
8. At this point each of the nodes has weight one, so now no matter what node we select next we would have to add one edge, so we arbitrarily select node *C* and add the edge between nodes {*A*, *E*}, as shown in Figure C.3(c).
9. This induces cluster *ACE*. A similar result holds at Figure C.3(d) where we remove node B and the process repeats itself for the remaining nodes in Figure C.3(e),(f),(g).

C.1.3   Step 3: Construct the Junction Tree

The final step is to connect the clusters obtained in the previous step in such a way that, when joined, they form a tree and not a graph. The tree property is essential for it to be useful for probabilistic inference, and we favor those junction trees that minimize the computational time required for inference.

A tree is a connected graph in which there are no cycles.

Following is the process for constructing the junction tree:

1. Remove any clusters that are not maximal (i.e., remove any cluster that is a subset of another cluster).
2. For the n remaining clusters, we can form a junction tree by iteratively inserting edges between pairs of clusters, until the clusters are connected by $(n-1)$ edges. We can also view this task as inserting $(n-1)$ separators between clusters. A separator between two clusters must be a common subset of the nodes in the two clusters.
3. We select candidate separators by choosing the separator with the highest number of members. If there is a tie (and if it is relevant to consider states of a node), then choose the separator with minimum weight (the product of the states of all nodes contained in that separator).

Example C.2

From our moral graph, we induced the following clusters:

GEH, GEC, DEF, ACE, ABD, ADE, AE, A

Cluster *A* and cluster *AE* are discarded since they are both subsets of *ADE*. This leaves six clusters:

GEH, GEC, DEF, ACE, ABD, ADE

Candidate separators include separator *EG* between clusters *CEG* and *GEH*, since *EG* is a subset of both. Separator *AE* is required between *ADE* and *ACE*. Separator *AD* is required between *ABD* and *ADE*, and separator *DE* between *ADE* and *DEF*. Separator CE is required between ACE and CEG. This gives us five separators, which is one less than the number of clusters, and therefore, clusters and separators combined, results in an optimal tree, as shown in Figure C.4.

This is equivalent to the following factorization order under variable elimination:

$$P(F) = \sum_D \left(\sum_E P(F|D,E) \left(\sum_A \left(\sum_C P(E|C)P(C|A)P(A) \Theta \right) \left(\sum_B P(D|D)P(B|A) \right) \right) \right)$$

where

$$\Theta = \sum_G P(G|C) \left(\sum_H P(H|E,G) \right).$$

Note that we are marginalizing out each variable only once. Other factorization orders are "of course" just as feasible.

Let's step through the factorization and compare the results against the junction tree, substituting the results from previous steps into later steps:

1. $\sum_H P(H|E,G) = P(E,G)$ cluster *GEH* and separator *EG*.

2. $\sum_G P(G|C)P(E,G) = P(E,C)$ cluster *EGC* and separator *CE*.

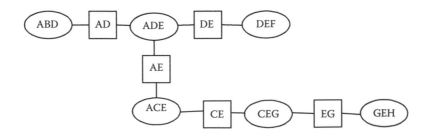

FIGURE C.4   Junction tree for example BN.

3. $\displaystyle\sum_B P(D\,|\,B)P(B\,|\,A)=P(A,D)$ cluster *ABD* and separator *AD*.

4. $\displaystyle\sum_C P(E\,|\,C)P(C\,|\,A)P(A)P(E,C)=P(A,E)$ cluster *ACE* and separator *AE*.

5. $\displaystyle\sum_A P(D,A)P(A,E)=P(D,E)$ cluster *ADE* and separator *DE*.

6. $\displaystyle\sum_D\left(\sum_E P(F\,|\,D,E)P(D,E)\right)$ cluster *DEF*

Notes:

1. As we have observed, there may, in general, be many ways to triangulate an undirected graph. An optimal triangulation is one that minimizes the sum of the state space sizes of the clusters of the triangulated graph. The task of finding an optimal triangulation is an example of a problem that mathematicians call NP-hard (i.e., there is no guarantee you have found the optimal one, from a computational perspective, but in practice a near optimal solution will still provide correct answers).
2. An undirected graph is triangulated if every cycle of length four or greater contains an edge that connects two non-adjacent nodes in the cycle. A *cycle* is a closed trail with at least one edge and with no repeated vertices except that the initial vertex is the terminal vertex. A *trail* is a walk with no repeated edges. A *walk* is an alternating sequence of vertices and edges, with each edge being incident to the vertices immediately preceding and succeeding it in the sequence.
3. The algorithm, that we describe next, for evidence propagation in a junction tree is called *sum-product* and is exact. There are other forms of exact algorithm such as *sum-max* and many types of approximate algorithms for use with really large models.

C.2   Algorithm for Evidence Propagation in a Junction Tree

The junction tree is now a series of pipeline graphs. We can now formulate an evidence propagation scheme using the algebra of NPTs (Appendix B).

Take the example model $P(A,B,C)=P(C\,|\,B)P(B\,|\,A)P(A)$ with BN graph and junction tree as shown in Figure C.5.

We assign NPTs to clusters as per the membership of nodes assigned. In some cases a node may belong to more than one cluster so we assign it to a cluster of the smallest size. So, in this case we assign $P(A)$ to the

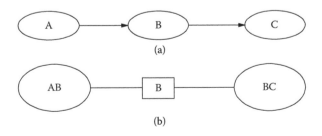

FIGURE C.5   BN and junction tree for propagation example.

cluster AB and using the notation $t(AB)$ to indicate a cluster composed from nodes belonging to that cluster we get the following cluster tables:

$$t(AB) = P(B \mid A)P(A)$$

$$t(BC) = P(C \mid B)$$

In the case of our separator, B, the node B has already been assigned to a cluster so we simply initialize it to 1:

$$t(B) = 1$$

Let's talk through the propagation one step at a time. When we update evidence at cluster BC, either on node B or node C, we produce an amended table for the cluster, $t(BC)$, which we will denote by $t^*(BC)$ with the asterisk (*) showing that the cluster table has changed. So in the case where we have evidence on node C:

$$t^*(BC) = P(C = c_1 \mid B)$$

We can marginalize to calculate the conditional change in node B given this change in C and use Bayes' theorem to convert $P(C = c_1 \mid B)$ to $P(B \mid C = c_1)$:

$$t^*(B) = \sum_C t^*(BC) = \sum_C P(B \mid C = c_1)$$

We now have a new separator $t^*(B)$, whereas before we had initialized this as $t(B) = 1$.

Next we want to update the cluster AB since node B's marginal has changed via the change in the separator B. We can do this by multiplying in the likelihood change in the separator, $t^*(B) / t(B)$:

$$t^*(AB) = t(AB)\left[\frac{t^*(B)}{t(B)}\right]$$

This is equivalent to changing the odds ratio and it should be clear that it simply involves uplifting state probabilities that have been favored by the new evidence and deflating those unfavored.

We can take a closer look at the results of multiplication to see how the cluster AB updates:

$$t^*(AB) = t(AB)\left[\frac{t^*(B)}{t(B)}\right]$$

$$= P(B \mid A)P(A)\frac{t^*(B)}{1}$$

$$= P(B \mid A)P(A)\sum_C P(B \mid C = c_1)$$

$$\Rightarrow t^*(AB) = P(A, B) = P(B \mid A)P(A)\sum_C P(B \mid C = c_1)$$

Thus, after evidence has been entered all cluster and separator tables are updated to reflect the new state of affairs, giving $t^*(AB)$, $t^*(BC)$ and $t^*(B)$.

The use of multiplication and marginalization at each stage should be pretty straightforward but the role of the likelihood change, $t^*(B)/t(B)$, may look a little puzzling at first; however, its use is key to understanding how local calculations, via separators, ensure the approach is successful.

Separators can be obtained by marginalizing from one of their neighboring clusters or from the other neighbor (there are only two neighbors of course because the clusters belong to a tree and it is easy to see how these two marginalizations are equivalent to collection and distribution). In this example these are

$$t(B) = \sum_C t(BC)$$

$$t(B) = \sum_A t(AB)$$

Let's use the asterisk again to show that these are different, where we have evidence on C, so

$$t^*(B) = \sum_C t^*(BC) \tag{C.1}$$

The trick is to realize that

$$t^*(B) = \sum_C t^*(BC) = \frac{t^*(B)}{t(B)} t(B) \tag{C.2}$$

And since

$$t(B) = \sum_A t(AB) \tag{C.3}$$

then we simply substitute Equation (C.3) into Equation (C.2):

$$t^*(B) = \sum_C t^*(BC) = \frac{t^*(B)}{t(B)} \sum_A t(AB)$$

We can then expand Equation (C.1) to

$$t^*(B) = \sum_C t^*(BC) = \sum_A \left[t(AB) \frac{t^*(B)}{t(B)} \right] = \sum_A t^*(AB)$$

So we have effectively proven that the propagated evidence from node C through the junction tree starting from cluster BC, via the separator, B, updates cluster AB. At each stage the separator table can be marginalized out from any separator or cluster and will be consistent with that obtained from any other cluster.

It should be relatively easy to see that propagation can also occur in the opposite direction using separator tables as messages:

$$t^*(B) = \sum_A t^*(AB) = \frac{t^*(B)}{t(B)} \sum_C t(BC) = \sum_C \left[t(BC) \frac{t^*(B)}{t(B)} \right] = \sum_C t^*(BC) = t^*(B)$$

To better illustrate the steps involved in propagation we shall carry out propagation from cluster AB to cluster BC:

1. First we initialize the separators to 1:

$t(B) =$	b_1	b_2
	1	1

2. Then we update cluster AB with new evidence.

$$t^*(AB)$$

3. Next we marginalize to get our separator, B.

$$t^*(B) = \sum_A t^*(AB)$$

4. We then calculate the likelihood change at the separator, B.

$$\frac{t^*(B)}{t(B)}$$

5. Finally we multiply the likelihood from the separator by the previous table for cluster BC to get a new table:

$$t^*(BC) = t(BC)\frac{t^*(B)}{t(B)}$$

Let's add some numbers to this to make it concrete:

$$t(AB) = P(A)P(B \mid A) \qquad\qquad t(BC) = P(C \mid B)$$

$$=\quad \begin{array}{c|cc} & a_1 & a_2 \\ \hline b_1 & 0.27 & 0.14 \\ b_2 & 0.03 & 0.56 \end{array} \qquad = \quad \begin{array}{c|cc} & b_1 & b_2 \\ \hline c_1 & 0.1 & 0.4 \\ c_2 & 0.9 & 0.6 \end{array}$$

First we initialize the separator and then take account of evidence, which is available on node C:

$$e = \quad \begin{array}{c|cc} & c_1 & c_2 \\ \hline & 0 & 1 \end{array}$$

Next we multiply this into cluster table BC:

$$t^*(BC) = t(BC) \times e = \quad \begin{array}{c|cc} & b_1 & b_2 \\ \hline c_1 & 0.1 & 0.4 \\ c_2 & 0.9 & 0.6 \end{array} \quad \times \quad \begin{array}{c|cc} & c_1 & c_2 \\ \hline & 0 & 1 \end{array} \quad = \quad \begin{array}{c|cc} & b_1 & b_2 \\ \hline c_1 & 0 & 0 \\ c_2 & 0.9 & 0.6 \end{array}$$

Then we marginalize out the separator:

$$t^*(B) = \sum_C t^*(BC) = \quad \begin{array}{c|cc} & b_1 & b_2 \\ \hline & 0.9 & 0.6 \end{array}$$

We can now calculate the likelihood ratio at the separator:

$$\frac{t^*(B)}{t(B)} = \frac{\begin{array}{cc} b_1 & b_2 \\ \hline 0.9 & 0.6 \end{array}}{\begin{array}{cc} b_1 & b_2 \\ \hline 1 & 1 \end{array}} = \begin{array}{cc} b_1 & b_2 \\ \hline 0.9 & 0.6 \end{array}$$

Finally we multiply this into the cluster table AB:

$$t^*(AB) = t(AB)\frac{t^*(B)}{t(B)} = \begin{array}{c|cc} & a_1 & a_2 \\ \hline b_1 & 0.27 & 0.14 \\ b_2 & 0.03 & 0.56 \end{array} \times \frac{\begin{array}{cc} b_1 & b_2 \\ \hline 0.9 & 0.6 \end{array}}{\begin{array}{cc} b_1 & b_2 \\ \hline 1 & 1 \end{array}} = \begin{array}{c|cc} & a_1 & a_2 \\ \hline b_1 & 0.243 & 0.126 \\ b_2 & 0.018 & 0.336 \end{array}$$

We can now query our model to get the marginals for nodes A and B (normalizing the result to sum to one along the way):

$$t^*(A) = \sum_B t^*(AB) = \begin{array}{cc} a_1 & a_2 \\ \hline 0.261 & 0.462 \end{array} = \begin{array}{cc} a_1 & a_2 \\ \hline 0.3610 & 0.6390 \end{array}$$

$$t^*(B) = \sum_A t^*(AB) = \begin{array}{cc} b_1 & b_2 \\ \hline 0.369 & 0.354 \end{array} = \begin{array}{cc} b_1 & b_2 \\ \hline 0.5104 & 0.4896 \end{array}$$

Note that some clusters will not have any NPTs assigned to them, such as cluster ADE in Figure C.4. How are they to be populated with probabilities then? This is done by initializing such clusters with 1's and then performing an initial propagation, using vacuous evidence (i.e., 1) allocated on any node belonging to any leaf cluster in the junction tree. The effect of this propagation is to update such clusters via their separators.

Box C.2 From Junction Tree Back to Joint Distribution

The joint distribution of a BN can be calculated by multiplying all of the clusters together and then dividing by the separators. For Figure C.4 we identified the clusters and separators:

1. Cluster GEH, $P(H \mid E,G)$ and separator EG, $P(E,G)$.
2. Cluster EGC, $P(G \mid C)P(E,G)$ and separator CE, $P(E,C)$.
3. Cluster ABD, $P(D \mid B)P(B \mid A)$ and separator AD, $P(A,D)$.
4. Cluster ACE, $P(E \mid C)P(C \mid A)P(A)P(E,C)$ and separator AE, $P(A,E)$.
5. Cluster ADE, $P(D,A)P(A,E)$ and separator DE, $P(D,E)$.
6. Cluster DEF, $P(F \mid D,E)P(D,E)$.

The joint probability distribution is therefore the clusters divided by the separators:

$$P(A,B,C,D,E,F,G,H)$$

$$= \frac{\alpha \times \beta}{P(E,G)P(E,C)P(D,A)P(A,E)P(D,E)}$$

$$= P(H \mid E,G)P(F \mid D,E)P(D \mid B)P(B \mid A)P(E \mid C)P(C \mid A)P(A)$$

$$= P(A,B,C,D,E,F,G,H)$$

where

$$\alpha = P(H \mid E,G)\ P(G \mid C)\ P(E,G)\ P(D \mid B)\ P(B \mid A)\ P(E \mid C)$$

and

$$\beta = P(C \mid A)\ P(A)\ P(E, C)\ P(D, A)\ P(A, E)\ P(F \mid D,E)\ P(D,E)$$

At the end of message passing through the junction tree the full joint probability distribution is therefore recovered.

Appendix D: Dynamic Discretization

The ability to include numeric variables in a Bayesian network (BN) without having to predefine a set of static states for each such variable is achieved through a dynamic discretization algorithm together with various heuristics to ensure a good balance between accuracy and efficiency. This appendix provides a formal description of the dynamic discretization algorithm and associated heuristics for readers with a keen interest in the underlying theory.

Our approach to dynamic discretization allows us to automatically estimate the conditional probability densities for all deterministic and statistical functions specified in the BN. No separate analytical calculation needs to be performed, no simulation methods are required, and no tables need to be populated. Because of this you do not need to use techniques like Monte Carlo simulation alongside BNs and can instead use this algorithm as the all-in-one solution.

In Section D.1 we introduce the notation and provide an overview of the dynamic discretization approach. The core algorithm is described in detail Section D.2. Section D.3 describes the methods used for approximating the statistical and deterministic functions we need to manipulate using the dynamic discretization algorithm. Section D.4 describes the various methods that are used in practice to ensure the algorithm works efficiently and accurately.

D.1 Overview of Dynamic Discretization Approach

Let X be a continuous numeric node in a BN. The range of X is denoted by Ω_X, and the probability density function (pdf) of X is denoted by f_X. The idea of discretization is to approximate f_X by, first, partitioning Ω_X into a set of intervals $\Psi_X = \{w_j\}$, and second, defining a locally constant function $\tilde{f}_X$ on the partitioning intervals. The task consists of finding an optimal discretization set $\Psi_X = \{\omega_i\}$ and optimal values for the discretized probability density function $\tilde{f}_X$. Discretization operates in much the same way when X takes integer values, but here we will focus on the case where X is continuous.

The approach to dynamic discretization described here searches Ω_X for the most accurate specification of the high-density regions (i.e., around the modes), given the model and the evidence by calculating a sequence of discretization intervals in Ω_X iteratively. At each stage in the iterative process a candidate discretization, Ψ_X, is tested to determine whether the resulting discretized probability density $\tilde{f}_X$ has converged to the true probability density f_X within an acceptable degree of precision. At convergence, f_X is then approximated by $\tilde{f}_X$.

By dynamically discretizing the model we achieve more accuracy in the regions that matter and incur less storage space compared with static discretizations. Moreover, we can adjust the discretization anytime in response to new evidence to achieve greater accuracy.

The approach to dynamic discretization presented here is influenced by the work of Kozlov and Koller on using nonuniform discretization in hybrid BNs. The key features of the approach include:

A BN containing both discrete and continuous numeric nodes is called a *hybrid BN* (HBN).

---

1. The relative entropy, or Kullback-Leibler (KL) measure, is the distance between two density functions f and g and can be used as a metric D of the error introduced by approximating the true (but unknown) function $f(x)$, by some approximate function, $g(x)$:

$$D(f \parallel g) = \int_S f(x) \log \frac{f(x)}{g(x)} dx$$

Under this metric, the optimal value for the discretized function $\tilde{f}$ is given by the mean of the function f in each of the intervals of the discretized domain.

2. Given that we are approximating the true function we use a bound on the KL distance as an estimate of the relative entropy error between a function f and its discretization $\tilde{f}$ based on the function mean $\overline{f}$, the function maximum f_{max}, and the function minimum f_{min} in the given discretization interval w_j:

$$E_j = \left[\frac{f_{max} - \overline{f}}{f_{max} - f_{min}} f_{min} \log \frac{f_{min}}{\overline{f}} + \frac{\overline{f} - f_{min}}{f_{max} - f_{min}} f_{max} \log \frac{f_{max}}{\overline{f}} \right] |w_j|$$

where $|w_j|$ denotes the length of the discretization interval w_j and the density values $\overline{f}$, f_{max}, f_{min} are approximated using the midpoint of an interval and its points of intersection with neighboring intervals.

Throughout this appendix when we talk of relative entropy error we mean the bound on the KL distance, but think it easier to employ this linguistic shortcut.

3. Evidence propagation uses an extension to standard BN inference algorithms such as the junction tree approach described in Appendix C.

KL distance and entropy are very important in quantum information theory as well as statistical mechanics.

For speed and convenience we choose to discretize univariate partitions (i.e., marginal densities only), which are simpler to implement, instead of tackling the problem of partitioning joint multivariate distributions such as clusters.

In outline, dynamic discretization follows these steps:

1. Choose initial discretizations for all continuous variables.
2. Calculate the discretized conditional probability density of each continuous node given the current discretization and propagate evidence through the BN.
3. Query the BN to get posterior marginals for each node and split those intervals with highest entropy error in each node.
4. Continue to iterate the process by recalculating the conditional probability densities and propagating the BN, querying to get the marginals, and then splitting the intervals with highest entropy error until the model converges to an acceptable level of accuracy.

We consider here a BN for a set of random variables, X, and partition X into the sets, $\mathbf{X}_Q$ and $\mathbf{X}_E$, consisting of the set of query variables and the set of observed variables, respectively.

D.2   The Dynamic Discretization Algorithm

Our approximate inference approach using dynamic discretization is based on the following algorithm:

1:   Initialize the discretization, $\Psi_X^{(0)}$, for each continuous variable $X \in \mathbf{X}$.
2:   Build a junction tree structure to determine the clusters, Φ, and sepsets.
3:   **for** $l=1$ to max_num_iterations
4:        Compute the NPTs, $P^{(l)}(X \mid pa\{X\})$, on $\Psi_X^{(l-1)}$ for all nodes $X \in \mathbf{X}_Q$ that have new discretization or that are children of parent nodes that have a new discretization
5:        Initialize the junction tree by multiplying the NPTs for all nodes into the relevant members of Φ
6:        Enter evidence, $\mathbf{X}_E = \mathbf{e}$, into the junction tree
7:        Perform global propagation on the junction tree
8:        **for** all nodes $X \in \mathbf{X}_Q$
9:             Marginalize/normalize to get the discretized posterior marginals $P^{(l)}(X \mid \mathbf{X}_E = \mathbf{e})$
10:            Compute the approximate relative entropy error $S_X^{(l)} = \sum_{w_j} E_j$, for $P^{(l)}(X \mid \mathbf{X}_E = \mathbf{e})$ over all intervals w_j in $\Psi_X^{(l-1)}$
11:            **If**

$$\left\{ 1 - \alpha \le \frac{S_X^{(l-k)}}{S_X^{(l-k+1)}} \le 1 + \alpha \text{ for } k = 1, 2, 3 \right\} \quad \text{\# Stable-entropy-error stopping rule \#}$$

or

$$\left\{ S_i^X < \beta \right\} \quad \text{\# Low-entropy-error stopping rule \#}$$

12:                 **then** stop discretization for node X
13:                 **else** create a new discretization $\Psi_X^{(l)}$ for node X:
14:                      Split into two halves the interval w_j in $\Psi_X^{(l-1)}$ with the highest entropy error, E_j.
15:                      Merge those consecutive intervals in $\Psi_X^{(l-1)}$ that have zero mass and zero entropy error
16:            **end if**
17:        **end for**
18:   **end for**

Iteratively updating the partitioning intervals using the dynamic discretization scheme involves the following steps:

1. Recalculate the NPT approximations over the current discretized domains.
2. Propagate the discrete BN to compute the approximate marginal posterior probability density function of each node.
3. Split/merge intervals according to the relative entropy criteria until the model converges.

This approach is able to obtain accurate approximations for the marginal probability density functions of conditionally deterministic variables defined by deterministic functions and general statistical distributions.

Next we turn to methods used for approximating the statistical and deterministic functions we need to manipulate using the dynamic discretization algorithm. First we will discuss the challenge of doing so using Monte Carlo and other techniques, and then describe how this problem is solved in AgenaRisk.

D.3 Methods for Approximating Functions

D.3.1 Why Monte Carlo Simulation Is Inadequate

Once a discretization has been defined at each step in the algorithm, we need to calculate the marginal probability for all X in the model by marginalization from the conditional distribution of X given its parents, $p(X \mid pa\{X\})$. For more complex conditional distributions, approximation techniques need to be used, and at first glance, the obvious way of approximating these is simply to simulate them using Monte Carlo methods to sample the distributions involved.

Consider, for instance, the case in which the conditional distributions $p(X \mid pa\{X\})$ involve a deterministic function f of random variables, for example, $X = f(pa\{X\})$. We could generate the local conditional probability table $p(X \mid pa\{X\})$ by sampling values, with Monte Carlo methods, from each parent interval in $\Omega_{pa\{X\}}$ for all parents of X and calculating the result $X = f(pa\{X\})$, then counting the frequencies with which the results fall within the static bins predefined for X, and finally normalizing the NPT.

Although simple this procedure is flawed. On the one hand, there is no guarantee that every interval in Ω_X will contain a probability density if the parents' node values are undersampled. The implication of this is that some regions of Ω_X might be void; they should have probability mass but do not. Any subsequent inference in the BN will return an inconsistency when it encounters either a valid observation in a zero mass interval in X or attempts inference involving X. The only way to counter this is to generate a large number of samples, which is expensive and cannot guarantee a solution.

Consider, for example, $Z = X + Y$ with $X,Y \sim N(10,100)$. Here

$$p(X,Y,Z) = p(Z \mid X,Y)p(X)p(Y) \text{ with } Z = f(X,Y).$$

The resulting marginal distribution $p(Z)$ using a static discretization is shown in Figure D.1 and we can clearly see that the interval [80.0, 90.1] has been undersampled resulting in a zero mass interval. Should we actually observe evidence $Z = 80.05$ then we will achieve a falsely inconsistent result, and any attempt at inference about the parents of Z will stall.

So, given that we approximate the conditional distribution, how do we do this in AgenaRisk? First we will describe how the approximation works for statistical distributions, and then we will describe how it works for deterministic functions.

D.3.2 Estimating Statistical Distribution Functions by Sampling Partitions

We calculate all conditional statistical distribution functions as follows:

1. Sample m equally spaced partitions from each parent node.
2. Generate the resulting distribution on the child node.
3. Normalize the result.

For $m = 2$ this is

1. Take the upper and lower bounds of each interval in $\Psi_{pa\{X\}}^{(l)}$.
2. Calculate the conditional density.
3. Use the appropriate distribution function for the current intervals in the child node X, $\Psi_X^{(l)}$.
4. Normalize the resulting densities.

Under dynamic discretization an increasing number of intervals are produced, resulting in many more interval combinations, which has the effect of fitting an equally weighted mixture of conditional

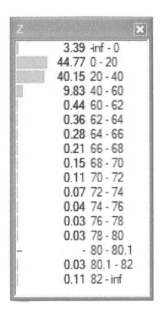

FIGURE D.1 $p(Z)$ with zero mass in the interval $]80.0, 80.1]$.

distributions to the continuous function, and resulting in a piecewise continuous function with no voids (at least theoretically but more on this later).

We describe how the approach works in the case where there are just two parents and where $m = 2$ (i.e., lower and upper bounds only):

1. Consider a model with three variables A, B, and C. We can define variable C as a statistical distribution function with A and B as parameters, $f(C \mid A,B)$, such as $f(C \mid A,B) \sim N(\mu = A, \sigma^2 = B)$.

2. Suppose the variables A and B have discretizations, Ψ_A and Ψ_B over their respective domains Ω_A and Ω_B. For each pair of intervals in the respective sets Ψ_A and Ψ_B, such as interval (a_1, a_2) in Ψ_A and interval (b_1, b_2) in Ψ_B, the approach computes the following distributions over the child node, C, $f(C \mid a_1, b_1), f(C \mid a_1, b_2), f(C \mid a_2, b_1)$ and $f(C \mid a_2, b_2)$.

3. If n is the set of all such pairs of parent intervals then we get a set of intervals $j = 1,\ldots,n$ each of which generates a set of statistical distributions $f(C \mid A = a_j, B = b_j)$. Assuming that Ψ_C consists of the intervals $\omega_1,\ldots\omega_n$ then the NPT for the conditional density function of the node C, $f(C \mid A = a_j, B = b_j)$, is approximated as

$$f(C \mid A = a_j, B = b_j) = \sum_{j=1}^{n} \left[\int_{\omega_k} f(C \mid A = a_j, B = b_j) dC \right]$$

4. We can generate the statistical distribution functions, $f(C \mid A = a_j, B = b_j)$, using standard libraries, which will return the probability mass in the interval required when supplied with valid parameter values.

The approach described here could be said to have more in common with stratified Monte Carlo sampling in the sense that more partitions that will be generated by the algorithm will tend to be in the high-density regions rather than spread randomly.

Note that the process used in dynamic discretization is not the same as Monte Carlo sampling because samples are not generated randomly over the multidimensional space. However, the approaches are similar in that they both aim to solve the problem of integration of a conditional probability function. In fact, at the limit, they will provide identical results since, as the number of partitions in the model increases, a greater number of conditional distributions are being generated and averaged, much in the same way that Monte Carlo generates sample points that are then averaged to estimate the function. However, the key difference is that dynamic discretization is generating densities over intervals rather than single point values. This generation of densities is crucial if we wish to solve the inverse problem and reason from effect to cause. To do otherwise means that evidence might exist in a perfectly valid region of the model that may remain unsampled under Monte Carlo, as mentioned earlier.

D.3.3   Approximating Deterministic Functions of Random Variables Using Mixtures of Uniform Distributions

In general, performing approximate inference on models that have conditionally deterministic functions of random variables represents a major challenge. By deterministic function we mean a function with no uncertainty, where the conditionally dependent variable is an arithmetical function, such as $Z = 5X + Z$.

We resolve all deterministic functions by modeling them as an approximate mixture of Uniform distributions. This involves the following steps:

1. Take the upper and lower bounds of each interval in $\Psi_{pa\{X\}}^{(l)}$.
2. Calculate all values using the deterministic function.
3. Calculate the min and max values, and enter these as the parameters in the Uniform distribution.

Under dynamic discretization an increasing number of intervals are produced, resulting in many more interval combinations, which has the effect of fitting a histogram composed of Uniform distributions to the continuous function, and resulting in a piecewise continuous function with no voids.

We describe how the approach works in the case where there are just two parents:

1. Consider a model with three variables A, B, and C. We can define variable C as a deterministic function of $C = g(A,B)$, such as $g(A,B) = A \times B$.
2. Suppose the variables A and B have discretizations, Ψ_A and Ψ_B over their respective domains Ω_A and Ω_B. For each pair of intervals in the respective sets Ψ_A and Ψ_B, such as interval (a_1, a_2) in Ψ_A and interval (b_1, b_2) in Ψ_B, the approach computes the minimum l and the maximum u for each of the set of values $g(a_1,b_1), g(a_1,b_2), g(a_2,b_1)$ and $g(a_2,b_2)$.
3. If I is the set of all such pairs of intervals then we get a set of intervals (l_i, u_i) for each $i \in I$, and this generates a Uniform probability density mass, $U(l_i, u_i)$, over the range of C. Assuming that Ψ_C consists of the intervals $\omega_1, ... \omega_n$ then the approximating NPT for the conditional density function of the node C, $\hat{f}(C \mid pa(C))$, is defined as a weighted Uniform distribution by

$$p_{l_i, u_i}(C \in \omega_k) \times U(C; l_i, u_i)$$

where $p_{l_i, u_i}(C \in \omega_k)$ represents the fraction of the Uniform mass $U(C; l_i, u_i)$ corresponding to the interval ω_k, that is,

$$p_{l_i, u_i}(C \in \omega_k) = \begin{cases} \displaystyle\int_{\omega_k} U(C; l_i, u_i) dC & \text{if } \omega_k \cap (l_i, u_i) \neq \varnothing \\ 0 & \text{otherwise} \end{cases}$$

For instance with $Z = X + Y$ we take the boundaries of each interval from X and Y and calculate the conditional probability for Z from these:

$$p(Z \mid X \in [x_l, x_u], Y \in [y_l, y_u]) = U(\min(x_l + y_l, x_l + y_u, x_u + y_l, x_u + y_u),$$

$$\max(x_l + y_l, x_l + y_u, x_u + y_l, x_u + y_u))$$

If calculating sums of random variables this approach is equivalent to an alternative approach called *convolution*. See Box D.1 for an example of convolution and a nonlinear deterministic case.

Box D.1 Deterministic Function Examples: Sum of Random Variables and Nonlinear Functions

Let us first consider the probability distribution for a sum of two independent random variables $Z = X + Y$, where $X \sim f_X$ and $Y \sim f_Y$, given by the convolution function:

$$f_Z(z) = f_X \times f_Y(z) = \int f_X(x) f_Y(z - x)\, dx$$

Calculating such a distribution represents a major challenge for most BN software. Traditional methods to obtain this function include fast Fourier transform (FFT) or Monte Carlo simulation. Here we compare an example and solution using AgenaRisk with the analytical solution produced by convolution of the density functions.

Consider the case $f_X = Uniform(-2, 2)$ and $f_Y = Triangular(0, 0, 2)$. The probability density for $Z = X + Y$ can be obtained analytically by

$$f_Z(z) = \int_0^{2+z} (1/4 + x/8) dx + \int_0^2 (1/4 + x/8) dx + \int_{z-2}^0 (1/4 + x/8) dx$$

The resulting mean and variance are $E[Z] = 0.667$ and $Var(Z) = 1.555$.

Using dynamic discretization over 25 iterations results in the set of marginal distributions for $f_Z = f_X \times f_Y$ as shown in Figure D.2. The summary statistics are $\mu_Z = 0.667$ and $\sigma_Z^2 = 1.567$, which are very accurate estimates of the analytical solution.

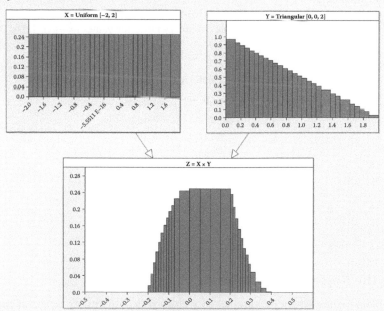

FIGURE D.2 Marginal distributions from function $f_Z = f_X \times f_Y$ after 25 iterations.

The sum entropy error for the estimate of $P(Z)$ is approximately 10^{-6}.

The second example here illustrates how our approach can estimate the distribution of variables that are non-linear deterministic functions of their continuous parents. The model consists of a variable, X, with distribution Beta(2.7,1.3,0,1), a variable Y that is a nonlinear deterministic function of its parent, $Y = -0.3X^3 + X^2$, and a variable Z with a Gaussian conditional distribution, $Z \sim N(2Y + 1, 1)$. Figure D.3 shows the posterior marginals for each of the variables before and after entering the evidence $Z = 0$.

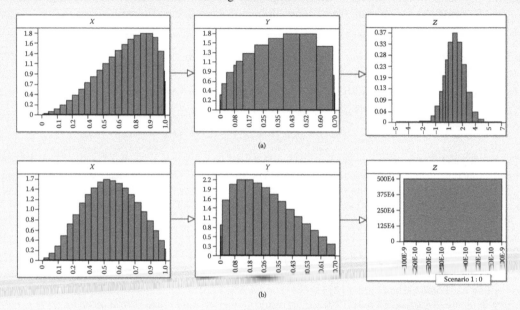

FIGURE D.3   Marginal distributions for X, Y, and Z after 1 iteration. (a) Before entering the evidence. (b) After entering the evidence $Z = 0$.

Running the model for 25 iterations results in the summary posterior values given in Table D.1.

Table D.1
Summary Posterior Values

	Mean	Variance
X	0.6747	0.0440
Y	0.3822	0.0334
Z	1.7685	1.1928
After Observing $Z = 0$		
X	0.5511	0.044
Y	0.2755	0.0288

D.3.4   Solving the Resolution Problem

The resolution problem occurs when the variance of a child node, Y, is many orders of magnitude smaller compared with the variance of the parent nodes. As a result, the mass generated by each sample taken on the parent node, X, is so small that the conditional probability distribution $f(Y \mid X)$ is sparse and contains areas without mass that should actually contain it.

Consider, for example, a model containing X and Y with ranges Ω_X and Ω_Y, respectively, and joint density function $f_{X,Y}(x,y) = f_Y(y \mid X = x)f_X(x)$, with distributions:

$$f_X(x) = N(0,1E8)$$

$$f_Y(y \mid X = x) = N(0,1E-8)$$

Here, $f_X(x)$ has extremely high variance, at $1E8$, and the conditional distribution $f_Y(y \mid X = x)$ has extremely low variance at $1E-8$. If we take two samples per parent region, we achieve a marginal posterior density for Y which is jagged, containing areas that are clearly under-sampled, as shown in Figure D.4. We should achieve a smooth function resembling $f(X)$ for Y given it is nearly an identity function. Likewise, the fact that the result on Y is jagged means that the DD algorithm treats the marginal distribution as multi-modal and therefore spends more computation time splitting intervals that are in fact smooth and locally linear.

What is happening here? Samples taken from X generate an extremely small part of the mass of the conditional probability distribution of Y and it would require a very large number of samples taken from X to generate mass such that it would adequately approximate the posterior range of Y. In this case, we roughly calculate how many samples would be needed to accurately cover the likelihood: Here, X has prior range $\Omega_X = \{-2 \times 10^{-5}, 2 \times 10^5\}$ and each sample would generate a probability mass with width $|\Omega_Y| = |10^{-3}|$, thus requiring $4 \times 10^5/10^{-3}$ samples, which equals 400 million samples in total. This is clearly too many.

We can solve this resolution problem by firstly tuning the sampling process to generate a high number of samples in the target region of the child node's conditional probability distribution, given the range of the parents and, additionally, by smoothing any under-sampled regions with probability mass, using an additional optimization procedure called ***uniform smoothing***.

Sample tuning is done by computing the resolution, r_i, for each sub-region, w_i, in the child node Y, as a function of the conditional probability mass generated in that sub-region by the states of the parent nodes, $pa\{Y\}$. This mass on the child node Y is computed from the expectations and standard deviations computed on Y from the parent states sampled on, $pa\{Y\}$ (to make the notation easier we will simply assume $pa\{Y\}$ means the sample values from the parent nodes of child Y). Thus, we compute two bounds using $E(Y \mid pa\{Y\}) \pm \text{s.d.}(Y \mid pa\{Y\})$ and from this determine: r_i:

$$r_i = \frac{|w_i|}{|E(Y \mid pa\{Y\}) - \text{s.d.}(Y \mid pa\{Y\}), \ E(Y \mid pa\{Y\}) + \text{s.d.}(Y \mid pa\{Y\})|}$$

If the mass bounds generated $(E(Y \mid pa\{Y\}) - \text{s.d.}(Y \mid pa\{Y\}), E(Y \mid pa\{Y\}) + \text{s.d.}(Y \mid pa\{Y\}))$ lie outside of the state range for sub-region w_i, then $r_i = 0$, since no mass is generated in that interval.

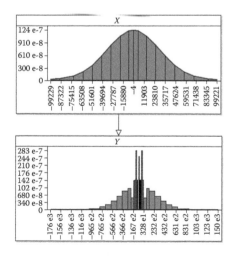

FIGURE D.4 The resolution problem: under sampling of X given Y.

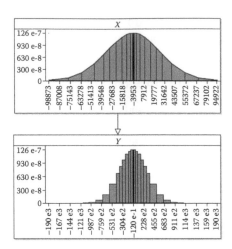

FIGURE D.5 The resolution problem solved using tuned sampling and uniform smoothing.

If $r_i < 0$ then the mass generated by the parent samples is wider than the state range for child interval, w_i; therefore, we can simply take a minimum number of samples. Otherwise, we take samples equal to r_i (up to some upper limit).

Clearly this is a two-stage process whereby we take an initial number of samples from the parent nodes and calculate r_i and use this to perform sampling again and generate the conditional probability distribution on each child node. However, even this can result in sample values that are too large and therefore computationally intensive. This is where we need uniform smoothing.

Uniform smoothing superimposes a series of piecewise constant approximations on $f(Y \mid pa\{Y\})$ and combines these with the sampled distributions, calculated based on sample tuning, and then averages the total resulting mass. Why use a uniform distribution for smoothing? When we approximate the conditional probability distribution we must be careful to ensure that we use a piecewise maximum entropy approximation over the range of the function and this is naturally the uniform distribution. Any other distribution might add additional information to the conditional probability distribution that should not be present.

Uniform smoothing proceeds by detecting the modes over the child node conditional probability distribution for Y by tuned sampling of the parents sampled, $pa\{Y\}$, then calculating a unit mass between these modes and adding this to the existing probability mass, then normalising. This process essentially smooths the conditional probability distribution to remove the kinds of spikes we get when we encounter the resolution problem. Reassuringly, as the number of tuned samples increases the width of each uniform smoothing component superimposed decreases and at the limit the effect of the uniform smoothing disappears. This is equivalent to the way we approximate conditionally deterministic functions, which effectively have a conditional distribution function with zero variance.

The effects of the sample tuning and uniform smoothing process on our example in Figure D.4 are shown in Figure D.5. As you can see the results are very much improved.

D.3.5 Approximating Mixture Distributions

The preceding cases deal with deterministic and statistical functions but we can also deal with more interesting and more challenging cases, such as mixture distributions where continuous nodes are conditionally dependent on labeled or Boolean nodes, as may be the case in hybrid BNs. Unlike other algorithms, dynamic discretization does not enforce any restriction on whether particular continuous or discrete nodes can be parents or children of others. Neither does the algorithm make any assumptions about whether some nodes in a HBN can or cannot receive evidence.

Mixture distributions are easily declared as hybrid models containing at least one labeled node that specifies the mixtures we wish to model along with the prior probabilities of each mixture indexing a continuous node containing the distributions we wish to mix. A mixture distribution is usually specified as a marginal distribution, indexed by discrete states in parent nodes A and B:

$$P(C \in \omega_k) = \sum_{j=1}^{n} \left[\int_{\omega_k} f(C \mid A = a_j, B = b_j) dC \right] P(A = a_j) P(B = b_j)$$

In the BN we therefore simply specify a set of conditional distribution functions partitioned by the labeled nodes we wish to use to define the mixture (in AgenaRisk we use partitioned expressions to declare these). Thus, for a mixture function where continuous variable, A, is conditioned on a discrete variable, B, with discrete states, $\{b_1, b_2, ..., b_n\}$, we could generate a different statistical or deterministic function for each state in the parent B:

$$f(A \mid B = \{b_1, b_2, ..., b_n\}) = \begin{cases} f(A \mid B = b_1) = N(0,10) \\ f(A \mid B = b_2) = Gamma(5,4) \\ \quad . \quad . \quad . \quad . \\ f(A \mid B = b_n) = TNormal(10,100,0,10) \end{cases}$$

Since dynamic discretization is flexible and agnostic about the underlying distributions chosen it will simply substitute the corresponding conditional distribution function into the NPT and execute, producing the mixture as if it was any other function.

Example D.1   Mixture of Normal Distributions

The Normal mixture distribution is an example of statistical models of continuous nodes that have discrete nodes as parents. Consider a mixture model with distributions

$$p(X = false) = p(X = true) = 0.5$$

$$p(Y \mid X) = \begin{cases} Normal(\mu_1, \sigma_1^2) & X = false \\ Normal(\mu_2, \sigma_2^2) & X = true \end{cases}$$

The marginal distribution of Y is a mixture of Normal distributions

$$P(Y) = \frac{1}{2} N\left(Y \mid \mu_1, \sigma_1^2\right) + \frac{1}{2} N\left(Y \mid \mu_2, \sigma_2^2\right)$$

with mean and variance given by this analytical solution:

$$E[Y] = \frac{1}{2}(\mu_1 + \mu_2)$$

$$Var[Y] = \frac{1}{2}\left[\left(\sigma_1^2 + \mu_1^2\right) + \left(\sigma_2^2 + \mu_2^2\right)\right] - \frac{1}{4}(\mu_1 + \mu_2)^2$$

Figure D.6 shows the resulting marginal distribution $p(Y)$ after 25 iterations for the mixture of $N(Y \mid 10,100)$ and $N(Y \mid 50,10)$, calculated under the static and the dynamic discretization approaches. While using the latter approach we are able to recover the exact values for the mean and variance, $E(Y) = 30$, $Var(Y) = 455$, the static case produces the approximated values $\mu_Y = 82.8$ and $\sigma_Y^2 = 12518$, showing clearly just how badly a static discretization performs.

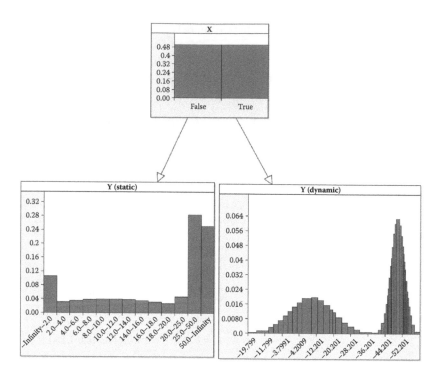

FIGURE D.6 Comparison of static and dynamic discretizations of the marginal distribution $p(Y)$ for the normal mixture model.

D.3.6 Approximating Comparative Functions

For conditionally deterministic functions we use Uniform distributions to smooth the conditional density function and also to gain a better approximation than can be achieved by sampling.

In AgenaRisk the currently implemented algorithm relies on sampling rather than the procedure described here.

We consider comparative statements of the general form:

$$X = True \text{ if } f(pa\{X\}) \otimes g(pa\{X\})$$

$$X = False \text{ otherwise}$$

where the operator $\otimes$ represents one of ($==$, $!=$, $\leq$, $\geq$, $<$, $>$).

In the continuous domain $==$, $!=$ always returns zero, but for integer values domains they are perfectly valid.

We take the upper and lower bounds of each interval in $\Psi^{(l)}_{pa\{X\}}$, calculate all values using the deterministic functions, $f(pa\{X\}$, $g(pa\{X\}$, and then calculate the min and max values, and then enter these as the parameters in the Uniform distribution. Under dynamic discretization an increasing number of intervals are produced, resulting in many more interval combinations, which has the effect of fitting a histogram composed of Uniform distributions to the continuous function, and resulting in a piecewise continuous function with no voids. Thus, we approximate the functions using:

$$f(pa\{X\}) \sim U(\min(f(pa\{X\})), \max(f(pa\{X\})))$$

$$g(pa\{X\}) \sim U(\min(g(pa\{X\})), \max(g(pa\{X\})))$$

Next we need to calculate the cumulative area under the curve where $f(pa\{X\})-g(pa\{X\}) \otimes 0$. The conditional probability $p(X = True \mid pa\{X\})$ is then the cumulative probability of the convolution of $f(pa\{X\}) - g(pa\{X\})$, which again can be approximated by a histogram:

$$p(X = True \mid pa\{X\}) =$$

$$\int_{\Omega_{f(pa\{X\})-g(pa\{X\})}} U[\min f(pa\{X\}) - \max g(pa\{X\}), \max f(pa\{X\}) - \min g(pa\{X\})]d(pa\{X\})$$

The relevant domain for $\Omega_{f(pa\{X\})-g(pa\{X\})}$ is $[\min(f(pa\{X\})-g(pa\{X\})), 0]$ for the $\leq, <$ operators and $[0, \max(f(pa\{X\})-g(pa\{X\}))]$ for the $\geq, >$ operators.

Note that some of the above results in invalid domains. For an invalid domain or a continuous domain with zero width, $p(X = True \mid pa\{X\}) = 0$. Additionally, for integer values we need to treat the $\leq, <$ and $\geq, >$ cases slightly differently since each pair of comparatives is not equivalent; practically this simply involves adding or subtracting one from the domain as appropriate. For labeled, as opposed to Boolean, nodes the approximation is relatively straightforward. In the case of nested comparatives things become a little trickier, but this is best solved by factorizing the comparatives first.

Example D.2   Comparative Function

Consider the case where $pa\{X\} = \{W,X,Y\}$ and the comparative statement is

$$Z = True \text{ if } W + X < Y$$

$$Z = False \text{ otherwise}$$

The intervals in $\Psi_{pa\{X\}}^{(l)}$ are W: [2, 5]; X: [5, 10]; and Y: [10, 12]; where

$$f(pa\{X\}) = W + X$$

$$g(pa\{X\}) = Y$$

The domains are $\Omega_{f(pa\{Z\})} = [7,15]$ and $\Omega_{g(pa\{Z\})} = [10,12]$
The relevant domain for $\Omega_{f(pa\{X\})-g(pa\{X\})}$ is

$$[\min(f(pa\{X\}) - g(pa\{X\})), 0]$$

which is [−5, 0].
Therefore

$$p(X = True \mid pa\{X\}) = \int_{-5}^{0} U(-5,5) = 1/2$$

$$p(X = False \mid pa\{X\}) = 1 - 1/2 = 1/2$$

D.3.7   Solving Nonlinear and Other Complex Functions

Solving nonlinear problems, by finding roots of continuous functions can be very challenging using BNs. There are a wide variety of root finding algorithms, many of which are iterative and aim to provide a successively more accurate approximation, but it isn't the role of this appendix to discuss them exhaustively. In AgenaRisk we use the simplest method, Bisection, which while relatively slow gives very good results in practice. The method is guaranteed to converge for a continuous function f, if on the interval $[a, b]$, $f(a)$ and $f(b)$ have opposite signs. So, provided the initial samples taken to evaluate the function are either side of the true roots of the function we are guaranteed to find them.

During dynamic discretisation we take partitions of the upper and lower bounds of states in the parents of the child and calculate the expectation of the nonlinear function. For a single child node, Y, with observation $Y = y$ and with parents, $pa\{Y\}$, we use bisection to find solutions to

$$E(Y \mid pa\{Y\}) - y = 0$$

Consider, for example, a model containing X and Y with joint density function

$$f_{X,Y}(x,y) = f_Y(y \mid X = x)f_X(x)$$

with distributions:

$$f_X(x) = Uniform(0,360)$$

$$f_Y(y \mid X = x) = N(\cos(X),0.01)$$

and with evidence $Y = 0.07$.

The resulting posterior marginal probability for X is shown in Figure D.7, which clearly shows a fine toothcomb mesh of roots in range [0, 360] with normal distributions around the roots.

Consider another more complex example model containing three parent nodes and three child nodes where the child nodes are simultaneous equations with nonlinear functions of two of the three parents. There are three parent nodes, X_i, where $i = 1, 2, 3$ with distribution:

$$f_{X_i}(X_i) = Normal(0,100000)$$

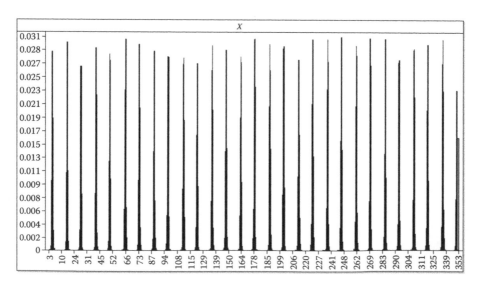

FIGURE D.7   Finding roots of cosine function.

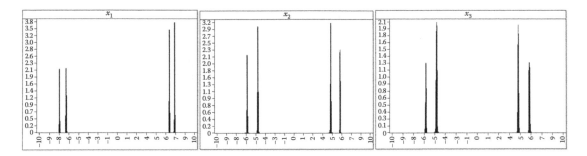

FIGURE D.8   Roots and associated regions for nonlinear simultaneous equations example.

There are three child nodes with conditional probability distribution functions:

$$f_{Y_1}\left(y_1 \,|\, X_1 = x_1, X_2 = x_2\right) = N\left(x_1 + x_2^2, 0.1\right)$$

$$f_{Y_2}\left(y_2 \,|\, X_1 = x_1, X_2 = x_2, X_3 = x_3\right) = N\left(x_1^2 + x_2, 0.1\right)$$

$$f_{Y_2}\left(y_3 \,|\, X_1 = x_1, X_3 = x_3\right) = N\left(x_1 + x_3^2, 0.1\right)$$

Given observations $y_1 = 30$, $y_2 = 50$ and $y_3 = 30$ we therefore need to solve these equations to find the roots:

$$x_1 + x_2^2 - 30 = 0$$

$$x_1^2 + x_2 - 50 = 0$$

$$x_1 + x_3^2 - 30 = 0$$

From dynamic discretisation we get the roots with associated posterior regions for each parent node X_i as shown in Figure D.8.

D.4   Ensuring Efficient Running of the Algorithm

Inevitably with any complex algorithm a number of programmatic tricks are needed to ensure smooth operation and accuracy, given the limits of modern computers. We start with some heuristics that are common to numerical approximation methods. We then describe the stopping rules that are used to determine when the algorithm has produced a good enough approximation. Finally we describe the method (binary factorization) that is used to overcome the complexity of nodes with multiple parents.

D.4.1   Heuristics

The following is a list of heuristics used in the implemented algorithm and recommendations for how best to use the resulting algorithm.

1. In order to control the growth of the resulting discretization sets, Ψ_X, after each iteration, we merge those consecutive intervals in Ψ_X with the lowest error estimate or those that have zero mass and zero entropy error. Merging intervals is difficult in practice because of a number of factors. First

we do not necessarily want to merge intervals because they have a zero relative entropy error, as is the case with Uniform distributions, since we want those intervals to help generate finer grained discretizations in any connected child nodes. Also, we wish to ensure that we only merge zero mass intervals with zero relative entropy error if they belong to sequences of zero mass intervals because some zero mass intervals might separate out local maxima in multimodal distributions. To resolve these issues we therefore apply a number of heuristics while merging.

2. To enter point values as evidence into a continuous node X, we assign a tolerance bound around the evidence, namely, $\delta(x)$, and instantiate X on the interval $(x - \delta(x), x + \delta(x))$. This bound can be set by the user in AgenaRisk. As this interval gets smaller there is a risk that the probability mass generated gets approximated as zero since at the limit the probability density of a point value is zero. Under these circumstances AgenaRisk assigns a very small nonzero probability to ensure the model will not fail.

3. Similar problems can occur when generating the statistical distribution functions over the parent intervals in circumstances where the probability for a child interval is vanishingly small, treated as zero by the library of standard approximations or indeed by the CPU of the machine, but should, at least theoretically, be nonzero. In this circumstance the zero values generated are replaced by linear interpolation between nonzero values generated over the NPT, thus ensuring that voids in the density are circumvented.

4. The current dynamic discretization algorithm is designed to work with simple array-based data structures that can be easily manipulated by the junction tree algorithm. It is not optimized for complex multidimensional models that cannot be factorized, such as a multidimensional Normal distribution, since the size of the data structures needed would be exponentially large. However, we believe that the style and type of problem covered in this book do not demand solutions of this kind.

5. Surprisingly, some problems that are easy to solve using classical statistical algorithms can be very difficult to solve using a BN. A well-known procedure is linear regression, which involves predicting a dependent variable from a linear sum of dependent variables of unknown parameters that require estimation. Although simple algorithms using least squares can solve this, it is very difficult in BNs. The reason for this is that a regression model captured as a BN will have very wide tree width in the junction tree; this induces large cluster sizes and these become Exponential in size. So, despite the ability of dynamic discretization algorithm to handle continuous nodes, further research is needed before it can be applied to this class of problem.

6. AgenaRisk supports the specification of mixed states for nodes containing continuous intervals and constant (integer and noninteger) values. Because the posterior probability of these constant values is not known at runtime, they are inserted into the state ranges of any node that requires them, as and when they are generated or encountered by the algorithm. Therefore, at its simplest, AgenaRisk can calculate exact arithmetical results alongside distributions in a way that other algorithms and tools currently struggle with. This capability is especially useful when computing deterministic functions where the result can include the absorbing state zero, enabling us to include a crisp zero in the distribution alongside the continuous state range, that is, $\Psi \in \{[0], (0-10], (10-20], \ldots.\}$.

7. AgenaRisk attempts to compute the upper and lower state boundaries for nodes based on the values of the parent nodes and automatically inserts these into the nodes active state range. For example, the model $f(X \mid Y, Z) \sim TNormal(\mu, \sigma^2, Y + Z, Z - Y)$ with state ranges $\Psi_X \in \{[0 - 10]\}$, $\Psi_Y \in \{[0 - 10]\}$ and $\Psi_Z \in \{[10 - 20]\}$ will run with the state range for X replaced with $\Psi_X \in \{[\min_{\Psi_{Y,Z}}\{Y + Z\} = 10, \max_{\Psi_{Y,Z}}\{Z - Y\} = 20]\}$. Indeed, under particular circumstances we can go further and examine all of the distributions and functions declared in the model and determine a reasonable initial state approximation based on where the prior probability mass of

the node lies. This will, of course, depend on the parameters declared in each node and its parent node state ranges. AgenaRisk propagates these ranges through the BN model in the first iteration, from parents to children, computing approximate initial ranges for each node. Then in subsequent iterations the dynamic discretization algorithm is used to refine the result. However running the model for one iteration provides for considerable speed improvements with little loss of accuracy in most cases.

8. We have already remarked in Chapter 10 about the problem where negative domain ranges are generated for nodes where logically this should be impossible. An example is where we observe Ψ_X that includes values $X < 0$ because $X = Y - Z$ in spite of the fact that the model is specified such that $Z \leq Y$. Propagating these sorts of logical constraints within the model and maintaining their consistency is a future research task, but practical means can ensure that domain ranges can remain logically consistent by using appropriate max and min bounds on conditionally deterministic functions.

9. The entropy error metric, E_j, can suffer from overflows and underflows where any of $f_{max} - \bar{f}$, $f_{max} - f_{min}$, $\bar{f} - f_{min}$ equals zero. For instance, this will happen where we have a modal region split into two neighboring equal width intervals, where $f_{max} = \bar{f}$. In this case we break the deadlock by using the arithmetical mean of the three intervals instead of the middle interval.

D.4.2   Using Stopping Rules and Convergence to Determine Accuracy

An important property of any algorithm is whether it converges to produce a stable answer. Stopping rules are then used to determine whether the answer currently approximated is good enough for the purposes at hand; if not the algorithm will continue.

There are three types of stopping rule implemented for the dynamic discretization algorithm in AgenaRisk. The first is the bluntest and this is the maximum number of iterations to run the algorithm for. This is run at the model level and when this stopping rule is exceeded the algorithm will stop regardless of whether the nodes are accurate enough.

The other two stopping rules are the *stable-entropy-error* stopping rule:

$$\left\{ 1 - \alpha \leq \frac{S_X^{(l-k)}}{S_X^{(l-k+1)}} \leq 1 + \alpha \text{ for } k = 1, 2, 3 \right\}$$

and the *low-entropy-error* stopping rule:

$$\left\{ S_i^X < \beta \right\}$$

The stable-entropy-error stopping rule determines over three iterations (k) whether the entropy has converged to a stable value within some limiting region defined by $(1 - \alpha, 1 + \alpha)$. The low-entropy-stopping rule determines whether a particular node has breached an absolute entropy error threshold. Both of these stopping rules apply at the node level, therefore during calculation some nodes will stop discretizing, should their stopping rule be triggered, whereas others will continue.

Let's use Example D.1 to illustrate how the convergence, interval splitting, and stopping rules work in AgenaRisk. Figure D.9 shows the posterior marginal probability distribution for $p(Y)$ after 2, 4, 6, and 25 iterations.

After two iterations in Figure D.9 the following intervals are candidates for splitting: [−100, 10], [10, 55], and [55, 100] with relative entropy error values of 4.1, 156 and 168, respectively, thus the interval [55, 100] is split with highest priority and then [10, 55]. The interval [−100, 10] is long and thin but has such small relative entropy value compared to the other intervals that it has very low priority.

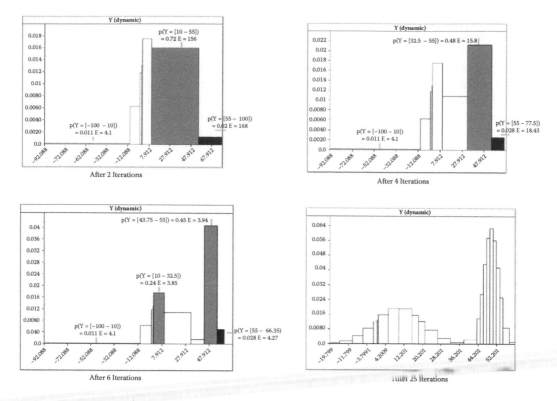

FIGURE D.9 Approximation of $p(Y)$ for Normal mixture problem over 2, 4, 6, and 25 iterations. (Graphs show 99.8 percentile range around the median.)

After four iterations, the interval [55.0, 77.5] has the highest relative entropy error of 18.43. The next highest relative entropy error, standing at 15.8, corresponds to the interval [32.5, 55.0]. Again the interval [−100, 10] is very low priority.

After 6 iterations the multi modal nature of $p(Y)$ is gradually being revealed and both modes are competing for attention: [10.0, 32.5] and [43.75, 55.0] each have very close relative entropy error values at 3.85 and 3.94, respectively. However, [55.0, 56.25] gives rise to a higher relative entropy error at 4.27 and so is the next interval to split.

After 25 iterations we can clearly see a very good approximation to the true multimodal distribution. Notice that the long-tail interval [−100, −10] has now been split so many times that it has dropped out of the displayed percentile range for the graph, thus producing an accurate discretization in the tail region.

For most problems each node in the model converges relatively quickly by converging according to one of the two stopping rules of the dynamic discretization algorithm. In Figure D.10 we show the resulting logarithm of the sum of the relative entropy errors for our example $p(Y)$ over 20 iterations. The low-entropy-error stopping rule used a very small threshold value to ensure it continued up to 50 iterations.

Clearly, from Figure D.10 we can see that the results are highly accurate after as few as 15 iterations and that the sum relative entropy error metric converges nicely. At iteration 15 some intervals merged resulting in a slight decrease in accuracy. The sum entropy error for the estimates of $p(Y)$ eventually converges to around 10^{-3}.

In practice the choice of stopping rule values needs to be traded off against computation time. For single nodes the increase in computation time is linear, but for larger networks with many parent nodes the increase is exponential.

For detailed information on entropy error metrics, the intervals, and resulting densities generated during BN execution, AgenaRisk provides an output log for visual inspection.

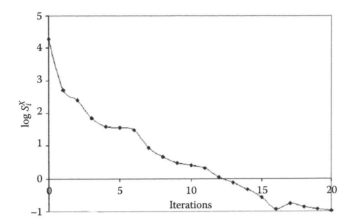

FIGURE D.10   Convergence of $p(Y)$ over 20 iterations.

D.4.3   Binary Factorization

Suppose that in your model you have a numeric node F that is the sum of, say, five numeric variables A, B, C, D, and E. It is very tempting to simply make each of these five nodes parents of F and define the NPT of F as $F = A + B + C + D + E$. The problem is that such a model will take a very long time to calculate even if the number of simulation iterations is set to a low number. That is because with dynamic discretization there will typically be dozens of states created for each node (the problem is usually even more acute for static discretization). If, say, just 20 states were created for each node, then the table for node F would require 20^6 entries (that is 64 million).

AgenaRisk automatically factorizes any arithmetical expression into factors containing at most two continuous nodes as parents (including comparatives). The binary factorization approach is described in Box D.2 where there are multiple parents.

Box D.2   Binary Factorization

Suppose that you have a situation where the node F is defined as the sum of nodes:

$$F = A + B + C + D + E$$

The strategy for reorganizing the structure here is to use binary factorization. In this the algorithm introduces synthetic nodes as shown in Figure D.11 where

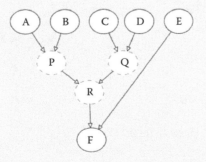

FIGURE D.11   Binary factorization.

$$P = A + B;\ Q = C + D;\ R = P + Q;\ F = R + E$$

and computes the BN based on this new factorized BN structure. This is done behind the scenes in AgenaRisk but the results can be viewed within the output log.

Binary factorization is performed on conditionally deterministic functions and statistical distribution functions, as illustrated by the following two examples.

Example D.3 Conditionally Deterministic Function

In the BN G the NPT for node B is shown in Figure D.12(a) as an expression, which is an arithmetical function, $B(Y_1, Y_2, Y_3) = Y_1Y_2 + Y_3$, of its parents, Y_1, Y_2, Y_3. This expression can be represented by a binary factorized network with five nodes, with new factors f_1 and B'. The factorized graph would then be defined with appropriate NPTs $B'(f_1, Y_3) = f_1 + Y_3$ and $f_1(Y_1, Y_2) = Y_1Y_2$ resulting in the new BN, $\mathbf{G}'$, as shown in Figure D.12(b).

Example D.4 Statistical Distribution Function

In BN G the CPT for node B, as shown in Figure D.13(a), is a Normal distribution whose mean and variance are arithmetical functions of its parents Y_1, Y_2, Y_3, Y_4; specifically, the mean is equal to $Y_1Y_2 + Y_3$ and variance is $Y_1 + Y_4$. Each expression declared on parameters for a statistical distribution needs to be parsed. Then it is factorized into f_1, f_2, f_3 and B'. The binary factorized network is the new BN $\mathbf{G}'$, which is a combination of two binary factorized networks each sharing a common parent node, Y_1, as shown in Figure D.13(b). Specifically:

$$f_1(Y_1, Y_2) = Y_1Y_2$$

$$f_2(f_1, Y_3) = Y_1 + Y_3 \quad f_3(Y_1, Y_4) = Y_1 + Y_4, \quad B'(f_2, f_3) = N(f_2, f_3)$$

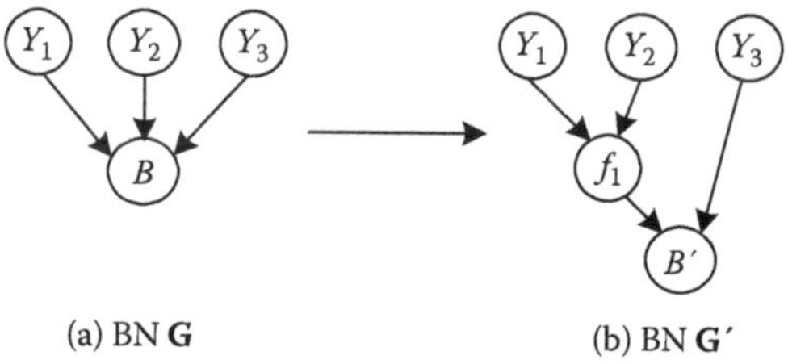

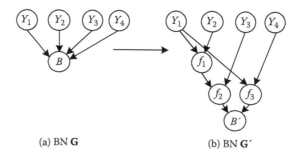

(a) BN **G**   (b) BN **G**′

FIGURE D.12   BNs for Example D.1. (a) BN **G**. (b) BN **G**′.

(a) BN **G**   (b) BN **G**′

FIGURE D.13   BNs for Example D.2. (a) BN **G**. (b) BN **G**′.

Appendix E: Statistical Distributions

E.1 Introduction

The continuous and discrete statistical distributions that are supported in AgenaRisk and approximated using dynamic discretization are listed in Table E.1. Each distribution function is described more fully in the following subsections. Continuous distribution types are only usable on continuous interval node types, with the exception of Uniform, Triangular, Normal and TNormal which can also be applied to discrete (integer) node types.

Table E.1
Statistical Distributions

Continuous	Discrete
Beta	Binomial
BetaPert	Exponential
Chi-Square	Geometric
Exponential	Hypergeometric
Extreme Value	Negative Binomial
Gamma	Normal
Log Normal	Poisson
Logistic	TNormal
Normal	Triangular
Student-*t*	Uniform
TNormal	
Triangular	
Uniform	
Weibull	

E.2 Continuous Distributions

E.2.1 Beta

Probability function: $P(X) = \frac{(1-x)^{\beta-1}x^{\alpha-1}}{B(\alpha,\beta)} = \frac{\Gamma(\alpha+\beta)}{\Gamma(\alpha)\Gamma(\beta)}(1-x)^{\beta-1}x^{\alpha-1}$

Domain: $0 \le X \le 1$

Parameter domain(s): $\alpha > 0$, $\beta > 0$

Mean: $E(X) = \frac{\alpha}{\alpha+\beta}$

Variance: $V(X) = \frac{\alpha\beta}{(\alpha+\beta)^2(\alpha+\beta+1)}$

Note: The domain of the Beta distribution can be extended to any finite range in the region $L \le X \le U$.

Example: Beta(3, 7, 0, 10)

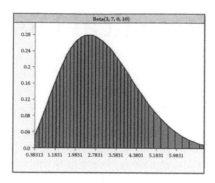

FIGURE E.1   Beta(3,7,0,10) distribution example.

E.2.2   BetaPERT

Probability function: $p(X) = \dfrac{(1-x)^{\beta-1} x^{\alpha-1}}{B(\alpha,\beta)} = \dfrac{\Gamma(\alpha+\beta)}{\Gamma(\alpha)\Gamma(\beta)} (1-x)^{\beta-1} x^{\alpha-1}$

The probability distribution is the same as the Beta distribution except the parameters are defined by mode (*m*), confidence (*lambda*), lower (*a*) and upper (*b*) bounds, where:

$$\alpha = \frac{(\mu-a)}{(b-m)}\left[\frac{(\mu-a)(b-\mu)}{\sigma^2} - 1\right]$$

$$\beta = \frac{(b-\mu)}{(b-a)}\left[\frac{(\mu-a)(b-\mu)}{\sigma^2} - 1\right]$$

Domain: $a \le m \le b$

Parameter domain(s): $\alpha > 0,\ \beta > 0$

Mean: $E(X) = \mu = \dfrac{a + \lambda m + b}{\lambda + 2}$

Variance: $V(X) = \sigma^2 = \dfrac{(b-a)^2}{(\lambda+2)^2}$

Example: BetaPERT(7, 6, –10, 10)

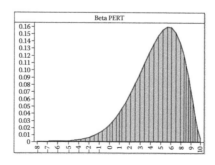

FIGURE E.2   BetaPERT(7,6,–10,10) distribution example.

E.2.3 Chi-Square

Probability function: $P(X) = \frac{2^{-v/2}}{\Gamma(v/2)} x^{(v/2)-1} e^{-v/2}$
Domain: $X > 0$
Parameter domain(s): $v > 1$, where v is the degrees of freedom.
Mean: $E(X) = v$
Variance: $V(X) = 2v$
Example: Chisquare(5)

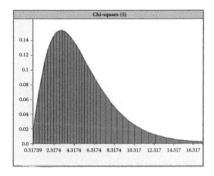

FIGURE E.3 Chi-Square(5) distribution example.

E.2.4 Exponential

Probability function: $P(X) = e^{-\lambda x}$
Domain: $X > 0$
Parameter domain(s): $\lambda > 0$
Mean: $E(X) = \lambda$
Variance: $V(X) = \lambda^2$
Example: Exponential(2)

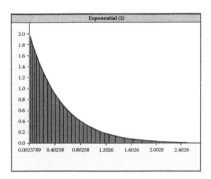

FIGURE E.4 Exponential(2) distribution example.

E.2.5 Extreme Value (Gumbel)

Cumulative probability function: $F(X) = e^{-e^{\frac{(x-\mu)}{\sigma}}}$ if $v = 0$ or $F(X) = e^{-(1+v\frac{(x-\mu)}{\sigma})^{-1/v}}$ if $v = 1$
Domain: $-\infty < X < \infty$

Parameter domain(s): $-\infty < \mu < \infty$, $\sigma > 0$, $\nu = \{0,1\}$
Order: $\nu = 0$ for maxima, $\nu = 1$ for minima
Location: μ
Scale: σ
Shape: ν
Example: Extreme Value($\nu = 0$, $\mu = 10$, $\sigma = 0$)

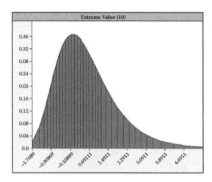

FIGURE E.5   Extreme Value(10) distribution example.

E.2.6   Gamma

Probability function: $P(X) = x^{\alpha-1} \dfrac{\beta^\alpha e^{-\beta x}}{\Gamma(\alpha)}$
Domain: $X > 0$

Parameter domain(s): $\alpha > 0$, $\beta > 0$ where $\beta = \frac{1}{\lambda}$ and λ is a rate parameter.
Mean: $E(X) = \alpha\beta$
Variance: $V(X) = \alpha\beta^2$
Example: Gamma(3, 20)

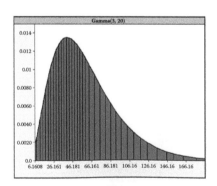

FIGURE E.6   Gamma(3,20) distribution example.

E.2.7   LogNormal

Probability function: $P(X) = \dfrac{1}{x\sigma\sqrt{2\pi}} e^{-(\ln x - \mu)^2/(2\sigma^2)}$
Domain: $X > 0$
Parameter domain(s): $-\infty < \mu < \infty$, $\sigma^2 > 0$

Mean: $E(X) = e^{(\mu + (1/2\sigma^2))}$
Variance: $V(X) = e^{2\mu}\, e^{\sigma^2}(e^{\sigma^2} - 1)$
Example: LogNormal(1.5, 2)

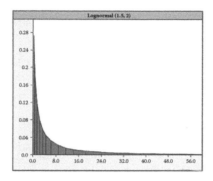

FIGURE E.7 LogNormal(1.5, 2) distribution example.

E.2.8 Logistic

Probability function: $P(X) = \dfrac{e^{-(x-\mu)/\beta}}{\beta[1+e^{-(x-\mu)/\beta}]^2}$
Domain: $-\infty < X < \infty$
Parameter domain(s): $v > 1$, where v is the degrees of freedom.
Mean: $E(X) = \beta$
Variance: $V(X) = \frac{1}{3}\pi^2\beta^2$
Note: In AgenaRisk the required parameters are "Mu" $= \mu$ and "Beta" $= \beta$.
Example: Logistic(2,3)

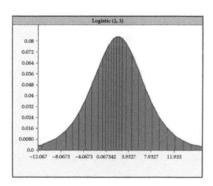

FIGURE E.8 Logistic(2,3) distribution example.

E.2.9 Normal

Probability function: $P(X) = \dfrac{1}{\sigma\sqrt{2\pi}} e^{-(x-\mu)^2/(2\sigma^2)}$
Domain: $-\infty < X < \infty$
Parameter domain(s): $-\infty < \mu < \infty$, $\sigma^2 > 0$

Mean: $E(X) = \mu$
Variance: $V(X) = \sigma^2$
Example: Normal(0, 100)

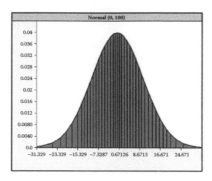

FIGURE E.9   Normal(0,100) distribution example.

E.2.10   Student-*t*

Probability function: $P(X) = \dfrac{\Gamma[\frac{1}{2}(r+1)]}{\sqrt{r\pi}\,\Gamma(r/2)(1+x^2/r)^{(r+1)/2}}$
Domain: $-\infty < X < \infty$
Parameter domain(s): $r > 0$, where $r = n - 1$ and n is the degrees of freedom.
Mean: $E(X) = 0$
Variance: $V(X) = \dfrac{r}{r-2}$
Example: Student(10)

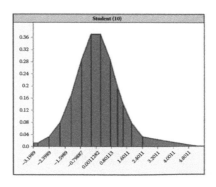

FIGURE E.10   Student-*t*(10) distribution example.

E.2.11   Truncated Normal

Probability function: $P(X) = \dfrac{1}{\sigma\sqrt{2\pi}}\,e^{-(x-\mu)^2/(2\sigma^2)}$
Domain: $L \le X \le U$
Parameter domain(s): $-\infty < \mu < \infty,\ \sigma^2 > 0$

Mean: $E(X) \approx \mu$

Variance: $V(X) \approx \sigma^2$

Note: The domain of the (doubly) truncated Normal distribution is restricted to the region $L \leq X \leq U$ and under these circumstances the mean and variance of the truncated distribution is only approximated by the mean and variance of the untruncated distribution. Depending on the truncation the true mean and variance may differ significantly from the supplied values.

Example: TNormal(0, 100, 0, 50)

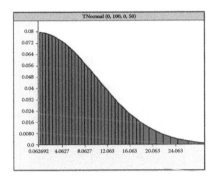

FIGURE E.11   Truncated Normal (0,100,0,50) distribution example.

E.2.12   Triangular

Probability function: $P(X) = \begin{cases} \frac{2(x-a)}{(b-a)(c-a)} & \text{for } a \leq x \leq c \\ \frac{2(b-x)}{(b-a)(c-a)} & \text{for } c < x \leq b \end{cases}$

Domain: $a \leq c \leq b$

Parameter domain(s): $b > c > a$

Mean: $E(X) \frac{a+b+c}{3}$

Variance: $V(X) \frac{a^2+b^2+c^2-ac-ab-cb}{18}$

Note: In AgenaRisk a is "Left" c is "Middle" and b is "Right."

Example: Triangular(5, 7, 10)

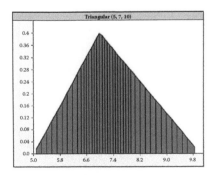

FIGURE E.12   Triangular(5,7,10) distribution example.

E.2.13   Uniform

Probability function: $P(X) = \begin{cases} 0 & \text{for } x < a \\ \dfrac{1}{(b-a)} & \text{for } a < x < b \\ 0 & \text{for } x > b \end{cases}$

Domain: $a < X < b$

Parameter domain(s): $b > a$

Mean: $E(X) = \frac{a+b}{2}$

Variance: $V(x) = \frac{(b-a)^2}{12}$

Note: In AgenaRisk a is "Lower Bound" and b is "Upper Bound."

Example: Uniform(0, 50)

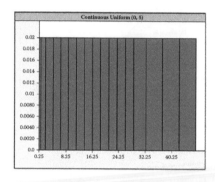

FIGURE E.13   Uniform(0,50) distribution example.

E.2.14   Weibull

Probability function: $P(X) = \alpha\beta^{-\alpha} x^{\alpha-1} e^{-(x/\beta)^{\alpha}}$

Domain: $X > 0$

Parameter domain(s): $\alpha > 0$, $\beta > 0$ [where α is the shape parameter and β is the scale parameter]

Mean: $E(X) = \beta^{1/\alpha}\, \Gamma(1 + 1/\alpha)$

Variance: $V(X) = \beta^{2/\alpha}\, [\Gamma(1 + 2/\alpha) - \Gamma^2(1 + 1/\alpha)]$

Example: Weibull(5, 3)

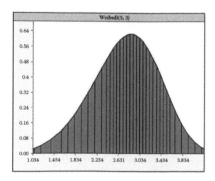

FIGURE E.14   Weibull(5,3) distribution example.

E.3 Discrete Distributions

E.3.1 Binomial

Probability function: $P(X = x) = \binom{N}{x} p^x (1-p)^{n-x}$

Domain: $x = 0, 1..., n$

Parameter domain(s): $n > 0$, where n is the number of trials. $0 < p < 1$, where p is the probability of success.

Mean: $E(X) = np$

Variance: $V(X) = np(1-p)$

Example: Binomial(0.2, 10)

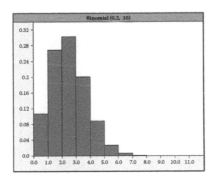

FIGURE E.15 Binomial(0.2, 10) distribution example.

E.3.2 Geometric

Probability function: $P(X = x) = p(1-p)^{x-1}$

Domain: $x = 1, 2, 3, ...$, where x is the number of trials until the first success.

Parameter domain(s): $0 < p < 1$, where p is the probability of success.

Mean: $E(X) = \frac{1}{p}$

Variance: $V(X) = \frac{1-p}{p^2}$

Example: Geometric(0.2)

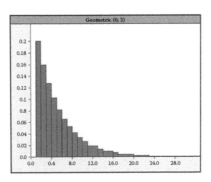

FIGURE E.16 Geometric(0.2) distribution example.

E.3.3   Hypergeometric

Probability function: $P(X = x) = \dfrac{\begin{pmatrix} r \\ x \end{pmatrix}\begin{pmatrix} m - r \\ n - x \end{pmatrix}}{\begin{pmatrix} m \\ n \end{pmatrix}}$

Domain: $x < m$, where x is the number of successful selections in a trial of size n from a total population of successes (type 1 observations), r, and failures, $m - r$, equal to a population of size m.

Parameter domain(s): $x < m$

Mean: $E(X) = \frac{nr}{m}$

Variance: $V(X) = \frac{nr(m-r)(m-n)}{m^2(m-1)}$

Example: Hypergeometric(100, 25, 10)

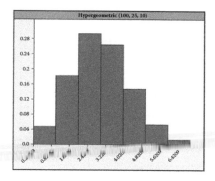

FIGURE E.17   Hypergeometric(100, 25, 10) distribution example.

E.3.4   Negative Binomial

Probability function: $P(X = x) = \begin{pmatrix} x + r - 1 \\ r - 1 \end{pmatrix} p^r (1 - p)^x$

Domain: $x = 0,1,2,...$, where x is the number of failed trials until the $(r-1)^{\text{th}}$ success in $(x + r - 1)$ trials

Parameter domain(s): $r > 0$ and $0 < p < 1$

Mean: $E(X) = r\frac{(1-p)}{p}$

Variance: $V(X) = r\frac{(1-p)}{p^2}$

Example: Negative Binomial(10, 0.3)

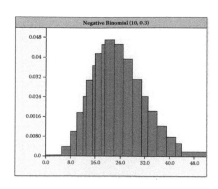

FIGURE E.18   Negative Binomial(10,0.3) distribution example.

E.3.5   Poisson

Probability function: $P(X = x) = \frac{1}{x!}\lambda^x e^{-\lambda}$
Domain: $x = 0,1,2,3...$
Parameter domain(s): $\lambda > 0$ where λ is the rate.
Mean: $E(X) = \lambda$
Variance: $V(X) = \lambda$
Example: Poisson(5)

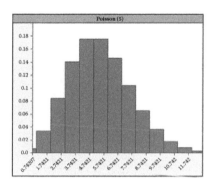

FIGURE E.19   Poisson(5) distribution example.

E.3.6   Uniform

Probability function: $P(X = x) = \frac{1}{N}$
Domain: $x = 0,1,..., N$
Parameter domain(s): $N > 0$
Mean: $E(X) = \frac{N+1}{2}$
Variance: $V(X) = \frac{(N-1)(N+1)}{12}$
Note: In AgenaRisk the required parameters are "Lower Bound," L, and "Upper Bound," U, and $N = (U - L) +1$.
Example: Uniform(0, 50)

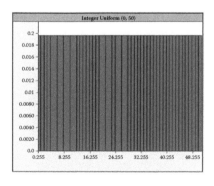

FIGURE E.20   Uniform(0,50) distribution example.

Index

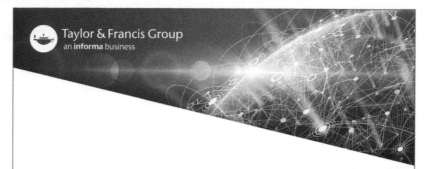